STUDENT SOLUTIONS MANUAL

GLORIA LANGER

A INTRODUCTORY LGEBRA FOR COLLEGE STUDENTS

ROBERT F. BLITZER

PRENTICE HALL, Englewood Cliffs, NJ 07632

Production Editor: *Joseph F. Tomasso*
Acquisitions Editor: *Melissa Acuña*
Supplement Acquisitions Editor: *Audra Walsh*
Production Coordinator: *Alan Fischer*

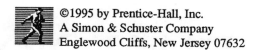 ©1995 by Prentice-Hall, Inc.
A Simon & Schuster Company
Englewood Cliffs, New Jersey 07632

Printed in the United States of America

10 9 8 7 6 5 4 3 2 1

ISBN 0-13-208927-0

Prentice-Hall International (UK) Limited, *London*
Prentice-Hall of Australia Pty. Limited, *Sydney*
Prentice-Hall Canada Inc., *Toronto*
Prentice-Hall Hispanoamericana, S.A., *Mexico*
Prentice-Hall of India Private Limited, *New Delhi*
Prentice-Hall of Japan, Inc., *Tokyo*
Simon & Schuster Asia Pte. Ltd., *Singapore*
Editora Prentice-Hall do Brasil, Ltda., *Rio de Janeiro*

STUDENT
SOLUTIONS
MANUAL

GLORIA LANGER

INTRODUCTORY
ALGEBRA
FOR COLLEGE STUDENTS

ROBERT F. BLITZER

Introductory Algebra Student's Solution Manual

Table of Contents

Chapter 1   The Real Number System

Section 1.1 The Real Numbers

Problem Set 1.1, pp. 12-15

1. $\{1, 2, 3\}$

3. $\{0, 1, 2, 3, 4, 5\}$

5. $\{-2, -1, 0, 1, \dots\}$

7. $\{-6, -5, -4, -3, \dots\}$

9. $\{7\}$

11. $\left\{-\dfrac{3}{4}\right\}$

13. $\{0\}$

15. $\left\{\dfrac{2}{3}, 1\right\}$

17. $\{\pi\}$

19. a. Natural numbers: $\left\{\sqrt{100}\right\}$ since $\sqrt{100} = 10$.

    b. Whole numbers: $\left\{0, \sqrt{100}\right\}$. The whole numbers consist of the natural numbers and 0.

    c. Integers: $\left\{-9, 0, \sqrt{100}\right\}$.

    d. Rational numbers: $\left\{-9, \dfrac{-4}{5}, 0, 0.25, 5\dfrac{1}{8}, 9.2, \sqrt{100}\right\}$.

        Each of the numbers can be expressed as the quotient of two integers. $-9 = \dfrac{-9}{1}$, $0 = \dfrac{0}{1}$, $0.25 = \dfrac{1}{4}$,

        $5\dfrac{1}{8} = \dfrac{41}{8}$, $9.2 = 9\dfrac{1}{5} = \dfrac{46}{5}$, $\sqrt{100} = 10 = \dfrac{10}{1}$.

    e. Irrational numbers: $\left\{\sqrt{3}, e\right\}$. In decimal form neither number terminates nor has a repeating pattern.

    f. Real numbers: $\left\{-9, \dfrac{-4}{5}, 0, 0.25, \sqrt{3}, e, 5\dfrac{1}{8}, 9.2, \sqrt{100}\right\}$.

        All the numbers in the set are real numbers.

21. A: -1

23. C: 0

25. E: $\dfrac{1}{2}$

27. G: 1

29. I: $1\dfrac{1}{2}$

31. K: 2

33. $\dfrac{1}{2}\boxed{<} 2$; $\dfrac{1}{2}$ is to the left of 2 so $\dfrac{1}{2}$ is less than 2.

35. $3\boxed{>} -\dfrac{5}{2}$, $-\dfrac{5}{2}$ is to the left of 3 so $-\dfrac{5}{2} < 3$ and $3 > -\dfrac{5}{2}$.

37. $-4\ \boxed{>}\ -6$; -6 is to the left of -4 so $-6 < -4$ and $-4 > -6$.

39. $-2.5\ \boxed{<}\ 1.5$; -2.5 is to the left of 1.5 so $-2.5 < 1.5$.

41. $-\dfrac{3}{4}\ \boxed{>}\ -\dfrac{5}{4}$, $-\dfrac{5}{4}$ is to the left of $-\dfrac{3}{4}$ so $-\dfrac{5}{4} < -\dfrac{3}{4}$ and $-\dfrac{3}{4} > -\dfrac{5}{4}$.

43. $-4.5\ \boxed{<}\ 3$; -4.5 is to the left of 3 so $-4.5 < 3$.

45. Opposite of 6: -6.

47. Opposite of -7: $-(-7) = 7$.

49. Opposite of $\dfrac{2}{3}$: $-\dfrac{2}{3}$.

51. Opposite of $-\dfrac{1}{4}$: $-\left(-\dfrac{1}{4}\right) = \dfrac{1}{4}$.

53. $\left|6\right| = 6$ because the distance between 6 and 0 on the number line is 6.

55. $\left|-7\right| = 7$ because the distance between -7 and 0 on the number line is 7.

57. $\left|\dfrac{2}{3}\right| = \dfrac{2}{3}$ because the distance between $\dfrac{2}{3}$ and 0 on the number line is $\dfrac{2}{3}$.

59. $\left|-\dfrac{1}{4}\right| = \dfrac{1}{4}$ because the distance between $-\dfrac{1}{4}$ and 0 on the number line is $\dfrac{1}{4}$.

61. when the number is negative; If $x < 0$, then the additive inverse is $-x\,(-x > 0)$ and the absolute value of x is $\left|x\right| = -x\,(-x > 0)$. Thus, the additive inverse of a number is equal to the absolute value of the number when the number is negative.

63. $\left|-8.03\right| = 8.03$

65. $-\left|8\right| = -8$

67. $-\left|-8\right| = -8$

69. $\left|-8\right| - \left|-2\right| = 8 - 2 = 6$

71. $\left|-\dfrac{3}{4}\right| - \left|-\dfrac{1}{2}\right| = \dfrac{3}{4} - \dfrac{1}{2} = \dfrac{1}{4}$

73. $\left|-9\right|$ $\boxed{=}$ $\left|9\right|$ since $\left|-9\right| = 9$ and $\left|9\right| = 9$.

75. -1.3 $\boxed{>}$ -1.36 since -1.3 is to the right of -1.36 on the number line.

77. $-\dfrac{1}{3}$ $\boxed{>}$ $-\dfrac{1}{2}$ since $-\dfrac{1}{3}$ is to the right of $-\dfrac{1}{2}$ on the number line.

79. $\left|-5\right|$ $\boxed{>}$ $\left|2\right|$ since $\left|-5\right| = 5$ and $\left|2\right| = 2$.

81. $\left|25\right|$ $\boxed{<}$ $\left|-30\right|$ since $\left|25\right| = 25$ and $\left|-30\right| = 30$.

83. $\left|\dfrac{3}{16}\right|$ $\boxed{<}$ $\left|-\dfrac{1}{2}\right|$ since $\left|\dfrac{3}{16}\right| = \dfrac{3}{16}$ and $\left|-\dfrac{1}{2}\right| = \dfrac{1}{2}$.

85. $-\left|-8\right|$ $\boxed{<}$ $\left|-8\right|$ since $-\left|-8\right| = -8$ and $\left|-8\right| = 8$.

87. $\left|-5.1\right|$ $\boxed{>}$ $-\left|-5.1\right|$ since $\left|-5.1\right| = 5.1$ and $-\left|-5.1\right| = -5.1$.

89. $-(-6)$ $\boxed{>}$ $-\left|-4\right|$ since $-(-6) = 6$ and $-\left|-4\right| = -4$.

91. Opposite of 20° below zero: +20; 20° above zero.

93. Opposite of 65 feet above sea level: –65; 65 feet below sea level.

95. 3; The distance between –1 and 3 is 4 which is twice as far as the distance between 3 and 5 which is 2.

97. –3 and 3 since $\left|-3\right| = 3$ and $\left|3\right| = 3$.

99. –5;

101. C is true; $-\dfrac{1}{2}$ is an example of a rational number which is not positive.

103. D is true; $\dfrac{7}{3} = 2\dfrac{1}{3}$ and the whole numbers less than $\dfrac{7}{3}$ are $\{0, 1, 2\}$.

105. B is true; $\left|-4\right| = 4$, $4 > 3$.

107. C is false; $\left|0\right| = 0$, $\left|-4\right| = 4$ and $0 < 4$.

109. B is true; $-8 = -8$, $-(-3) = 3$ and $-8 < 3$.

111–114: Answers will vary. Samples given.

111. $\{x \mid x$ is a positive real number but not an integer$\}$: $\left\{ \dfrac{1}{2}, \dfrac{5}{2}, \dfrac{7}{2} \right\}$.

113. $\{x \mid x$ is a rational number but not an integer$\}$: $\left\{ \dfrac{1}{2}, \dfrac{5}{2}, \dfrac{7}{2} \right\}$.

115. THINKING: (20, 18): $20 > 18$, $20 = $ T; $(-8, -12)$: $-8 > -12$, $\left| -8 \right| = 8 = $ H

(0, 9): $9 > 0$, $9 = $ I; $(-14, -17)$: $-14 > -17$, $\left| -14 \right| = 14 = $ N

(0, 11): $11 > 0$, $11 = $ K; $(-17, 9)$: $9 > -17$, $9 = $ I

$(-16, -14)$: $14 > -16$, $14 = $ N; $(-22, 7)$: $7 > -22$, $7 = $ G

117. $-\sqrt{12} \approx -3.464$ **119.** $2 - 3\sqrt{5} \approx -4.708$

Section 1.2 Addition of Real Numbers

Problem Set 1.2, pp. 24-26

1. $-7 + (-5) = \left(\left| -7 \right| + \left| -5 \right| \right) = -(7 + 5) = -12$ **3.** $12 + (-8) = +\left(\left| 12 \right| - \left| -8 \right| \right) = +(12 - 8) = 4$

5. $6 + (-9) = -\left(\left| -9 \right| - \left| 6 \right| \right) = -(9 - 6) = -3$ **7.** $-9 + (+4) = -\left(\left| -9 \right| - \left| +4 \right| \right) = -(9 - 4) = -5$

9. $-0.4 + (-0.9) = -1.3$ **11.** $-3.6 + 2.1 = -1.5$

13. $-9 + (-9) = -18$ **15.** $9 + (-9) = 0$

17. $-\dfrac{7}{10} + \left(\dfrac{-3}{10} \right) = -\dfrac{10}{10} = -1$ **19.** $\dfrac{9}{10} + \left(-\dfrac{3}{5} \right) = \dfrac{9}{10} + \left(-\dfrac{6}{10} \right) = \dfrac{3}{10}$

21. $3\dfrac{1}{2} + \left(-4\dfrac{1}{4} \right) = -\left(\left| -4\dfrac{1}{4} \right| - \left| 3\dfrac{1}{2} \right| \right) = -\left(3\dfrac{5}{4} - 3\dfrac{2}{4} \right) = -\dfrac{3}{4}$

23. $-8.74 - 8.74 = -17.48$ **25.** $5.86 + (-7.89) = -\left(\left| -7.89 \right| - \left| 5.86 \right| \right)$
$= -(7.89 - 5.86) = -2.03$

27. $5 + [13 + (-6)] = 5 + 7 = 12$ **29.** $7 + [-3 + (-1)] = 7 + (-4) = 3$

31. $-4 + [5 + (-1)] = -4 + 4 = 0$ **33.** $-15 + [9 + (-3)] = -15 + 6 = -9$

35. $[(-7) + (-11)] + 10 = (-18) + 10 = -8$ **37.** $[-3 + (-5)] + [7 + (-10)] = (-8) + (-3) = -11$

39. $-5.7 + (-7.2 + 6.6 = -12.9 + 6.6 = -6.3$ **41.** $-7.5 + 6.9 + (-3.7) = -0.6 + (-3.7) = -4.3$

43. $-12 + (-5) + 11 + (-7) = -17 + 11 + (-7) = -6 + (-7) = -6 + (-7) = -13$

45. $-13 + [7 + (-4)] = -13 + 3 = -10$

47. $9 + [(-2 + 4) + (-11)] = 9 + [2 + (-11)] = 9 + (-9) = 0$

49. $-20 + [(-8 + 2) + (-5)] = -20 + [(-6) + (-5)] = -20 + (-11) = -31$

51. $\left| -3 + (-5) \right| + \left| 2 + (-6) \right| = \left| -8 \right| + \left| -4 \right| = 8 + 4 = 12$

53. $[-4 + (-7)] + [(-3) + (-9)] + (13 + (-12)] = (-11) + (-12) + (1) = -23 + 1 = -22$

55. $-\dfrac{4}{5} + \dfrac{8}{5} + \left(-\dfrac{9}{5} \right) = \dfrac{4}{5} + \left(-\dfrac{9}{5} \right) = -\dfrac{5}{5} = -1$

57. $\left| -8 + (-2) \right| + (-13) = \left| -10 \right| + (-13) = 10 + (-13) = -3$

59. $-20 + \left[-\left| 15 + (-25) \right| \right] = -20 + \left[-\left| -10 \right| \right] = -20 + (-10) = -30$

61. Elevation of Dead Sea + elevation of person standing $= -1312 + 712 = -600$

The elevation of the person is $\boxed{600 \text{ feet below sea level}}$.

63. Temperature at 8:00 A.M. + rise 15°F by noon + fall 5°F by 4 P.M. $= -7°F + 15°F - 5°F = 3°F$

The temperature at 4:00 P.M. was $\boxed{3°F}$.

65. Start at 27-yard line + 4-yard gain + 2-yard loss + 8-yard gain + 12-yard loss

$= 27 + 4 - 2 + 8 - 12 = 39 - 14 = 25$

The location of the football at the end of the fourth play is at the $\boxed{25\text{--yard line}}$.

67. Starting price + rose $1\dfrac{1}{2}$ points + fell $\dfrac{1}{4}$ point + fell $\dfrac{1}{2}$ point

$= 35\dfrac{1}{2} + 1\dfrac{1}{2} - \dfrac{1}{4} - \dfrac{1}{2} = 37 - \dfrac{3}{4} = 36\dfrac{1}{4}$

The final price for the stock is $\boxed{36\dfrac{1}{4} \text{ per share}}$.

69. Answers will vary.

71. The sum of a number and 6: $x + 6$.

73. Five more than a number: $x + 5$.

75. a number increased by 17: $x + 17$.

77. Four added to a number: $x + 4$.

79. The sum of -7 and 4 and a number: $(-7 + 4) + x = -3 + x$.

81. The opposite of a number added to the sum of -14 and -22: $-x + [-14 + (-22)] = -x + (-36)$ or $-x - 36$.

83. The sum of the absolute value of –4 and the additive inverse of 6, increased by the absolute value of a number:

$$\left[\left|-4\right| + (-6)\right] + \left|x\right| = [4 + (-6)] + \left|x\right| = -2 + \left|x\right|.$$

85. C is true; example:

let $a = 1$, $b = 2$, then $\left|a\right| + \left|b\right| = \left|1\right| + \left|2\right| = 3$, $\left|a + b\right| = \left|1 + 2\right| = 3$, and $\left|a\right| + \left|b\right| = \left|a + b\right|$

let $a = -1$, $b = -2$, then $\left|a\right| + \left|b\right| = \left|-1\right| + \left|-2\right| = 3$, $\left|a + b\right| = \left|-1 - 2\right| = 3$ and $\left|a\right| + \left|b\right| = \left|a + b\right|$

If two numbers are both positive or both negative, then the absolute value of their sum equals the sum of their absolute values.

87. $6 + [2 + (-13)] \boxed{?} -3 + [4 + (-8)]$

$6 + (-11) \boxed{?} -3 + (-4)$

$-5 \boxed{>} -7$

89. $\frac{2}{3} + \left(-\frac{1}{6}\right) \boxed{?} \frac{9}{10} + \left(-\frac{3}{5}\right)$

$\frac{4}{6} + \left(-\frac{1}{6}\right) \boxed{?} \frac{9}{10} + \left(-\frac{6}{10}\right)$

$\frac{3}{6} \boxed{?} \frac{3}{10}$

$\frac{1}{2} \boxed{>} \frac{3}{10}$ since $\frac{1}{2} = \frac{5}{10}$ and $\frac{5}{10} > \frac{3}{10}$

91. $9.873 + (-11.732) + (-16.999) + 72.405 = 53.547.$

Review Problems

95. $\{x \mid x$ is an integer but not a natural number$\}$

$\{\ldots, -3, -2, -1, 0\}$

96. a. Natural numbers: $\left\{\sqrt{9}\right\}$.

   b. Whole numbers: $\left\{0, \sqrt{9}\right\}$.

   c. Integers: $\left\{-17, 0, \sqrt{9}\right\}$.

   d. Rational numbers: $\left\{-17, -\frac{2}{3}, 0.\overline{3}, \sqrt{9}, 10\frac{1}{7}\right\}$.

   e. Irrational numbers: $\left\{\sqrt{5}, \pi, \sqrt{7}\right\}$.

f. Real numbers: $\left\{ -17, -\frac{2}{3}, 0, \overline{3}, \sqrt{5}, \pi, \sqrt{7}, \sqrt{9}, 10\frac{1}{7} \right\}$.

97. $\left|-\frac{3}{7}\right| \boxed{?} \left|\frac{-3}{10}\right|$

$\frac{3}{7} \boxed{>} \frac{3}{10}$ since $\frac{3}{7} = \frac{30}{70} > \frac{3}{10} = \frac{21}{70}$.

Section 1.3 Subtraction of Real Numbers

Problem Set 1.3, pp. 32-34

1. $13 - 8 = 13 + (-8) = 5$

3. $8 - 15 = 8 + (-15) = -7$

5. $4 - (-10) = 4 + 10 = 14$

7. $-6 - (-17) = -6 + 17 = 11$

9. $-12 - (-3) = -12 + 3 = -9$

11. $-11 - 17 = -28$

13. $\frac{1}{5} - \left(-\frac{3}{5}\right) = \frac{1}{5} + \frac{3}{5} = \frac{4}{5}$

15. $-\frac{4}{5} - \left(-\frac{1}{5}\right) = -\frac{4}{5} + \frac{1}{5} = -\frac{3}{5}$

17. $\frac{1}{2} - \left(-\frac{1}{4}\right) = \frac{1}{2} + \frac{1}{4} = \frac{3}{4}$

19. $-4.4 - 9.3 = -13.7$

21. $-7.2 - (-5.1) = -7.2 + 5.1 = -2.1$

23. $3.1 - (-6.03) = 3.1 + 6.03 = 9.13$

25. $13 - 2 - (-8) = 13 - 2 + 8 = 11 + 8 = 19$

27. $9 - 8 + 3 - 7 = 1 + 3 - 7 = 4 - 7 = -3$

29. $-6 - 2 + 3 - 10 = -8 + 3 - 10 = -5 - 10 = -15$

31. $-10 - (-5) + 7 - 2 = -10 + 5 + 7 - 2 = -5 + 7 - 2 = 2 - 2 = 0$

33. $-23 - 11 - (-7) + (-25) = -23 - 11 + 7 + (-25) = -34 + 7 + (-25) = -27 + (-25) = -52$

35. $-823 - 146 - 50 - (-832) = -823 - 146 - 50 + 832 = -969 - 50 + 832 = -1019 + 832 = -187$

37. $1 - \frac{2}{3} - \left(-\frac{5}{6}\right) = 1 - \frac{2}{3} + \frac{5}{6} = \frac{1}{3} + \frac{5}{6} = \frac{2}{6} + \frac{5}{6} = \frac{7}{6} = 1\frac{1}{6}$

39. $-30 - 14 + 11 - (-9) - (-6) + 17$
$= -30 - 14 + 11 + 9 + 6 + 17$
$= -44 + 11 + 9 + 6 + 17$
$= -33 + 9 + 6 + 17$
$= -24 + 6 + 17$
$= -18 + 17$
$= -1$

41. $-0.16 - 5.2 - (-0.87) = -0.16 - 5.2 + 0.87 = -5.36 + 0.87 = -4.49$

43. $-\frac{3}{4} - \frac{1}{4} - \left(-\frac{5}{8}\right) = -\frac{3}{4} - \frac{1}{4} + \frac{5}{8} = -1 + \frac{5}{8} = -\frac{3}{8}$

45. $3 - (7 - 15) = 3 - (-8) = 3 + 8 = 11$

47. $-6 - [-6 - (-3)] = -6 - [-6 + 3] = -6 - (-3) = -6 + 3 = -3$

49. $[3 - (-5)] - (-8 + 10) = (3 + 5) - (2) = 8 + (-2) = 6$

51. $-(25 - 88) - 33 = -(-63) - 33 = 63 - 33 = 30$

53. $25 - [(-15) - (-15)] = 25 - [-15 + 15] = 25 - 0 = 25$

55. $-8 - \{8 - [6 - (-3)]\} = -8 - \{5 - [6 + 3]\} = -8 - (5 - 9) = -8 - (-4) = -8 + 4 = -4$

57. $-20 - [-(3 - 6) - 2] - 7 = -20 - [-(-3) - 2] - 7 = -20 - [3 - 2] - 7 = -20 - 1 - 7 = -21 - 7 = -28$

59. $-\left|-9 - (-6)\right| - (-12) = -\left|-9 + 6\right| + 12 = -\left|-3\right| + 12 = -3 + 12 = 9$

61. $(5.2 - 2.5) - (5 - 5.45) = 2.7 - (-0.45) = 2.7 + 0.45 = 3.15$

63. $12.5 - \{2.4 - [10.2 - (-2.4)]\}$
$= 12.5 - \{2.4 - [10.2 + 2.4]\}$
$= 12.5 - (2.4 - 12.6)$
$= 12.5 - (-10.2)$
$= 12.5 + 10.2$
$= 22.7$

65. $\dfrac{5}{8} - \left(\dfrac{1}{2} - \dfrac{3}{4}\right) = \dfrac{5}{8} - \left(-\dfrac{1}{4}\right) = \dfrac{5}{8} + \dfrac{2}{8} = \dfrac{7}{8}$

67. $-40 - \{-(6 - [(-21) + 5])\} = -40 - \{-(6 - [-16])\} = -40 - \{-(6 + 16)\} = -40 - (-22) = -40 + 22 = -18$

69. $\left|-9 - (-3 + 7)\right| - \left|-17 - (-2)\right|$
$= \left|-9 - (4)\right| - \left|-17 + 2\right|$
$= \left|-13\right| - \left|-15\right|$
$= 13 - 15 = -2$

71. Elevation of peak of Mount Whitney – elevation of Death Valley
$= 14{,}494 - (-282)$
$= 14{,}494 + 282$
$= 14{,}776.$

The peak of Mount Whitney is $\boxed{14{,}776 \text{ feet}}$ above Death Valley.

73. High temperature – low temperature
$= -4°F - (-17°F)$
$= -4°F + 17°F$
$= 13°F.$

The difference between the high and the low temperature is $\boxed{13°F}$.

75. Latitude of Sacramento – latitude of Lima $= +39 - (-12) = 39 + 12 = 51$

The difference in latitudes between Sacramento and Lima is $\boxed{51 \text{ degrees}}$.

77. Let "+" represent *north* and "−" represent *south*.

distance north + distance south
= +125 + (−316)
= −191

The car is $\boxed{191 \text{ miles south}}$ from its starting point.

79. Let "−" represent B.C. and "+" represent A.D.

usual way: golden age of India − golden age of Athens
= +500 − (−212)
= 500 + 212
= 712

Since there was no year 0, you must modify the difference by subtracting 1 year.

$712 - 1 = 711$

The difference between the years in the usual way is $\boxed{712 \text{ years}}$. However, the number of years that elapsed between these dates is $\boxed{711 \text{ years}}$.

81. The difference between 9 and −2: $9 - (-2) = 9 + 2 = 11$.

83. 7 less than 4: $4 - 7 = -3$.

85. −6 subtracted from the sum of 4 and −9
$= [4 + (-9)] - (-6)$
$= (-5) + 6$
$= 1$

87. The sum of 17 and the difference between −8 and 4
$= 17 + (-8 - 4) = 17 + (-12) = 5$

89. 6 less than the difference between −9 and −5
$= [-9 - (-5)] - 6 = (-9 + 5) - 6 = -4 - 6 = -10$

91. 24, decreased by 6 less than 15
$24 - (15 - 6) = 24 - 9 = 15$

93. A number decreased by 17: $x - 17$.

95. 6 less than a number: $x - 6$.

97. −14 subtracted from the sum of 7 and a number: $(7 + x) - (-14) = 7 + x + 14 = x + 21$.

99. The length of the remaining piece: $x - 3$ feet.

101. D is true;
       a subtracted from b: $b - a$
       sum of b and the opposite of a: $b + (-a) = b - a$
    The result is the same.

Review Problems

105. **a.** Natural numbers: $\left\{\sqrt{1}\right\}$.

  b. Whole numbers: $\left\{0,\sqrt{1}\right\}$.

  c. Integers: $\left\{-123, 0, \sqrt{1}\right\}$.

  d. Rational numbers: $\left\{-123, -\frac{3}{9}, 0, 0.45, \sqrt{1}, 8\frac{1}{5}\right\}$.

  e. Irrational numbers: $\left\{\sqrt{7}, e\right\}$.

  f. Real numbers: $\left\{-123, -\frac{3}{9}, 0, 0.45, \sqrt{1}, \sqrt{7}, e, 8\frac{1}{5}\right\}$.

106. First reading + increase of 4°F + decrease of 17°F + decrease of 2°F
= 12°F + 4°F − 17°F − 2°F
= −3°F

The temperature of the final reading is $\boxed{-3°F}$.

107. $-\frac{1}{2}\;\boxed{?}\;-\frac{1}{10}$

$-\frac{5}{10}\;\boxed{<}\;-\frac{1}{10}$

Section 1.4 Multiplication of Real Numbers

Problem Set 1.4, pp. 43-45

1. $6(-9) = -(6 \cdot 9) = -54$

3. $(-7)(-3) = +(7 \cdot 3) = 21$

5. $(-2)(6) = -12$

7. $(-13)(-1) = 13$

9. $0(-5) = 0$

11. $\frac{1}{2}(-14) = -7$

13. $\left(-\frac{3}{4}\right)(-20) = \frac{3 \cdot 20}{4 \cdot 1} = 15$

15. $-\frac{3}{5}\left(-\frac{4}{7}\right) = \frac{3 \cdot 4}{5 \cdot 7} = \frac{12}{35}$

17. $-\frac{7}{9} \cdot \frac{2}{3} = -\frac{7 \cdot 2}{9 \cdot 3} = -\frac{14}{27}$

19. $\left(\frac{4}{15}\right)\left(-1\frac{1}{4}\right) = \left(\frac{4}{15}\right)\left(-\frac{5}{4}\right) = -\frac{4 \cdot 5}{15 \cdot 4} = -\frac{1}{3}$

21. $(-4.1)(0.03) = -(4.1)(0.03) = -0.123$

23. $(-3.8)(-2.4) = +(3.8)(2.4) = 9.12$

25. $(3.08)(-0.25) = -0.77$

27. $(-5)(-2)(-3)(4) = 10(-3)(-4) = -30(4) = -120$

29. $-2(-3)(-4)(-1) = 6(-4)(-1) = -24(-1) = 24$

31. $-3\left(-\frac{1}{6}\right)(-50) = \frac{1}{2}(-50) = -25$

33. $(-3)(-1)(-2)\left(-\frac{1}{2}\right)(-4) = 3(-2)\left(-\frac{1}{2}\right)(-4) = 3(-4) = -12$

35. $-\frac{1}{8}(-24)\left(-\frac{1}{2}\right)(-6) = 3\left(-\frac{1}{2}\right)(-6) = 3(3) = 9$

37. $(-5)(-5)(-5) = 25(-5) = -125$

39. $(-6)(-4) - (-4)(2) = -24 - (-8) = -24 + 8 = -16$

41. $-8 - (-2) \cdot 4 = -8 + 8 = 0$ **43.** $4(-3) - (-6) = -12 + 6 = -6$

45. $7(-2)(-5) - (-11) = 70 + 11 = 81$ **47.** $-15 - (-3)(-4)(-2) = -15 + 24 = 9$

49. $(-4)(2)(-1) - (-5)(3)(-2) = 8 - 30 = -22$ **51.** $-6(-8 - 2) = -6(-10) = 60$

53. $6 - 4(2 - 10) = 6 - 4(-8) = 6 + 32 = 38$

55. $4(2 + 5) - 5(7 + 3) = 4(7) - 5(10) = 28 - 50 = -22$

57. $2(8 - 10) - 3(-6 + 4) = 2(-2) - 3(-2) = -4 + 6 = 2$

59. $(4 - 11)(6 - 10) = (-7)(-4) = 28$

61. $(-3 - 2)(-6 + 10) = (-5)(4) = -20$

63. $(-4 - 6)(-3) + 5 = (-10)(-3) + 5 = 30 + 5 = 35$

65. $-3(-5) + 7(-1) = 15 + (-7) = 8$

67. $4(3) - 5(-2) + 7(-3) = 12 + 10 - 21 = 22 - 21 = 1$

69. $-3[-6 - 4(-3)] = -3[-6 + 12] = -3(6) = -18$

71. $6 - 2[-5 - 3(-1)] = 6 - 2[-5 + 3] = 6 - 2(-2) = 6 + 4 = 10$

73. $8 - 3[4(-2 - 2) - 6(-1 - 1)] = 8 - 3[4(-4) - 6(-2)] = 8 - 3[-16 + 12] = 8 - 3[-4] = 8 + 12 = 20$

75. $(7 - 4)[2(-3 - 5) - 8] = (3)[2(-8) - 8] = 3[-16 - 8] = 3(-24) = -72$

77. $3\left|-3 - 12\right| = 3\left|-15\right| = 3(15) = 45$

79. $2\left|-8 + 6\right| - 4\left|-3 + 5\right| = 2\left|-2\right| - 4\left|2\right| = 2(2) - 4(2) = 4 - 8 = -4$

81. $-2\left|3 - 10\right| - 6\left|2 - 5\right| = -2\left|-7\right| - 6\left|-3\right| = -2(7) - 6(3) = -14 - 18 = -32$

83. The product of 5 and the sum of 7 and -15:

$5[7 + (-15)] = 5(-8) = -40.$

85. 8 added to the product of 4 and -10:

$4(-10) + 8 = -40 + 8 = -32.$

87. Twice the difference between 6 and -18:

$2[6 - (-18)] = 2[6 + 18] = 2(24) = 48.$

89. One-half of the sum of –2 and –24:

$$\frac{1}{2}[(-2) + (-24)] = \frac{1}{2}(-26) = -13$$

91. 60% of the sum of –4 and 44:

$$0.60[(-4) + 44] = 0.60(40) = 24$$

93. The product of –9 and –3 decreased by –3:

$$(-9)(-3) - (-3) = 27 + 3 = 30$$

95. The difference between –10 and the product of –1 and 7:

$$-10 - (-1)(7) = -10 + 7 = -3$$

97. The value in cents of x nickels:
the value of one nickel times number of nickels
$5¢ \cdot x$
$5x$ cents

99. The amount of acid in x liters of a solution that is 40% acid:
40% of number of liters in the solution
$0.40 \cdot x$
$0.40x$ liters

101. The distance covered by a car traveling at 50 miles per hour for x hours:

50 miles per hour times x hours
$50 \cdot x$
$50x$ miles

103. The total cost of a computer that sells for x dollars plus 8% tax:

cost of computer plus tax
$x + 0.08x$
$1.08x$

105. The dollar amount of the discount received for a 30% discount on an item priced at x dollars:

percent discount times dollar amount
$0.30 \cdot x$
$0.30x$ dollars

107. The reduced price of an item priced at x dollars with a 30% discount:

price of item minus discount
$x - 0.30x$
$0.70x$ dollars

109. The amount of weight in an elevator that is carrying a person who weighs 155 pounds plus x bags of cement weighing 75 pounds each:

weight of person + weight of each bag times number of bags
$155 + 75x$ pounds

111. The annual salary of a person who gets paid x dollars per week:

   weeks per year times salary per week
   52 weeks $\cdot$ x dollars per week
   $52x$ dollars

113. The total fee charged at a campground that charges \$35 for two adults and \$4 for each of x children:

   fee for two adults plus fee per child times number of children
   $35 + 4x$ dollars

115. The total hourly earnings for an employee earning \$5.50 per hour plus 25 cents for each of x units of a product manufactured during the hour:

   earnings per hour plus earnings per unit times number of units
   $5.50 + 0.25x$ dollars

117. The consecutive integer that follows any integer represented by x: $x + 1$.

119. The cost (in cents) for a phone call lasting x minutes if the company charges 15 cents for the first minute and 5 cents for each additional minute:

   cost for first minute plus cost for each additional minute times the number of additional minutes (which is 1 less than the total number of minutes)
   $15 + 5(x - 1)$ cents
   $15 + 5x - 5$
   $10 + 5x$ cents

121. B is true; If x is positive $(x > 0)$, then $-x$ is negative $(x < 0)$.

   $(-x) \cdot (-x) = +x \cdot x$ which is always positive.

123. $a + b = ab$ when a is a whole number and b is a mixed number, $1\frac{1}{c}$, where c is one less than the whole number a.

   $$b = 1\frac{1}{a-1} \quad \text{or} \quad b = \frac{a}{a-1}$$

125. $(-0.92532)(73.052)(-2.0041) = 135.4700988$

Review Problems

126. $\left\{ x \middle| \sqrt{x} \text{ is irrational and } x \text{ is a natural number between 2 and 10, not including 2 and not including 10} \right\}$:

   $\left\{ \sqrt{3}, \sqrt{5}, \sqrt{6}, \sqrt{7}, \sqrt{8} \right\}$

127. $\left| (-7) + 3 \right| \boxed{?} \left| -7 \right| + \left| 3 \right|$

   $\left| -4 \right| \boxed{?} 7 + 3$

   $4 \boxed{<} 10$

128. **a.** Natural numbers: $\{1492\}$.

b. Whole numbers: $\{0, 1492\}$.

c. Integers: $\left\{-\sqrt{25}, 0, 1492\right\}$.

d. Rational numbers: $\left\{-\sqrt{25}, 0, \frac{17}{125}, 1492\right\}$.

e. Irrational numbers: $\left\{-\sqrt{2}, \frac{\pi}{2}\right\}$.

f. Real numbers: $\left\{-\sqrt{25}, -\sqrt{2}, 0, \frac{17}{125}, \frac{\pi}{2}, 1492\right\}$.

Section 1.5 Exponents; Division of Real Numbers

Problem Set 1.5, pp. 54-57

1. $7^2 = 7 \cdot 7 = 49$

3. $4^3 = 4 \cdot 4 \cdot 4 = 64$

5. $(-4)^2 = (-4)(-4) = 16$

7. $(-4)^3 = (-4)(-4)(-4) = -64$

9. $(-2)^4 = (-2)(-2)(-2)(-2) = 16$

11. $-2^4 = -2 \cdot 2 \cdot 2 \cdot 2 = -16$

13. $2^6 = 2 \cdot 2 \cdot 2 \cdot 2 \cdot 2 \cdot 2 = 64$

15. $\left(\frac{2}{3}\right)^2 = \frac{2}{3} \cdot \frac{2}{3} = \frac{4}{9}$

17. $\left(-\frac{1}{3}\right)^3 = \left(-\frac{1}{3}\right)\left(-\frac{1}{3}\right)\left(-\frac{1}{3}\right) = -\frac{1}{27}$

19. $\left(-\frac{3}{4}\right)^3 = \left(-\frac{3}{4}\right)\left(-\frac{3}{4}\right)\left(-\frac{3}{4}\right) = -\frac{27}{64}$

21. $\left(-\frac{2}{3}\right)^4 = \left(-\frac{2}{3}\right)\left(-\frac{2}{3}\right)\left(-\frac{2}{3}\right)\left(-\frac{2}{3}\right) = \frac{16}{81}$

23. $-\left(\frac{2}{3}\right)^4 = -\frac{2}{3} \cdot \frac{2}{3} \cdot \frac{2}{3} \cdot \frac{2}{3} = -\frac{16}{81}$

25. $-\left(-\frac{1}{2}\right)^3 = -\left(-\frac{1}{2}\right)\left(-\frac{1}{2}\right)\left(-\frac{1}{2}\right) = \frac{1}{8}$

27. $(-1)^{17} = \underbrace{(-1)(-1)(-1)\ldots(-1)(-1)}_{17 \text{ times}} = -1$

29. $-(-1)^{13} = -\underbrace{(-1)(-1)(-1)\ldots(-1)(-1)}_{13 \text{ times}} = -(-1) = 1$

31. $-(-1)^{12} = -(-1)(-1)(-1)(-1)(-1)(-1)(-1)(-1)(-1)(-1)(-1)(-1) = -1$

33. $(-1.2)^3 = (-1.2)(-1.2)(-1.2) = -1.728$

35. $\frac{1}{4^3} = \frac{1}{4 \cdot 4 \cdot 4} = \frac{1}{64}$

37. $-\frac{12}{4} = -12 \cdot \frac{1}{4} = -3$

39. $\frac{21}{-3} = 21\left(-\frac{1}{3}\right) = -7$

41. $\frac{-90}{-3} = -90\left(-\frac{1}{3}\right) = 30$

43. $\frac{0}{-7} = 0\left(-\frac{1}{7}\right) = 0$

45. $\frac{-7}{0}$ is undefined; zero has no reciprocal.

47. $(-480) \div 24 = -480 \cdot \frac{1}{24} = -20$

49. $(465) \div (-15) = 465 \cdot \left(-\frac{1}{15}\right) = -31$

51. $\dfrac{-15 \cdot 9}{0.003} = -5300$

53. $\dfrac{8.25}{-0.05} = 165$

55. $4.06 \div (-0.7) = 4.06 \cdot \left(-\dfrac{1}{0.7}\right) = -5.8$

57. $-\dfrac{14}{9} \div \dfrac{7}{8} = -\dfrac{14}{9} \cdot \dfrac{8}{7} = -\dfrac{16}{9}$ or $-1\dfrac{7}{9}$

59. $\dfrac{3}{8} \div \left(-\dfrac{3}{4}\right) = \dfrac{3}{8} \cdot \left(-\dfrac{4}{3}\right) = -\dfrac{1}{2} \cdot \dfrac{1}{1} = -\dfrac{1}{2}$

61. $-\dfrac{4}{3} \div \left(-\dfrac{16}{9}\right) = -\dfrac{4}{3} \cdot \left(-\dfrac{9}{16}\right) = \dfrac{1}{3} \cdot \dfrac{9}{4} = \dfrac{3}{4}$

63. $0 \div \left(-\dfrac{3}{7}\right) = 0 \cdot \left(-\dfrac{7}{3}\right) = 0$

65. $-\dfrac{3}{7} \div 0 = -\dfrac{3}{7} \cdot \dfrac{1}{0}$ is undefined.

67. $-\dfrac{5}{7} \div \left(-\dfrac{5}{7}\right) = -\dfrac{5}{7} \cdot \left(-\dfrac{7}{5}\right) = 1$

69. $-\dfrac{5}{7} \div \dfrac{5}{7} = -\dfrac{5}{7} \cdot \left(\dfrac{7}{5}\right) = -1$

71. $6 \div \left(-\dfrac{2}{5}\right) = 6 \cdot \left(-\dfrac{5}{2}\right) = -15$

73. $-1\dfrac{2}{3} \div \left(-\dfrac{2}{9}\right) = -\dfrac{5}{3} \cdot \left(-\dfrac{9}{2}\right) = \dfrac{15}{2}$ or $7\dfrac{1}{2}$

75. $\dfrac{7}{12} \div (-7) = \dfrac{7}{12} \cdot \left(-\dfrac{1}{7}\right) = -\dfrac{1}{12}$

77. $\left(3 - 4\dfrac{1}{3}\right) \div \left(-\dfrac{2}{3} + \dfrac{5}{6}\right) = \left(-1\dfrac{1}{3}\right) \div \left(\dfrac{1}{6}\right) = -\dfrac{1}{4} \cdot \dfrac{6}{1} = -8$

79. $\dfrac{-80}{6 - (-2)} = -\dfrac{80}{6 + 2} = -\dfrac{80}{8} = -10$

81. $\dfrac{-25 - (-7)}{-4} = \dfrac{-25 + 7}{-4} = \dfrac{-18}{-4} = \dfrac{9}{2}$ or $4\dfrac{1}{2}$

83. $\dfrac{10(-3) - 20}{5(-2)} = \dfrac{-30 - 20}{-10} = \dfrac{-50}{-10} = 5$

85. $\dfrac{27 - 4(-2)}{-5(3)} = \dfrac{27 + 8}{-15} = \dfrac{35}{-15} = -\dfrac{7}{3}$ or $-2\dfrac{1}{3}$

87. $\dfrac{4(-3) + 6(-5)}{10 - 6} = \dfrac{-12 - 30}{4} = \dfrac{-42}{4} = -\dfrac{21}{2}$ or $-10\dfrac{1}{2}$

89. $\dfrac{4(-6) - 2(7 - 10)}{-3 - 5 - 6} = \dfrac{-24 - 2(-3)}{-14} = \dfrac{-24 + 6}{-14} = \dfrac{-18}{-14} = \dfrac{9}{7}$ or $1\dfrac{2}{7}$

91. $\dfrac{\left|3 - 10\right| - \left|4 - 6\right|}{11 - 6} = \dfrac{\left|-7\right| - \left|-2\right|}{5} = \dfrac{7 - 2}{5} = \dfrac{5}{5} = 1$

93. $\dfrac{5\left|-2 + 6\right| - 6\left|-5 + 3\right|}{-2 - 4} = \dfrac{5\left|4\right| - 6\left|-2\right|}{-6} = \dfrac{5(4) - 6(2)}{-6} = \dfrac{20 - 12}{-6} = \dfrac{8}{-6} = -\dfrac{4}{3}$ or $-1\dfrac{1}{3}$

95. $\dfrac{-2(7 - 11) - 3(12 - 9)}{3(-6) - 2(-6)} = \dfrac{-2(-4) - 3(3)}{-18 + 12} = \dfrac{8 - 9}{-6} = \dfrac{-1}{-6} = \dfrac{1}{6}$

97. $\dfrac{7^2 - 4^2}{-7 + 4} = \dfrac{49 - 16}{-3} = \dfrac{33}{-3} = -11$

99. $\dfrac{2(-5) + [7(-2) - (-4)]}{-4 - (-2)} = \dfrac{-10 + [-14 + 4]}{-4 + 2} = \dfrac{-10 + (-10)}{-2} = \dfrac{-20}{-2} = 10$

101. $\dfrac{(-2)(-9) - [(-5)(-3) - 4]}{-3(2) - 2(2)} = \dfrac{18 - (15 - 4)}{-6 - 4} = \dfrac{18 - 11}{-10} = -\dfrac{7}{10}$

103. The quotient of –18 and the sum of –15 and 12:

$$\frac{-18}{-15+12}=\frac{-18}{-3}=6.$$

105. The sum of –14 and the quotient of –21 and –7:

$$-14+\frac{-21}{-7}=-14+3=-11.$$

107. The difference between –6 and the quotient of 12 and –4 :

$$-6-\frac{12}{(-4)}=-6+3=-3.$$

109. The product of 6 and –3, divided by the difference between –7 and –9:

$$\frac{6(-3)}{(-7)-(-9)}=\frac{-18}{-7+9}=\frac{-18}{2}=-9.$$

111.   –7 subtracted from the quotient of 15 and $-\frac{1}{2}$:

$$15\div\left(-\frac{1}{2}\right)-(-7)=15\cdot\left(-\frac{2}{1}\right)+7=-30+7=-23.$$

113. The quotient of the product of –8 and –3, and $\frac{1}{3}$:

$$(-8)(-3)\div\left(\frac{1}{3}\right)=24\cdot\left(\frac{3}{1}\right)=72.$$

115. The quotient of the opposite of a number x and 16:

$$-x\div 16 \text{ or } \frac{-x}{16}.$$

117. The monthly salary for a person earning x dollars per year:

yearly salary ÷ 12 months per year

$$\frac{x}{12}\text{ dollars}.$$

119. The quotient of 6 and the consecutive integer that follows the integer x:

$$\frac{6}{x+1}.$$

121. The cost per orange when x oranges cost \$5 :

total cost ÷ number of oranges

$$\frac{5}{x}\text{ dollars}.$$

123. The height (in feet) of a person who is c inches tall:

height in inches ÷ 12 inches per foot

$\dfrac{c}{12}$ feet.

125. The average of 12 and x:

sum of items ÷ number of items

$\dfrac{12 + x}{2}$.

127. The length (in meters) of a line segment that measures x centimeters:

length in centimeters ÷ 100 centimeter per meter

$\dfrac{x}{100}$ meters.

129. The cost per calculator when all but one of x calculators are sold for \$50:

total cost ÷ number of calculators

$\dfrac{50}{x-1}$ dollars .

131. The fraction of the data above horizontal line l and to the right of vertical line m if the total number of dots is represented by x:

$\dfrac{\text{number of dots above line } l \text{ and to the right of } m}{\text{total dots}} = \dfrac{20}{x}$.

133. The number of minutes a person spends washing each dish if that person can wash dishes at a rate that is twice as fast as another person who can wash x dishes per minute:

1 dish ÷ rate (which is twice x)

1 dish ÷ $2x$ dishes per minute

$\dfrac{1}{2x}$ minutes.

135. The number of campaign workers needed to distribute x boxes of fliers so that each campaign worker gets $\dfrac{1}{2}$ of a box:

number of boxes ÷ amount per worker

x boxes ÷ $\left(\dfrac{1}{2} \text{box/worker} \right)$

$2x$ workers.

137. B is true: Since a is negative and c is positive, then $a - c$ is negative. Since b is positive, then $\frac{a-c}{b}$ must be negative because a negative divided by a positive is a negative.

139. Yes; 1 since $\left|1\right| = \frac{1}{1} = 1.$

141. Reciprocal of $\frac{3}{4}$: $1.\overline{3}.$

143. Reciprocal of $-\frac{13}{17}$: $-1.\overline{307692}.$

145. Reciprocal of -0.02: $-50.$

Review Problems

149. $\{x \mid x \text{ is a whole number less than 6 and a positive number}\}$

$\{1, 2, 3, 4, 5\}$

150. $-3 \ ? \ \dfrac{3(5+1)-6}{-5(7-5)+4}$

$-3 \ ? \ \dfrac{3(6)-6}{-5(2)+4}$

$-3 \ ? \ \dfrac{18-6}{-10+4}$

$-3 \ ? \ \dfrac{12}{-6}$

$-3 < -2, \ \boxed{\text{true}}$

151. $12°F - (-16°F) = 12 °F + 16°F = 28°F$

Section 1.6 Order of Operations; Mathematical Models

Problem Set 1.6, pp. 65-70

1. $-45 \div 5 \cdot 3 = -9 \cdot 3 = -27$

3. $-3 + 5(1 - 4)^3 = -3 + 5(-3)^3 = -3 + 5(-27) = -3 - 135 = -138$

5. $36 - 24 \div 2^3 \cdot 3 - 1 = 36 - 24 \div 8 \cdot 3 - 1 = 36 - 3 \cdot 3 - 1 = 36 - 9 - 1 = 26$

7. $(15 - 3^3)^2 = (15 - 27)^2 = (-12)^2 = 144$

9. $16 - (-3)(-12) \div 9 = 16 - (36) \div 9 = 16 - 4 = 12$

11. $[7 + 3(2^3 - 1)] \div 21 = [7 + 3(8 - 1)] \div 21 = [7 + 3(7)] \div 21 = (7 + 21) \div 21 = 28 \div 21 = 28 \cdot \frac{1}{21} = \frac{4}{3} \text{ or } 1\frac{1}{3}$

13. $\dfrac{37 + 15 \div (-3)}{16} = \dfrac{37 + (-5)}{16} = \dfrac{32}{16} = 2$

15. $\frac{3}{5}\left(\frac{2}{3}-\frac{3}{4}\right)=\frac{3}{5}\left(\frac{8-9}{12}\right)=\frac{3}{5}\left(-\frac{1}{12}\right)=-\frac{1}{20}$

17. $4(3-6)^2-2(3-4)=4(-3)^2-2(-1)=4(9)+2=36+2=38$

19. $\frac{5(4-6)}{2}-\frac{27}{-3}=\frac{5(-2)}{2}+9=-5+9=4$

21. $5-5\div5\cdot5-5^2=5-5\div5\cdot5-25=5-1\cdot5-25=5-5-25=-25$

23. $\frac{4^2-3^2}{(4-3)^2}=\frac{16-9}{1^2}=\frac{7}{1}=7$

25. $3(-2)^3-5(-2)+4=3(-8)+10+4=-24+10+4=-10$

27. $[5+3(-2)]^7=(5-6)^7=(-1)^7=-1$

29. $\left(\frac{3}{2}\right)^2\div\left(-\frac{3}{4}\right)=\frac{9}{4}\cdot\left(-\frac{4}{3}\right)=-3$

31. $6(6-7)^3-9(3-6)^2=6(-1)^3-9(-3)^2=6(-1)-9(9)=-6-81=-87$

33. $\frac{(-11)(-4)+2(-7)}{7-(-3)}=\frac{44-14}{7+3}=\frac{30}{10}=3$

35. $-2^2+4[16\div(3-5)]=-4+4[16\div(-2)]=-4+4(-8)=-4-32=-36$

37. $24\div\frac{3^2}{8-5}-(-6)=24\div\frac{9}{3}+6=24\div3+6=8+6=14$

39. $40\div10[8+(2-6)^2]-3^3$
$=40\div10[8+(-4)^2]-27$
$=40\div10[8+16]-27$
$=40\div10(24)-27$
$=4(24)-27$
$=96-27$
$=69$

41. $18\div(9-2^3)-(-4)=18\div(9-8)+4=18\div(1)+4=18+4=22$

43. $18\div9-2^3-(-4)=2-8+4=-2$

45. $(0.2)^2(-0.5)+1.92=(0.04)(-0.5)+1.92=-0.02+1.92=1.9$

47. $\frac{12(9-25)}{5^2-3^2}\div(-12)=\frac{12(-16)}{25-9}\div(-12)=\frac{12(-16)}{16}\div-12=-12\div(-12)=1$

49. $\frac{2}{3}-4(3+5)\div\left(-\frac{1}{2}\right)\left(-\frac{1}{8}\right)=\frac{2}{3}-4(8)\div\left(-\frac{1}{2}\right)\left(-\frac{1}{8}\right)=\frac{2}{3}+64\left(-\frac{1}{8}\right)=\frac{2}{3}-8=-7\frac{1}{3}$ or $-\frac{22}{3}$

51. $2\cdot(4-5)^3+12\div6+(5-8)^2=2(-1)=2+(-3)^2=2(-1)+2+9=-2+2+9=9$

53. $\frac{7-\{6-[3-(7-8)]\}}{19-2[4-(-3)]}=\frac{7-\{6-[3-(-1)]\}}{19-2[4+3]}=\frac{7-[6-(3+1)]}{19-2(7)}=\frac{7-(6-4)}{19-14}=\frac{7-2}{5}=\frac{5}{5}=1$

55. $\left[\frac{1}{3}-\left(\frac{1}{2}\right)^2\right]=\left[\frac{1}{3}-\frac{1}{4}\right]^2=\left(\frac{1}{12}\right)^2=\frac{1}{144}$

57. $\dfrac{(7-5)^2+2}{42-(7+3)}=\dfrac{2^2+2}{16-10}=\dfrac{4+2}{6}=\dfrac{6}{6}=1$

59. $\dfrac{2}{3}-\dfrac{1}{2}\left(\dfrac{1}{6}-\dfrac{7}{9}\right)=\dfrac{2}{3}-\dfrac{1}{2}\left(\dfrac{3-14}{18}\right)=\dfrac{2}{3}-\dfrac{1}{2}\left(-\dfrac{11}{18}\right)=\dfrac{2}{3}+\dfrac{11}{36}=\dfrac{24+11}{36}=\dfrac{35}{36}$

61. $\dfrac{\left|-14\right|-\left|3-8\right|}{-15}=\dfrac{14-\left|-5\right|}{-15}=\dfrac{14-5}{-15}=\dfrac{9}{-15}=-\dfrac{3}{5}$

63. $\left|2^3-3^2\right|-\left|8-11\right|=\left|8-9\right|-\left|-3\right|=\left|-1\right|-3=1-3=-2$

65. $(2^3-3^2)-(8-11)=(8-9)-(-3)=-1+3=2$

67. $\dfrac{(1+5)^2+(-1)^9}{12\div3\cdot2-3}=\dfrac{6^2-1}{4\cdot2-3}=\dfrac{36-1}{8-3}=\dfrac{35}{5}=7$

69. $\dfrac{-3(3-8)^2+5(2^3)}{-\left|10-(-10)\right|-5}=\dfrac{-3(-5)^2+5(8)}{-\left|20\right|-5}=\dfrac{-3(25)+40}{-25}=\dfrac{-75+40}{-25}=\dfrac{-35}{-25}=\dfrac{7}{5}\text{ or }1\dfrac{2}{5}$

71. $3.5(2.3-1.2)^2-(4.6-7.2)=3.5(1.1)^2-(-2.6)=3.5(1.21)+2.6=4.235+2.6=6.835$

73. $\dfrac{4\cdot3+3^3}{3\cdot13-2^2}+\dfrac{2^4+5\cdot3}{7^2-14}=\dfrac{12+27}{39-4}+\dfrac{16+15}{49-14}=\dfrac{39}{35}+\dfrac{31}{35}=\dfrac{70}{35}=2$

75. $\dfrac{2(2^2\cdot3-15)}{5^2-3^2}+\dfrac{11+12\div4\cdot9}{-2(5-13)}$

$=\dfrac{2(4\cdot3-15)}{25-9}+\dfrac{11+3\cdot9}{-2(-8)}$

$=\dfrac{2(12-15)}{16}+\dfrac{11+27}{16}$

$=\dfrac{2(-3)}{16}+\dfrac{38}{16}$

$=\dfrac{-6}{16}+\dfrac{38}{16}$

$=\dfrac{32}{16}$

$=2$

77. $\dfrac{1}{9}\div\left(\dfrac{1}{3}-2\right)^2=\dfrac{1}{9}\div\left(\dfrac{25}{9}\right)^2=\dfrac{1}{9}\cdot\dfrac{9}{25}=\dfrac{1}{25}$

79. $H=0.8(200-A)$
$=0.9(200-150)$ substitute 150 for A
$=0.8(50)$
$=40$ The handicap is 40.

$150+40=190$. The final score is $\boxed{190}$.

81. $T = \frac{1}{4}C + 37$

$= \frac{1}{4}(180) + 37$ $\qquad\qquad C = 180$

$= 45 + 37$

$= 82$ $\qquad\qquad\qquad$ The estimated temperature is $\boxed{82°\text{F}}$.

83. $P = \frac{25t^2 + 125t}{t^2 + 1}$

$= \frac{25(3)^2 + 125(3)}{3^2 + 1}$ $\qquad\qquad t = 3$ seconds

$= \frac{25(9) + 375}{9 + 1}$

$= \frac{225 + 375}{10}$

$= \frac{600}{10}$

$= 60$ $\qquad$ The systolic pressure 3 seconds after blood moves from the heart is $\boxed{60 \text{ millimeters of mercury}}$.

85. $W = \frac{60(x + 1)}{x + 5}$

$= \frac{60(10 + 1)}{10 + 5}$ $\qquad\qquad x = 10$ weeks

$= \frac{60(11)}{15}$

$= 44$ $\qquad$ After 10 weeks a person could type $\boxed{44 \text{ words per minute}}$.

Try $x = 100$

$W = \frac{60(101)}{105} \approx 57.7$

Try $x = 1000$

$W \; \frac{60(1,001)}{1005} \approx 59.76$

As x becomes larger, W approaches a limit of $\boxed{60 \text{ words per minute}}$.

87. $D = \frac{14,400}{x^2 + 10x}$

For $x = 10$,

$D = \frac{14,400}{(10)^2 + 10(10)}$

$= \frac{14,400}{100 + 100}$

$= \frac{14,400}{200}$

$= 72$ people at $10/calculator

For $x = 15$,

$$D = \frac{14{,}400}{15^2 + 10(5)}$$
$$= \frac{14{,}400}{225 + 150}$$
$$= \frac{14{,}400}{375}$$
$$= 38.4 \text{ people at } \$15/\text{calculator}$$

The difference is

$$72 - 38.4 = 33.6 \approx 34.$$

$\boxed{34 \text{ people}}$ more people are willing to purchase calculators at a price of \$10 than at a price of \$15.

89. $m = \dfrac{2x}{1 - x}$

Percentage of Contaminant Removed	$m = \dfrac{2x}{1-x}$	Required Amount of Money in Monetary Pool
50% ($x = 0.5$)	$m = \dfrac{2(0.5)}{1 - 0.5} = \dfrac{1}{0.5} = 2$	\$2 million
60% ($x = 0.6$)	$m = \dfrac{2(0.6)}{1 - 0.6} = \dfrac{1.2}{0.4} = 3$	\$3 million
70% ($x = 0.7$)	$m = \dfrac{2(0.7)}{1 - 0.7} = \dfrac{1.4}{0.3} = 4.\overline{6}$	\$4.$\overline{6}$ million
80% ($x = 0.8$)	$m = \dfrac{2(0.8)}{1 - 0.8} = \dfrac{1.6}{0.2} = 8$	\$8 million
90% ($x = 0.9$)	$m = \dfrac{2(0.9)}{1 - 0.9} = \dfrac{1.8}{0.1} = 18$	\$18 million
95% ($x = 0.95$)	$m = \dfrac{2(0.95)}{1 - 0.95} = \dfrac{1.9}{0.05} = 38$	\$38 million
99% ($x = 0.99$)	$m = \dfrac{2(0.99)}{1 - 0.99} = \dfrac{1.98}{0.01} = 198$	\$198 million

As the percent of contaminant gets closer and closer to 100% the cost $\boxed{\text{increases rapidly}}$.

91. $P = 100t^2 + 10{,}000$
$= 100(6^2) + 10{,}000$ $t = 6$
$= 100(36) + 10{,}000$
$= 3600 + 10{,}000$
$= 13{,}600$

$C = 0.5P + 2$
$= 0.5(13{,}600) + 2$ $P = 13{,}600$
$= 6{,}800 + 2$
$= 6{,}802$

The average daily level of carbon monoxide 6 years from now is $\boxed{6802 \text{ parts/million}}$. The relationship implies that $\boxed{\text{air pollution increases with time}}$.

93. $V = C\left(1 - \dfrac{n}{N}\right)$

$\qquad = 10{,}000\left(1 - \dfrac{7}{20}\right) \qquad\qquad C = 10{,}000,\ n = 7,\ N = 20$

$\qquad = 10000\left(\dfrac{13}{20}\right)$

$\qquad = 6500$

The value is $\boxed{\$6{,}500}$.

95. a. $x = -2,\ y = -3$

$\qquad 5x^2 - 4y^2 = 5(-2)^2 - 4(-3)^2 = 5(4) - 4(9) = 20 - 36 = -16$

b. $x = -3,\ y = 5$

$\qquad 5x^2 - 4y^2 = 5(-3)^2 - 4(5)^2 = 5(9) - 4(25) = 45 - 100 = -55$

97. a. $x = -1,\ y = -4$

$\qquad y - 3(y - 5x) = -4 - 3[-4 - 5(-1)] = -4 - 3[-4 + 5] = -4 - 3(1) = -4 - 3 = -7$

b. $x = 3,\ y = -2$

$\qquad y - 3(y - 5x) = -2 - 3[-2 - 5(3)] = -2 - 3[-2 - 15] = -2 - 3(-17) = -2 + 51 = 49$

99. a. $a = 2,\ b = -5,\ c = -1$

$\qquad b^2 - 4ac = (-5)^2 - 4(2)(-1) = 25 + 8 = 33$

b. $a = -3,\ b = 4,\ c = -2$

$\qquad b^2 - 4ac = 4^2 - 4(-3)(-2) = 16 - 24 = -8$

101. a. $x = 5,\ y = 3$

$\qquad \dfrac{x - 3y}{-x + 4y} = \dfrac{5 - 3(3)}{-5 + 4(3)} = \dfrac{5 - 9}{-5 + 12} = \dfrac{-4}{7} = -\dfrac{4}{7}$

b. $x = -5,\ y = -3$

$\qquad \dfrac{x - 3y}{-x + 4y} = \dfrac{-5 - 3(-3)}{-(-5) + 4(-3)} = \dfrac{-5 + 9}{5 - 12} = \dfrac{4}{-7} = -\dfrac{4}{7}$

103. a. $x = -4,\ y = 2$

$\qquad \dfrac{-3x}{x^2 - y^3} = \dfrac{-3(-4)}{(-4)^2 - 2^3} = \dfrac{12}{16 - 8} = \dfrac{12}{8} = \dfrac{3}{2} \text{ or } 1\dfrac{1}{2}$

b. $x = 4,\ y = -2$

$\qquad \dfrac{-3x}{x^2 - y^3} = \dfrac{-3(4)}{4^2 - (-2)^3} = \dfrac{-12}{17 - (-8)} = \dfrac{-12}{16 + 8} = \dfrac{-12}{24} = -\dfrac{1}{2}$

105.   **a.** $x = -1,\ y = -2$

$(x - y)(-3x) = [(-1) - (-2)][-3(-1)] = (-1 + 2)(3) = (1)(3) = 3$

b. $x = 0,\ y = -700$

$(x - y)(-3x) = [0 - (-700)][-3(0)] = 700(0) = 0$

107.   **a.** $x = -\dfrac{1}{2},\ y = -2$

$(2x - 3y)(4x + y) = \left[2\left(\dfrac{-1}{2}\right) - 3(-2)\right]\left[4\left(-\dfrac{1}{2}\right) + (-2)\right] = (-1 + 6)(-2 - 2) = 5(-4) = -20$

b. $x = -5,\ y = 0$

$(2x - 3y)(4x + y) = [2(-5) - 3(0)][4(-5) + 0] = (-10 - 0)(-20 + 0) = (-10)(-20) = 200$

109   **a.** $x = 5,\ y = -2$

$4\left|y - x\right| - 3(y - x) = 4\left|-2 - 5\right| - 3(-2 - 5) = 4\left|-7\right| - 3(-7) = 4(7) + 21 = 28 + 21 = 49$

b. $x = -5,\ y = -2$

$4\left|y - x\right| - 3(y - x) = 4\left|-2 - (-5)\right| - 3[-2 - (-5)]$
$= 4\left|-2 + 5\right| - 3(-2 + 5) = 4(3) - 3(3) = 12 - 9 = 3$

111.   **a.** $x = 0,\ y = 4$

$\dfrac{x}{-4 - y} = \dfrac{0}{-4 - 4} = \dfrac{0}{-8} = 0$

b. $x = 4,\ y = -4$

$\dfrac{x}{-4 - y} = \dfrac{4}{-4 - (-4)} = \dfrac{4}{-4 + 4} = \dfrac{4}{0}$ is undefined

113.   D is true;

$\dfrac{\left|3 - 7\right| - 2^3}{(-2)(-3)} = \dfrac{\left|-4\right| - 8}{6} = \dfrac{4 - 8}{6} = -\dfrac{4}{6} = -\dfrac{2}{3}$

and $-\dfrac{1}{3} - \dfrac{1}{3} = -\dfrac{2}{3}.$

Challenge Problems

115.   **a.** $(2 \cdot 3 + 3) \cdot 5 = 45$

b. $\left(2 \cdot 5 - \dfrac{1}{2} \cdot 10\right) \cdot 9 = 45$

c. $4^2 \div \dfrac{1}{4} - 3 \cdot 5 - 2^2 = 45$

117. $1 - 2 \div 3 + 4 \times 5 = 20\frac{1}{3}$

119. **a.**   yes; $9 \div (17 - 14) - (4 \cdot 2) = 1$

     b. $(17 - 14 + 9) \div (4 + 2) = 2$

     c. $(9 + 14 - 17) \div (4 - 2) = 3$

121. $7(4)^2 - (-5) = 117$

123. $\dfrac{\frac{3}{8} - 4}{12 + 3 - 4} = -0.329\overline{54}$ or $-\dfrac{29}{88}$

125. $\left(3 + \dfrac{3}{4} - \dfrac{1}{2}\right)^4 = 111.56640625$ or $\dfrac{28{,}561}{256}$

127. $\dfrac{-3.02(1.7)^3 - (-16.3)}{(17.5 + 8.6)\left(\frac{1}{2} - \frac{2}{5}\right)} = 0.5604367816$

Review Problems

131 **a.**   Natural numbers: $\left\{\sqrt{100}\right\}$.
     b.   Whole numbers: $\left\{\sqrt{100}\right\}$.
     c.   Integers: $\left\{\sqrt{100}\right\}$.
     d.   Rational numbers: $\left\{-\dfrac{2}{3}, 0.\overline{6}, \sqrt{100}\right\}$.
     e.   Irrational numbers: $\left\{-\dfrac{\pi}{3}, e, \sqrt{99}\right\}$.
     f.   Real numbers: $\left\{-\dfrac{\pi}{3}, -\dfrac{2}{3}, 0.\overline{6}, e, \sqrt{99}, \sqrt{100}\right\}$.

132. D is true;

$$(-4)^2 + 3^2 = (-5)^2$$
$$16 + 9 = 25$$
$$25 = 25.$$

133. A is true;

$$12{,}006 \in \{1, 2, 3, \ldots\}.$$

Section 1.7 **Properties of Real Numbers**

Problem Set 1.7, pp. 80-83

1. $6(-9) = (-9)6$

Commutative Property of Multiplication

3. $(3 + y) + z = 3 + (y + z)$

Associative Property of Addition

5. $\pi \cdot 1 = \pi$

Multiplicative Identity Property

7. $4(3y + 8r) = 12y + 32r$

Distributive Property

9. $(3 + y) + z = (y + 3) + z$

Commutative Property of Addition

11. $-17 + 0 = -17$

Additive Identity Property

13. $5\left(\dfrac{1}{5}\right) = 1$

Multiplicative Inverse Property

15. $-13\left(\dfrac{-1}{13}\right) = 1$

Multiplicative Inverse Property

17. $3(5x) = (3 \cdot 5)x$

Associative Property of Multiplication

19. $\dfrac{7}{8} + \left(-\dfrac{7}{8}\right) = 0$

Additive Inverse Property

21. $7 + (5 + x) = (7 + 5) + x = 12 + x$

23. $(y + 3) + (-11) = y + [3 + (-11)] = y - 8$

25. $7(4x) = (7 \cdot 4)x = 28x$

27. $-3(-5x) = (-3)(-5)x = 15x$

29. $\dfrac{1}{3}(-6z) = \left(\dfrac{1}{3}\right)(-6)z = -2z$

31. $\dfrac{1}{5}(5w) = \left(\dfrac{1}{5} \cdot 5\right)w = w$

33. $\dfrac{2}{3}\left(\dfrac{3}{2}z\right) = \left(\dfrac{2}{3} \cdot \dfrac{3}{2}\right)z = z$

35. $-\dfrac{1}{2}(2m) = \left(-\dfrac{1}{2}\right)(2)m = -m$

37. $\dfrac{3}{11}\left(\dfrac{-11}{3}s\right) = \left(\dfrac{3}{11}\right)\left(-\dfrac{11}{3}\right)s = -s$

39. $(5x) \cdot 3 = 3 \cdot (5x) = (3 \cdot 5)x = 15x$

41. $(6y)(-5) = (-5)(6y) = (-5 \cdot 6)y = -30y$

43. $(4w)\left(\dfrac{1}{4}\right) = \left(\dfrac{1}{4}\right)(4w) = \left(\dfrac{1}{4} \cdot 4\right)w = w$

45. $\left(\dfrac{2}{5}r\right)\left(-\dfrac{5}{2}\right) = \left(-\dfrac{5}{2}\right)\left(\dfrac{2}{5}r\right) = -r$

47. $(3 + y) + (-23) = (-23) + (3 + y) = (-23 + 3) + y = -20 + y$

49. $(-6 + r) + (-15) = (-15) + (-6 + r) = [-15 + (-6)] + r = -21 + r$

51. $3(x + 5) = 3(x) + 3(5) = 3x + 15$

53. $8(2x + 3) = 8(2x) + 8(3) = 16x + 24$

55. $-4(y + 3) = -4(y) + (-4)(3) = -4y - 12$

57. $-11\left(2y + \dfrac{1}{11}\right) = -11(2y) + (-11)\left(\dfrac{1}{11}\right) = -22y - 1$

59. $\dfrac{1}{3}(12 + 6r) = \dfrac{1}{3}(12) + \dfrac{1}{3}(6r) = 4 + 2r$

61. $5(x + y) = 5x + 5y$

63. $3(y - 7) = 3y + 3(-7) = 3y - 21$

65. $10(7x - 3) = 10(7x) + 10(-3) = 70x - 30$

67. $\dfrac{1}{2}(5x - 12) = \dfrac{1}{2}(5x) + \dfrac{1}{2}(-12) = \dfrac{5}{2}x - 6$

69. $-4(y - 6) = -4y + (-4)(-6) = -4y + 24$

71. $(m + 7)5 = m(5) + 7(5) = 5m + 35$

73. $(6 - k)(-3) = 6(-3) + (-k)(-3) = -18 + 3k$

75. $(5x - 7y)(-10) = 5x(-10) - 7y(-10) = -50x + 70y$

77. $-3(2x + 3y - 5) = -3(2x) - 3(3y) - 3(-5) = -6x - 9y + 15$

79. $(9r - 12s - 15)\left(-\dfrac{1}{3}\right) = 9r\left(-\dfrac{1}{3}\right) - 12s\left(-\dfrac{1}{3}\right) - 15\left(-\dfrac{1}{3}\right) = -3r + 4s + 5$

81. $4(x + 5) + 30 = 4x + 4(5) + 30 = 4x + 20 + 30 = 4x + 50$

83. $-5(2x) - 5(-3) + 15 = -10x + 15 + 15 = -10x + 30$

85. $(6 - 4x)\left(-\dfrac{1}{2}\right) + 7 = 6\left(-\dfrac{1}{2}\right) - 4x\left(-\dfrac{1}{2}\right) + 7 = -3 + 2x + 7 = 2x - 3 + 7 = 2x + 4$

87. $-(3x + 12) = -3x - 12$

89. $-(12y - 2) = -12y + 2$

91. $-(-5x + 17) = 5x - 17$

93. $-(-3x + 14) + 12 = 3x - 14 + 12 = 3x - 2$

95. $-(-4r - 2) - (-6) = 4r + 2 + 6 = 4r + 8$

97. $-(7w - 3) + (-3) = -7w + 3 + (-3) = -7w$

99. Additive inverse of 7: -7.

101. Additive inverse of $\left|-7\right|$: $-\left|-7\right| = -7$.

103. Additive inverse of $-\sqrt{2}$: $-\left(-\sqrt{2}\right) = \sqrt{2}$.

105. Additive inverse of $-(-13)$: $-[-(-13)] = -13$.

107. Multiplicative inverse of 7: $\dfrac{1}{7}$.

109. Multiplicative inverse of $\left|-7\right|$: $\dfrac{1}{\left|-7\right|} = \dfrac{1}{7}$.

111. Multiplicative inverse of $-\sqrt{2}$: $\dfrac{1}{-\sqrt{2}} = -\dfrac{1}{\sqrt{2}}$.

113. Multiplicative inverse of $-(-13)$: $\dfrac{1}{-(-13)} = \dfrac{1}{13}$.

115. Multiplicative inverse of $\dfrac{2}{8}$: $\dfrac{1}{2/8} = \dfrac{8}{2}$ or 4.

117. Multiplicative inverse of $-\left|-\dfrac{3}{7}\right| = \dfrac{1}{-\left|-3/7\right|} = -\dfrac{1}{3/7} = -\dfrac{7}{3}$.

119. Multiplicative inverse of $2\dfrac{2}{3}$: $\dfrac{1}{2\frac{2}{3}} = \dfrac{1}{\frac{8}{3}} = \dfrac{3}{8}$.

121. $5x + 0 = 5x$
Additive identity property

123. $3 + 5y = 5y + 3$
Commutative property of addition

125. $1x + 19 = x + 19$
Multiplicative identity property

127. $-3(y - 2) = -3y + 6$
Distributive property

129. $7x + (-7x) = 0$
Additive inverse property

131. $4(x + 3) = 4x + \boxed{12}$ **133.** $9(x + y) = 9x + \boxed{9}y$

135. $7 \cdot 0 = \boxed{0}$ **137.** $7 + (-7) = \boxed{0}$

139. $3(2 \cdot 5) = (3 \cdot 2)\boxed{5}$ **141.** $5 \cdot \dfrac{1}{5} = \boxed{1}$

143. D is true;

Multiplicative inverse of -0.02: $\dfrac{1}{-0.02} = -\dfrac{1}{0.02} = -50$.

145. B is true; $\dfrac{1}{1} = 1$

147. $\begin{aligned}7 + 8(y + 1) &= 7 + 8y + 8\\ &= 7 + 8 + 8y\\ &= (7 + 8) + 8y\\ &= 15 + 8y\\ &= 8y + 15\end{aligned}$ Distributive Property
Commutative Property of Addition
Associative Property of Addition
Addition
Commutative Property of Addition

149. $\begin{aligned}\dfrac{1}{a}(ba) &= \dfrac{1}{a}(ab) \quad a \neq 0\\[2mm] &= \left(\dfrac{1}{a} \cdot a\right)b\\[2mm] &= (1)b\\ &= b\end{aligned}$ Commutative Property of Multiplication

Associative Property of Multiplication

Multiplicative Inverse Property
Multiplicative Identity

151. Commutative   **153.** Commutative

155. Answers will vary.

157. a. yes;

$a * b = a + b + 3$
$b * a = b + a + 3 = a + b + 3$
$a * b = b * a$

b. yes;

$\begin{aligned}(a * b) * c &= (a + b + 3) * c\\ &= (a + b + 3) + c + 3\\ &= a + b + c + 6\\ a * (b * c) &= a * (b + c + 3)\\ &= a + (b + c + 3) + 3\\ &= a + b + c + 6\end{aligned}$

$(a * b) * c = a * (b * c)$

159. No; Example: Let $a = 1$, $b = 2$, $c = 3$ then

$$a + bc = 1 + 6 = 7$$
$$a(b + c) = 1(2 + 3) = 5$$
$$7 \neq 5$$

Review Problems

163. $3(4 - 7) - 32 \div (11 - 9) = 3(-3) - 32 \div (2) = -9 - 16 = -25$

164. $-24 \div 8 \cdot 3 + 28 \div (-7) = -3 \cdot 3 - 4 = -9 - 4 = -13$

165. $\quad y^3 - x(2z^2 - xy^2) = (-3)^3 - 2[2(-2)^2 - 2(-3)^2] \qquad x = 2,\ y = -3,\ z = -2$
$$\begin{aligned}
&= -27 - 2[2(4) - 2(9)] \\
&= -27 - 2[2(4) - 2(9)] \\
&= -27 - 2[8 - 18] \\
&= -27 - 2(-10) \\
&= -27 + 20 \\
&= -7
\end{aligned}$$

Section 1.8 Simplifying Algebraic Expressions

Problem Set 1.8, pp. 89-91

1. $7x + 10x = (7 + 10)x = 17x$

3. $10z + 20z + 5z = (10 + 20 + 5)z = 35z$

5. $8w^2 + w^2 = 9w^2$

7. $20xy^2 - 15xy^2 = (20 - 15)xy^2 = 5xy^2$

9. $-7w - (-10w) = -7w + 10w = (-7 + 10)w = 3w$

11. $-17wz - (-4wz) = -17wz + 4wz = -13wz$

13. $9r^2 + r + = 10r^2$

15. $y^2 - 8y^2 = (1 - 8)y^2 = -7y^2$

17. $-y^2 - (-7y^2) = -y^2 + 7y^2 = 6y^2$

19. $9b - b + 5b = (9 - 1 + 5)b = 13b$

21. $5x - 3 + 6x = (5x + 6x) - 3 = 11x - 3$

23. $3z - 8z + 5 = -5z + 5$

25. $2x - 5 + 7x + 4 = 2x + 7x - 5 + 4 = 9x - 1$

27. $3b - 2 + b + 5 = 3b + b - 2 + 5 = 4b + 3$

29. $-4x - 8 - x + 8 = -4x - x - 8 + 8 = -5x$

31. $7xy - 3x - 6xy - x = 7xy - 6xy - 3x - x = xy - 4x$

33. $5(3x - 2) + 4 = 15x - 10 + 4 = 15x - 6$

35. $2 + 5(3x - 2) = 2 + 15x - 10 = 15x + 2 - 10 = 15x - 8$

37. $-7(3y - 1) + 4 = -21y + 7 + 4 = -21y + 11$

39. $6 - 2(y + 1) = 6 - 2y - 2 = -2y + 6 - 2 = -2y + 4$

41. $4(2z - 3) - 5z = 8z - 12 - 5z = 8z - 5z - 12 = 3z - 12$

43. $5(x + y) + 7(y - 4x) = 5x + 5y + 7y - 28x = 5x - 28x + 5y + 7y = -23x + 12y$

45. $-3(2x - y) - 4(y + 5x) = -6x + 3y - 4y - 20x = -6x - 20x + 3y - 4y = -26x - y$

47. $8 - 3(2x - 5) = 8 - 6x + 15 = -6x + 8 + 15 = -6x + 23$

49. $-4(3w - 1) - (8 - 14w) = -12w + 4 - 8 + 14w = -12w + 14w + 4 - 8 = 2w - 4$

51. $-9 - 6(2 - r) + 4 = -9 - 12 + 6r + 4 = 6r - 17$

53. $-10(2y - 1) - (-7y - 11) = -20y + 10 + 7y + 11 = -13y + 21$

55. $-3x^2 - (-4x^2 - 10x^3) = -3x^2 + 4x^2 + 10x^3 = x^2 + 10x^3$

57. $7(y - 4) + 3y - (6 - y) = 7y - 28 + 3y - 6 + y = 7y + 3y + y - 28 - 6 = 11y - 34$

59. $w + 3(7 - w) - (2w - 4) - w^2 = w + 21 - 3w - 2w + 4 - w^2 = 21 + 4 + w - 3w - 2w - w^2 = 25 - 4w - w^2$

61. $-5(y - 6) + 3(y - 4) + 2y - 3 = -5y + 30 + 3y - 12 + 2y - 3 = -5y + 3y + 2y + 30 - 12 - 3 = 15$

63. $-\dfrac{2}{3}(12x - 15) + 18 = -8x + 10 + 18 = -8x + 28$

65. $\dfrac{3}{8}(x - 4) - \dfrac{7}{2} = \dfrac{3}{8}x - \dfrac{3}{2} - \dfrac{7}{2} = \dfrac{3}{8}x - 5$

67. $3[6 - (y + 1)] = 3(6 - y - 1) = 3(5 - y) = 15 - 3y$

69. $7 - 4[3 - (-4y - 5)] = 7 - 4(3 + 4y + 5) = 7 - 4(8 + 4y) = 7 - 32 - 16y = -25 - 16y$

71. $3[6x - 2y - 4(-5x - y)] = 3(6x - 2y + 20x + 4y) = 3(26x + 2y) = 78x + 6y$

73. $12x - 4[6x - 8y - (2x - 4y)]$
$= 12x - 4(6x - 8y - 2x + 4y)$
$= 12x - 4(4x - 4y)$
$= 12x - 16x + 16y$
$= -4x + 16y$

75. $3x - 8x + 2 \quad = 3(-3) - 8(-3) + 2 \qquad x = -3$
$\qquad\qquad\qquad = -9 + 24 + 2$
$\qquad\qquad\qquad = 17$

$\quad 3x - 8x + 2 \quad = -5x + 2$
$\qquad\qquad\qquad = -5(-3) + 2 \qquad\qquad x = -3$
$\qquad\qquad\qquad = 15 + 2$
$\qquad\qquad\qquad = 17$

77. $8x - 2 - x - 9 \quad = 8(-3) - 2 - (-3) - 9 \qquad x = -3$
$\qquad\qquad\qquad\quad = -24 - 2 + 3 - 9$
$\qquad\qquad\qquad\quad = -32$

$\quad\; 8x - 2 - x - 9 \quad = 7x - 11$
$\qquad\qquad\qquad\quad = 7(-3) - 11 \qquad\qquad x = -3$
$\qquad\qquad\qquad\quad = -21 - 11$
$\qquad\qquad\qquad\quad = -32$

79. $4(3x + 1) + 5x$ $= 4[3(-3) + 1] + 5(-3)$ $x = -3$
$= 4(-9 + 1) - 15$
$= 4(-8) - 15$
$= -32 - 15$
$= -47$

$4(3x + 1) + 5x = 12x + 4 + 5x$
$= 17x + 4$
$= 17(-3) + 4$ $x = -3$
$= -51 + 4$
$= -47$

81. $7 - 5(1 - 4x) - 19x$ $= 7 - 5[1 - 4(-3)] - 19(-3)$ $x = -3$
$= 7 - 5(1 + 12) + 57$
$= 7 - 5(13) + 57$
$= 7 - 65 + 57$
$= -1$

$7 - 5(1 - 4x) - 19x = 7 - 5 + 20x - 19x$
$= x + 2$
$= -3 + 2$ $x = -3$
$= -1$

83. The sum of a number and −4, decreased by 6:

   a. $[x + (-4)] - 6$
   b. $x - 4 - 6 = x - 10$
   c. the difference between a number and 10

85. The difference between the sum of a number and 4, and two times the number:

   a. $(x + 4) - 2x$
   b. $x + 4 - 2x = 4 - x$
   c. the difference between 4 and a number

87. The product of three times a number and −2, decreased by four times the number:

   a. $(3x)(-2) - 4x$
   b. $-6x - 4x = -10x$
   c. the product of a number and −10

89. Six times a number subtracted from ten times the number, with the result subtracted from eight times the number:

   a. $8x - (10x - 6x)$
   b. $8x - 10x + 6x = 4x$
   c. the product of 4 and a number

91. The difference between a number and six times the number, with the result subtracted from the sum of 10 and the product of 9 and the number:

   a. $(10 + 9x) - (x - 6x)$
   b. $10 + 9x - x + 6x = 10 + 14x$
   c. the sum of 10 and the product of 14 and a number

93. The sum of 4 multiplied by the sum of three times a number and 5, and the difference between 8 and the number:

   a. $4(3x + 5) + (8 - x)$
   b. $12x + 20 + 8 - x = 11x + 28$
   c. the sum of the product of 11 and a number, and 28

95. The product of the identity element of addition and three times a number decreased by 4:

   a. $(0)(3x - 4)$
   b. 0
   c. the number zero

97. The product of the identity element of addition and three times a number, with the result decreased by 4:
   a. $0(3x) - 4$
   b. $0 - 4 = -4$
   c. the number (-4)

99. Cost of airfare + cost per day × number of days

$$\boxed{500 + 60(x - 3) \text{ dollars}}$$

$$500 + 60x - 180$$

$$\boxed{320 + 60x \text{ dollars}}$$

101. Amount of paper now − paper used per week × number of weeks

$$\boxed{7000 - \frac{350}{4}(x - 2) \text{ tons}}$$

$$7000 - \frac{175}{2}(x - 2)$$

$$7000 - \frac{175}{2}x + 175$$

$$\boxed{7175 - 87.5x \text{ tons}}$$

103. D is true;

$$x - 0.02(x + 200) = x - 0.02x - 4 = 0.98x - 4.$$

105. $7.3 - 1.8(x - 3.7) - 9.62 = 7.3 - 1.8x + 6.66 - 9.62 = -1.8x + 4.34$

Review Problems

111. $\begin{aligned}4(x + 3) &= 4x + 12 \\ &= 12 + 4x\end{aligned}$ Distributive Property and
                                        Commutative Property of Addition

112. $-15 \cdot 3 \div 9 - 6 \div (-3) = -45 \div 9 + 2 = -5 + 2 = -3$

113.   4 nickels + 2 dimes + x quarters

$$\boxed{4(0.05) + 2(0.10) + x(0.25) \text{ dollars}}$$

$$= 0.2 + 0.2 + 0.25x$$

$$= \boxed{0.40 + 0.25x \text{ dollars}}$$

Chapter 1 Review Problems

Chapter 1 Review Problems, pp. 94-96

1. $\{x \mid x$ is a whole number that is less than 6$\}$

  $\{0, 1, 2, 3, 4, 5\}$

2. $\{x \mid x$ is an integer that is greater than $-3\}$

  $\{-2, -1, 0, 1, 2 \dots \}$

3.   **a.**   Natural numbers: $\left\{ \sqrt{81} \right\}$.
  b.   Whole numbers: $\left\{ 0, \sqrt{81} \right\}$.
  c.   Integers: $\left\{ -17, 0, \sqrt{81} \right\}$
  d.   Rational numbers: $\left\{ -17, -\dfrac{9}{13}, 0, 0.75, 5\tfrac{1}{4}, \sqrt{81} \right\}$.
  e.   Irrational numbers: $\left\{ \sqrt{2}, \pi \right\}$.
  f.   Real numbers: $\left\{ -17, -\dfrac{9}{13}, 0, 0.75, \sqrt{2}, \pi, 5\tfrac{1}{4}, \sqrt{81} \right\}$.

4.   17 $\boxed{>}$ 5

5. $-|-3.2| \ \boxed{\phantom{<}} \ -(-3.2)$

  $-3.2 \ \boxed{<} \ 3.2$

6. $0 \boxed{>} -\dfrac{1}{3}$

7. $-\dfrac{1}{4} \ \boxed{<} \ -\dfrac{1}{5}$

8. $6 + [-7 + (-3)] = 6 + (-10) = -4$

9. $[-11 + (-5)] + [-8 + (-4)] = (-16) + (-12) = -28$

10. $-8.6 - (-3.4) = -8.6 + 3.4 = -5.2$

11. $(-9 - 6) - (-10 - 7) = (-15) - (-17) = -15 + 17 = 2$

12. $-5 + [(-8 + 2) - (-2 - 11)] = -5 + [-6 - (-13)] = -5 + (-6 + 13) = -5 + 7 = 2$

13. $\left(\frac{1}{2}-\frac{1}{3}\right)-\left(-\frac{1}{12}\right)=\frac{1}{6}+\frac{1}{12}=\frac{3}{12}=\frac{1}{4}$

14. $(-5.2)(-0.3) = 1.56$

15. $(-10)(-6) - (-8)(4) = 60 + 32 = 92$

16. $-8[-4 - 5(-3)] = -8(-4 + 15) = -8(11) = -88$

17. $(-8 - 4)(-3) - 10 = (-12)(-3) - 10 = 36 - 10 = 26$

18. $\left|-4(-10)\right| - \left|-7\right| = 40 - 7 = 33$

19. $-\frac{2}{3} \div \left(-\frac{8}{9}\right) = -\frac{2}{3} \cdot \left(-\frac{9}{8}\right) = \frac{3}{4}$

20. $-14.8 \div 0.8 = -18.5$

21. $\frac{9(-3)}{6-(-3)} = \frac{-27}{6+3} = -\frac{27}{9} = -3$

22. $\frac{6(-10+3)}{2(-15)-9(-3)} = \frac{6(-7)}{-30+27} = \frac{-42}{-3} = 14$

23. $\frac{5[-8-(-3)]}{2(-13)+(-4+1)(-6-1)} = \frac{5(-8+3)}{-26+(-3)(-7)} = \frac{5(-5)}{-26+21} = \frac{-25}{-5} = 5$

24. $\frac{6^2-8}{4(-1)+2(2)} = \frac{36-8}{-4+4} = \frac{28}{0}$ is undefined

25. $\frac{-7[8-(-6+11)]}{-6[3-(-2)]-3(-3)} = \frac{-7(8-5)}{-6(3+2)+9} = \frac{-7(3)}{-6(5)+9} = \frac{-21}{-30+9} = \frac{-21}{-21} = 1$

26. $8^2 - 36 \div 3^2 \cdot 4 - (-7) = 64 - 36 \div 9 \cdot 4 + 7 = 64 - 4 \cdot 4 + 7 = 64 - 16 + 7 = 55$

27. $-\frac{1}{2}(-16)(-5) = 8(-5) = -40$

28. $(-3)^3 \cdot 2^2 = -27 \cdot 4 = -108$

29. $25 \div \left(\frac{11+9}{2^3 \cdot 3}\right) - 6 = 25 \div \left(\frac{20}{8 \cdot 3}\right) - 6 = 25 \div \frac{20}{24} - 6 = 30 - 6 = 24$

30. $-2^4 + (-2)^4 = -16 + 16 = 0$

31. $3 + (9 \div 3)^3 - 25 \div 5 \cdot 2 - (4-1)^3 = 3 + 3^3 - 5 \cdot 2 - 3^3 = 3 + 27 - 10 - 27 = -7$

32. $\left[\left(\frac{1}{2}\right)^2 -\frac{1}{2}\right]^2 = \left(\frac{1}{4}-\frac{1}{2}\right)^2 = \left(-\frac{1}{4}\right)^2 = \frac{1}{16}$

33. $\frac{5}{12} - \frac{11}{12} \div \left(\frac{1}{6}-\frac{3}{8}\right) = \frac{5}{12} - \frac{11}{12} \div \left(\frac{-5}{24}\right) = \frac{5}{12} - \frac{11}{12} \cdot \left(-\frac{24}{5}\right) = \frac{5}{12} + \frac{22}{5} = \frac{25+264}{60} = \frac{289}{60}$

34. The sum of -14 and -12, decreased by -20:

$[-14 + (-12)] - (-20) = -26 + 20 = -6.$

35. -5 decreased by 8 less than -3:

$-5 - (-3 - 8) = -5 - (-11) = -5 + 11 = 6$

36. The difference between -8 and the product of -7 and 6:

$-8 - (-7)(6) = -8 + 42 = 34.$

37. 70% of the difference between 32 and -8:

$0.7[32 - (-8)] = 0.7(32 + 8) = 0.7(40) = 28.$

38. -10 added to seven-eighths of the difference between -3 and 5:

$\frac{7}{8}(-3 - 5) + (-10) = \frac{7}{8}(-8) + (-10) = -7 - 10 = -17.$

39. The quotient of the sum of 12 and -8 and the difference between -8 and -6:

$\frac{12 + (-8)}{-8 - (-6)} = \frac{4}{-8 + 6} = \frac{4}{-2} = -2.$

40. The quotient of the difference between 20 and -16 and the product of -4 and 3:

$\frac{20 - (-16)}{(-4)(3)} = \frac{20 + 16}{-12} = \frac{36}{-12} = -3.$

41. The quotient of 25 and -5 less than the quotient of -16 and 4:

$\frac{25}{(-16) \div (4) - (-5)} = \frac{25}{-4 + 5} = 25.$

42. Elevation of Dead Sea + elevation of person

$-1312 + 512 = -800$

The person's elevation is $\boxed{\text{800 feet below sea level}}$.

43. elevation of plane $-$ elevation of submarine

$26,500 - (-650) = 26,500 + 650 = 27,150$

The difference in elevation is $\boxed{\text{27,150 feet}}$.

44. The value in cents of x 50-cent pieces:

value of each 50-cent piece $\times$ number of 50-cent pieces

$50x$

45. The amount of weight in an elevator that is carrying a 200-pound operator plus x bags of concrete weighing 45 pounds each:

weight of operator plus weight of each bag times number of bags

$200 + 45x$

46. The reduced price of an item priced at x dollars with a 35% discount:

price of item – discount

$x - 0.35x$

$0.65x$

47. The cost per pencil when x pencils are purchased for $6:

total cost ÷ by number of pencils

$\dfrac{6}{x}$ dollars

48. The weekly salary for a person who earns x dollars per month

salary per year ÷ by number of weeks per year

$\dfrac{x \text{ dollars per month} \times 12 \text{ months per year}}{52 \text{ weeks per year}}$

$\dfrac{12x}{52}$ dollars per week

$\dfrac{3x}{13}$ dollars per week

49. $C = \dfrac{5}{9}(F - 32)$

    a. $C = \dfrac{5}{9}(32 - 32)$ $F = 32°$

            $= \dfrac{5}{9}(0)$

            $= \boxed{0°C}$

    b. $C = \dfrac{5}{9}(14 - 32)$ $F = 14°$

            $= \dfrac{5}{9}(-18)$

            $= \boxed{-10°C}$

50. $N = 14W^3 - 17W^2 - 16W + 34$
$= 14(2^3) - 17(2^2) - 16(2) + 34$
$= 14(8) - 17(4) - 32 + 34$
$= 112 - 68 - 32 + 34$
$= 46$

A moth with an abdominal width of 2 millimeters has $\boxed{46 \text{ eggs}}$.

51. $T = 3(A - 20)^2 \div 50 + 10$

Time for 40-year-old:
$$\begin{aligned} T &= 3(40 - 20)^2 \div 50 + 10 \\ &= 3(20)^2 \div 50 + 10 \\ &= 3(400) \div 50 + 10 \\ &= 1,200 \div 50 + 10 \\ &= 24 + 10 \\ &= 34 \end{aligned}$$

Time for 30-year-old:
$$\begin{aligned} T &= 3(30 - 20)^2 \div 50 + 10 \\ &= 3(10)^2 \div 50 + 10 \\ &= 3(100) \div 50 + 10 \\ &= 300 \div 50 + 10 \\ &= 6 + 10 \\ &= 16 \end{aligned}$$

difference in time: $34 - 16 = 18$

The difference in time is $\boxed{18 \text{ seconds}}$.

52. After 1 hour:
$$\begin{aligned} T &= 10t + 40 \\ &= 10(1) + 40 \qquad t = 1 \\ &= 10 + 40 \\ &= 50 \end{aligned}$$

One hour after the experiment begins, the temperature of the environment is 50°F.

$$\begin{aligned} N &= -2T^2 + 240T - 5400 \\ &= -2(50)^2 + 240(50) - 5,400 \qquad T = 50 \\ &= -2(2,500) + 12,000 - 5,400 \\ &= -5,000 + 12,000 - 5,400 \\ &= 1,600 \end{aligned}$$

There are $\boxed{1600 \text{ bacteria}}$ present one hour after the experiment began.

At the beginning:
$$\begin{aligned} T &= 10(0) + 40 \qquad t = 0 \\ &= 0 + 40 \\ &= 40 \end{aligned}$$

$$\begin{aligned} N &= -2(40)^2 + 240(40) - 5,400 \\ &= -2(1,600) + 9,600 - 5,400 \\ &= -3,200 + 9,600 - 5,400 \\ &= 1,000 \end{aligned}$$

There are $\boxed{1000 \text{ bacteria}}$ present at the beginning.

53. $\begin{aligned} -6x^2 - 2y^2 &= -6(-1)^2 - 2(-4)^2 \qquad x = -1, y = -4 \\ &= -6(1) - 2(16) \\ &= -6 - 32 \\ &= -38 \end{aligned}$

54. $\begin{aligned} y - 3(y - 6x) &= -4 - 3[(-4) - 6(-1)] \qquad x = -1, y = -4 \\ &= -4 - 3[-4 + 6] \\ &= -4 - 3(2) \\ &= -4 - 6 \\ &= -10 \end{aligned}$

55. $\dfrac{y-4x}{x^2+y^3} = \dfrac{(-4)-4(-1)}{(-1)^2+(-4)^3}$ $x=-1,\ y=-4$

$$= \dfrac{-4+4}{1+(-64)}$$

$$= \dfrac{0}{-63}$$

$$= 0$$

56. $\dfrac{|x-y|}{6y-xy} = \dfrac{|(-1)-(-4)|}{6(-4)-(-1)(-4)}$ $x=-1,\ y=-4$

$$= \dfrac{|-1+4|}{-24-4}$$

$$= \dfrac{3}{-28}$$

$$= -\dfrac{3}{28}$$

57. $\dfrac{2x}{-4-y} = \dfrac{2(-1)}{-4-(-4)}$ $x=-1,\ y=-4$

$$= \dfrac{-2}{0}\ \text{is undefined.}$$

58. $(7+5)+(-2)=(5+7)+(-2)$

Commutative Property of Addition

59. $1x=x$

Multiplicative Identity Property

60. $5(4x-3)=20x-15$

Distributive Property

61. $-2(3x)=(-2\cdot 3)x=-6x$

Associative Property of Multiplication

62. $17+(-17)=0$

Additive Inverse Property

63. $3y+8y-y-15y+6=-5y+6$

64. $5s-9-(-10)-6s-(-45)=5s-9+10-6s+4s=3s+1$

65. $7(3x-2y)-4(y-6x)=21x-14y-4y+24x=45x-18y$

66. $7(3z-2)-(-4z+9)=21z-14+4z-9=25z-23$

67. $-6y^2-(-3y^2-10y^3)=-6y^2+3y^2+10y^3=-3y^2+10y^3$

68. $7(3w-1)-5(1-2w)-(8w+3)=21w-7-5+10w-8w-3=23w-15$

69. $7x-4[x-3(2x-1)]=7x-4(x-6x+3)=7x-4(-5x+3)=7x+20x-12=27x-12$

70. $-4[-3(x+5y)-2(y-x)]=-4(-3x-15y-2y+2x)=-4(-x-17y)=4x+68y$

71. The difference between the sum of a number and –6, and three times the number:

    a. $[x + (-6)] - 3x$
    b. $x - 6 - 3x = -2x - 6$
    c. the difference between the product of a number and –2, and 6

72. The product of five times a number and –2, decreased by –3 times the number:

    a. $(5x)(-2) - (-3x)$
    b. $-10x + 3x = -7x$
    c. the product of –7 and a number

73. The difference between a number and four times the number, with the result subtracted from the sum of 6 and the product of 5 and the number:

    a. $(6 + 5x) - (x - 4x)$
    b. $6 + 5x - x + 4x = 8x + 6$
    c. the sum of the product of 8 and a number, and 6

74. The sum of 5 multiplied by the sum of three times a number and 7, and the difference between 6 and the number:

    a. $5(3x + 7) + (6 - x)$
    b. $15x + 35 + 6 - x = 14x + 41$
    c. the sum of the product of 14 and a number, and 41

75. The difference between 11 and four times the sum of $\frac{1}{2}$ and double a number:

    a. $11 - 4\left(\frac{1}{2} + 2x\right)$
    b. $11 - 2 - 8x = 9 - 8x$
    c. the difference of 9 and 8 times a number

Chapter 2   Linear Equations and Inequalities in One Variable

Section 2.1 The Addition Property of Equality

Problem Set 2.1, pp. 109–110

1.
$$\begin{aligned}
x - 7 &= 13 \\
x - 7 + 7 &= 13 + 7 \\
x + 0 &= 20 \\
\textit{Check: } x - 7 &= 13 \\
20 - 7 &= 13 \\
13 &= 13 \ \checkmark
\end{aligned}$$
$\boxed{\{20\}}$

3.
$$\begin{aligned}
z + 5 &= -12 \\
x + 5 - 5 &= -12 - 5 \\
z &= -17 \\
\textit{Check: } z + 5 &= -12 \\
-17 + 5 &= -12 \\
-12 &= -12 \ \checkmark
\end{aligned}$$
$\boxed{\{-17\}}$

5.
$$\begin{aligned}
3.2 + x &= 7.5 \\
3.2 + x - 3.2 &= 7.5 - 3.2 \\
x &= 4.3 \\
\textit{Check: } 3.2 + 4.3 &= 7.5 \\
7.5 &= 7.5 \ \checkmark
\end{aligned}$$
$\boxed{\{4.3\}}$

7.
$$\begin{aligned}
x - \frac{3}{4} &= \frac{9}{2} \\
x - \frac{3}{4} + \frac{3}{4} &= \frac{9}{2} + \frac{3}{4} \\
x &= \frac{21}{4} \\
\textit{Check: } \frac{21}{4} - \frac{3}{4} &= \frac{9}{2} \\
\frac{18}{4} &= \frac{9}{2} \\
\frac{9}{2} &= \frac{9}{2} \ \checkmark
\end{aligned}$$
$\boxed{\left\{\dfrac{21}{4}\right\}}$

9.
$$\begin{aligned}
5 &= -13 + y \\
5 + 13 &= y \\
18 &= y \\
\textit{Check: } 5 &= -13 + 18 \\
5 &= 5 \ \checkmark
\end{aligned}$$
$\boxed{\{18\}}$

11.
$$\begin{aligned}
-\frac{3}{5} &= -\frac{3}{2} + s \\
-\frac{3}{5} + \frac{3}{2} &= s \\
-\frac{6}{10} + \frac{15}{10} &= s \\
\frac{9}{10} &= s \\
\textit{Check: } -\frac{3}{5} &= -\frac{3}{2} + \frac{9}{10} \\
-\frac{6}{10} &= -\frac{15}{10} + \frac{9}{10} \\
-\frac{6}{10} &= -\frac{6}{10} \ \checkmark
\end{aligned}$$
$\boxed{\left\{\dfrac{9}{10}\right\}}$

13.
$$830 + y = 520$$
$$y = 520 - 830$$
$$y = -310$$
Check: $830 - 310 = 520$
$$520 = 520 \ \checkmark$$

$$\boxed{\{-310\}}$$

15.
$$r + 3.7 = 8$$
$$r = 8 - 3.7$$
$$r = 4.3$$
Check: $4.3 + 3.7 = 8$
$$8 = 8 \ \checkmark$$

$$\boxed{\{4.3\}}$$

17.
$$3\tfrac{2}{5} + x = 5\tfrac{2}{5}$$
$$x = 5\tfrac{2}{5} - 3\tfrac{2}{5}$$
$$x = 2$$
Check: $3\tfrac{2}{5} + 2 = 5\tfrac{2}{5}$
$$5\tfrac{2}{5} = 5\tfrac{2}{5}$$

$$\boxed{\{2\}}$$

19.
$$-3.7 + m = -3.7$$
$$m = -3.7 + 3.7$$
$$m = 0$$
Check: $-3.7 + 0 = -3.7$
$$-3.7 = -3.7 \ \checkmark$$

$$\boxed{\{0\}}$$

21.
$$6y + 3 - 5y = 14$$
$$y + 3 = 14$$
$$y = 14 - 3$$
$$y = 11$$
Check: $6(11) + 3 - 5(11) = 14$
$$66 + 3 - 55 = 14$$
$$14 = 14 \ \checkmark$$

$$\boxed{\{11\}}$$

23.
$$6y + 5 - 5y + 2 = 4(4)$$
$$y + 7 = 4(4)$$
$$y = 16 - 7$$
$$y = 9$$
Check: $6(9) + 5 - 5(9) + 2 = 4(4)$
$$54 + 5 - 45 + 2 = 16$$
$$16 = 16 \ \checkmark$$

$$\boxed{\{9\}}$$

25.
$$7 - 5x + 8 + 2x + 4x - 3 = 2 + 3 \cdot 5$$
$$x + 12 = 2 + 15$$
$$x = 17 - 12$$
$$x = 5$$
Check: $7 - 5(5) + 8 + 2(5) + 4(5) - 3 = 2 + 3 \cdot 5$
$$7 - 25 + 8 + 10 + 20 - 3 = 2 + 15$$
$$45 - 28 = 17$$
$$17 = 17 \ \checkmark$$

$$\boxed{\{5\}}$$

27.
$$7y + 4 = 6y - 9$$
$$7y - 6y + 4 = -9$$
$$y = -9 - 4$$
$$y = -13$$
Check: $7(-13) + 4 = 6(-13) - 9$
$$-91 + 4 = -78 - 9$$
$$-87 = -87 \ \checkmark$$

$$\boxed{\{-13\}}$$

29.
$$3(2x - 4) = 10 + 5x$$
$$6x - 12 = 10 + 5x$$
$$x - 12 = 10$$
$$x = 22$$
Check: $3[2(22) - 4] = 10 + 5(22)$
$$3[44 - 4] = 10 + 110$$
$$3(40) = 120$$
$$120 = 120 \checkmark$$

$$\boxed{\{22\}}$$

31.
$$18 - 7x = 12 - 6x$$
$$18 = 12 + x$$
$$6 = x$$
Check:
$$18 - 7(6) = 12 - 6(6)$$
$$18 - 42 = 12 - 36$$
$$-24 = -24 \ \checkmark$$

$$\boxed{\{6\}}$$

33.
$$2(y + 3) = y + 4$$
$$2y + 6 = y + 4$$
$$y + 6 = 4$$
$$y = -2$$
Check:
$$2(-2 + 3) = -2 + 4$$
$$2(1) = 2$$
$$2 = 2 \ \checkmark$$

$$\boxed{\{-2\}}$$

35.
$$-3(y - 4) + 7 = 3 - 4y$$
$$-3y + 12 + 7 = 3 - 4y$$
$$-3y + 19 = 3 - 4y$$
$$y + 19 = 3$$
$$y = -16$$
Check:
$$-3(-16 - 4) + 7 = 3 - 4(-16)$$
$$-3(-20) + 7 = 3 + 64$$
$$60 + 7 = 67$$
$$67 = 67 \ \checkmark$$

$$\boxed{\{-16\}}$$

37.
$$5(2y + 1) + 6 = 8 + 9y$$
$$10y + 5 + 6 = 8 + 9y$$
$$10y + 11 = 8 + 9y$$
$$y + 11 = 8$$
$$y = -3$$
Check:
$$5[2(-3) + 1] + 6 = 8 + 9(-3)$$
$$5(-6 + 1) + 6 = 8 - 27$$
$$5(-5) + 6 = -19$$
$$-25 + 6 = -19$$
$$-19 = -19 \ \checkmark$$

$$\boxed{\{-3\}}$$

39.
$$-(y + 3) - 1 = 6 - 2y$$
$$-y - 3 - 1 = 6 - 2y$$
$$-y - 4 = 6 - 2y$$
$$y - 4 = 6$$
$$y = 10$$
Check:
$$-(10 + 3) - 1 = 6 - 2(10)$$
$$-13 - 1 = 6 - 20$$
$$-14 = -14 \ \checkmark$$

$$\boxed{\{10\}}$$

41.
$$4x + 2 = 3(x - 6) + 8$$
$$4x + 2 = 3x - 18 + 8$$
$$4x + 2 = 3x - 10$$
$$x + 2 = -10$$
$$x = -12$$
Check:
$$4(-12) + 2 = 3(-12 - 6) + 8$$
$$-48 + 2 = 3(-18) + 8$$
$$-42 = -54 + 8$$
$$-42 = -42 \ \checkmark$$

$$\boxed{\{-12\}}$$

43.
$$2(3y + 1) + 10 = 5(y + 2) + 1$$
$$6y + 2 + 10 = 5y + 10 + 1$$
$$6y + 12 = 5y + 11$$
$$y + 12 = 11$$
$$y = -1$$
Check:
$$2[3(-1) + 1] + 10 = 5(-1 + 2) + 1$$
$$2(-2) + 10 = 5(1) + 1$$
$$-4 + 10 = 5 + 1$$
$$6 = 6 \ \checkmark$$

$$\boxed{\{-1\}}$$

45.
$$-3(2x - 9) + 9 = 12 - 7(x - 4)$$
$$-6x + 27 + 9 = 12 - 7x + 28$$
$$-6x + 36 = 40 - 7x$$
$$x + 36 = 40$$
$$x = 4$$
Check:
$$-3[2(4) - 9] + 9 = 12 - 7(4 - 4)$$
$$-3(8 - 9) + 9 = 12 - 7(0)$$
$$-3(-1) + 9 = 12 - 0$$
$$3 + 9 = 12$$
$$12 = 12 \ \checkmark$$

$$\boxed{\{4\}}$$

47.

$$
\begin{aligned}
3(2-5y)+4(7y-1) &= 4(5+3y) \\
6-15y+28y-4 &= 20+12y \\
2+13y &= 20+12y \\
2+y &= 20 \\
y &= 18
\end{aligned}
$$

Check:
$$
\begin{aligned}
3[2-5(18)]+4[7(18)-1] &= 4[5+3(18)] \\
3(2-90)+4(126-1) &= 4(5+54) \\
3(-88)+4(125) &= 4(59) \\
-264+500 &= 236 \\
236 &= 236 \ \sqrt{}
\end{aligned}
$$

$$\boxed{\{18\}}$$

49.

$$
\begin{aligned}
2(2z+3)-6 &= -3(4-7z)-2(8z+7) \\
4z+6-6 &= -12+21z-16z-14 \\
4z &= 5z-26 \\
0 &= z-26 \\
26 &= z
\end{aligned}
$$

Check:
$$
\begin{aligned}
2[2(26)+3]-6 &= 3[4-7(26)]-2[8(26)+7] \\
2(52+3)-6 &= 3(4-182)-2(208+7) \\
2(55)-6 &= -3(-178)-2(215) \\
110-6 &= 534-430 \\
104 &= 104 \ \sqrt{}
\end{aligned}
$$

$$\boxed{\{26\}}$$

51. Four times a number is five more than three times the number:

$$
\begin{aligned}
4x &= 5+3x \\
x &= \boxed{5}
\end{aligned}
$$

53. If five times a number is added to three times the number, the result is the sum of seven times the number and 11:

$$
\begin{aligned}
5x+3x &= 7x+11 \\
8x &= 7x+11 \\
x &= \boxed{11}
\end{aligned}
$$

55. The sum of five times a number and 8 is ten less than four times the number:

$$
\begin{aligned}
5x+8 &= 4x-10 \\
x+8 &= -10 \\
x &= \boxed{-18}
\end{aligned}
$$

57. Twelve minus six times a number is equal to the difference of 13 and five times the number:

$$
\begin{aligned}
12-6x &= 13-5x \\
12 &= 13+x \\
\boxed{-1} &= x
\end{aligned}
$$

59. The sum of five and three-halves times a number is the sum of nine and half the number:

$$
\begin{aligned}
5+\frac{3}{2}x &= 9+\frac{1}{2}x \\
5+x &= 9 \\
x &= \boxed{4}
\end{aligned}
$$

61.
$$\begin{aligned} C + M &= S \\ C + 95.20 &= 698.75 \\ C &= 698.75 - 95.20 \\ C &= 603.55 \end{aligned}$$
The cost (to the retailer) of the television is $\boxed{\$603.55}$.

63. The width of a rectangle is 2.7 meters more than its length.
$$\begin{aligned} W &= 2.7 + L \\ 8.3 &= 2.7 + L \\ 5.6 &= L \end{aligned}$$
The length is $\boxed{5.6 \text{ meters}}$.

65.
$$\begin{aligned} x - a &= b \\ x &= \boxed{a + b} \end{aligned}$$

67.
$$\begin{aligned} a + b &= \frac{c}{2} \\ a &= \boxed{\frac{c}{2} - b} \end{aligned}$$

69.
$$\begin{aligned} -2 &= a + b \\ -2 - b &= a \\ a &= \boxed{-2 - b} \end{aligned}$$

71.
$$\begin{aligned} 4x + 3a &= 3x - 2a \\ x + 3a &= -2a \\ x &= \boxed{-5a} \end{aligned}$$

73.
$$\begin{aligned} 4(2x - 3b) &= 7x + 5b \\ 8x - 12b &= 7x + 5b \\ x - 12b &= 5b \\ x &= \boxed{17b} \end{aligned}$$

75.
$$\begin{aligned} 3y(y - 1) &= 2y + 2 \\ 3\left(-\frac{1}{3}\right)\left(-\frac{1}{3} - 1\right) &= 2\left(-\frac{1}{3}\right) + 2 \\ -1\left(-\frac{4}{3}\right) &= -\frac{2}{3} + 2 \\ \frac{4}{3} &= \frac{4}{3} \ \surd \end{aligned}$$
$\boxed{\text{yes}}$

77.
$$\begin{aligned} y(y + 1) &= 2.6 - 2y \\ (-1.8)(-1.8 + 1) &= 2.6 - 2(-1.8) \\ (-1.8)(-0.8) &= 2.6 + 3.6 \\ 1.44 &\neq 6.2 \end{aligned}$$
$\boxed{\text{no}}$

79. $\boxed{\text{C}}$ is true; "eight less than a number (x) gives 15":
$$x - 8 = 15$$

81.
$$\begin{aligned} 2(y + 0.08) &= 5y - 2.6 \\ y &= 4.4 \end{aligned}$$
Using a calculator:
$$\begin{aligned} \text{LHS:} &\quad 8.96 \\ \text{RHS:} &\quad 19.4 \\ \text{LHS} &\neq \text{RHS} \end{aligned}$$
$\boxed{\text{no}}$

Review Problems

87. $-16 - (50 \div 5^2) = -16 - (50 \div 25) = -16 - 2 = -18$

88. $3y(y - 5) + 5y(y + 2) = 3y^2 - 15y + 5y^2 + 10y = 8y^2 - 5y$

89. $d = t^2 + 7t = 5^2 + 7(5)$ $t = 5$ seconds
$$= 25 + 35$$
$$= 60$$

The shock wave travels $\boxed{60 \text{ miles}}$ in 5 seconds.

Section 2.2 The Multiplication Property of Equality

Problem Set 2.2, pp. 121–124

1.
$$5x = 45$$
$$\frac{5x}{5} = \frac{45}{5}$$
$$x = 9$$
Check: $5(9) = 45$
$$45 = 45 \; \checkmark$$
$\boxed{\{9\}}$

3.
$$7b = 56$$
$$\frac{7b}{7} = \frac{56}{7}$$
$$b = 8$$
Check: $7(8) = 56$
$$56 = 56 \; \checkmark$$
$\boxed{\{8\}}$

5.
$$8r = -24$$
$$\frac{8r}{8} = -\frac{24}{8}$$
$$r = -3$$
Check: $8(-3) = -24$
$$-24 = -24 \; \checkmark$$
$\boxed{\{-3\}}$

7.
$$-3y = -15$$
$$\frac{-3y}{-3} = \frac{-15}{-3}$$
$$y = 5$$
Check: $-3(5) = -15$
$$-15 = -15 \; \checkmark$$
$\boxed{\{5\}}$

9.
$$-8m = 2$$
$$\frac{-8m}{-8} = \frac{2}{-8}$$
$$m = -\frac{1}{4}$$
Check: $-8\left(-\frac{1}{4}\right) = 2$
$$2 = 2 \; \checkmark$$
$\boxed{\left\{-\dfrac{1}{4}\right\}}$

11.
$$7y = 0$$
$$\frac{7y}{7} = \frac{0}{7}$$
$$y = 0$$
Check: $7(0) = 0$
$$0 = 0 \; \checkmark$$
$\boxed{\{0\}}$

13.
$$\frac{y}{3} = 4$$
$$3\left(\frac{y}{3}\right) = 3(4)$$
$$y = 12$$
Check: $\frac{12}{3} = 4$
$$4 = 4 \; \checkmark$$
$\boxed{\{12\}}$

15.
$$-\frac{x}{5} = 11$$
$$(-5)\left(-\frac{x}{5}\right) = (-5)(11)$$
$$x = -55$$
Check: $-\frac{(-55)}{5} = 11$
$$11 = 11 \; \checkmark$$
$\boxed{\{-55\}}$

17.

$$-\frac{x}{5} = -10$$
$$(-5)\left(-\frac{x}{5}\right) = (-5)(-10)$$
$$x = 50$$

Check:
$$-\frac{50}{5} = -10$$
$$-10 = -10 \ \checkmark$$

$$\boxed{\{50\}}$$

19.

$$\frac{2}{3}y = 8$$
$$\frac{3}{2}\left(\frac{2}{3}y\right) = \frac{3}{2}(8)$$
$$y = 12$$

Check:
$$\frac{2}{3}(12) = 8$$
$$8 = 8 \ \checkmark$$

$$\boxed{\{12\}}$$

21.

$$-\frac{2}{5}a = \frac{6}{15}$$
$$-\frac{5}{2}\left(-\frac{2}{5}a\right) = -\frac{5}{2}\left(\frac{6}{15}\right)$$
$$a = -1$$

Check:
$$-\frac{2}{5}(-1) = \frac{6}{15}$$
$$\frac{2}{5} = \frac{2}{5} \ \checkmark$$

$$\boxed{\{-1\}}$$

23.

$$-\frac{7}{2}x = -21$$
$$-\frac{2}{7}\left(-\frac{7}{2}x\right) = -\frac{2}{7}(-21)$$
$$x = 6$$

Check:
$$-\frac{7}{2}(6) = -21$$
$$-21 = -21 \ \checkmark$$

$$\boxed{\{6\}}$$

25.

$$-r = 7$$
$$(-1)(-r) = (-1)(7)$$
$$r = -7$$

Check:
$$-(-7) = 7$$
$$7 = 7 \ \checkmark$$

$$\boxed{\{-7\}}$$

27.

$$-15 = -y$$
$$(-1)(-15) = (-1)(-y)$$
$$15 = y$$

Check:
$$-15 = -(15)$$
$$-15 = -15 \ \checkmark$$

$$\boxed{\{15\}}$$

29.

$$-4y - 2y = 24$$
$$-6y = 24$$
$$y = -4$$

Check:
$$-4(-4) - 2(-4) = 24$$
$$16 + 8 = 24$$
$$24 = 24 \ \checkmark$$

$$\boxed{\{-4\}}$$

31.

$$5y + 3y - 4y = 10 + 2$$
$$4y = 12$$
$$y = 3$$

Check:
$$5(3) + 3(3) - 4(3) = 10 + 2$$
$$15 + 9 - 12 = 12$$
$$12 = 12 \ \checkmark$$

$$\boxed{\{3\}}$$

33.

$$-6 - 2 = 5y + 3y - 10y$$
$$-8 = -2y$$
$$\frac{-8}{-2} = y$$
$$y = 4$$

Check:
$$-6 - 2 = 5(4) + 3(4) - 10(4)$$
$$-8 = 20 + 12 - 40$$
$$-8 = -8 \ \checkmark$$

$$\boxed{\{4\}}$$

35.

$$3y - 2 = 9$$
$$3y = 9 + 2$$
$$3y = 11$$
$$y = \frac{11}{3}$$

Check:
$$3\left(\frac{11}{3}\right) - 2 = 9$$
$$11 - 2 = 9$$
$$9 = 9 \ \checkmark$$

$$\boxed{\left\{\frac{11}{3}\right\}}$$

37.
$$\begin{aligned} 2a + 1 &= 7 \\ 2a &= 6 \\ a &= 3 \end{aligned}$$
Check:
$$\begin{aligned} 2(3) + 1 &= 7 \\ 6 + 1 &= 7 \\ 7 &= 7 \ \checkmark \end{aligned}$$

$\boxed{\{3\}}$

39.
$$\begin{aligned} -2y + 5 &= 7 \\ -2y &= 2 \\ y &= -1 \end{aligned}$$
Check:
$$\begin{aligned} -2(-1) + 5 &= 7 \\ 2 + 5 &= 7 \\ 7 &= 7 \ \checkmark \end{aligned}$$

$\boxed{\{-1\}}$

41.
$$\begin{aligned} -2y - 5 &= 7 \\ -2y &= 7 + 5 \\ -2y &= 12 \\ y &= -6 \end{aligned}$$
Check:
$$\begin{aligned} -2(-6) - 5 &= 7 \\ 12 - 5 &= 7 \\ 7 &= 7 \ \checkmark \end{aligned}$$

$\boxed{\{-6\}}$

43.
$$\begin{aligned} 12 &= 4m + 3 \\ 12 - 3 &= 4m \\ 9 &= 4m \\ \frac{9}{4} &= m \end{aligned}$$
Check:
$$\begin{aligned} 12 &= 4\left(\frac{9}{4}\right) + 3 \\ 12 &= 9 + 3 \\ 12 &= 12 \ \checkmark \end{aligned}$$

$\boxed{\left\{\dfrac{9}{4}\right\}}$

45.
$$\begin{aligned} 0.03x + 21 &= 27 \\ 0.03x &= 27 - 21 \\ 0.03x &= 6 \\ x &= \frac{6}{0.03} \\ x &= 200 \end{aligned}$$
Check:
$$\begin{aligned} 0.03(200) + 21 &= 27 \\ 6 + 21 &= 27 \\ 27 &= 27 \ \checkmark \end{aligned}$$

$\boxed{\{200\}}$

47.
$$\begin{aligned} -x - 3 &= 3 \\ -x &= 3 + 3 \\ -x &= 6 \\ x &= -6 \end{aligned}$$
Check:
$$\begin{aligned} -(-6) - 3 &= 3 \\ 6 - 3 &= 3 \\ 3 &= 3 \ \checkmark \end{aligned}$$

$\boxed{\{-6\}}$

49.
$$\begin{aligned} -x - \frac{1}{3} &= \frac{2}{3} \\ -x &= \frac{2}{3} + \frac{1}{3} \\ -x &= 1 \\ x &= -1 \end{aligned}$$
Check:
$$\begin{aligned} -(-1) - \frac{1}{3} &= \frac{2}{3} \\ 1 - \frac{1}{3} &= \frac{2}{3} \\ \frac{2}{3} &= \frac{2}{3} \ \checkmark \end{aligned}$$

$\boxed{\{-1\}}$

51.
$$\begin{aligned} 6y &= 2y - 12 \\ 6y - 2y &= -12 \\ 4y &= -12 \\ y &= -3 \end{aligned}$$
Check:
$$\begin{aligned} 6(-3) &= 2(-3) - 12 \\ -18 &= -6 - 12 \\ -18 &= -18 \ \checkmark \end{aligned}$$

$\boxed{\{-3\}}$

53.

$$
\begin{aligned}
3x &= -2x - 15 \\
3x + 2x &= -15 \\
5x &= -15 \\
x &= \frac{-15}{5} \\
x &= -3
\end{aligned}
$$

Check:
$$
\begin{aligned}
3(-3) &= -2(-3) - 15 \\
-9 &= 6 - 15 \\
-9 &= -9 \; v
\end{aligned}
$$

$\boxed{\{-3\}}$

55.

$$
\begin{aligned}
-5y &= -2y - 12 \\
-5y + 2y &= -12 \\
-3y &= -12 \\
y &= \frac{-12}{-3} \\
y &= 4
\end{aligned}
$$

Check:
$$
\begin{aligned}
-5(4) &= -2(4) - 12 \\
-20 &= -8 - 12 \\
-20 &= -20 \; \surd
\end{aligned}
$$

$\boxed{\{4\}}$

57.

$$
\begin{aligned}
8y + 4 &= 2y - 5 \\
8y - 2y + 4 &= -5 \\
6y + 4 &= -5 \\
6y &= -5 - 4 \\
6y &= -9 \\
y &= -\frac{9}{6} \\
y &= -\frac{3}{2}
\end{aligned}
$$

Check: $\quad 9\left(-\dfrac{3}{2}\right) + 4 \quad 2\left(-\dfrac{3}{2}\right) - 5$

$$
\begin{aligned}
-12 + 4 &= -3 - 5 \\
-8 &= -8 \; \surd
\end{aligned}
$$

$\boxed{\left\{-\dfrac{3}{2}\right\}}$

59.

$$
\begin{aligned}
6x - 5 &= x + 5 \\
6x - x - 5 &= 5 \\
5x &= 5 + 5 \\
5x &= 10 \\
x &= 2
\end{aligned}
$$

Check:
$$
\begin{aligned}
6(2) - 5 &= 2 + 5 \\
12 - 5 &= 7 \\
7 &= 7 \; \surd
\end{aligned}
$$

$\boxed{\{2\}}$

61.

$$
\begin{aligned}
6x + 14 &= 2x - 2 \\
6x - 2x + 14 &= -2 \\
4x &= -2 - 14 \\
4x &= -16 \\
x &= -4
\end{aligned}
$$

Check:
$$
\begin{aligned}
6(-4) + 14 &= 2(-4) - 2 \\
-24 + 14 &= -8 - 2 \\
-10 &= -10 \; \surd
\end{aligned}
$$

$\boxed{\{-4\}}$

63.

$$
\begin{aligned}
-3y - 1 &= 5 - 2y \\
-3y + 2y - 1 &= 5 \\
-y &= 5 + 1 \\
-y &= 6 \\
y &= -6
\end{aligned}
$$

Check:
$$
\begin{aligned}
-3(-6) - 1 &= 5 - 2(-6) \\
18 - 1 &= 5 + 12 \\
17 &= 17 \; \surd
\end{aligned}
$$

$\boxed{\{-6\}}$

65.

$$
\begin{aligned}
-3(y - 6) &= 2 - y \\
-3y + 18 &= 2 - y \\
-3y + y + 18 &= 2 \\
-2y &= 2 - 18 \\
-2y &= -16 \\
y &= 8
\end{aligned}
$$

Check:
$$
\begin{aligned}
-3(8 - 6) &= 2 - 8 \\
-3(2) &= 2 - 8 \\
-6 &= -6 \; \surd
\end{aligned}
$$

$\boxed{\{8\}}$

67.

$$
\begin{aligned}
10 + 5r &= 3(2r - 5) \\
10 + 5r &= 6r - 15 \\
10 + 5r - 6r &= -15 \\
10 - r &= -15 \\
-r &= -15 - 10 \\
-r &= -25 \\
r &= 25
\end{aligned}
$$

Check:
$$
\begin{aligned}
10 + 5(25) &= 3[2(25) - 5] \\
10 + 125 &= 3(50 - 5) \\
135 &= 3(45) \\
135 &= 135 \; \surd
\end{aligned}
$$

$\boxed{\{25\}}$

69.
$$4(y + 3) = 2(y - 6)$$
$$4y + 12 = 2y - 12$$
$$4y - 2y + 12 = -12$$
$$2y + 12 = -12$$
$$2y = -12 - 12$$
$$2y = -24$$
$$y = -12$$

Check:
$$4(-12 + 3) = 2(-12 - 6)$$
$$4(-9) = 2(-18)$$
$$-36 = -36 \ \checkmark$$

$$\boxed{\{-12\}}$$

71.
$$\frac{1}{5}y - 4 = -6$$
$$5\left(\frac{1}{5}y - 4\right) = 5(-6)$$
$$5\left(\frac{1}{5}y - 4\right) = 5(-6)$$
$$6\left(\frac{1}{5}y\right) - 5(4) = -30$$
$$y - 20 = -30$$
$$y = -30 + 20$$
$$y = -10$$

Check:
$$\frac{1}{5}(-10) - 4 = -6$$
$$-2 - 4 = -6$$
$$-6 = -6 \ \checkmark$$

$$\boxed{\{-10\}}$$

73.
$$\frac{2}{3}y - 5 = 7$$
$$3\left(\frac{2}{3}y - 5\right) = 3(7)$$
$$3\left(\frac{2}{3}y\right) - 3(5) = 21$$
$$2y - 15 = 21$$
$$2y = 21 + 15$$
$$2y = 36$$
$$y = 18$$

Check:
$$\frac{2}{3}(18) - 5 = 7$$
$$12 - 5 = 7$$
$$7 = 7 \ \checkmark$$

$$\boxed{\{18\}}$$

75.
$$y + \frac{1}{2} = \frac{1}{4}y - \frac{5}{8}$$
$$8\left(y + \frac{1}{2}\right) = 8\left(\frac{1}{4}y - \frac{5}{8}\right) \qquad \text{LCM of 2, 4, 8 is 8.}$$
$$8y + 8\left(\frac{1}{2}\right) = 8\left(\frac{1}{4}y\right) - 8\left(\frac{5}{8}\right)$$
$$8y + 4 = 2y - 5$$
$$8y - 2y + 4 = -5$$
$$6y + 4 = -5$$
$$6y = -5 - 4$$
$$6y = -9$$
$$y = -\frac{9}{6}$$
$$y = -\frac{3}{2}$$

Check: $$-\frac{3}{2}+\frac{1}{2} = \frac{1}{4}\left(-\frac{3}{2}\right)-\frac{5}{8}$$

$$-\frac{2}{2} = -\frac{3}{8}-\frac{5}{8}$$

$$-1 = -\frac{8}{8}$$

$$-1 = -1 \ \sqrt{}$$

$$\boxed{\left\{-\frac{3}{2}\right\}}$$

77. $$\frac{1}{12}x+\frac{5}{12} = \frac{1}{2}x-\frac{5}{12}$$

$$12\left(\frac{1}{12}x+\frac{5}{12}\right) = 12\left(\frac{1}{2}x-\frac{5}{12}\right) \quad \text{LCM of 2 and 12 is 12.}$$

$$12\left(\frac{1}{12}x\right)+12\left(\frac{5}{12}\right) = 12\left(\frac{1}{2}x\right)-12\left(\frac{5}{12}\right)$$

$$x+5 = 6x-5$$

$$5 = 6x-x-5$$

$$5 = 5x-5$$

$$5+5 = 5x$$

$$10 = 5x$$

$$2 = x$$

Check: $$\frac{1}{12}(2)+\frac{5}{12} = \frac{1}{2}(2)-\frac{5}{12}$$

$$\frac{2}{12}+\frac{5}{12} = \frac{12}{12}-\frac{5}{12}$$

$$\frac{7}{12} = \frac{7}{12} \ \sqrt{}$$

$$\boxed{\{2\}}$$

79. $$\frac{1}{2}x-\frac{1}{2} = \frac{3}{2}x+\frac{7}{2}$$

$$2\left(\frac{1}{2}x-\frac{1}{2}\right) = 2\left(\frac{3}{2}x+\frac{7}{2}\right)$$

$$2\left(\frac{1}{2}x\right)-2\left(\frac{1}{2}\right) = 2\left(\frac{3}{2}x\right)+2\left(\frac{7}{2}\right)$$

$$x-1 = 3x+7$$

$$-1 = 3x-x+7$$

$$-1 = 2x+7$$

$$-1-7 = 2x$$

$$-8 = 2x$$

$$-4 = x$$

Check: $$\frac{1}{2}(-4)-\frac{1}{2} = \frac{3}{2}(-4)+\frac{7}{2}$$

$$-\frac{4}{2}-\frac{1}{2} = -\frac{12}{2}+\frac{7}{2}$$

$$-\frac{5}{2} = -\frac{5}{2} \ \sqrt{}$$

$$\boxed{\{-4\}}$$

81.
$$\frac{2}{3}y - 4 = \frac{1}{6}y$$
$$6\left(\frac{2}{3}y - 4\right) = 6\left(\frac{1}{6}y\right) \quad \text{LCM of 3 and 6 is 6.}$$
$$6\left(\frac{2}{3}y\right) - 6(4) = y$$
$$4y - 24 = y$$
$$4y - y = 24$$
$$3y = 24$$
$$y = 8$$

Check:
$$\frac{2}{3}(8) - 4 = \frac{1}{6}(8)$$
$$\frac{16}{3} - \frac{12}{3} = \frac{4}{3}$$
$$\frac{4}{3} = \frac{4}{3} \; \checkmark$$

$$\boxed{\{8\}}$$

83.
$$\frac{Y}{6} - \frac{Y}{8} = \frac{1}{12}$$
$$24\left(\frac{Y}{6} - \frac{Y}{8}\right) = 24\left(\frac{1}{12}\right) \quad \text{LCM of 6, 8, and 12 is 12.}$$
$$4Y - 3Y = 2$$
$$Y = 2$$

Check:
$$\frac{2}{6} - \frac{2}{8} = \frac{1}{12}$$
$$\frac{1}{3} - \frac{1}{4} = \frac{1}{12}$$
$$\frac{4 - 3}{12} = \frac{1}{12}$$
$$\frac{1}{12} = \frac{1}{12} \; \checkmark$$

$$\boxed{\{2\}}$$

85.
$$ay = r$$
$$\frac{ay}{a} = \frac{r}{a}$$
$$\boxed{y = \frac{r}{a}}$$

87.
$$\frac{A}{W} = L$$
$$W\left(\frac{A}{W}\right) = W(L)$$
$$\boxed{A = LW}$$

89.
$$P = \frac{I}{rt}$$
$$P(rt) = \left(\frac{I}{rt}\right)(rt)$$
$$Prt = I \quad \text{or} \quad \boxed{I = Prt}$$

91.
$$3ax = 2ax + b$$
$$3ax - 2ax = b$$
$$ax = b$$
$$\frac{ax}{a} = \frac{b}{a}$$
$$\boxed{x = \frac{b}{a}}$$

93.
$$5ax = 3ax - b$$
$$5ax - 3ax = -b$$
$$2ax = -b$$
$$\frac{2ax}{2a} = \frac{-b}{2a}$$
$$\boxed{x = -\frac{b}{2a}}$$

95. $P = 28$ pounds per square inch
$A = 24$ squares inches
$$\frac{W}{4A} = P$$
$$\frac{W}{4(24)} = 28$$
$$(4)(24)\left[\frac{W}{4(24)}\right] = 4(24)(28)$$
$$W = 2688$$
The weight of the car is $\boxed{2688 \text{ pounds}}$.

97.
$$p = 15 + \frac{15d}{33}$$
$$165 = 15 + \frac{15d}{33} \qquad p = 165 \text{ pounds per square inch}$$
$$165 - 15 = \frac{15d}{33}$$
$$150 = \frac{15d}{33}$$
$$\frac{33}{15}(150) = \frac{33}{15}\left(\frac{15d}{33}\right)$$
$$330 = d$$
The diver is $\boxed{330 \text{ feet}}$ below the surface.

99.
$$H = 1.2L + 27.8$$
$$65 = 1.2L + 27.8 \qquad H = 65 \text{ inches}$$
$$65 - 27.8 = 1.2L$$
$$37.2 = 1.2L$$
$$\frac{37.2}{1.2} = L$$
$$31 = L$$
The approximate length of this primate's humerus is $\boxed{31 \text{ inches}}$.

101. Four increased by three times a number is 13. Let x equal the number.
$$\boxed{4 + 3x = 13}$$
$$3x = 9$$
$$x = 3$$
Check: $\quad 4 + 3(3) = 13$
$$4 + 9 = 13$$
$$13 = 13 \ \checkmark$$
The number is $\boxed{3}$.

103. Four increased by five times a number is –41. Let x equal the number.
$$\boxed{4 + 5x = -41}$$
$$5x = -45$$
$$x = -9$$
Check: $\ 4 + 5(-9) = -41$
$$4 - 45 = -41$$
$$-41 = -41 \ \checkmark$$
The number is $\boxed{-9}$.

105. When 18 is subtracted from six times a certain number, the result is 96. Let x equal the number.

$$\boxed{6x - 18 \;=\; 96}$$

$$6x \;=\; 114$$
$$x \;=\; 19$$

Check: $6(19) - 18 \;=\; 96$
$$114 - 18 \;=\; 96$$
$$96 \;=\; 96\,\checkmark$$

The number is $\boxed{19}$.

107. Adding one-third of a number to the number itself results in 48. Let x equal the number.

$$\boxed{x + \frac{1}{3}x \;=\; 48}$$

$$\frac{4}{3}x \;=\; 48$$

$$x \;=\; \frac{3}{4}(48)$$

$$x \;=\; 36$$

Check: $36 + \frac{1}{3}(36) \;=\; 48$

$$36 + 12 \;=\; 48$$
$$48 \;=\; 48\,\checkmark$$

The number is $\boxed{36}$.

109. A number divided by -9 is 12. Let x equal the number.

$$\boxed{\frac{x}{-9} \;=\; 12}$$

$$-9\left(\frac{x}{-9}\right) \;=\; -9(12)$$

$$x \;=\; -108$$

Check: $\dfrac{-108}{-9} \;=\; 12$

$$12 \;=\; 12\,\checkmark$$

The number is $\boxed{-108}$.

111. If you double a number and then add 85, you get three-fourths of the original number. Let x equal the number.

$$\boxed{2x + 85 \;=\; \frac{3}{4}x}$$

$$2x - \frac{3}{4}x + 85 \;=\; 0$$

$$\frac{5}{4}x \;=\; -85$$

$$x \;=\; \frac{4}{5}(-85)$$

$$x \;=\; -68$$

Check: $2(-68) + 85 \;=\; \frac{3}{4}(-68)$

$$-136 + 85 \;=\; -51$$
$$-51 \;=\; -51\,\checkmark$$

The original number is $\boxed{-68}$.

113. If twice the weight of a truck is increased by 1500 pounds, the truck will weigh 9000 pounds. Let W equal the truck's weight.

$$2W + 1,500 = 9,000$$
$$2W = 7,500$$
$$W = 3,750$$

Check: $2(3,750) + 1,500 = 9,000$
$$7,500 + 1,500 = 9,000$$
$$9,000 = 9,000 \checkmark$$

The weight of the truck is $\boxed{3,750 \text{ pounds}}$.

115. 80% of what number is 28? Let x equal the number.

$$\boxed{0.8x = 28}$$
$$x = \frac{28}{0.8}$$
$$x = 35$$

Check: $0.8(35) = 28$
$$28 = 28 \checkmark$$

The number is $\boxed{35}$.

117. A person spends 30% of monthly income on food. The amount she spent on food was $315. Let x equal her monthly income for January.

$$\boxed{0.3x = 315}$$
$$x = \frac{315}{0.3}$$
$$x = 1050$$

Check: $0.3(1050) = 315$
$$315 = 315 \checkmark$$

The person's monthly income for January was $\boxed{\$1,050}$.

119. A person paid seven-tenths of the original price for a computer, paying $1400 at the discounted price. Let x equal the original price.

$$\boxed{\frac{7}{10}x = 1,400}$$
$$x = \frac{10}{7}(1,400)$$
$$x = 2,000$$

Check: $\frac{7}{10}(2,000) = 1,400$
$$1,400 = 1,400 \checkmark$$

The original price of the computer was $2,000.

121.
$$4x - 3 = 4x + 5$$
$$4x - 3 - 4x = 4x + 5 - 4x$$
$$-3 = 5$$

A contradiction exists since $-3 \neq 5$. There is no solution.

123.

$$4x + 6 \ = \ 2(x + 3) + 2x$$

Try $x = 0$:
$$4(0) + 6 \ = \ 2(0 + 3) + 2(0)$$
$$0 + 6 \ = \ 2(3) + 0$$
$$6 \ = \ 6$$

Try $x = 1$:
$$4(1) + 6 \ = \ 2(1 + 3) + 2(1)$$
$$4 + 6 \ = \ 2(4) + 2$$
$$10 \ = \ 8 + 2$$
$$10 \ = \ 10$$

$$4x + 6 \ = \ 2x + 6 + 2x$$
$$4x + 6 \ = \ 4x + 6$$
$$4x + 6 - 4x \ = \ 4x + 6 - 4x$$
$$6 \ = \ 6$$
$$6 - 6 \ = \ 6 - 6$$
$$0 \ = \ 0$$

Both sides of the equation are equal to the same value. The equation is true for all real numbers.

Writing in Mathematics

125.
$$7x \ = \ 21$$

| Divide *not* subtract. |

$$\frac{7x}{7} \ = \ \frac{21}{7}$$

| $x \ = \ 3$ |

127.
$$3\left|x\right| + 6 \ = \ 12$$
$$3\left|x\right| + 6 - 6 \ = \ 12 - 6$$
$$3\left|x\right| \ = \ 6$$
$$\frac{3\left|x\right|}{3} \ = \ \frac{6}{3}$$
$$\left|x\right| \ = \ 2$$

| $x \ = \ 2$ or -2 |

| $\{-2, 2\}$ is the solution set |

129.
$$0x \ = \ 17$$

| Division by zero is undefined; x is undefined. |

131.
$$a(x - 2) \ = \ b$$
$$ax - 2a \ = \ b$$
$$ax \ = \ b + 2a$$
$$x \ = \ \frac{b + 2a}{a}$$

Review Problems

132. $\dfrac{1}{3} - \left(-\dfrac{1}{4}\right) = \dfrac{1}{3} + \dfrac{1}{4} = \dfrac{7}{12}$

133. $4x + 1 = \ 1 + 4x$

Commutative Property of Addition

134. At the start of the party: 256 eligible contestants

After one-half hour: $\frac{1}{4}(256)$ eliminated

$\frac{3}{4}(256)$ still eligible

$\frac{3}{4}(256) = 192$

After one hour: $\frac{3}{4}(192)$ still eligible

$\frac{3}{4}(192) = 144$

After one hour $\boxed{144}$ contestants are still eligible.

Let n equal the number of half-hour periods after the start of the party.

Then the number of eligible contestants after n half-hour periods is $\left(\frac{3}{4}\right)^n(256)$.

After 1 hour, $n = 2$: $\left(\frac{3}{4}\right)^2(256) = \frac{9}{16}(256) = 144$.

Section 2.3 Solving Linear Equations

Problem Set 2.3, pp. 133–135

1.
$$
\begin{aligned}
3(2y + 3) &= -3y - 9 \\
6y + 9 &= -3y - 9 \\
6y + 9 - 9 &= -3y - 9 - 9 \\
6y &= -3y - 18 \\
6y + 3y &= -3y - 18 + 3y \\
9y &= -18 \\
\frac{9y}{9} &= \frac{-18}{7} \\
y &= -2
\end{aligned}
$$
$\boxed{\{-2\}}$

3.
$$
\begin{aligned}
3(y + 3) &= -2(2y - 1) \\
3y + 9 &= -4y + 2 \\
3y + 9 - 9 &= -4y - 7 \\
3y &= -4y - 7 \\
3y + 4y &= -4y - 7 + 4y \\
7y &= -7 \\
\frac{7y}{7} &= \frac{-7}{7} \\
y &= -1
\end{aligned}
$$
$\boxed{\{-1\}}$

5.
$$
\begin{aligned}
8(y + 2) &= 2(3y + 4) \\
8y + 16 &= 6y + 8 \\
8y + 16 - 16 &= 6y + 8 - 16 \\
8y &= 6y - 8 \\
8y - 6y &= 6y - 7 - 6y \\
2y &= -8 \\
y &= -4
\end{aligned}
$$
$\boxed{\{-4\}}$

7.
$$
\begin{aligned}
10x - 2(4 + 5x) &= -8 \\
10x - 8 - 10x &= -8 \\
-8 &= -8
\end{aligned}
$$
all real numbers
$\boxed{\{x \mid x \in R\}}$

9.
$$3(y + 1) = 7(y - 2) - 3$$
$$3y + 3 = 7y - 14 - 3$$
$$3y + 3 = 7y - 17$$
$$3y + 3 - 3 = 7y - 17 - 3$$
$$3y = 7y - 20$$
$$3y - 7y = 7y - 20 - 7y$$
$$-4y = -20$$
$$\frac{-4y}{-4} = \frac{-20}{-4}$$
$$y = 5$$

$$\boxed{\{5\}}$$

11.
$$2(10y - 3) = 5(4y - 1)$$
$$20y - 6 = 20y - 5$$
$$-6 \neq -5$$
no solution

$$\boxed{\varnothing}$$

13.
$$5(2z - 8) - 2 = 5(z - 3) + 3$$
$$10z - 40 - 2 = 5z - 15 + 3$$
$$10z - 42 = 5z - 12$$
$$10z - 42 + 42 = 5z - 12 + 42$$
$$10z = 5z + 30$$
$$10z - 5z = 5z + 30 - 5z$$
$$5z = 30$$
$$\frac{5z}{5} = \frac{30}{5}$$
$$z = 6$$

$$\boxed{\{6\}}$$

15.
$$17(x + 3) = 13 + 4(x - 10)$$
$$17x + 51 = 13 + 4x - 40$$
$$17x + 51 = 4x - 27$$
$$17x + 51 - 51 = 4x - 27 - 51$$
$$17x = 4x - 78$$
$$17x - 4x = 4x - 78 - 4x$$
$$13x = -78$$
$$\frac{13x}{13} = \frac{-78}{13}$$
$$x = -6$$

$$\boxed{\{-6\}}$$

17.
$$6 = -4(1 - x) + 3(x + 1)$$
$$6 = -4 + 4x + 3x + 3$$
$$6 = -1 + 7x$$
$$6 + 1 = -1 + 7x + 1$$
$$7 = 7x$$
$$\frac{7}{7} = \frac{7x}{7}$$
$$1 = x$$

$$\boxed{\{1\}}$$

19.
$$6(2y + 1) - 3 = 8(y - 3) + 4y$$
$$12y + 6 - 3 = 8y - 24 + 4y$$
$$12y + 3 = 12y - 24$$
$$12y + 3 - 12y = 12y - 24 - 12y$$
$$3 \neq -24$$

no solution

$$\boxed{\varnothing}$$

21.
$$10(y + 4) - 4(y - 2) = 3(y - 1) + 2(y - 3)$$
$$10y + 40 - 4y + 8 = 3y - 3 + 2y - 6$$
$$6y + 48 = 5y - 9$$
$$6y + 48 - 48 = 5y - 9 - 40$$
$$6y = 5y - 57 - 5y$$
$$y = -57$$

$$\boxed{\{-57\}}$$

23.
$$9 - 6(2z + 1) = 3 - 7(z - 1)$$
$$9 - 12z - 6 = 3 - 7z + 7$$
$$3 - 12z = 10 - 7z$$
$$3 - 12z - 3 = 10 - 7z - 3$$
$$-12z = 7 - 7z$$
$$-5z = 7$$
$$\frac{-5z}{-5} = \frac{7}{-5}$$
$$z = -\frac{7}{5}$$

$$\boxed{\left\{-\frac{7}{5}\right\}}$$

25.
$$3(2x - 1) = 5[3x - (2 - x)]$$
$$6x - 3 = 5(3x - 2 + x)$$
$$6x - 3 = 5(4x - 2)$$
$$6x - 3 = 20x - 10$$
$$6x - 3 + 3 = 20x - 10 + 3$$
$$6x = 20x - 7$$
$$6x - 20x = 20x - 7 - 20x$$
$$-14x = -7$$
$$\frac{-14x}{-14} = \frac{-7}{-14}$$
$$x = \frac{1}{2}$$

$$\boxed{\left\{ \frac{1}{2} \right\}}$$

27.
$$18(-2 + z) = 9(2z - 4)$$
$$-36 + 18z = 18z - 36$$
$$-36 + 18z - 18z = 18z - 36 - 18z$$
all real numbers
$$\boxed{\{ z | z \in R \}}$$

29.
$$2(3x + 4) = 3x + 2[3(x - 1) + 2]$$
$$6x + 8 = 3x + 2(3x - 3 + 2)$$
$$6x + 8 = 3x + 2(3x - 1)$$
$$6x + 8 = 3x + 6x - 2$$
$$6x + 8 - 8 = 9x - 2 - 8$$
$$6x = 9x - 10$$
$$6x - 9x = 9x - 10 - 9x$$
$$-3x = -10$$
$$\frac{-3x}{-3} = \frac{-10}{-3}$$
$$x = \frac{10}{3}$$

$$\boxed{\left\{ \frac{10}{3} \right\}}$$

31.
$$\frac{x}{3} + \frac{x}{2} = \frac{5}{6}$$
$$6\left(\frac{x}{3} + \frac{x}{2} \right) = 6\left(\frac{5}{6} \right) \quad \text{LCM} = 6$$
$$2x + 3x = 5$$
$$5x = 5$$
$$\frac{5x}{5} = \frac{5}{5}$$
$$x = 1$$

$$\boxed{\{1\}}$$

33.
$$20 - \frac{z}{3} = \frac{z}{2}$$
$$6\left(20 - \frac{z}{3} \right) = 6\left(\frac{z}{2} \right) \quad \text{LCM} = 6$$
$$120 - 2z = 3z$$
$$120 - 2z + 2z = 3z + 2z$$
$$120 = 5z$$
$$\frac{120}{5} = \frac{5z}{5}$$
$$24 = z$$

$$\boxed{\{24\}}$$

35.
$$\frac{3x}{4} - 3 = \frac{x}{2} + 2$$
$$4\left(\frac{3x}{4} - 3 \right) = 4\left(\frac{x}{2} + 2 \right) \quad \text{LCM} = 4$$
$$3x - 12 = 2x + 8$$
$$3x - 12 + 12 = 2x + 8 + 12$$
$$3x = 2x + 20$$
$$3x - 2x = 2x + 20 - 2x$$
$$x = 20$$

$$\boxed{\{20\}}$$

37.
$$\frac{3x}{5} - x = \frac{x}{10} - \frac{5}{2}$$

$$10\left(\frac{3x}{5} - x\right) = 10\left(\frac{x}{10} - \frac{5}{2}\right) \quad \text{LCM} = 10$$

$$6x - 10x = x - 25$$
$$-4x = x - 25$$
$$-4x - x = x - 25 - x$$
$$-5x = -25$$
$$\frac{-5x}{-5} = \frac{-25}{-5}$$
$$x = 5$$

$\boxed{\{5\}}$

39.
$$\frac{5z - 1}{7} - \frac{3z - 2}{5} = 1$$

$$35\left(\frac{5z-1}{7}\right) - 35\left(\frac{3z-2}{5}\right) = 35(1)$$

$$5(5z - 1) - 7(3z - 2) = 35$$
$$25x - 5 - 21z + 14 = 35$$
$$4z + 9 = 35$$
$$4z + 9 - 9 = 35 - 9$$
$$4z = 26$$
$$\frac{4z}{4} = \frac{26}{4}$$
$$z = \frac{13}{2}$$

$\boxed{\left\{\dfrac{13}{2}\right\}}$

41.
$$\frac{z - 3}{4} - 1 = \frac{z}{2}$$

$$4\left(\frac{z-3}{4} - 1\right) = 4\left(\frac{z}{2}\right) \quad \text{LCM} = 4$$

$$z - 3 - 4 = 2z$$
$$z - 7 = 2z$$
$$z - 7 - z = 2z - z$$
$$-7 = z$$

$\boxed{\{-7\}}$

43.
$$\frac{2y - 3}{9} + \frac{y - 3}{2} = \frac{y + 5}{6} - 1$$

$$18\left(\frac{2y-3}{9}\right) + 18\left(\frac{y-3}{2}\right) = 18\left(\frac{y+5}{6}\right) - 18(1)$$

$$2(2y - 3) + 9(y - 3) = 3(y + 5) - 18$$
$$4y - 67 + 9y - 27 = 3y + 15 - 18$$
$$13y - 33 = 3y - 3$$
$$13y - 33 + 33 = 3y - 3 + 33$$
$$13y = 3y + 30$$
$$13y - 3y = 3y + 30 - 3y$$
$$10y = 30$$
$$\frac{10y}{10} = \frac{30}{10}$$
$$y = 3$$

$\boxed{\{3\}}$

45.

$$\frac{1}{3}(6x+9) = -2(x-1)$$
$$2x+3 = -2x+2$$
$$2x = -2x-1$$
$$4x = -1$$
$$x = -\frac{1}{4}$$

$$\boxed{\left\{-\frac{1}{4}\right\}}$$

47.

$$12\left(\frac{y}{3}-\frac{1}{2}\right) = 8\left(\frac{y}{2}-1\right)$$
$$4y-6 = 4y-8$$
$$4y-6-4y = 4y-8-4y$$
$$-6 \neq -8$$

no solution

$$\boxed{\varnothing}$$

49.

$$\frac{1}{2}(y+2)-\frac{1}{4}(y-12) = 4+y$$
$$4\left(\frac{1}{2}\right)(y+2)-4\left(\frac{1}{4}\right)(y-12) = 4(4+y)$$
$$2(y+2)-(1)(y-12) = 16+4y$$
$$2y+4-y+12 = 16+4y$$
$$y+16 = 16+4y$$
$$y+16-16 = 16+4y-16$$
$$y = 4y$$
$$y-y = 4y-y$$
$$0 = 3y$$
$$\frac{0}{3} = \frac{3y}{3}$$
$$0 = y$$

$$\boxed{\{0\}}$$

51.

$$0.05y = 0.10(y+60)-12$$
$$100(0.05y) = 100(0.10)(y+60)-100(12)$$
$$5y = 10(y+60)-1200$$
$$5y = 10y+600-1200$$
$$5y = 10y-600$$
$$5y+600 = 10y-600+600$$
$$5y+600 = 10y$$
$$5y+600-5y = 10y-5y$$
$$600 = 5y$$
$$\frac{600}{5} = \frac{5y}{5}$$
$$120 = y$$

$$\boxed{\{120\}}$$

53.

$$0.06x+0.04(x-5,000) = 200$$
$$100(0.06x)+100(0.04)(x-5,000) = 100(200)$$
$$6x+4(x-5,000) = 100(200)$$
$$6x+4x-20,000 = 20,000$$
$$10x-20,000 = 20,000$$
$$10x-20,000+20,000 = 20,000+20,000$$
$$10x = 40,000$$
$$\frac{10x}{10} = \frac{40,000}{10}$$
$$x = 4,000$$

$$\boxed{\{4,000\}}$$

55.
$$
\begin{aligned}
0.025(y + 5000) &= 0.03y + 100 \\
1,000(0.025)(y + 5000) &= 1,000(0.03y + 100) \\
25(y + 5000) &= 30y + 100,000 \\
25y + 125000 &= 30y + 100,000 \\
25y + 125,000 - 100,000 &= 30y + 100,000 - 100,000 \\
25y + 25000 &= 30y \\
25y + 25,000 - 25y &= 30y - 25y \\
25,000 &= 5y \\
5,000 &= y
\end{aligned}
$$

$\boxed{\{5000\}}$

57.
$$
\begin{aligned}
0.1(y + 80) &= 14 - 0.2y \\
0.1y + 8 &= 14 - 0.2y \\
0.1y &= 6 - 0.2y \\
0.3y &= 6 \\
y &= \frac{6}{0.3} \\
y &= 20
\end{aligned}
$$

$\boxed{\{20\}}$

59.
$$
\begin{aligned}
3.2y - 2.2 &= 4.9y + 5.9 \\
3.2y - 2.2 + 2.2 &= 4.9y + 5.9 + 2.2 \\
3.2y &= 4.9y + 8.1 \\
3.2y - 4.9y &= 4.9y + 8.1 - 4.9y \\
-1.7y &= 8.1 \\
y &= \frac{8.1}{-1.7} \\
y &= -\frac{81}{17}
\end{aligned}
$$

$\boxed{\left\{-\dfrac{81}{17}\right\}}$

61.
$$
\begin{aligned}
0.03(y + 200) &= 86 - 0.05y \\
100(0.03)(y + 200) &= 100(86) - 100(0.05y) \\
3(y + 200) &= 8,600 - 5y \\
3y + 600 &= 8,600 - 5y \\
3y + 600 - 600 &= 8,600 - 5y - 600 \\
3y &= -5y + 8,000 \\
3y + 5y &= -5y + 8,000 + 5y \\
8y &= 8,000 \\
y &= 1,000
\end{aligned}
$$

$\boxed{\{1000\}}$

63.
$$
\begin{aligned}
0.05(y - 300) &= 105 - 0.1y \\
0.05y - 15 &= 105 - 0.1y \\
0.05y &= 120 - 0.1y \\
0.15y &= 120 \\
y &= \frac{120}{0.15} \\
y &= 800
\end{aligned}
$$

$\boxed{\{800\}}$

65.
$$
\begin{aligned}
y - 0.2y &= 72 \\
0.8y &= 72 \\
y &= \frac{72}{0.8} \\
y &= 90
\end{aligned}
$$

$\boxed{\{90\}}$

67. When one–third of a number is added to one–fifth of the number, the sum is 16:

$$
\boxed{\frac{1}{3}x + \frac{1}{5}x = 16}
$$

$$
\begin{aligned}
15\left(\frac{1}{3}x + \frac{1}{5}x\right) &= 15(16) \\
5x + 3x &= 240 \\
8x &= 240 \\
x &= 30
\end{aligned}
$$

The number is $\boxed{30}$.

69. When three–eighths of a number is subtracted from two–fifths of the number, the result is 1:

$$\boxed{\frac{2}{5}x - \frac{3}{8}x = 1}$$

$$40\left(\frac{2}{5}x\right) - 40\left(\frac{3}{8}x\right) = 40(1)$$

$$16x - 15x = 40$$

$$x = 40$$

The number is $\boxed{40}$.

71. When 3 is subtracted from three–fourths of a number, the result is equal to one–half of the number:

$$\boxed{\frac{3}{4}x - 3 = \frac{1}{2}x}$$

$$4\left(\frac{3}{4}x\right) - 4(3) = 4\left(\frac{1}{2}x\right)$$

$$3x - 12 = 2x$$

$$3x = 2x + 12$$

$$x = 12$$

The number is $\boxed{12}$.

73. When 2 is added to 60% of a number, the sum is 20:

$$\boxed{0.60x + 2 = 20}$$

$$0.60x = 18$$

$$x = \frac{18}{0.6}$$

$$x = 30$$

The number is $\boxed{30}$.

75. Thirty–five percent of the sum of a number and 16 is 5.35:

$$\boxed{0.35(x + 16) = 5.35}$$

$$0.35x + 5.6 = 5.35$$

$$0.35x = -0.25$$

$$x = \frac{-0.25}{0.35}$$

$$x = -\frac{5}{7}$$

The number is $\boxed{-\frac{5}{7}}$.

77. Twenty–five percent of the difference between a number and 28 is 13.05:

$$\boxed{0.25(x - 28) = 13.05}$$

$$0.25x - 7 = 13.05$$

$$0.25x = 20.05$$

$$x = \frac{20.05}{0.25}$$

$$x = 80.2$$

The number is $\boxed{80.2}$.

79. $\boxed{x - 6 \;=\; \dfrac{1}{3}x + \dfrac{1}{5}x + \dfrac{1}{6}x + \dfrac{1}{4}x}$

$$
\begin{aligned}
60(x - 6) &= 60\left(\dfrac{1}{3}x + \dfrac{1}{5}x + \dfrac{1}{6}x + \dfrac{1}{4}x\right) \\
60x - 360 &= 20x + 12x + 10x + 15x \\
60x - 360 &= 57x \\
60x &= 57x + 360 \\
3x &= 360 \\
x &= 120
\end{aligned}
$$

The total number of lilies is $\boxed{120}$.

81. C is true;

$$
\begin{aligned}
2 - 3y &= 11 \\
-3y &= 9 \\
y &= -3
\end{aligned}
$$

$y^2 + 2y - 3 = (-3)^2 + 2(-3) - 3 = 9 - 6 - 3 = 0$

neither positive nor negative

83.
$$
\begin{aligned}
8.05x + 2.03x &= 17.06 - 4.3 \\
10.08x &= 12.76 \\
x &= 1.2658730
\end{aligned}
$$
$\boxed{\{1.2658733\}}$

85.
$$
\begin{aligned}
8497x + 7947 &= -5689x - 8576 \\
8497x + 5689x &= -8576 - 7949 \\
14186x &= -16523 \\
x &= -1.1647399
\end{aligned}
$$
$\boxed{\{-1.1647399\}}$

87.
$$
\begin{aligned}
3.7y - 15.1 &= 9y - 6.2 \\
3.7y - 9y &= -6.2 + 15.1 \\
-5.3y &= -8.9 \\
y &= -1.679245
\end{aligned}
$$
$\boxed{\{-1.679245\}}$

89.
$$
\begin{aligned}
0.003x - 0.1297 &= 1.43x + 8.5 \\
0.003x - 1.43x &= 0.1297 + 8.5 \\
-1.427x &= 8.6297 \\
x &= -6.0474422
\end{aligned}
$$
$\boxed{\{-6.0474422\}}$

Writing in Mathematics

91. Examples will vary.

$$
\begin{aligned}
4(x + 2) + 2 &= 2(x + 1) - 3 \\
4x + 8 + 2 &= 2x + 2 - 3 \\
4x + 10 &= 2x - 1 \\
2x &= -11 \\
\\
x &= -\dfrac{11}{2}
\end{aligned}
$$

– Simplify each side of the equation.

– Collect all like terms with the specified variables on one side and all other terms on the other side.

– Isolate the specified variables by dividing both sides of the equation by the coefficient of the variables.

$$
\begin{aligned}
4\left(-\dfrac{11}{2} + 2\right) + 2 &= 2\left(-\dfrac{11}{2} + 1\right) - 3 \\
4\left(-\dfrac{7}{2}\right) + 2 &= 2\left(-\dfrac{9}{2}\right) - 3 \\
-14 + 2 &= -9 - 3 \\
-12 &= -12 \;\checkmark
\end{aligned}
$$

– Check the solution in the original equation.

93.

	number of solutions	example
1	one	$2x + 3 = 7$ $x = 2$ $\{2\}$
2	no solution	$2x + 3 = 2x + 4$ $3 \neq 4$ $\emptyset$ the empty set
3	infinitely many solutions	$2x + 3 = 2x + 3$ $3 = 3$ all real numbers $\{x \mid x \in R\}$ (equation is an identity)

Review Problems

94. $-(x - 7) - 3(y - x) = -x + 7 - 3y + 3x = 2x - 3y + 7$

95. $\dfrac{\frac{1}{4} - \frac{3}{2}}{\frac{3}{8}} = \dfrac{\frac{1}{4} - \frac{6}{4}}{\frac{3}{8}} = \dfrac{-\frac{5}{4}}{\frac{3}{8}} = \dfrac{-5}{4} \cdot \dfrac{8}{3} = -\dfrac{10}{3}$

96. Let n = number of minutes.
cost for first minute + cost per minute × number of minutes
$= 1.20 + 0.85(n - 1)$
$= 1.20 + 0.85n - 0.85$
$= 0.35 + 0.85n \qquad\qquad n = 10$
$= 0.35 + 0.85(10)$
$= 0.35 + 8.50$
$= 8.85$
The cost for a 10–minute call is $\boxed{\$8.85}$.

Section 2.4 An Introduction to Problem Solving

Problem Set 2.4, pp. 148–152

1. When 6 is subtracted from 12 times a number, the result is 7 times the number increased by 24. Let x equal the number.

$$
\begin{aligned}
12x - 6 &= 7x + 24 \\
12x &= 7x + 30 \\
5x &= 30 \\
x &= 6
\end{aligned}
$$

Check:
$$
\begin{aligned}
12(6) - 6 &= 7(6) + 24 \\
72 - 6 &= 42 + 24 \\
66 &= 66 \ \checkmark
\end{aligned}
$$

The number is $\boxed{6}$.

3. When 3 is multiplied by 5 less than a number, the result is double the sum of 1 and twice that number. Let x equal the number.

$$
\begin{aligned}
3(x - 5) &= 2(1 + 2x) \\
3x - 15 &= 2 + 4x \\
3x - 17 &= 4x \\
-17 &= x
\end{aligned}
$$

Check:
$$
\begin{aligned}
3(-17 - 5) &= 2[1 + 2(-17)] \\
3(-22) &= 2(1 - 34) \\
-66 &= 2(-33) \\
-66 &= -66 \ \checkmark
\end{aligned}
$$

The number is $\boxed{-17}$.

5. When 45 is added to 3 times a number, the result is 8 times the number. Let x equal the number.

$$
\begin{aligned}
3x + 45 &= 8x \\
45 &= 5x \\
9 &= x
\end{aligned}
$$

Check:
$$
\begin{aligned}
3(9) + 45 &= 9(9) \\
27 + 45 &= 72 \\
72 &= 72 \ \checkmark
\end{aligned}
$$

The number is $\boxed{9}$.

7. Let x equal the number.

$$
\begin{aligned}
3(x - 2) + x &= 2x - 4 \\
3x - 6 + x &= 2x - 4 \\
4x - 6 &= 2x - 4 \\
4x &= 2x + 2 \\
2x &= 2 \\
x &= 1
\end{aligned}
$$

Check:
$$
\begin{aligned}
3(1 - 2) + 1 &= 2(1) - 4 \\
3(-1) + 1 &= 2 - 4 \\
-3 + 1 &= -2 \\
-2 &= -2 \ \checkmark
\end{aligned}
$$

The number is $\boxed{1}$.

9. The sum of three consecutive integers is 30.
Let x equal the first integer.
$x + 1$ and $x + 2$ are the next two consecutive integers.

$$
\begin{aligned}
x + (x + 1) + (x + 2) &= 30 \\
3x + 3 &= 30 \\
3x &= 27 \\
x &= 9 \\
x + 1 &= 10 \\
x + 2 &= 11
\end{aligned}
$$

Check:
$$
\begin{aligned}
9 + 10 + 11 &= 30 \\
30 &= 30 \ \checkmark
\end{aligned}
$$

The integers are $\boxed{9, \ 10, \text{ and } 11}$.

11. The sum of three consecutive even integers is 198. Let x equal the first even integer. $x + 2$ and $x + 4$ are the next two consecutive even integers.

$$
\begin{aligned}
x + (x + 2) + (x + 4) &= 198 \\
3x + 6 &= 198 \\
3x &= 192 \\
x &= 64 \\
x + 2 &= 66 \\
x + 4 &= 68 \\
\textit{Check:} \quad 64 + 66 + 68 &= 198 \\
198 &= 198 \ \sqrt{}
\end{aligned}
$$

The integers are $\boxed{64, 66, \text{ and } 68}$.

13. The sum of four consecutive odd integers is 216. Let x equal the first odd integer. $x + 2, x + 4$, and $x + 6$ are the next three consecutive odd integers.

$$
\begin{aligned}
x + (x + 2) + (x + 4) + (x + 6) &= 216 \\
4x + 12 &= 216 \\
4x &= 204 \\
x &= 51 \\
x + 2 &= 53 \\
x + 4 &= 55 \\
x + 6 &= 57 \\
\textit{Check:} \quad 51 + 53 + 55 + 57 &= 216 \\
216 &= 216 \ \sqrt{}
\end{aligned}
$$

The integers are $\boxed{51, 53, 55, \text{ and } 57}$.

The check for the remainder of the odd numbered problems is left to the reader.

15. The largest of three consecutive even integers is 6 less than twice the smallest. Let x equal the first even integer. $x + 2$ equals the next even integer. $x + 4$ equals the largest of three consecutive even integers.

$$
\begin{aligned}
x + 4 &= 2x - 6 \\
x + 10 &= 2x \\
10 &= x \\
x + 2 &= 12 \\
x + 4 &= 14
\end{aligned}
$$

The integers are $\boxed{10, 12, \text{ and } 14}$.

17. Let x equal the first even integer. $x + 2$ equals the second even integer. $x + 4$ equals the third even integer.

$$
\begin{aligned}
x + (x + 4) &= \frac{1}{2}(x + 2) + 15 \\
2x + 4 &= \frac{1}{2}x + 1 + 15 \\
2x + 4 &= \frac{1}{2}x + 16 \\
2x &= \frac{1}{2}x + 12 \\
\frac{4x}{2} - \frac{1}{2}x &= 12 \\
\frac{3x}{2} &= 12 \\
x &= \frac{2}{3}(12) \\
x &= 8
\end{aligned}
$$

$$x + 2 = 10$$
$$x + 4 = 12$$

The integers are $\boxed{8, 10, \text{ and } 12}$.

19. Let x equal the smaller number. $4x + 1$ equals the larger number.

$$3(4x + 1) - 2x = 43$$
$$12x + 3 - 2x = 43$$
$$10x + 3 = 43$$
$$10x = 40$$
$$x = 4$$
$$4x + 1 = 4(4) + 1 = 16 + 1 = 17$$

The numbers are $\boxed{4 \text{ and } 17}$.

21. Let x equal Dale's share. $2x + 5,000$ equals Jackie's share. $6x - 15,000$ equals Maggie's share.

$$x + (2x + 5,000) + (6x - 15,000) = 80,000$$
$$9x - 10,000 = 80,000$$
$$9x = 90,000$$
$$x = 10,000$$
$$2x + 5,000 = 2(10,000) + 5,000 = 20,000 + 5,000 = 25,000$$
$$6x - 15,000 = 6(10,000) - 15,000 = 60,000 - 15,000 = 45,000$$

$\boxed{\text{Dale's share, \$10,000; Jackie's share, \$25,000; Maggie's share, \$45,000}}$

23. Let x equal the number of atoms of oxygen. $2x + 1$ equals the number of atoms of carbon.
$2x + 1 - 1 = 2x$ equals the number of atoms of hydrogen.

$$x + (2x + 1) + (2x) = 21$$
$$5x + 1 = 21$$
$$5x = 20$$
$$x = 4$$
$$2x + 1 = 2(4) + 1 = 8 + 1 = 9$$

There are $\boxed{9 \text{ atoms of carbon}}$.

25. Let x equal the number of Republicans. $x - 8$ equals the number of Democrats. $\dfrac{x-8}{2}$ equals the number of Independents.

$$x + (x - 8) + \left(\frac{x-8}{2}\right) = 68$$
$$x + x - 8 + \frac{x}{2} - 4 = 68$$
$$\frac{5}{2}x - 12 = 68$$
$$\frac{5}{2}x = 80$$
$$x = \frac{2}{5}(80)$$
$$x = 32$$
$$x - 8 = 32 - 8 = 24$$
$$\frac{x-8}{2} = \frac{24}{2} = 12$$

$\boxed{\text{Republicans, 32; Democrats, 24; Independents, 12}}$

27. Let x equal the number of inhabitants per square mile in Canada. Canada is x. The United States is $9x - 1$. Australia is $x - 2$. England is $10(9x - 1) - 9 = 90x - 10 - 9 = 90x - 19$. England exceeds the sum of the others by 537.

$$\begin{aligned} 90x - 19 &= x + (9x - 1) + (x - 2) + 537 \\ 90x - 19 &= 11x + 534 \\ 79x &= 553 \\ x &= 7 \end{aligned}$$

Population density:

Canada: 7 inhabitants per square mile

United States: $9(7) - 1 = 63 - 1 = $ 62 inhabitants per square mile

Australia: $7 - 2 = $ 5 inhabitants per square mile

England: $90(7) - 19 = 630 - 19 = $ 611 inhabitants per square mile

29. Let x equal the number of days of rain per year in London. London is x. New York is $x + 7$. Vienna is $\dfrac{x+7}{2} + 36$.

$$\begin{aligned} (\text{New York}) + (\text{London}) &= (\text{Vienna}) + 137 \\ (x + 7) + (x) &= \left(\frac{x+7}{2} + 36 \right) + 137 \\ 2x + 7 &= \frac{x}{2} + \frac{7}{2} + 173 \\ 2(2x) &= 2 \left(\frac{x}{2} + \frac{7}{2} + 166 \right) \\ 4x &= x + 7 + 332 \\ 3x &= 339 \\ x &= 113 \end{aligned}$$

London: 113 days of rain per year

New York: $113 + 7 = $ 120 days of rain per year

Vienna: $\dfrac{120}{2} + 36 = 60 + 36 = $ 96 days of rain per year

31. Let x equal the number of Republicans. $5x$ equals the number of Democrats.

$$\begin{aligned} (\text{Democrats}) &= 2 \times (\text{Republicans} + 12) \\ 5x &= 2(x + 12) \\ 5x &= 2x + 24 \\ 3x &= 24 \\ x &= 8 \\ 5x &= 5(8) = 40 \end{aligned}$$

There are 40 Democrats at the gathering.

33. Let x equal Barry's present age. $4x$ equals Simon's present age. Barry's age in 4 years is $x + 4$. Simon's age in 4 years is $4x + 4$. In 4 years:

$$\begin{aligned} (\text{Simon}) &= 2(\text{Barry}) \\ 4x + 4 &= 2(x + 4) \\ 4x + 4 &= 2x + 8 \\ 2x &= 4 \\ x &= 2 \\ 4x &= 4(2) = 8 \end{aligned}$$

Present ages: Barry, 2 years old; Simon, 8 years old .

35. Let x equal the number of years ago. Rita's age x years ago is $23 - x$. Joel's age x years ago is $15 - x$.

$$23 - x = 3(15 - x)$$
$$23 - x = 45 - 3x$$
$$2x = 22$$
$$x = 11$$

$\boxed{11 \text{ years ago}}$

37. Let x equal the number of years from now. Pen drawing age x years from now: $8 + x$. Pencil drawing age x years from now: $22 + x$.

$$22 + x = 2(8 + x)$$
$$22 + x = 16 + 2x$$
$$6 = x$$

$\boxed{6 \text{ years from now}}$

39. Let x equal the age of the painting now. $x + 74$ equals the age of the rug now. The painting's age 30 years ago is $x - 30$. The rug's age 30 years ago is $(x + 74) - 30 = x + 44$.

$$x + 44 = 3(x - 30)$$
$$x + 44 = 3x - 90$$
$$134 = 2x$$
$$67 = x$$
$$x + 74 = 67 + 74 = 141$$

$\boxed{\text{painting now, 67 years old; rug now, 141 years old}}$

41. Let x equal the age of Tweedledum now. $16 - x$ is the age of Tweedledee now. The age of Tweedledum 4 years ago is $x - 4$. Age of Tweedledee 4 years ago: $16 - x - 4 = 12 - x$.

Four years ago:

$$\text{(Tweedledum)} = 3\text{(Tweedledee)}$$
$$x - 4 = 3(12 - x)$$
$$x - 4 = 36 - 3x$$
$$4x = 40$$
$$x = 10$$
$$16 - x = 16 - 10 = 6$$

$\boxed{\text{Tweedledum's present age, 10 years old; Tweedledee's present age, 6 years old}}$

43. Let x equal the present age of the canoe. $24 - x$ is the present age of the rowboat. The age of the canoe in 6 years: $x + 6$. The age of the rowboat 4 years ago: $24 - x - 4 = 20 - x$.

$$x + 6 = 20 - x$$
$$2x = 14$$
$$x = 7$$
$$24 - x = 17$$

$\boxed{\text{present age of canoe, 7 years old; present age of rowboat, 17 years old}}$

45. Let x equal the man's final age.

childhood: $\frac{1}{9}x$ ⎫
                              ⎬ extraneous facts
youth: $\frac{1}{12}x$ ⎭

childless marriage: $\frac{1}{3}x$; 10 years

$$\frac{1}{3}x = 10$$
$$x = 30$$

$\boxed{\text{man's final age, 30 years old}}$

47. Let p equal the price of the car. $0.12p$ equals the price reduction.

$$
\begin{aligned}
p - 0.12p &= 17{,}600 \\
0.88p &= 17{,}600 \\
p &= \frac{17{,}600}{0.88} \\
p &= 20{,}000
\end{aligned}
$$

price of car, $20,000

49. Let p equal the price of the car. $0.065p = $ sales tax.

$$
\begin{aligned}
p + 0.65p &= 17{,}466 \\
1.065p &= 17{,}466 \\
p &= 16{,}400
\end{aligned}
$$

price of car, $16,400

51. Let x equal the dealer's cost of the refrigerator.

$$
\begin{aligned}
0.25x &= \text{markup} \\
x + 0.25x &= 584 \\
1.25x &= 584 \\
x &= 467.20
\end{aligned}
$$

dealer's cost, $467.20

53. Let x equal the original price of the sofa.

$$
\begin{aligned}
\frac{2}{7}x &= \text{reduction} \\
x - \frac{2}{7}x &= 235 \\
\frac{5}{7}x &= 235 \\
x &= \frac{7}{5}(235) \\
x &= 329
\end{aligned}
$$

sofa's price, $329

55. Let x equal the original price. $0.30x$ equals the first reduction.
Price after the first reduction:
$$x - 0.30x = 0.70x.$$
Second reduction: $0.40(0.70x) = 0.28x.$
Price after the second reduction: $0.70x - 0.28x = 0.42x.$

$$
\begin{aligned}
0.42x &= 372.40 \\
x &= 886.67
\end{aligned}
$$

original price of VCR, $886.67

57. Let x equal the original price.
Price increased by 50%: $x + 0.50x = 1.50x.$
Then decreased by 50%: $1.50x - 0.50(1.50x) = 0.50(1.50x) = 0.75x.$
The final price is 280.

$$
\begin{aligned}
0.75x &= 280 \\
x &= 373.33
\end{aligned}
$$

original price, $373.33

The net reduction is $1 - 0.75 = 0.25 = $ 25% .

59. Let x = amount of money you had originally

amount of money spent: $\frac{1}{5}x$

amount left: $x - \frac{1}{5}x = \frac{4}{5}x$

amount of money lost: $\frac{1}{3}\left(\frac{4}{5}x\right) = \frac{4}{15}x$

amount left: $\frac{4}{5}x - \frac{1}{3}\left(\frac{4}{5}x\right) = \frac{2}{3}\left(\frac{4}{5}x\right) = \frac{8}{15}x$

$$\frac{8}{15}x = 96$$

$$x = 180$$

$\boxed{\text{original amount of money, \$180}}$

61. Let x equal the length of call. Cost of first minute is \$0.75. Cost of each additional minute is \$0.60.

$$
\begin{aligned}
0.75 + 0.60(x - 1) &= 12.15 \\
0.75 + 0.60x - 0.60 &= 12.15 \\
0.60x + 0.15 &= 12.15 \\
0.60x &= 12.00 \\
x &= 20
\end{aligned}
$$

$\boxed{\text{length of call, 20 minutes}}$

63. Let x equal the number of advertisements distributed.

$$
\begin{aligned}
45 + 0.05x &= 82.50 \\
0.05x &= 37.50 \\
x &= 750
\end{aligned}
$$

$\boxed{\text{750 advertisements}}$ were distributed.

65. Let x equal the amount of money brought in by ticket sales.

$$
\begin{aligned}
3,500 + \frac{1}{5}x &= 5,200 \\
\frac{1}{5}x &= 1,700 \\
x &= 8,500
\end{aligned}
$$

The total ticket sales were $\boxed{\$8,500}$.

67. Let x equal the length of call in minutes.

$$
\begin{aligned}
0.43 + 0.32(x - 1) + 2,110 &= 5.73 \\
0.43 + 0.32x - 0.32 + 2.10 &= 5.73 \\
0.32x + 2.21 &= 5.73 \\
0.32x &= 3.52 \\
x &= 11
\end{aligned}
$$

The person talked $\boxed{\text{11 minutes}}$.

69. Let x equal the number of A's.
$x + 2$ equals the number of B's.
$3x - 4$ equals the number of C's.
$x - 4$ equals the number of D's.

$$4x + 3(x + 2) + 2(3x - 4) + 1(x - 4) \qquad = 50$$
$$4x + 3x + 6 + 6x - 8 + x - 4 \qquad\qquad = 50$$
$$14x - 6 \ = \ 50$$
$$14x \ = \ 56$$

A: $\qquad\qquad\qquad x \ = \ 4$
B: $\qquad\qquad x + 2 \ = \ 6$
C: $\qquad\qquad 3x - 4 \ = \ 12 - 4 = 8$
D: $\qquad\qquad x - 4 \ = \ 4 - 4 = 0$

Janelle has $\boxed{4 \text{ A's, } 6 \text{ B's, } 8 \text{ C's, and no D's}}$.

71. Let x equal the amount of mulch the smaller truck can carry in one load.
$x + 2$ equals the load of larger truck.

$$9(\text{larger}) + 5(\text{smaller}) \ = \ 55$$
$$8(x + 2) + 5x \ = \ 55$$
$$8x + 16 + 5x \ = \ 55$$
$$13x \ = \ 39$$
$$x \ = \ 3$$

The smaller truck can carry $\boxed{3 \text{ cubic feet}}$ of mulch.

73. Let x equal the number of hours of darkness.
$x + 6$ hr 6 min equals the number of hours of daylight.

$$6 \text{ hr } 6 \text{ min} \ = \ 6.1 \text{ hours}$$
$$x + (x + 6.1) \ = \ 24$$
$$2x \ = \ 17.9$$
$$x \ = \ 8.95$$
$$8.95 \text{ hours} \ = \ 8 \text{ hours} + 0.95(60) \text{ minutes} = 8 \text{ hours } 57 \text{ minutes}$$

Darkness is experienced $\boxed{8 \text{ hours } 57 \text{ minutes}}$.

75. Let x equal the weight of Kay's orange.

$$x \ = \ \frac{9}{10}x + \frac{9}{10}$$
$$\frac{1}{10}x \ = \ \frac{9}{10}$$
$$x \ = \ 9$$

The weight of Kay's orange is $\boxed{9 \text{ lb}}$.

77. A is true; $(x + 7) - x = 7$.

79. Cost of article: $30.
Price for profit of 20%: $(1 + 0.2)(30) = 1.2(30) = 36$.
Let x equal tag price.
Discounted price 10% discount: $x - 0.1x = 0.9x$.

$$(\text{discounted price}) \ = \ (\text{price for profit of 20\%})$$
$$0.9x \ = \ 36$$
$$x \ = \ 40$$

$\boxed{\text{price on tag: } \$40}$

81. Let x equal the number of oranges in each of the original three piles.
$3x$ = total number of oranges.
$3x - 2$ = total after 2 are thrown away.

$$\frac{3x - 2}{2} = 32$$

$$3x - 2 = 64$$

$$3x = 66$$

$$x = 22$$

There are $\boxed{22 \text{ oranges}}$ in each of the original piles.

83. Let x equal Tom's age when daughter was born.
$2x - 44$ = Tom's wife's age when daughter born.
Sum of ages of Tom and wife at daughter's birth was 64.

$$x + (2x - 44) = 64$$

$$3x - 44 = 64$$

$$3x = 108$$

$$x = 36$$

Tom was born in 1952.
His daughter was born in $1952 + 36 = 1988$.
In 2005, she will be $(2005 - 1988 = 17)$

$\boxed{17 \text{ years old}}$

85. Let x equal the last four digits of Sophie's telephone number.
Sophie's age: 30.

$$30 + 70 = 3x - 5939$$

$$100 = 3x - 5939$$

$$6039 = 3x$$

$$2013 = x$$

The last 4 digits of telephone number.
Her phone number is $\boxed{279\text{--}2013}$.

87. Let x equal the number of condominiums Reuben owns.

Miami: $\frac{1}{2}x$

Atlanta: $\frac{1}{8}x$

New York: $\frac{1}{12}x$

Chicago: $\frac{1}{20}x$

San Francisco: $\frac{1}{30}x$

Boulder: 50

$$x = \frac{1}{2}x + \frac{1}{8}x + \frac{1}{12}x + + \frac{1}{20}x + \frac{1}{30}x + 50$$

Multiplying both sides by 120:

$$120x = 60x + 15x + 10x + 6x + 4x + 6000$$

$$120x - 95x = 6000$$

$$25x = 6000$$

$$x = 240$$

Reuben owns $\boxed{240 \text{ condominiums}}$.

89. Let x equal the amount Nathan had before finding \$2.

$$\text{(amount with \$2)} = 5(\text{amount would have had if lost \$2})$$

$$
\begin{aligned}
x + 2 &= 5(x - 2) \\
x + 2 &= 5x - 10 \\
12 &= 4x \\
3 &= x
\end{aligned}
$$

Nathan had $\boxed{\$3}$ originally.

91. Let x equal the number of plants originally stolen.

To first guard, he gives: $\quad\quad\quad\quad\quad\quad\quad\quad \frac{1}{2}x + 2$

After first guard, plants remaining: $\quad x - \left(\frac{1}{2}x + 2\right) = \frac{1}{2}x - 2$

To second guard, he gives: $\quad\quad\quad \frac{1}{2}\left(\frac{1}{2}x - 2\right) + 2 = \frac{1}{4}x - 1 + 2 = \frac{1}{4}x + 1$

After second guard, plants remaining: $\left(\frac{1}{2}x - 2\right) - \left(\frac{1}{4}x + 1\right) = \frac{1}{2}x - 2 - \frac{1}{4}x - 1 = \frac{1}{4}x - 3$

To third guard, he gives: $\quad\quad\quad\quad \frac{1}{2}\left(\frac{1}{4}x - 3\right) + 2 = \frac{1}{8}x - \frac{3}{2} + 2 = \frac{1}{8}x + \frac{1}{2}$

After third guard, plants remaining: $\left(\frac{1}{4}x - 3\right) - \left(\frac{1}{8}x + \frac{1}{2}\right) = \frac{1}{4}x - 3 - \frac{1}{8}x - \frac{1}{2} = \frac{1}{8}x - \frac{7}{2}$

He leaves with 1 plant.

$$
\begin{aligned}
\frac{1}{8}x - \frac{7}{2} &= 1 \\
8\left(\frac{1}{8}x - \frac{7}{2}\right) &= 8(1) \\
x - 28 &= 8 \\
x &= 36
\end{aligned}
$$

The thief originally stole $\boxed{36 \text{ plants}}$.

Review Problems

94. $3 + (7 + y) = (3 + 7) + y$
Associative Property of Addition

95. $4 - 2(2 - y) = 4 - 4 + 2y = 2y$

96. Income tax $= 4{,}200 + 0.38(9{,}500 - 8{,}000) = 4{,}200 + 0.38(1{,}500) = 4{,}200 + 570 = 4{,}770$
Income tax is $\boxed{\$4{,}770}$.

Section 2.5 Solving Linear Inequalities

Problem Set 2.5, pp. 162–164

1.
$$x - 3 > 2$$
$$x - 3 + 3 > 2 + 3$$
$$x > 5$$
$$\{x : x > 5\}$$

3.
$$x + 4 \leq 9$$
$$x + 4 - 4 \leq 9 - 4$$
$$x \leq 5$$
$$\{x \mid x \leq 5\}$$

5.
$$y - 3 < 0$$
$$y - 3 + 3 < 0 + 3$$
$$y < 3$$
$$\{y \mid y < 3\}$$

7.
$$4x < 20$$
$$\frac{4x}{4} < \frac{20}{4}$$
$$x < 5$$
$$\{x \mid x < 5\}$$

9.
$$3x \geq -15$$
$$x \geq -5$$
$$\{x \mid x \geq -5\}$$

11.
$$-3x < 15$$
$$\frac{-3x}{-3} > \frac{15}{-3}$$
$$x > -5$$
$$\{x \mid x > -5\}$$

13.
$$-3x \geq -15$$
$$\frac{-3x}{-3} \leq \frac{-15}{-3}$$
$$x \leq 5$$
$$\{x \mid x \leq 5\}$$

15.
$$2y - 3 > 7$$
$$2y - 3 + 3 > 7 + 3$$
$$2y > 10$$
$$\frac{2y}{2} > \frac{10}{2}$$
$$y > 5$$
$$\{y \mid y > 5\}$$

17.
$$3(x - 1) < 9$$
$$\frac{3(x - 1)}{3} < \frac{9}{3}$$
$$x - 1 < 3$$
$$x < 4$$
$$\{x \mid x < 4\}$$

19.
$$-2x - 3 < 3$$
$$-2x < 6$$
$$x > -3$$
$$\{x \mid x > -3\}$$

21.
$$3 - 7y \leq 17$$
$$-7y \leq 14$$
$$y \geq -2$$
$$\{y \mid y \geq -2\}$$

23.
$$-x < 4$$
$$(-1)(-x) > (-1)4$$
$$x > -4$$
$$\{x \mid x > -4\}$$

25.
$$5 - y \leq 1$$
$$-y \leq -4$$
$$y \geq 4$$

$$\{y \mid y \geq 4\}$$

27.
$$2y - 5 > -y + 6$$
$$2y > -y + 11$$
$$3y > 11$$
$$y > \frac{11}{3}$$

$$\left\{ y \mid y > \frac{11}{3} \right\}$$

29.
$$2y - 5 < 5y - 11$$
$$2y < 5y - 6$$
$$-3y < -6$$
$$y > 2$$

$$\{y \mid y > 2\}$$

31.
$$3(x + 1) - 5 < 2x + 1$$
$$3x + 3 - 5 < 2x + 1$$
$$3x - 2 < 2x + 1$$
$$3x < 2x + 3$$
$$x < 3$$

$$\{x \mid x < 3\}$$

33.
$$8x + 3 > 3(2x + 1) - x + 5$$
$$8x + 3 > 6x + 3 - x + 5$$
$$8x + 3 > 5x + 8$$
$$8x > 5x + 5$$
$$3x > 5$$
$$x > \frac{5}{3}$$

$$\left\{ x \mid x > \frac{5}{3} \right\}$$

35.
$$7(y + 4) - 13 < 12 + 13(3 + y)$$
$$7y + 28 - 13 < 12 + 39 + 13y$$
$$7y + 15 < 51 + 13y$$
$$7y < 36 + 13y$$
$$-6y < 36$$
$$y > -6$$

$$\{y \mid y > -6\}$$

37.
$$\frac{x}{4} - \frac{3}{8} < 2$$
$$8\left(\frac{x}{4} - \frac{3}{8}\right) < 8(2)$$
$$2x - 3 < 16$$
$$2x < 19$$
$$x < \frac{19}{2}$$

$$\left\{ x \mid x < \frac{19}{2} \right\}$$

39.
$$\frac{y}{3} + \frac{y}{4} \geq 1$$
$$12\left(\frac{y}{3} + \frac{y}{4}\right) \geq 12(1)$$
$$4y + 3y \geq 12$$
$$7y \geq 12$$
$$y \geq \frac{12}{7}$$

$$\left\{ y \mid y \geq \frac{12}{7} \right\}$$

41.
$$\frac{2y}{3} + 4 < 2$$
$$\frac{2y}{3} < -2$$
$$y < -2\left(\frac{3}{2}\right)$$
$$y < -3$$

$$\{y \mid y < -3\}$$

43.
$$-\frac{1}{5}x - \frac{1}{3} \geq \frac{2}{3}$$
$$15\left(-\frac{1}{5}x - \frac{1}{3}\right) \geq 15\left(\frac{2}{3}\right)$$
$$-3x - 5 \geq 10$$
$$-3x \geq 15$$
$$x \leq -5$$

$$\{x \mid x \leq -5\}$$

45.
$$4 - \frac{5}{6}x > -11$$
$$-\frac{5}{6}x > -15$$
$$x < -\frac{6}{5}(-15)$$
$$x < 18$$
$$\{x \mid x < 18\}$$

47.
$$\frac{1}{3}x - \frac{1}{2} < \frac{5}{6}x + \frac{1}{2}$$
$$6\left(\frac{1}{3}x - \frac{1}{2}\right) < 6\left(\frac{5}{6}x + \frac{1}{2}\right)$$
$$2x - 3 < 5x + 3$$
$$-3x < 6$$
$$x > -2$$
$$\{x \mid x > -2\}$$

49.
$$-0.4y + 2 > -1.2y - 0.4$$
$$-0.4y > -1.2y - 2.4$$
$$0.8y > -2.4$$
$$y > -3$$
$$\{y \mid y > -3\}$$

51.
$$0.3(y - 1) > 0.1y - 0.5$$
$$0.3y - 0.3 > 0.1y - 0.5$$
$$0.3y > 0.1y - 0.2$$
$$0.2y > -0.2$$
$$y > -1$$
$$\{y \mid y > -1\}$$

53.
$$5 < y - 3 < 7$$
$$5 + 3 < y - 3 + 3 < 7 + 3$$
$$8 < y < 10$$
$$\{y \mid 8 < y < 10\}$$

55.
$$5 < 2x - 3 < 11$$
$$5 + 3 < 2x < 11 + 3$$
$$8 < 2x < 14$$
$$4 < x < 7$$
$$\{x \mid 4 < x < 7\}$$

57.
$$-9 < 2x - 3 \leq 5$$
$$-6 < 2x \leq 8$$
$$-3 < x \leq 4$$
$$\{x \mid -3 < x \leq 4\}$$

59.
$$-4 \leq 5 - x < 7$$
$$-4 - 5 \leq 5 - x - 5 < 7 - 5$$
$$-9 \leq -x < 2$$
$$(-1)(-9) \geq (-1)(-x) > (-1)(2)$$
$$9 \geq x > -2$$
$$-2 < x \leq 9$$
$$\{x \mid -2 < x \leq 9\}$$

61.
$$-4 \leq \frac{2y - 3}{3} < 4$$
$$3(-4) \leq \frac{3(2y - 3)}{3} < 3(4)$$
$$-12 \leq 2y - 3 < 12$$
$$-9 \leq 2y < 15$$
$$-\frac{9}{2} \leq y < \frac{15}{2}$$
$$\left\{y \mid -\frac{9}{2} \leq y < \frac{15}{2}\right\}$$

63.
$$-8 \leq 2 - 3y \leq 10$$
$$-10 \leq -3y \leq 8$$
$$\frac{10}{3} \geq y \geq -\frac{8}{3}$$
$$-\frac{8}{3} \leq y \leq \frac{10}{3}$$
$$\left\{y \mid -\frac{8}{3} \leq y \leq \frac{10}{3}\right\}$$

65. Let x equal monthly sales.
$$0.30(x - 1000) > 700$$
$$0.3x - 300 > 700$$
$$0.3x > 1,000$$
$$x > 3333.33$$
$\boxed{\text{monthly sales greater than \$3333.33}}$

67. Let x equal the number of bags of cement.
$$245 + 95x \leq 300$$
$$95x \leq 2,755$$
$$x \leq 29$$
$\boxed{\text{29 or less}}$ bags of cement can be safely lifted

on the elevator in one trip.

69. Let x equal the score on third test.
$$\frac{44 + 72 + x}{3} \geq 60$$
$$116 + x \geq 180$$
$$x \geq 64$$
The maximum score is 100: $x \leq 100$
$$64 \leq x \leq 100$$
The score must be $\boxed{\text{at least 64}}$ or between and including 64 and 100.

71. Let x equal the number.
$$4 + 5x \leq 11 - 2x$$
$$5x \leq 7 - 2x$$
$$7x \leq 7$$
$$x \leq 1$$
range of numbers: $\boxed{\text{numbers less than or equal to 1}}$

73. Let x equal the number.
$$4 - x \leq 5$$
$$-x \leq 1$$
$$x \geq -1$$
range of numbers: $\boxed{\text{numbers greater than or equal to } -1}$

75. Let x equal the number.
$$15 - 3x \geq 6$$
$$-3x \geq -9$$
$$x \leq 3$$
range of numbers: $\boxed{\text{numbers less than or equal to 3}}$

77. Let x equal the number of customers.
$$\text{Profit} = 40x - 200$$
$$40x - 200 \geq 12,000$$
$$40x \geq 12,200$$
$$x \geq 305$$
$\boxed{\text{The company must have 305 or more customers}}$ or it will be sold by the stockholders.

79. Let x equal the pounds of uranium.
$$2x - 6 < 18$$
$$2x < 24$$
$$x < 12$$
$\boxed{\text{less than 12 pounds}}$ of uranium

81. Let x equal the number of advertisements.
$$1600 + 25x \geq 3560$$
$$25x \geq 1960$$
$$x \geq 78.4$$
The minimum number of advertisements is $\boxed{79}$.

83. Let x equal the number of hours working out.
Cost for first club: $500 + 1(x) = 500 + x$
Cost for second club: $440 + 1.75x$
$$
\begin{aligned}
500 + x &< 440 + 1.75x \\
x &< -60 + 1.75x \\
-0.75x &< -60 \\
x &> 80
\end{aligned}
$$
The person must work out $\boxed{\text{more than 80 hours}}$.

85. Let x equal the number.
$$
\begin{aligned}
5 &< 2x + 1 \le 9 \\
4 &< 2x \le 8 \\
2 &< x \le 4
\end{aligned}
$$
Range of numbers: $\boxed{\text{numbers greater than 2 and less than or equal to 4}}$.

87. Let x equal the number.
$$
\begin{aligned}
1 &\le 3x - 2 \le 7 \\
3 &\le 3x \le 9 \\
1 &\le x \le 3
\end{aligned}
$$
Range of numbers: $\boxed{\text{numbers greater than or equal to 1 or less than or equal to 3}}$.

89. Let x equal the first even integer.
$x + 2$ and $x + 4$ are the remaining two integers.
$$
\begin{aligned}
24 &\le x + (x + 2) + (x + 4) \le 36 \\
24 &\le 3x + 6 \le 36 \\
18 &\le 3x \le 30 \\
6 &\le x \le 10 \\
x &= 6, 8, 10 \text{ since } x \text{ is even} \\
x + 2 &= 8, 10, 12 \\
x + 4 &= 10, 12, 14
\end{aligned}
$$
The permissible values of the three integers are $\boxed{6, 8, 10; 8, 10, 12; 10, 12, 14}$.

91. B is true; Let x equal Mia's age. $2x + 3$ equals Eleanor's age.
$$
\begin{aligned}
x + (2x + 3) &\ge 24 \\
3x + 3 &\ge 24 \\
3x &\ge 21 \\
x &\ge 7 \\
2x + 3 &\ge 17
\end{aligned}
$$
Mia is at least 7.
Eleanor is at least 17.
A is false: Mia can be 8.
B is true: Eleanor can be 19.
C is false: Eleanor is at least 17 so she can be 17.
D is false: Mia cannot be 6 since she is at least 7.

93. D is true.
A is false: zero is greater than both negative numbers.
B is false:

$$
\begin{aligned}
3y - 2 &\le y \\
2y &< 2 \\
y &< 1
\end{aligned}
\qquad \text{and} \qquad
\begin{aligned}
y &< 3y - 2 \\
-2y &< -2 \\
y &> 1
\end{aligned}
$$

have different solutions.

C is false: $x > -3$ and $-4 < -3$.
D is true: $5 < x$ and $x > 5$ are the same.

95. B is true.

A is false:
$$-2x + 5 \geq 13$$
$$-2x \geq 8$$
$$x \leq -4$$

The largest integer is –4.

B is true:
$$-\frac{x}{3} > -7 \quad \text{is equivalent to } x < 21$$
$$-x > -21$$
$$x < 21$$

C is false: 8,000% of 10 = 80(10) = 800 not 80.

D is false: $8x > 4x$ is equivalent to $4x > 0$ or $x > 0$.

Review Problems

98. $b^2 - 4ac$
$= (-2)^2 - 4(-1)(3) \quad a = -1, b = -2, x = 3$
$= 4 + 12$
$= 16$

99. $2[10 - (y - 1)] = 2(10 - y + 1) = 2(11 - y) = 22 - 2y$

100.   Cost: $20

Reduction: $\frac{1}{5}(20) = \$4$

Reduced Price: $20 - \$4 = \$16 \left[\text{or } \frac{4}{5}(\$20) = \$16 \right]$

Cost of 25 calculators: 25($16) = $400

Section 2.6 Mathematical Models

Problem Set 2.6, pp. 170–173

1.
$$A = P(1 + rt)$$
$A = \$16,640, r = 9\% = 0.09, t = 12 \text{ years}, P = ?$
$$16,640 = P(2.08)$$
$$8,000 = P$$
invest $\boxed{\$8,000}$

3.
$$A = P(1 + rt)$$
$P = \$8,000, A = \$11,360, r = 7\% = 0.07, t = ?$
$$11360 = 8000(1 + 0.07t)$$
$$11360 = 8000 + 560t$$
$$3360 = 560t$$
$$6 = t$$
time: $\boxed{6 \text{ years}}$

5.
$$A = P(1 + rt)$$
$P = \$8000, A = \$13,760, t = 6 \text{ years}, r = ?$
$$13760 = 8,000(1 + r6)$$
$$13760 = 8,000 + 48,000r$$
$$5760 = 48,000r$$
$$0.12 = r$$
interest rate: $\boxed{12\%}$

7.
$$PB = A$$
$P = 18\% = 0.18, B = 40, A = ?$
$$0.18(40) = A$$
$$\boxed{7.2} = A$$

9.
$$PB = A$$
$P = ?, B = 15, A = 3$
$$P(15) = 3$$
$$P = \frac{1}{5} = 0.20 = \boxed{20\%}$$

11. $P = ?, B = 3, A = 15$
$$P(3) = 15$$
$$P = 5 = 5.00 = \boxed{500\%}$$

13. $P = 60\% = 0.60$, $B = ?$, $A = 3$

$$0.60B = 3$$
$$B = \boxed{5}$$

15.
$$PB = A$$
$P = 18\% = 0.18$, $B = ?$, $A = \$2,970$
$$0.18B = 2,970$$
$$B = 16,500$$
Your salary is $\boxed{\$16,500}$ for that year.

17.
$$PB = A$$
$P = ?$, $B = \$60,000$, $A = \$7,500$
$$P(60,000) = 7,500$$
$$P = 0.125 = 12.5\%$$
$\boxed{12.5\%}$ of goal raised.

19.
$$S = R - rR$$
$S = ?$, $R = \$760$, $r = 45\% = 0.45$
$$S = 760 - 0.45(760) = 760 - 342 = 418$$
Sale price: $\boxed{\$418}$.

21. $S = R - rR = R(1 - r)$
$S = ?$, $R = \$36$, $r = 25\% = 0.25$
$$S = 36(1 - 0.25) = 36(0.75) = 27$$
Sale price: $\boxed{\$27}$

23. $S = R - rR$
$S = \$27.30$, $R = \$42.00$, $r = ?$
$$27.30 = 42 - r(42)$$
$$-14.70 = -42r$$
$$0.35 = r$$
discount rate: $\boxed{35\%}$

25. $E = 250x + 100$

$E = 1,100$ watts, $x = ?$
$$1100 = 250x + 100$$
$$1000 = 250x$$
$$4 = x$$
Runner's rate of speed: $\boxed{4 \text{ meters per second}}$.

27. $R = -\dfrac{35}{2}L + 195$

$R = 125$ per minute, $L = ?$
$$125 = -\frac{35}{2}L + 195$$
$$-70 = -\frac{35}{2}L$$
$$4 = L$$
The wool length is $\boxed{4 \text{ centimeters}}$.

29. $C = \dfrac{1}{600}S + 0.11$

$C = \$0.16$ per mile, $S = ?$
$$0.16 = \frac{1}{600}S + 0.11$$
$$0.05 = \frac{1}{600}S$$
$$30 = S$$
$\boxed{\text{Keep the speed of the truck less than or equal to 30 miles per hour}}$.

31. $C = 0.25x + 0.6$
$R = 0.5x$
$R = C$
$$0.5x = 0.25x + 0.6$$
$$0.25x = 0.6$$
$$x = 2.4$$
Chrysler had to sell $\boxed{2.4 \text{ million cars}}$ to reach the break-even point.

33. $P = \dfrac{V}{180} + \dfrac{80}{9}$

$P = 40$ watts, $V = ?$

$$40 = \dfrac{V}{180} + \dfrac{80}{9}$$

$$180(40) = 180\left(\dfrac{V}{180} + \dfrac{80}{9}\right)$$

$$7{,}200 = V + 1600$$

$$5{,}600 = V$$

Volume of the room: $\boxed{5{,}600 \text{ cubic feet}}$.

35. $D = -4x + 800$

$D > 200$ watts, $x = ?$

$$-4x + 800 = D$$

$$-4x + 800 > 200$$

$$-4x > -600$$

$$x < 150$$

Price range of product: $\boxed{\text{less than } \$150}$.

37. $p = \dfrac{5}{11}d + 15$

$p > 60$ pounds per square inch, $d = ?$

$$\dfrac{5}{11}d + 15 = p$$

$$\dfrac{5}{11}d + 15 > 60$$

$$\dfrac{5}{11}d > 45$$

$$d > \dfrac{11}{5}(45)$$

$$d > 99$$

Depth: $\boxed{\text{depths greater than 99 feet}}$.

39. $C = 40x + 2000$

$R = 80x$

Profit: $R > C$

$$80x > 40x + 2000$$

$$40x > 2000$$

$$x > 50$$

The company will make a profit when $\boxed{\text{more than 50}}$ desks are sold.

41. Car rental agency A: $C = 0.15x + 4$ dollars

Car rental agency B: $C = 0.05x + 20$ dollars

(agency A) < (agency B)

$$0.15x + 4 < 0.05x + 20$$

$$0.10x < 16$$

$$x < 160$$

Car rental agency A is a better deal: $\boxed{\text{less than 160 kilometers}}$.

43. $w = \dfrac{11}{2}h - 230$

62 inches $\le h \le 68$ inches

$h = 62$: $w = \dfrac{11}{2}(62) - 230 = 341 - 230 = 111$

$h = 68$: $w = \dfrac{11}{2}(68) - 230 = 374 - 230 = 144$

$111 \le w \le 144$

weight range:

$\boxed{111 \text{ pounds to 144 pounds inclusively}}$.

45. $A = LW$ for L

$$\dfrac{A}{W} = L$$

$$\boxed{L = \dfrac{A}{W}}$$

47. $A = \dfrac{1}{2}bh$ for b

$$\dfrac{1}{2}bh = A$$

$$bh = 2A$$

$$\boxed{b = \dfrac{2A}{h}}$$

49. $Prt = I$ for P

$$\boxed{P = \dfrac{I}{rt}}$$

51. $E = mc^2$ for m

$$mc^2 = E$$

$$\boxed{m = \dfrac{E}{c^2}}$$

53.
$$y = mx + b \text{ for } m$$
$$mx + b = y$$
$$mx = y - b$$
$$\boxed{m = \frac{y-b}{x}}$$

55.
$$A = \frac{1}{2}(a+b) \text{ for } a$$
$$\frac{1}{2}(a+b) = A$$
$$a + b = 2A$$
$$\boxed{a = 2A - b}$$

57.
$$S = P + Prt \text{ for } r$$
$$P + Prt = S$$
$$Prt = S - P$$
$$\boxed{r = \frac{S-P}{Pt} \text{ or } \frac{S}{pt} - \frac{1}{t}}$$

59.
$$I = \frac{E}{R}$$
$$IR = E$$
$$\boxed{R = \frac{E}{I}}$$

61.
$$L = a + (n-1)d \text{ for } n$$
$$a + (n-1)d = L$$
$$a + nd - d = L$$
$$nd = L + d - a$$
$$\boxed{n = \frac{L+d-a}{d} \text{ or } \frac{L-a}{d} + 1}$$

63.
$$R = -\frac{35}{2}L + 195$$
$$\frac{35}{2}L = 195 - R$$
$$\boxed{L = \frac{390 - 2R}{35}}$$
$$L = \frac{390 - 2(125)}{35} \qquad R = 125$$
$$L = \frac{140}{35} = 4$$

The wool length is $\boxed{4 \text{ cm}}$.

65.
$$F = \frac{9}{5}C + 32$$
$$\frac{9}{5}C = F - 32$$
$$C = \frac{5}{9}(F - 32)$$
$$C = \frac{5}{9}(98.6 - 32) \quad F = 98.6$$
$$C = \frac{5}{9}(66.6)$$
$$C = 37 \quad \rightarrow \quad 37°C$$
$$K = C + 273$$
$$K = 37 + 273$$
$$K = 310$$

The normal body temperature of 98.6 °F corresponds to $\boxed{310 \text{ K}}$.

67.
$$A = P(1 + 0.08t)$$
The investment is doubled when $A = 2P$.
$$2P = P(1 + 0.08t)$$
$$2 = 1 + 0.08t$$
$$1 = 0.08t$$
$$t = \frac{1}{0.08} = 12.5 \quad \rightarrow \quad \boxed{12.5 \text{ years}}$$

69. B is true;

A is not true: $x = -\dfrac{b}{a}$

B is true: $ax = b + 2a \;\rightarrow\; x = \dfrac{b+2a}{a}$

C is not true: $C = 80 + 0.40x$

D is not true: $A = \dfrac{1}{2}bh \;\rightarrow\; b = \dfrac{2A}{h}$

Review Problems

71. $4ab^2c - 3abc = 4(2)(-4)^2(3) - 3(2)(-4)(3)$ $a=2,\,b=-4,\,c=3$
$$= 384 + 72$$
$$= 456$$

72. $\dfrac{4}{5} - \dfrac{\frac{2}{3}}{2-\frac{5}{3}} \div \dfrac{5}{6} = \dfrac{4}{5} - \dfrac{\frac{2}{3}}{\frac{1}{3}} \div \dfrac{5}{6} = \dfrac{4}{5} - 2 \cdot \dfrac{6}{5} = \dfrac{4}{5} - \dfrac{12}{5} = -\dfrac{8}{5}$

73. Weight maintenance is accomplished by 15 times the body weight in calories. To lose 1 pound per week, you would need to reduce the calorie intake by $\frac{1}{3}$, which is the same as an intake of $\frac{2}{3}$ of 15 times **the** body's weight in calories:

$$= \frac{2}{3}(15) = 10 \text{ times body's weight}$$

$$= 10(150) = \boxed{1,500 \text{ calories per day}} \text{ for a person who weighs 150 pounds.}$$

Chapter 2 Review Problems

Chapter 2 Review Problems, pp. 175–177

1. $2y - 5 = 7$
$2y = 12$
$y = 6$
$\boxed{\{6\}}$

2. $5z + 20 = 3z$
$2z = -20$
$z = -10$
$\boxed{\{-10\}}$

3. $7(y-4) = y+2$
$7y - 28 = y + 2$
$6y = 30$
$y = 5$
$\boxed{\{5\}}$

4. $1 - 2(6-y) = 3y + 2$
$1 - 12 + 2y = 3y + 2$
$-11 + 2y = 3y + 2$
$-13 = y$
$\boxed{\{-13\}}$

5. $2(y-4) + 3(y+5) = 2y - 2$
$2y - 8 + 3y + 15 = 2y - 2$
$5y + 7 = 2y - 2$
$3y = -9$
$y = -3$
$\boxed{\{-3\}}$

6. $2z - 4(5z+1) = 3z + 17$
$2z - 20z - 4 = 3z + 17$
$-18z - 4 = 3z + 17$
$-21z = 21$
$z = -1$
$\boxed{\{-1\}}$

7.
$$\frac{2}{3}x = \frac{1}{6}x + 1$$
$$4x = x + 6 \quad (\times 6)$$
$$3x = 6$$
$$x = 2$$
$$\boxed{\{2\}}$$

8.
$$\frac{1}{2}y - \frac{1}{10} = \frac{1}{5}y + \frac{1}{2}$$
$$5y - 1 = 2y + 5 \quad (\times 10)$$
$$3y = 6$$
$$y = 2$$
$$\boxed{\{2\}}$$

9.
$$0.2y - 0.3 = 0.8y - 0.3$$
$$-0.6y = 0$$
$$y = 0$$
$$\boxed{\{0\}}$$

10.
$$0.04z + 0.06(100 - z) = 4.6$$
$$4z + 6(100 - z) = 460 \quad (\times 100)$$
$$4z + 600 - 6z = 460$$
$$-2z = -140$$
$$z = 70$$
$$\boxed{\{70\}}$$

11.
$$-(3y - 2) = -2 - (6y - 2)$$
$$-2y + 8 - 3y + 2 = -2 - 6y + 2$$
$$-5y + 10 = 0 - 6y$$
$$y = -10$$
$$\boxed{\{-10\}}$$

12.
$$\frac{x}{4} = 2 + \frac{x - 3}{3}$$
$$3x = 24 + 4(x - 3) \quad (\times 12)$$
$$3x = 24 + 4x - 12$$
$$-x = 12$$
$$x = -12$$
$$\boxed{\{-12\}}$$

13.
$$3x + 15 = 3(x + 4)$$
$$3x + 15 = 3x + 12$$
$$15 \neq 12$$
no solution; $\boxed{\emptyset}$

14.
$$7x + 5 = 7(x + 2) - 9$$
$$7x + 5 = 7x + 14 - 9$$
$$7x + 5 = 7x + 5$$
$$5 = 5$$
all real numbers; $\boxed{\{x \mid x \in R\}}$

15. Six times a number, decreased by 20, is four times a number.
$$6n - 20 = 4n$$
$$2n = 20$$
$$n = 10$$
The number is $\boxed{10}$.

The check for exercises 15–30 is left to the reader.

16. Let n equal the smaller number. $2n + 3$ equals the larger number. The sum of the numbers is 39.
$$n + (2n + 3) = 39$$
$$3n + 3 = 39$$
$$3n = 36$$
$$n = 12$$
$$2n + 3 = 2(12) + 3 = 27$$
The numbers are $\boxed{12 \text{ and } 27}$.

17. Let x equal the number.
Fifteen percent of the number is 39.75.
$$0.15x = 39.75$$
$$x = 265$$
The number is $\boxed{265}$.

18. Let x equal the number. 8 added to 60% of the number is 332.

$$0.60x + 8 = 332$$
$$0.60x = 324$$
$$x = 540$$

The number is $\boxed{540}$.

19. Let x equal the number. When seven times a number decreased by 1, the result is nine more than five times the number.

$$7x - 1 = 5x + 9$$
$$2x = 10$$
$$x = 5$$

The number is $\boxed{5}$.

20. Let $x, x + 2, x + 4$ be three consecutive even integers.

$$x + (x + 2) + (x + 4) = 84$$
$$3x + 6 = 84$$
$$3x = 78$$
$$x = 26$$
$$x + 2 = 28$$
$$x + 4 = 30$$

The integers are $\boxed{26, 28 \text{ and } 30}$.

21. Let $x, x + 1, x + 2$ represent three consecutive integers. The sum of the first two integers exceeds the third integer by 12.

$$x + (x + 1) = (x + 2) + 12$$
$$2x + 1 = x + 14$$
$$x = 13$$
$$x + 1 = 14$$
$$x + 2 = 15$$

The integers are $\boxed{13, 14 \text{ and } 15}$.

22. Let x equal the total amount of money. $\frac{1}{2}x$ equals the amount in bank. $\frac{1}{10}x$ equals the amount in stock. $\frac{1}{20}x$ equals the amount in equipment. Remainder equals $35000 in savings.

$$x - \left(\frac{1}{2}x + \frac{1}{10}x + \frac{1}{20}x\right) = 35,000$$
$$x - \frac{13}{20}x = 35,000$$
$$\frac{7x}{20} = 35,000$$
$$x = 100,000$$

The total amount this person has is $\boxed{\$100,000}$.

23. Let x equal the number of Independents. $x + 1$ equals the number of Republicans. $5x + 1$ equals the number of Democrats.

$$x + (x + 1) + (5x + 1) = 23$$
$$7x + 2 = 23$$
$$7x = 21$$
$$x = 3$$
$$x + 1 = 4$$
$$5x + 1 = 5(3) + 1 = 16$$

$\boxed{3 \text{ Independents}; 4 \text{ Republicans}; 16 \text{ Democrats}}$

24. Let t equal the number of years ago.

$$
\begin{aligned}
95 - t &= 2(55 - t) \\
95 - t &= 110 - 2t \\
t &= 15
\end{aligned}
$$

The rug was twice as old as the dresser $\boxed{15 \text{ years ago}}$.

25. Let x equal Bill's present age. $x + 28$ equals Deborah's present age. $x + 6$ equals Bill's age in 6 years. $x + 34$ equals Deborah's age in 6 years.

$$
\begin{aligned}
x + 34 &= 2(x + 6) \\
x + 34 &= 2x + 12 \\
22 &= x \\
x + 28 &= 22 + 28 = 50
\end{aligned}
$$

Present age: $\boxed{\text{Bill, 22 years old; Deborah, 50 years old}}$.

26. Let x equal the price before reduction.

$$
\begin{aligned}
x - 0.45x &= 247.50 \\
0.55x &= 247.50 \\
x &= 450
\end{aligned}
$$

The price of the VCR before reduction was $\boxed{\$450}$.

27. Let x equal the yearly income.

$$
\begin{aligned}
x - 0.35x &= 32,630 \\
0.65x &= 32,630 \\
x &= 50,200
\end{aligned}
$$

Yearly income for all recreational needs to be satisfied is $\boxed{\$50,200}$.

28. Let x equal the number of days that the book was on loan.

$$
\begin{aligned}
1.25 + 0.55(x - 1) &= 10.65 \\
1.25 + 0.55x - 0.55 &= 10.05 \\
0.55x + 0.70 &= 10.05 \\
0.55x &= 9.35 \\
x &= 17
\end{aligned}
$$

The book was on loan for $\boxed{17 \text{ days}}$.

29. Let x equal the number of kilowatt–hours used that month.

$$
\begin{aligned}
17.50 + 0.18x &= 100.98 \\
0.18x &= 93.48 \\
x &= 463.7\overline{7}
\end{aligned}
$$

The energy used that month was about $\boxed{463.78 \text{ kilowatt–hours}}$.

30. Let x equal the present age of math club president.

$$
\begin{aligned}
x + 10 &= 2x - 5 \\
15 &= x
\end{aligned}
$$

$\boxed{\text{No}}$, Dora does not date the math club president. The president is only 15 years old, younger than 18.

31.
$$
\begin{aligned}
2y - 5 &< 3 \\
2y &< 8 \\
y &< 4
\end{aligned}
$$
$\{y \mid y < 4\}$

32.
$$
\begin{aligned}
3 - 5x &\le 18 \\
-5x &\le 15 \\
x &\ge -3
\end{aligned}
$$
$\{x \mid x \ge -3\}$

33.
$$4x + 6 \; < \; 5x$$
$$-x \; < \; -6$$
$$x \; > \; 6$$
$$\{x \mid x > 6\}$$

34.
$$9(z - 1) \; \geq \; 10(z - 2)$$
$$9z - 9 \; \geq \; 10z - 20$$
$$-z \; \geq \; -11$$
$$z \; \leq \; 11$$
$$\{z \mid z \leq 11\}$$

35.
$$-3(4 - x) \; < \; 4x + 3 + x$$
$$-12 + 3x \; < \; 5x + 3$$
$$-2x \; < \; 15$$
$$x \; > \; -\frac{15}{2}$$
$$\left\{ x \mid x > -\frac{15}{2} \right\}$$

36.
$$4y - (y - 4) \; \leq \; -3(2y - 7)$$
$$4y - y + 3 \; \leq \; -6y + 21$$
$$3y + 3 \; \leq \; -6y + 21$$
$$9y \; \leq \; 18$$
$$y \; \leq \; 2$$
$$\{y \mid y \leq 2\}$$

37.
$$\frac{5y}{4} - \frac{1}{4} \; \leq \; \frac{6y}{5} + \frac{1}{5}$$
$$\frac{5y}{4} - \frac{6y}{5} \; \leq \; \frac{1}{5} + \frac{1}{4}$$
$$25y - 24y \; \leq \; 4 + 5 \quad (\times 20)$$
$$y \; \leq \; 9$$
$$\{y \mid y \leq 9\}$$

38.
$$1.1y - 0.2 \; \leq \; 1.0 - 0.4y$$
$$1.5y \; \leq \; 1.2$$
$$y \; \leq \; \frac{4}{5}$$
$$\left\{ y \mid y \leq \frac{4}{5} \right\}$$

39.
$$-2x \; \geq \; 0$$
$$x \; \leq \; 0$$
$$\{x \mid x \leq 0\}$$

40.
$$-5 \; \leq \; 2x - 3 \; < \; 1$$
$$-2 \; \leq \; 2x \; < \; 4$$
$$-1 \; \leq \; x \; < \; 2$$
$$\{x \mid -1 \leq x < 2\}$$

41.
$$-3 \; < \; 3y - 6 \; \leq \; 0$$
$$3 \; < \; 3y \; \leq \; 6$$
$$1 \; < \; y \; \leq \; 2$$
$$\{y \mid 1 < y \leq 2\}$$

42. Let x equal the score on third test.
$$\frac{42 + 74 + x}{3} \; \geq \; 60$$
$$116 + x \; \geq \; 180$$
$$x \; \geq \; 64$$

The student needs $\boxed{\text{at least } 64}$ on the third test.

43. Let x equal the number.
$$16 - 4x \; \leq \; 24$$
$$-4x \; \leq \; 8$$
$$x \; \geq \; -2$$

The range of numbers is $\boxed{\text{at least } -2}$ or greater than or equal to -2.

44. Let x equal the number of customers.

$$\begin{aligned} \text{profit} &= 90n - 300 \\ 90n - 300 &\le 150000 \\ 90n &\le 150300 \\ n &\le 1670 \end{aligned}$$

The number of customers can be $\boxed{\text{at most 1670}}$ or the company will be nationalized.

45. Let x equal the number of half miles.

$$\begin{aligned} 0.75 + 0.35(x - 1) &< 7.40 \\ 0.35x + 0.40 &< 7.40 \\ 0.35x &< 7.00 \\ x &< 20 \end{aligned}$$

The person can travel $\boxed{\text{less than 20 half–miles}}$ or $\boxed{\text{less than 10 miles}}$.

46. Let x equal the number of small dogs.

$$\begin{aligned} 240 + 160 + 25x &\le 1000 \\ 400 + 25x &\le 1000 \\ 25x &\le 600 \\ x &\le 24 \end{aligned}$$

The number of dogs is $\boxed{\text{at most 24}}$.

47.

$$\begin{aligned} 2H - 7R &= 166 \\ 2H &= 7R + 166 \\ H &= \frac{7}{2}R + 83 \\ H &= \frac{7}{2}(24) + 83 \quad R = 24 \\ H &= 84 + 83 \\ H &= 167 \end{aligned}$$

The person is $\boxed{\text{167 centimeters}}$ tall.

48.

$$A = P(1 + rt)$$
$r = 12\% = 0.12,\ A = 40880,\ t = 16,\ P = ?$
$$\begin{aligned} 40880 &= P[1 + 0.12(16)] \\ 40880 &= P(1 + 1.92) \\ 2.92P &= 40880 \\ P &= 14000 \end{aligned}$$

$\boxed{\$14000}$

49.

$$A = P(1 + rt)$$
$P = 12000,\ A = 23880,\ t = 9,\ r = ?$
$$\begin{aligned} 23880 &= 12000(1 + 9r) \\ 12000(1 + 9r) &= 23880 \\ 12000 + 108000r &= 21880 \\ 108000r &= 11880 \\ r &= 0.11 \end{aligned}$$

$\boxed{11\%}$

50.

$$\begin{aligned} R &= 1.62t + 9.85 \\ 32.67 &= 1.63t + 9.85 \quad R = 32.67 \text{ billion} \\ 22.82 &= 1.63t \\ 14 &= t \end{aligned}$$

14 years after 1965 or $\boxed{1979}$.

51.

$$\begin{aligned} S &= R - rR \\ rR &= R - S \\ r &= 1 - \frac{S}{R} \end{aligned}$$

$R = 1250,\ S = 812.50,\ r = ?$
$$r = 1 - \frac{812.50}{1250} = 1 - 0.65 = 0.35$$

$\boxed{r = 35\%}$

52. $C = 400(3x - 1) + 700$
$R = 1250x$
break–even point:
$$\begin{aligned} R &= C \\ 1250x &= 400(3x-1) + 700 \\ 1250x &= 1200x - 400 + 700 \\ 50x &= 300 \\ x &= 6 \end{aligned}$$

$\boxed{6 \text{ units}}$ must be produced to reach the break–even point.

53.
$$\begin{aligned} N &= 600 - 5x \\ N &< 50S \\ 600 - 5x &< 50S \\ -5x &< -9S \\ x &> 19 \end{aligned}$$
$\boxed{\text{after 19 days}}$

54.
$$H = \frac{8T + 211}{3}$$
$$169 \le H \le 185$$
$$169 \le \frac{8T+211}{3} \le 185$$
$$507 \le 8T + 211 \le 555$$
$$296 \le 8T \le 344$$
$$37 \le T \le 43$$
$\boxed{37 \text{ cm} \le T \le 43 \text{ cm}}$

55. a.
$$\begin{aligned} F &= \tfrac{1}{4}C + 40 \\ 4F &= C + 160 \end{aligned}$$
$\boxed{C = 4F - 160}$

b.
$$\begin{aligned} 200 &= 4F - 160 \quad C = 200/\text{minute} \\ 4F &= 360 \\ F &= 90 \end{aligned}$$
The temperature must be $\boxed{90°F}$.

56.
$$A = \tfrac{1}{2}bh \text{ for } h$$
$$bh = 2A$$
$\boxed{h = \dfrac{2A}{b}}$

57.
$$\begin{aligned} Ax + By &= C \text{ for } y \\ By &= C - Ax \end{aligned}$$
$\boxed{y = \dfrac{C - Ax}{B}}$

58.
$$\begin{aligned} F &= f(1 - M) \text{ for } M \\ F &= f - fM \\ fM &= f - F \end{aligned}$$
$\boxed{M = \dfrac{f - F}{f} \text{ or } 1 - \dfrac{F}{f}}$

59.
$$\begin{aligned} P &= \frac{RT}{V} \text{ for } V \\ PV &= RT \end{aligned}$$
$\boxed{V = \dfrac{RT}{P}}$

60. No;
$$\begin{aligned} 3H - 8T &= 211 \\ 3(196) - 8(37) &= 211 \\ 588 - 296 &= 211 \\ 292 &\ne 211 \end{aligned}$$

61.
$$\begin{aligned} 1.8(0.3y + 7.4) &= 9.2y - 3 \\ 1.8(0.3)y + 1.8(7.4) &= 9.2y - 3 \\ 1.8(0.3)y - 9.2y &= -1.8(7.4) - 3 \\ y &= \frac{-1.8(7.4) - 3}{1.8(0.3) - 9.2} \\ y &\approx 1.884526559 \end{aligned}$$
$\boxed{\{1.884526559\}}$

62.

$$
\begin{aligned}
0.078(0.76y + 0.007) &= 0.084(0.96y + 0.009) \\
0.078(0.76)y + 0.078(0.007) &= 0.084(0.96)y + 0.084(0.009) \\
y[0.078(0.76) - 0.084(0.96)] &= 0.084(0.009) - 0.078(0.007) \\
y &= \frac{0.084(0.009) - 0.078(0.007)}{0.078(0.76) - 0.084(0.96)} \\
y &\approx -0.009831460674
\end{aligned}
$$

$$\boxed{\{-0.009831460674\}}$$

63.

$$
\begin{aligned}
3.5\left(\frac{1}{3}y + \frac{2}{7}\right) &= \frac{4}{5}\left(2 + \frac{2}{5}y\right) \\
\frac{3.5}{3}y + 3.5\left(\frac{2}{7}\right) &= \frac{8}{5} + \frac{8}{25}y \\
y\left(\frac{3.5}{3} - \frac{8}{25}\right) &= \frac{8}{5} - 1 \\
y\left(\frac{35}{30} - \frac{8}{25}\right) &= \frac{3}{5} \\
y &= \frac{\frac{3}{5}}{\frac{7}{6} - \frac{8}{25}} \\
y &= \frac{\frac{3}{5}}{\frac{127}{150}} = \frac{90}{127} \approx 0.7086614173
\end{aligned}
$$

$$\boxed{\left\{\frac{90}{127}\right\} \text{ or } \{0.7086614173\}}$$

Chapter 3   Problem Solving

Section 3.1 Critical Thinking

Problem Set 3.1, pp. 189-194

1. Let n equal the number.
$$7(n-11) = 588$$
$$7n-77 = 588$$
$$7n = 665$$
$$n = \boxed{95}$$

3. Let n equal the number.
$$7n-11 = 290$$
$$7n = 301$$
$$n = \boxed{43}$$

5. Let x equal the maximum price of dinner.
$$x+0.15x = 32.20$$
$$1.15x = 32.20$$
$$x = 28$$
$$\boxed{\$28}$$

7. Let x equal the selling price.
$$x-0.07x = 111600$$
$$0.93x = 111600$$
$$x = 120000$$
$$\boxed{\$120000}$$

9. Let x equal per capita income in Canada.
$x + 420$ equals per capita income in the U.S.
$2x - 40560$ equals per capita income in Mexico.
$$x+(x+420)+(2x-40560) = 47780$$
$$4x-40140 = 47780$$
$$4x = 87920$$
$$x = 21980$$
$$x+420 = 22400$$
$$2x-40560 = 3400$$
$$\boxed{\text{Canada, \$21980; U.S., \$22400; Mexico, \$3400}}$$

11. Let x equal the number of hours to repair the car.
$$\text{parts}+\text{labor} = 448$$
$$63+35x = 448$$
$$35x = 385$$
$$x = 11$$
$$\boxed{\text{11 hours}}$$

13. Let x equal the fixed rate.
$$0.35+(26-1)x = 6.60$$
$$0.35+25x = 6.60$$
$$25x = 6.25$$
$$x = 0.25$$
Fixed rate: $\boxed{25¢ \text{ per day}}$

15. Let x equal regular hourly salary.
$$40x+(50-40)\left(1\tfrac{1}{2}\right)(x) = 462$$
$$40x+10\left(\tfrac{3}{2}\right)x = 462$$
$$40x+15x = 462$$
$$55x = 462$$
$$x = 8.4$$
Regular rate: $\boxed{\$8.40 \text{ per hour}}$

17. Let x equal the number of gallons of oil.
$20x$ equals the number of gallons of gasoline.

$$
\begin{aligned}
x + 20x &= 42 \\
21x &= 42 \\
x &= 2 \\
20x &= 40
\end{aligned}
$$

$\boxed{\text{2 gallons of oil, 40 gallons of gasoline}}$

19. Let x equal the amount of each payment.

$$
\begin{aligned}
100 + 12x &= 202 \\
12x &= 102 \\
x &= 8.5
\end{aligned}
$$

payment: $\boxed{\$8.50 \text{ per month}}$

21. Let x equal the number of cans of soda sold.

$$
\begin{aligned}
7.50 + 1.25x &= 86.25 \\
1.25x &= 78.75 \\
x &= 63
\end{aligned}
$$

$\boxed{\text{63 cans of soda sold}}$

23. Let x equal the number of nickels.
$x + 5$ equals the number of quarters.

$$
\begin{aligned}
0.05x + 0.25(x + 5) &= 2.15 \\
0.05x + 0.25x + 1.25 &= 2.15 \\
0.3x &= 0.90 \\
x &= 3 \\
x + 5 &= 8
\end{aligned}
$$

$\boxed{\text{3 nickels, 8 quarters}}$

25. Let x equal the number of half–dollars.
$x - 3$ equals the number of quarters.
$3x + 2$ equals the number of dimes.

$$
\begin{aligned}
0.50x + 0.25(x - 3) + 0.10(3x + 2) &= 6.80 \\
0.50x + 0.25x - 0.75 + 0.30x + 0.20 &= 6.80 \\
1.05x - 0.55 &= 6.80 \\
1.05x &= 7.35 \\
x &= 7 \\
x - 3 &= 4 \\
3x + 2 &= 23
\end{aligned}
$$

$\boxed{\text{7 half–dollars, 4 quarters, 23 dimes}}$

27. Let x equal the number of \$5 bills.
$x + 6$ equals the number of \$10 bills.
$4x - 3$ equals the number \$20 bills.

$$
\begin{aligned}
5x + 10(x + 6) + 20(4x - 3) &= 665 \\
5x + 10x + 60 + 80x - 60 &= 665 \\
95x &= 665 \\
x &= 7 \\
4x - 3 &= 28 - 3 = 25
\end{aligned}
$$

$\boxed{\text{25 \$20 bills}}$

29. Let x equal the number of balcony tickets.
$2x$ equals the number of orchestra tickets.

$$
\begin{aligned}
4.50x + 6(2x) &= 1815 \\
4.50x + 12x &= 1815 \\
16.50x &= 1815 \\
x &= 110 \\
2x &= 220
\end{aligned}
$$

$\boxed{\text{110 balcony tickets, 220 orchestra tickets}}$

31. Let x equal the sample size.

$$
\begin{aligned}
0.005x &= 4 \\
x &= 800
\end{aligned}
$$

$\boxed{\text{800 parts}}$

33. Let x equal the number of tennis balls worth \$3.
$15 - x$ equals the number of tennis balls worth \$4.

$$
\begin{aligned}
3x + 4(15 - x) &= 49 \\
3x + 60 - 4x &= 49 \\
-x &= -11 \\
x &= 11 \\
15 - x &= 15 - 11 = 4
\end{aligned}
$$

$\boxed{\text{eleven \$3 balls, four \$4 balls}}$

35. Let x equal the number of ducks.
$20 - x$ equals the number of horses.
$$2x + 4(20 - x) = 64$$
$$2x + 80 - 4x = 64$$
$$-2x = -16$$
$$x = 8$$
$$20 - x = 20 - 8 = 12$$
$\boxed{\text{8 ducks, 12 horses}}$

37. Let x equal the smaller number.
$x + 5$ equals the larger number.
9 larger numbers $-$ 5 smaller numbers $= 53$
$$9(x + 5) - 5x = 53$$
$$9x + 45 - 5x = 53$$
$$4x = 8$$
$$x = 2$$
The smaller number is $\boxed{2}$.

39. Let x equal the weight of the body.
9 lb equals the weight of the tail.
$\frac{1}{2}x + 9$ equals the weight of the head.

weight of head + weight of tail = weight of body
$$\left(\frac{1}{2}x + 9\right) + 9 = x$$
$$\frac{1}{2}x + 18 = x$$
$$\frac{1}{2}x = 18$$
$$x = 36$$
$$\frac{1}{2}x + 9 = \frac{1}{2}(36) + 9 = 27$$

weight of fish = weight of head + weight of body + weight of tail = $27 + 36 + 9 = 72$
$\boxed{\text{72 pounds}}$

41. Let x equal the number of nickels.
$$0.10x = 0.05x + 0.45$$
$$0.05x = 0.45$$
$$x = 9$$
$\boxed{\text{9 nickels}}$

43. Let x equal the total number of votes cast.
$0.40x + 40$ equals the number of votes for candidate A.
$0.50(\text{candidate A}) - 40 = 0.50(0.40x - 40) = 0.20x + 20 = $ number of votes for candidate B.
$0.50(\text{candidate B}) + 20 = 0.50(0.20x + 20) + 20 = 0.10x + 10 + 20 = 0.10x + 30 = $ number of votes cast for candidate C.
$180 = $ number of votes for candidate D.
$$x - (\text{candidate A} + \text{candidate B} + \text{candidate C}) = (\text{candidate D})$$
$$x - [(0.40x + 40) + (0.20x + 20) + (0.10x + 30)] = 180$$
$$x - (0.70x + 90) = 180$$
$$0.30x - 90 = 180$$
$$0.30x = 270$$
$$x = 900$$

candidate A: $0.40x + 40 = 360 + 40 = 400$
candidate B: $0.20x + 20 = 180 + 20 = 200$
candidate C: $0.10x + 30 = 90 + 30 = 120$
candidate D: 180
$\boxed{\text{900 votes cast; candidate A won}}$

45-50. Answers may vary. Samples are given.

45. 2, 8, 14, 20, ___, ___, ___
$\boxed{\text{Add 6 repeatedly: } 26, 32, 38}$

47. 15, 11, 7, 3, ___, ___, ___
$\boxed{\text{Subtract 4 repeatedly: } -1, -5, -9}$

49. Answers will vary.

3, 4, 7, 11, 18, 29, ___, ___, ___

| Add the two previous numbers to obtain the next number in the sequence: 47, 76, 123 |

$18 + 29 = 47$
$29 + 47 = 76$
$47 + 76 = 123$

51.

$$1 + 3 \ = \ 4$$
$$1 + 3 + 5 \ = \ 9$$
$$1 + 3 + 5 + 7 \ = \ 16$$
$$1 + 3 + 5 + 7 + 9 \ = \ 25$$

The sum of the first n odd numbers is n^2. $1 + 3 + \ldots + 99 + 101 + \ldots + 197 + 199 = \boxed{100^2 = 10000}$

53.

	1 point free throw		2 point field goal		3 point long shot		total
	no.	points	no.	points	no.	points	
1.					5	15	15
2.	1	1	1	2	4	12	15
3.	3	3			4	12	15
4.			3	6	3	9	15
5.	2	2	2	4	3	9	15
6.	4	4	1	2	3	9	15
7.	6	6			3	9	15
8.	1	1	4	8	2	6	15
9.	3	3	3	6	2	6	15
10.	5	5	2	4	2	6	15
11.	7	7	1	2	2	6	15
12.	9	9			2	6	15
13.			6	12	1	3	15
14.	2	2	5	10	1	3	15
15.	4	4	4	8	1	3	15
16.	6	6	3	6	1	3	15
17.	8	8	2	4	1	3	15
18.	10	10	1	2	1	3	15
19.	12	12			1	3	15
20.	1	1	7	14			15
21.	3	3	6	12			15
22.	5	5	5	10			15
23.	7	7	4	8			15
24.	9	9	3	6			15
25.	11	11	2	4			15
26.	13	13	2	4			15
27.	15	15					15

There are $\boxed{\text{27 ways}}$.

55. We know the sum is $2 + 5 + 2 = 9$ since one of the area codes is 252.
No area code has a first digit of 4.
One area code begins with 6: $6 + a + b = 9$
Another area code ends with 1: $c + d + 1 = 9$
No area code contains a digit that is in one of the other area codes.
$a + b = 3$ only 0 and 3 are possible
$c + d = 8$ only 7 and 1 are possible
Thus, the number can be $\boxed{\text{711 or 171}}$.

57. pattern: $1, 9, 9 + 8 \;=\; 17, 17 + 8 = 25, \ldots$

$\qquad\qquad\qquad\;\; = \; 1, 1 + 8(1), 1 + 2(8), 1 + 3(8), \ldots$

$\qquad\qquad\qquad\;\; = \; 1, 1 + 8(2 - 1), 1 + 8(3 - 1), 1 + 8(4 - 1), \ldots$

number of dots in 10th term: $1 + 8(10 - 1) = 1 + 8(9) = \boxed{73}$

number of dots in n^{th} term: $1 + 8(n - 1) = 1 + 8n - 8 = \boxed{8n - 7}$

59. 1st trianglar.

Number	2nd	3rd	4th	5th	6th	12th	n^{th}
1	3	6	10	$\boxed{15}$	$\boxed{21}$	$\boxed{78}$	$\boxed{\dfrac{n(n+1)}{2}}$

Note pattern: $\dfrac{1(1+1)}{2} = 1$

$\qquad\qquad\quad \dfrac{2(2+10)}{2} = 3$

$\qquad\qquad\quad \dfrac{3(3+1)}{2} = 6$

$\qquad\qquad\quad \dfrac{4(4+1)}{2} = 10$

$\qquad\qquad\quad \dfrac{5(5+1)}{2} = 15$

$\qquad\qquad\quad \dfrac{6(6+1)}{2} = 21$

$\qquad\qquad\quad \dfrac{12(12+1)}{2} = 78$

	1st term $(n=1)$	2nd term $(n=2)$	3rd term $(n=3)$	4th term $(n=4)$	5th $(n=5)$	Formula for the nth term.
61.	1	3	5	$\boxed{7}$	$\boxed{9}$	$2n - 1$
63.	5	7	9	11	$\boxed{13}$	$\boxed{2n + 3}$
65.	2	$\boxed{8}$	18	$\boxed{32}$	$\boxed{50}$	$2n^2$

67. $\boxed{4}$ 13 $\boxed{9}$ since $4 + 9 \;=\; 13$

$\qquad\;\; 16 \;\; 21 \qquad\qquad\qquad\; 4 + 12 \;=\; 16$

$\qquad\;\;\; \boxed{12} \qquad\qquad\qquad\;\;\; 9 + 12 \;=\; 21$

69. 2 people in 3 days $\rightarrow$ $2(3) = 6$ people days

$\qquad$ 4 people in $\boxed{1\dfrac{1}{2}\text{ days}}$ $\rightarrow$ $4\left(\dfrac{3}{2}\right) = 6$ people days

71. a.   The Farey sequence F_5 contains all proper fractions where denominators are 2, 3, 4, and 5.

$\qquad \dfrac{1}{2}, \dfrac{1}{3}, \dfrac{2}{3}, \dfrac{1}{4}, \dfrac{3}{4}, \dfrac{1}{5}, \dfrac{2}{5}, \dfrac{3}{5}, \dfrac{4}{5},$ are arranged in order

$\qquad \dfrac{1}{5}, \dfrac{1}{4}, \dfrac{1}{3}, \dfrac{2}{5}, \dfrac{1}{2}, \dfrac{3}{5}, \dfrac{2}{3}, \dfrac{3}{4}, \dfrac{4}{5}$

b. F_6: (denominators 2, 3, 4, 5, 6)

$\qquad \dfrac{1}{6}, \dfrac{1}{5}, \dfrac{1}{4}, \dfrac{1}{3}, \dfrac{2}{5}, \dfrac{1}{2}, \dfrac{3}{5}, \dfrac{2}{3}, \dfrac{3}{4}, \dfrac{4}{5}, \dfrac{5}{6}$

c.

	Farey Sequence	A Fraction in the Sequence	Fractions Appearing Before and After Their Fraction	Sum of Numerator and Denominator of Fractions in Previous Column
F_3:	$\dfrac{1}{3}, \dfrac{1}{2}, \dfrac{2}{3}$	$\dfrac{1}{2}$	$\dfrac{1}{3}, \dfrac{2}{3}$	$\dfrac{1+2}{3+3} = \dfrac{3}{6} = \dfrac{1}{2}$
F_4:	$\dfrac{1}{4}, \dfrac{1}{3}, \dfrac{1}{2}, \dfrac{2}{3}, \dfrac{3}{4}$	$\dfrac{2}{3}$	$\dfrac{1}{2}, \dfrac{3}{4}$	$\dfrac{1+3}{2+4} = \dfrac{4}{6} = \dfrac{2}{3}$
F_5:	$\dfrac{1}{5}, \dfrac{1}{4}, \dfrac{1}{3}, \dfrac{2}{5}, \dfrac{1}{2}, \dfrac{3}{5}, \dfrac{2}{3}, \dfrac{3}{4}, \dfrac{4}{5}$	$\dfrac{1}{4}$	$\dfrac{1}{5}, \dfrac{1}{3}$	$\dfrac{1+1}{5+3} = \dfrac{2}{8} = \dfrac{1}{4}$
F_6:	$\dfrac{1}{6}, \dfrac{1}{5}, \dfrac{1}{4}, \dfrac{1}{3}, \dfrac{2}{5}, \dfrac{1}{2}, \dfrac{3}{5}, \dfrac{2}{3}, \dfrac{3}{4}, \dfrac{4}{5}, \dfrac{5}{6}$	$\dfrac{4}{5}$	$\dfrac{3}{4}, \dfrac{5}{6}$	$\dfrac{3+5}{4+6} = \dfrac{8}{10} = \dfrac{4}{5}$

d. The reduced form of the fraction formed by adding the numerator and the denominator of the fractions before and after the fractions is the fraction.

e. F_3: $\dfrac{1}{3} + \dfrac{2}{3} = 1$

F_4: $\dfrac{1}{4} + \dfrac{3}{4} = 1$

F_5: $\dfrac{1}{5} + \dfrac{4}{5} = 1$

F_6: $\dfrac{1}{6} + \dfrac{5}{6} = 1$

The sum is 1.

73. none; the hole is empty

75. decimal: $6 < 6.7 < 7$

77. Given:
$$3x + 1 = 16$$
$$3x = 15$$
$$x = 5$$
Conclusion: the number is 5.

True; argument is $\boxed{\text{valid}}$.

79. Given: Two numbers are both even. (number could be 8 and 12 or 8 and 10) Conclusion: The numbers differ by 2.

Not necessarily true; argument is $\boxed{\text{invalid}}$.

81. Given: My number: 8
Your number: 10
Fred's number: $8 + 3 = 11$
Conclusion: Fred's number is $10 - 1 = 9$.

False; argument is $\boxed{\text{invalid}}$.

83. Given: Tony's number is divisible by 5. Maria's number is divisible by 3.
Conclusion: Tony's number and Maria's number cannot be the same.

False; 15 is one example of a number that is divisible by 3 and 5; argument is $\boxed{\text{invalid}}$.

85. Given: $16 <$ my number < 19
$12 <$ your number < 20
Conclusion: Out number could be the same.

True; 17, 18 are two examples; argument is $\boxed{\text{valid}}$.

87. Given: (winning team – (losing team) = 4

Let x equal the points scored by losing team. $x + 4$ equals points scored by winning team.

$$2(\text{winning team}) = 3(\text{losing team}) - 14$$
$$2(x + 4) = 3x - 14$$
$$2x + 8 = 3x - 14$$
$$22 = x$$
$$x + 4 = 26$$

Conclusion: The final score of the game was 26 to 22.

True; argument is $\boxed{\text{valid}}$.

89. Given: Let x equal Ana's weight.

$230 - x$ equals Juan's weight.

Ana weighs less than Juan: $x < 230 - x$
$$2x < 230$$
$$x < 115 \quad \text{or (Ana)} < 115$$

$274 - x$ equals Mike's weight.
$$x = 274 - (\text{Mike})$$
$$274 - (\text{Mike}) < 115$$
$$-(\text{Mike}) < -159$$
$$(\text{Mike}) > 159$$

Ana's weight is less than 115 pounds. Mike's weight is greater than 159 pounds.

Conclusion: Mike weighs the most.

True; argument is $\boxed{\text{valid}}$.

91. Given: $50 \div \dfrac{1}{2} = 50 \cdot 2 = 100$

$$100 + 10 = 110$$

Conclusion: The resulting number is 35.

False; the argument is $\boxed{\text{invalid}}$.

93. Given: $V = E - IR$
$$25 = 28 - IR \quad E = 28, \ V = 25$$
$$IR = 3$$

If $R = 0.05$ ohm, $I = \dfrac{3}{R} = \dfrac{3}{0.05} = 60$

Conclusion: If $R = 0.05$ ohms, then $I = 60$ amps.

True; the argument is $\boxed{\text{valid}}$.

95. Let x equal the number.
$$3x + 2 = 14$$
The number is greater than zero – not necessary

$\boxed{\text{Too much information.}}$

97.
$$I = \dfrac{E}{R}$$
$$E = IR$$

Given: $I = 0.5$ ampere, $R = 440$ ohms

Find E.

$\boxed{\text{Just the right amount of information}}$ given.

99. Let x equal Lorna's weight.
$x + 10$ equals Ada's weight.
$(x + 10) + 10 = x + 20$ equals Ana's weight.
More information needs to be given such as the sum of weights in order to solve the individual weight.
Too little information.

101. Let x equal the number of nickels.
$x + 7$ equals the number of dimes.
$0.05x + 0.10(x + 7) = 235$
$29 - x = $ *number of dimes*
Too much information.

103. Let x equal the number of cans at $0.28.
$6 - x$ equals the number of cans at $0.32.
$0.28x + 0.32(6 - x) = 1.76$
Just the right amount of information   given.

105. Let x equal the length of the middle–sized piece.
$x - 16$ equals the length of the shortest piece.
$x + 12$ equals the length of the longest piece.
$x + (x - 16) + (x + 12) = 56$
Find $x - 16$.
Just the right amount of information   given.

107. Let x equal the number of videos purchased.
$19x + 6 = 100 - 37$
Find x.
Just the right amount of information   given.

109. Let x equal the number.
$4(x - 5) = 24$
When a number is decreased by 5 and the entire quantity is multiplied by 4, the result is 24. Find the number.

111. $x + (x + 1) + (x + 2) = 12$
The sum of three consecutive numbers is 12. Find the numbers.

113. Let x equal the number of nickels.
$5x + 10(3x - 1) = 60$
A collection of dimes and nickels is worth 60 cents. The number of dimes is one less than three times the number of nickels. How many of each king of coin are in the collection?

115. Let x equal a number.
$(x + 1)^2 = 36$
If a number is increased by one and the resulting number is squared, 36 is obtained. Find the number.

Review Problems

119.
$$
\begin{aligned}
5(2-y)+3 &= 3+4(3-y) \\
10-5y+3 &= 3+12-4y \\
13-5y &= 15-4y \\
-y &= 2 \\
y &= -2
\end{aligned}
$$
$\boxed{\{-2\}}$

120.
$$
\begin{aligned}
2y-(y+7) &< 3(y+2)-5 \\
2y-y-7 &< 3y+6-5 \\
y-7 &< 3y+1 \\
-2y &< 8 \\
y &> -4
\end{aligned}
$$
$\boxed{\{y\mid y>-4\}}$

121.
$$
\begin{aligned}
P &= 2s+b \text{ for } s \\
2s+b &= P \\
2s &= P-b \\
s &= \frac{P-b}{2}
\end{aligned}
$$
$\boxed{s = \dfrac{P-b}{2}}$

Section 3.2 Ratios and Proportions

Problem Set 3.2, pp. 205-207

1. $\dfrac{24\text{ feet}}{36\text{ feet}} = \dfrac{24}{36} = \boxed{\dfrac{2}{3}}$

3. $\dfrac{3\frac{1}{2}\text{ yards}}{5\text{ yards}} = \dfrac{3\frac{1}{2}}{5} = \dfrac{\frac{7}{2}}{5} = \dfrac{7}{2}\cdot\dfrac{1}{5} = \boxed{\dfrac{7}{10}}$

5. $\dfrac{4\text{ inches}}{3\text{ feet}} = \dfrac{4\text{ inches}}{3(12)\text{ inches}} = \dfrac{4}{36} = \boxed{\dfrac{1}{9}}$

7. $\dfrac{3\text{ gallons}}{2\text{ quarts}} = \dfrac{3(4\text{ quarts})}{2\text{ quarts}} = \dfrac{12}{2} = \boxed{6}$

9. $\dfrac{30\text{ centimeters}}{1\text{ meter}} = \dfrac{30\text{ centimeters}}{100\text{ centimeters}} = \dfrac{30}{100} = \boxed{\dfrac{3}{10}}$

11. $\dfrac{2000\text{ pounds}}{6\text{ tons}} = \dfrac{1\text{ ton}}{6\text{ tons}} = \boxed{\dfrac{1}{6}}$

13. $\dfrac{\text{hours studied}}{\text{hours spent in class}} = \dfrac{30\text{ hours}}{15\text{ hours}} = \dfrac{30}{15} = 2 \text{ or } \boxed{2:1}$

15. PE ratio: $\dfrac{\text{selling price}}{\text{earnings per share}} = \dfrac{\$68}{\$5.50} = \dfrac{136}{11} = \dfrac{12.\overline{36}}{1}$ or $\boxed{136:11 \text{ or } 12.\overline{36}:1}$

17. $\dfrac{\text{smokers}}{\text{nonsmokers}} = \dfrac{2}{48} = \dfrac{1}{24}$ or $\boxed{1:24}$

19. Let $3x$ equal the length of one part.
$4x$ equals the length of other part.

$$3x + 4x = 70$$
$$7x = 70$$
$$x = 10$$

length of one part $= 3x = 3 \cdot 10 = \boxed{30 \text{ feet}}$

length of other part $= 4x = 4 \cdot 10 = \boxed{40 \text{ feet}}$

21. Let $3x$ equal the smaller number. $7x$ equals the larger number.

$$7x = 3x + 12$$
$$4x = 12$$
$$x = 3$$

smaller number $= 3x = 3 \cdot 3 = \boxed{9}$

larger number $= 7x = 7 \cdot 3 = \boxed{21}$

23. Let $2x$ equal the amount of acid. $3x$ equals the amount of water.

$$2x + 3x = 12\frac{1}{2}$$
$$5x = \frac{25}{2}$$
$$x = \frac{5}{2}$$

amount of acid $= 2x = 2\left(\frac{5}{2}\right) = \boxed{5 \text{ quarts of acid}}$

amount of water $= 3x = 3\left(\frac{5}{2}\right) = \boxed{7\frac{1}{2} \text{ quarts of water}}$

25. unit price equals $\dfrac{\text{total price}}{\text{total units}}$

10–ounce size: $\dfrac{\$1.85}{10 \text{ oz}} = \boxed{\$0.185/\text{oz}}$

16–ounce size: $\dfrac{\$2.78}{16 \text{ oz}} = \boxed{\$0.17375/\text{oz}}$

best buy: $\boxed{16\text{–ounce size}}$

27.
$$\frac{24}{x} = \frac{12}{7}$$
$$7 \cdot 24 = 12x$$
$$12x = 7(24)$$
$$x = \frac{7(24)}{12} = 7(2) = 14$$

$\boxed{\{14\}}$

29.
$$\frac{y}{6} = \frac{18}{4}$$
$$4y = 6(18)$$
$$y = \frac{6(18)}{2 \cdot 2} = 3 \cdot 9 = 27$$

$\boxed{\{27\}}$

31.
$$\frac{y}{3} = -\frac{3}{4}$$
$$4y = -9$$
$$y = -\frac{9}{4}$$
$$\boxed{\left\{-\frac{9}{4}\right\}}$$

33.
$$\frac{x-2}{5} = \frac{3}{10}$$
$$10(x-2) = 15$$
$$10x - 20 = 15$$
$$10x = 35$$
$$x = \frac{7}{2}$$
$$\boxed{\left\{\frac{7}{2}\right\}}$$

35.
$$\frac{y+12}{34} = \frac{y}{10}$$
$$10(y+12) = 34y$$
$$10y + 120 = 34y$$
$$120 = 24y$$
$$5 = y$$
$$\boxed{\{5\}}$$

37.
$$\frac{y-5}{2} = \frac{y+6}{3}$$
$$3(y-5) = 2(y+6)$$
$$3y - 15 = 2y + 12$$
$$y = 27$$
$$\boxed{\{27\}}$$

39.
$$\frac{5\frac{1}{4}}{2\frac{1}{2}} = \frac{x}{3\frac{1}{8}}$$
$$\frac{21}{4} \cdot \frac{25}{8} = \frac{5}{2}x$$
$$\frac{5}{2}x = \frac{21}{4} \cdot \frac{25}{8}$$
$$x = \frac{2}{5} \cdot \frac{21}{4} \cdot \frac{25}{8} = \frac{105}{16} = 6\frac{9}{16}$$
$$\boxed{\left\{6\frac{9}{16}\right\} \text{ or } \left\{\frac{105}{16}\right\}}$$

41.
$$\frac{2}{3y+10} = \frac{10}{28-7y}$$
$$2(28-7y) = 10(3y+10)$$
$$56 - 14y = 30y + 100$$
$$-44 = 44y$$
$$-1 = y$$
$$\boxed{\{-1\}}$$

43.
$$\frac{0.8}{0.5} = \frac{0.3}{r}$$
$$0.8r = 0.3(0.5)$$
$$r = \frac{0.3(0.5)}{0.8}$$
$$r = \frac{3}{8}\left(\frac{1}{2}\right) = \frac{3}{16}$$
$$r = 0.1875$$
$$\boxed{\left\{\frac{3}{16}\right\}} \text{ or } \boxed{\{0.1875\}}$$

45.
$$\frac{\$45000}{180000 \text{ sq ft}} = \frac{x}{4800 \text{ sq ft}}$$
$$180000x = 4800(\$45000)$$
$$x = \frac{4800(45000)}{180000}$$
$$x = \boxed{\$1200}$$

47.
$$\frac{1}{40} = \frac{x}{\$38000}$$
$$40x = \$38000$$
$$x = \frac{38000}{40} = \boxed{\$950}$$

49.
$$\frac{x}{4963} = \frac{900}{218}$$
$$218x = 900(4963)$$
$$x = \frac{900(4963)}{218}$$
$$x = 20489.45$$
$$\approx \boxed{\text{about } 20500} \text{ fur seal pups}$$

51.
$$\frac{300 \text{ mg morphine}}{360 \text{ cubic cm solution}} = \frac{x}{250 \text{ cubic cm solution}}$$
$$360x = 300(250) \text{ mg}$$
$$x = \frac{300(250)}{360} \text{ mg}$$
$$x = \boxed{208\frac{1}{3} \text{ mg}} \text{ of morphine}$$

53. Let x equal the number of women. $x + 20$ equals the number of men.
$$\frac{3}{2} = \frac{x + 20}{x}$$
$$3x = 2(x + 20)$$
$$3x = 2x + 40$$
$$x = \boxed{40 \text{ women}}$$
$$x + 20 = \boxed{60 \text{ men}}$$

55. 5 : 1 ratio

	front: rear ratio	to have a 5 : 1 ratio	keep	change
rear	20 teeth		20 teeth	12 teeth
front	60 teeth		60 teeth	100 teeth

since $\dfrac{60 \text{ teeth}}{12 \text{ teeth}} = \dfrac{5}{1}$ and $\dfrac{100 \text{ teeth}}{20 \text{ teeth}} = \dfrac{5}{1}$

$\boxed{\text{change rear sprocket to 12 teeth or change front sprocket to 100 teeth}}$

57. Let x equal the age of friend in dog years.
$$\frac{\text{dog years}}{\text{human years}}: \quad \frac{x}{44} = \frac{7}{56}$$
$$56x = 7(44)$$
$$x = \frac{308}{56}$$
$$x = 5.5$$

My friend would be $\boxed{5.5 \text{ years}}$ old in dog years.

59. Let x equal the number of additional games.
$$\frac{8 + x}{20 + x} = 0.60$$
$$8 + x = 0.60(20 + x)$$
$$8 + x = 12 + 0.6x$$
$$0.4x = 4$$
$$x = 10$$

$\boxed{10}$ consecutive games must be won to raise the winning record to 60%.

61. ☐D☐ is true;

A. $\dfrac{3\text{ years}}{4\text{ feet}}=\dfrac{3(3\text{ feet})}{4\text{ feet}}=\dfrac{9}{4}\ne\dfrac{3}{4}$

B. $\dfrac{4}{3}=\dfrac{30}{\text{women}}\;\rightarrow\;\text{women}=\dfrac{90}{4}\approx22\ne40$

C. $\dfrac{4}{y}=\dfrac{5}{7}\;\rightarrow\;5y=28\ not\ 7y=20$

D. $\dfrac{y-4}{y}=\dfrac{3}{4}\;\rightarrow\;4y-16=3y;\ \text{true}$

63. Total invested by the three people: $A+B+B=A+2B$ dollars.
ratio of earnings:

first person: $\dfrac{A\text{ dollars}}{A+2B\text{ dollars}}$

Amount of profit from \$1000 for the first person: $\dfrac{A(\$1000)}{A+2B}=\boxed{\dfrac{1000A}{A+2B}}$ dollars

65.
$$\dfrac{2\text{ cm}}{13.47\text{ mi}}=\dfrac{9.85\text{ cm}}{x}$$
$$2x=9.85(13.47)\text{ mi}$$
$$x=\dfrac{9.85(13.47)}{2}\text{ mi}$$
$$x\approx\boxed{66.34\text{ miles}}$$

Review Problems

68.
$$\tfrac{1}{2}x+7=13-\tfrac{1}{4}x$$
$$\tfrac{3}{4}x=6$$
$$x=\tfrac{4}{3}(6)$$
$$x=8$$
$$\boxed{\{8\}}$$

69.
$$-3<2x-3\le5$$
$$0<2x\le8$$
$$0<x\le4$$
$$\boxed{\{x\,|\,0<x\le4\}}$$

70. Let $x;\ x+2;\ x+4$ represent three consecutive even integers.
$$x+(x+2)+(x+4)=234$$
$$3x+6=234$$
$$3x=228$$
$$x=76$$
$$x+2=78$$
$$x+4=80$$
The integers are $\boxed{76,\ 78\text{ and }80}$.

Section 3.3 Geometry Problems

Problem Set 3.3, pp. 220-228

1. $m\angle DAB=40°$; acute angle

3. $m\angle CAH=180°$; straight angle

5. $(2x + 50) + (4x + 10)\ =\ 90$
$6x + 60\ =\ 90$
$6x\ =\ 30$
$\boxed{x\ =\ 5}$

$(2x + 50)^\circ = (2 \cdot 5 + 50)^\circ = \boxed{60^\circ}$

$(4x + 10)^\circ = (4 \cdot 5 + 10)^\circ = \boxed{30^\circ}$

7. $(3x + 134) + (6x + 10)\ =\ 180$
$9x + 144\ =\ 180$
$9x\ =\ 36$
$\boxed{x\ =\ 4}$

$(3x + 134)^\circ = (3 \cdot 4 + 134)^\circ = (12 + 134)^\circ = \boxed{146^\circ}$

$(6x + 10)^\circ = (6 \cdot 4) + 10)^\circ = \boxed{34^\circ}$

9. $25 + x\ =\ 90$
$\boxed{x\ =\ 65}$

The measures of the angles are $\boxed{25^\circ \text{ and } 65^\circ}$.

11. $2x\ =\ x + 43$
$\boxed{x\ =\ 43}$

$(2x)^\circ = (2 \cdot 43)^\circ = \boxed{86^\circ}$

$(x + 43)^\circ = (43 + 43)^\circ = \boxed{86^\circ}$

13. $12x - 3\ =\ 10x + 15$
$2x\ =\ 18$
$\boxed{x\ =\ 9}$

$(12x - 3)^\circ = (12 \cdot 9 - 3)^\circ = (108 - 3)^\circ = \boxed{105^\circ}$

$(10x + 15)^\circ = (10 \cdot 9 + 15)^\circ = (90 + 15)^\circ = \boxed{105^\circ}$

15. $\boxed{m \angle a\ =\ 105^\circ}$
$b + 105\ =\ 180$
$b\ =\ 75$
$c + 105\ =\ 180$
$c\ =\ 75$
$d + 105\ =\ 180$
$d\ =\ 75$
$d\ =\ e$
$\boxed{m \angle f\ =\ 105^\circ}$

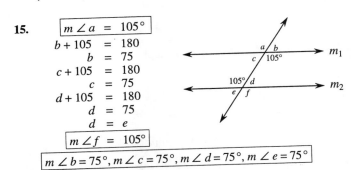

$\boxed{m \angle b = 75^\circ, m \angle c = 75^\circ, m \angle d = 75^\circ, m \angle e = 75^\circ}$

$c + 105 = 75 + 105 = 180$
$d + 105 = 75 + 105 = 180$

$\boxed{\text{The sum of the measures of the interior angles on the same side of the transveral of two parallel lines is } 180^\circ.}$

17. $(4x) + (3x + 4) + (2x + 5)\ =\ 180$
$9x + 9\ =\ 180$
$9x\ =\ 171$
$\boxed{x\ =\ 19}$

$(4x)^\circ = (4 \cdot 19)^\circ = \boxed{76^\circ}$

$(3x + 4)^\circ = (3 \cdot 19 + 4)^\circ = (57 + 4)^\circ = \boxed{61^\circ}$

$(2x + 5)^\circ = (2 \cdot 19 + 5)^\circ = (38 + 5)^\circ = \boxed{43^\circ}$

19. $61 + x\ =\ 90 + 50$
$x + 61\ =\ 140$
$\boxed{x\ =\ 79}$

$x^\circ\ =\ \boxed{79^\circ}$

21. $m \angle PQT = 180 - (70 + 60) = 180 - 130 = 50°$
$m \angle SQR = 180 - (50 + 50) = 180 - 100 = 80°$
$m \angle R = x° = 180 - (80 + 30) = 180 - 110 = \boxed{70°}$

23. Let x equal the measure of the angle. $90 - x$ equals the measure of its complement.
$$\begin{aligned} x &= (90 - x) + 60 \\ 2x &= 150 \\ x &= 75 \end{aligned}$$
The angle measures $\boxed{75°}$.

25. Let x equal the measure of the angle. $180 - x$ equals the measure of its supplement.
$$\begin{aligned} x &= (180 - x) - 41 \\ 2x &= 139 \\ x &= 695 \end{aligned}$$
The angle measures $\boxed{69.5°}$.

27. Let x equal the measure of the angle. $90 - x$ equals the measure of its complement.
$$\begin{aligned} x &= 2(90 - x) \\ x &= 180 - 2x \\ 3x &= 180 \\ x &= 60 \end{aligned}$$
The angle measures $\boxed{60°}$.

29. Let x equal the measure of the angle. $180 - x$ equals the measure of its supplement.
$$\begin{aligned} x &= 180 - x \\ 2x &= 180 \\ x &= 90 \end{aligned}$$
The angle measures $\boxed{90°}$.

31. Let x equal the measure of the angle. $90 - x$ equals the measure of its complement.
$$\begin{aligned} x &= 4(90 - x) - 25 \\ x &= 360 - 4x - 25 \\ 5x &= 335 \\ x &= 67 \end{aligned}$$
The angle measures $\boxed{67°}$.

33. Let x equal the measure of the angle. $180 - x$ equals the measure of its supplement.
$$\begin{aligned} x &= 2(180 - x) + 81 \\ x &= 360 - 2x + 81 \\ 3x &= 441 \\ x &= 147 \end{aligned}$$
The angle measures $\boxed{147°}$.

35. Let x equal the measure of the angle. $90 - x$ equals the measure of its complement. $180 - x$ equals the measure of its supplement.
$$\begin{aligned} 180 - x &= 3(90 - x) + 10 \\ 180 - x &= 270 - 3x + 10 \\ 2x &= 100 \\ x &= 50 \end{aligned}$$
The angle measures $\boxed{50°}$.

37. Let x equal the measure of the angle. $90 - x$ equals the measure of its complement. $180 - x$ equals the measure of its supplement.

$$
\begin{aligned}
2(90 - x) &= 16 + [(180 - x) - (90 - x)] \\
180 - 2x &= 16 + 180 - x - 90 + x \\
74 &= 2x \\
37 &= x
\end{aligned}
$$

The angle measures $\boxed{37^\circ}$.

39. Let x equal the measure of the smallest angle. $2x$ equals the measure of the second angle. $x + 20$ equals the measure of the third angle.

$$
\begin{aligned}
x + 2x + (x + 20) &= 180 \\
4x + 20 &= 180 \\
4x &= 160 \\
x &= 40 \\
2x &= 80 \\
x + 20 &= 60
\end{aligned}
$$

The angles of the triangle measure $\boxed{40^\circ, 80^\circ, \text{ and } 60^\circ}$.

41. Let x equal the measure of the first angle. $4x + 6$ equals the measure of the second angle. $12x - 13$ equals the measure of the third angle.

$$
\begin{aligned}
x + (4x + 6) + (12x - 13) &= 180 \\
17x - 7 &= 180 \\
17x &= 187 \\
x &= 11 \\
4x + 6 &= 44 + 6 = 50 \\
12x - 13 &= 132 - 13 = 119
\end{aligned}
$$

The angles of the triangle measure $\boxed{11^\circ, 50^\circ, \text{ and } 119^\circ}$.

43. Let $2x$ equal the measure of the first angle. $5x$ equals the measure of the second angle.

$$
\begin{aligned}
5x - 2x &= 15 \\
3x &= 15 \\
x &= 5 \\
5x &= 5(5) = 25 \\
2x &= 2(5) = 10 \\
\text{Third angle} &= 180 - (25 + 10) = 180 - 35 = 145
\end{aligned}
$$

The measure of the largest angle of the triangle is $\boxed{145^\circ}$.

45. Let x equal the measure of one of the equal angles. $4x$ equals the measure of the non–equal angle.

$$
\begin{aligned}
4x + x + x &= 180 \\
6x &= 180 \\
x &= 30 \\
4x &= 120
\end{aligned}
$$

The angles of the triangle measure $\boxed{30^\circ, 30^\circ, \text{ and } 120^\circ}$.

47. Let x equal the measure of the first angle. $3x$ equals the measure of the second angle. $(x + 3x) - 25 = 4x - 25$ equals the measure of the third angle.

$$
\begin{aligned}
x + 3x + (4x - 25) &= 180 \\
8x - 25 &= 180 \\
8x &= 205 \\
x &= 25.625
\end{aligned}
$$

The measure of the angle is $\boxed{25.625^\circ}$.

49. Let w equal width. $25 + w$ equals length.
$$\begin{aligned}
2w + 2(25 + w) &= 310 \\
2w + 50 + 2w &= 310 \\
4w &= 260 \\
w &= 65 \\
w + 25 &= 90
\end{aligned}$$
The dimensions of the lot are $\boxed{65\,\text{m} \times 90\,\text{m}}$.

51. Let w equal the width. $3w - 1$ equals length.
$$\begin{aligned}
2w + 2(3w - 1) &= 90 - 12 \quad \leftarrow 90\text{ yd} - 12\text{ yd needed to enclose lot.} \\
2w + 6w - 2 &= 78 \\
8w &= 80 \\
w &= 10 \\
3w - 1 &= 30 - 1 = 29
\end{aligned}$$
The dimensions of the lot are $\boxed{10\text{ yd} \times 29\text{ yd}}$

53. Let x equal the width of the frame. Width of the framed picture: $12 + 2x$. Length of the framed picture: $18 + 2x$.
$$\begin{aligned}
2(12 + 2x) + 2(18 + 2x) &= 84 \\
24 + 4x + 36 + 4x &= 84 \\
8x + 60 &= 84 \\
8x &= 24 \\
x &= 3
\end{aligned}$$
The width of the frame is $\boxed{3\text{ inches}}$.

55. Let x equal the length of the second side. $2x - 1$ equals the length of the first side. $2x + 1$ equals the length of the third side.
$$\begin{aligned}
x + (2x - 1) + (2x + 1) &= 30 \\
5x &= 30 \\
x &= 6 \\
2x - 1 &= 11 \\
2x + 1 &= 13
\end{aligned}$$
The length of the sides of the triangle are $\boxed{6\text{ inches, }11\text{ inches, and }13\text{ inches}}$.

57. Let x equal the length of the side of the triangle. $x - 3$ equals the height of the rectangle. x equals the width of the rectangle.
$$\begin{aligned}
\text{width of rectangle} &= \text{length of side of triangle} \\
\text{perimeter of figure} &= 2 \text{ length of side of triangle} + 2 \text{ height of rectangle} + \text{width of rectangle} \\
34 &= 2x + 2(x - 3) + x \\
34 &= 2x + 2x - 6 + x \\
40 &= 5x \\
8 &= x
\end{aligned}$$
The length of a side of the triangle is $\boxed{8\text{ meters}}$.

59. Let x equal the width of the original rectangle. $2x - 1$ equals the length of the original rectangle. $x - 1$ equals the width of the new rectangle. $2x - 1 - 2 = 2x - 3$ equals the length of the new rectangle.
$$\begin{aligned}
2(x - 1) + 2(2x - 3) &= 16 \\
2x - 2 + 4x - 6 &= 16 \\
6x - 8 &= 16 \\
6x &= 24 \\
x &= 4 \\
2x - 1 &= 8 - 1 = 7
\end{aligned}$$
The dimensions of the original rectangle are $\boxed{4\text{ cm} \times 7\text{ cm}}$.

61.
$$LW = A$$
$$L(75) = 9375$$
$$L = 125$$
The length of the plot is $\boxed{125 \text{ feet}}$.

63.
$$\frac{1}{2}(21)h = 147$$
$$h = \frac{2}{21}(147)$$
$$h = 14$$
The height is $\boxed{14 \text{ m}}$.

65. Let L equal the length of the garden.
$$LW = A$$
$$L(2) = 12$$
$$L = 6 \text{ yd}$$
Area of garden and the path combined $= (6+2)(2+2) = 8(4) = \boxed{32 \text{ square yards}}$.

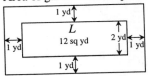

67. area of square + area of triangle = 1020
$$30(30) + \frac{1}{2}(30)h = 1020$$
$$900 + 15h = 1020$$
$$15h = 120$$
$$h = 8$$
The height of the triangle is $\boxed{8 \text{ ft}}$.

69.
$$B = b + 4$$
$$A = \frac{1}{2}h(B + b)$$
$$A = \frac{1}{2}h(2b + 4)$$
$$A = h(b + 2) \quad h = 8 \text{ m}, A = 136 \text{ sq m}$$
$$136 = 8(b + 2)$$
$$17 = b + 2$$
$$b = 15$$
$$b + 4 = 19$$
The length of the bases are $\boxed{15 \text{ and } 19 \text{ m}}$.

71. area of trapezoid + area of triangle = area of combined figure

$$\frac{1}{2}x(13 + 6) + \frac{1}{2}(5)(12) = 68$$

$$\frac{1}{2}x(19) + \frac{1}{2}(60) = 68$$

$$\frac{19x}{2} + 30 = 68$$

$$\frac{19x}{2} = 38$$

$$x = \frac{2(38)}{19} = 4$$

The height of the trapezoid is $\boxed{4 \text{ ft}}$.

73.
$$\frac{\pi 8^2}{\pi 4^2} = \frac{x}{3}$$

$$\frac{64}{16} = \frac{x}{3}$$

$$4 = \frac{x}{3}$$

$$12 = x$$

The charge for an 8 inch pizza is $\boxed{\$12}$.

75.
$$\frac{\frac{4}{3}\pi 2^3}{\frac{4}{3}\pi 1^3} = \frac{x}{12}$$

$$\frac{8}{1} = \frac{x}{12}$$

$$96 = x$$

The cost is $\boxed{\$96}$.

77. Let $x, x + 1, x + 2$ represent the measures of the angles of a triangle.

$$\begin{aligned}
x + (x + 1) + (x + 2) &= 180 \\
3x + 3 &= 180 \\
3x &= 177 \\
x &= 59 \\
x + 1 &= 60 \\
x + 2 &= 61
\end{aligned}$$

The measures of the angles of the triangles are $\boxed{59°, 60°, \text{ and } 61°}$.

79. Let x equal the length of the second rectangle. $x + 3$ equals the length of the first rectangle.

$$\begin{aligned}
4(x + 3) + 7x &= 45 \\
11x + 12 &= 45 \\
11x &= 33 \\
x &= 3 \\
x + 3 &= 6
\end{aligned}$$

The length of the first rectangle is $\boxed{6 \text{ m}}$ and the length of the second rectangle is $\boxed{3 \text{ m}}$.

81.
$$\frac{18}{9} = \frac{10}{x}$$

$$2 = \frac{10}{x}$$

$$2x = 10$$

$$x = \boxed{5}$$

83.
$$\frac{20}{15} = \frac{x}{12}$$

$$15x = 20(12)$$

$$x = \frac{240}{15} = \boxed{16}$$

85. C is true;

    A.
$$
\begin{aligned}
2w + 2(8x) &= 48x \\
2w &= 32x \\
w &= 16x \neq 32x
\end{aligned}
$$

    B. Square of side 4 has a perimeter numerically equal to its area.

    C. Let x equal the measure of the angle. $90 - x$ equals the measure of its complement. $180 - x$ equals the measure of **its** supplement.

        $(180 - x) - (90 - x) = 90$, true

    D. $x + 2(90 - x) = 180 - x = $ measure of its supplement

87. Let $x, x + 2, x + 4$ represent the length of the triangle. The length of the rectangle equals $x + (x + 4) = 2x + 4$. Width of the rectangle equals $x + 2$.

$$
\begin{aligned}
\text{perimeter of rectangle} &= \text{perimeter of triangle} + 30 \\
2(2x + 4) + 2(x + 2) &= x + (x + 2) + (x + 4) + 30 \\
4x + 8 + 2x + 4 &= 3x + 36 \\
6x + 12 &= 3x + 36 \\
3x &= 24 \\
x &= 8 \\
x + 2 &= 10 \\
2x + 4 &= 2(8) + 4 = 20
\end{aligned}
$$

The dimensions of the rectangle are $\boxed{10\text{ m} \times 20\text{ m}}$.

89.
$$
\begin{aligned}
\text{area of garden} + \text{sidewalk} &= \text{area of garden} + \text{area of sidewalk} \\
(8 + 2)(12 + 2) &= 8(12) + A \\
10(14) &= 96 + A \\
140 - 96 &= A \\
A &= \boxed{44\text{ sq yds}}
\end{aligned}
$$

91. Length of edges $= 4x + 4(x + 1) + 4(x + 2) = 4x + 4x + 4 + 4x + 8 = \boxed{12x + 12}$.

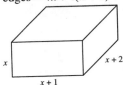

93. The sides of the equilateral triangle are each k.

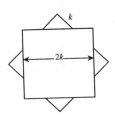

Let a represent the portion of the square extending outside the equilateral triangle.

$$2a + k = 2k$$
$$a = \frac{k}{2}$$

The figure has 4 equal sides, each equal to $a + k + k + a = 2k + 2a = 2k + 2\left(\frac{k}{2}\right) = 3k$.

The perimeter is $4(3k) = \boxed{12k}$.

95.

base	number of triangles
3	9
4	16
pattern	
n	n^2
12	$12^2 = \boxed{144 \text{ triangles}}$

97.

Polygon	number of sides	number of triangles	sum of the angles measures
	4	2	$2 \times 180 = \boxed{360}$
	5	3	$3 \times 180 = \boxed{540}$
	6	4	$4 \times 180 = \boxed{720}$

The sum of the angle measures of a polygon of n sides is $\boxed{(n-2)180}$.

98-101. Answers will vary. Sample given.

99. The length of the nonequal side of an isosceles triangle is 3 less than the length of an equal side. The perimeter of the triangle is 69 inches. Find the length of each side of the triangle.

101. The measure of one angle of a triangle is $10°$ more than that of the first angle. The measure of the third angle is $20°$ more than that of the first angle. Find the measure of each angle of the triangle.

Review Problems

103. Let $x, x + 2, x + 4$ represent three consecutive odd integers.
$$x + (x + 2) = (x + 4) + 15$$
$$2x + 2 = x + 19$$
$$x = 17$$
$$x + 2 = 19$$
$$x + 4 = 21$$
The integers are $\boxed{17, 19, \text{ and } 21}$.

104. Let x equal the age of Riff now. $x + 20$ equals the age of Tony now. $x + 6$ equals the age of Riff in 6 years. $x + 26$ equals the age of Tony in 6 years.
$$x + 26 = 3x + 18$$
$$8 = 2x$$
$$4 = x$$
$$x + 20 = 24$$
Present ages: $\boxed{\text{Riff, 4 years old; Tony, 24 years old}}$.

105. Let x equal a number.

$$3x + 2 \ \le \ 11$$
$$3x \ \le \ 9$$
$$\boxed{x \ \le \ 3}$$

Section 3.4 Classic Algebraic Word Problems

Problem Set 3.4, pp. 238-240

1.

50 mph →   55 mph ←

315 miles

Let t equal time.

$$50t + 55t \ = \ 315$$
$$105t \ = \ 315$$
$$t \ = \ 3$$

The vehicles meet after $\boxed{3 \text{ hours}}$.

3.

r → $r-5$ ←

110 miles

r = rate of one car
$r - 5$ = rate of the other car
t = 2 hours

$$rt + (r - 5)t \ = \ 110$$
$$2r + (r - 5)2 \ = \ 110$$
$$2r + 2r - 10 \ = \ 120$$
$$4r \ = \ 120$$
$$r \ = \ 30$$
$$r - 5 \ = \ 25$$

The speeds of the cars
are $\boxed{30 \text{ mph and } 25 \text{ mph}}$.

5.

10 mph ←   12 mph
66 miles

$t = ?$

$$10t + 12t \ = \ 66$$
$$22t \ = \ 66$$
$$t \ = \ 3$$

$\boxed{3 \text{ hours}}$

7.

t
80 km/h → going
60 km/h ← returning
$t + \frac{3}{4}$

$$45 \text{ minutes} \ = \ \frac{3}{4} \text{ hour}$$
$$\text{distance going} \ = \ \text{distance returning}$$
$$80t \ = \ 60\left(t + \frac{3}{4}\right)$$
$$80t \ = \ 60t + 45$$
$$20t \ = \ 45$$
$$t \ = \ 2.25$$
$$d = 80t = 80(2.25) = \ \boxed{180 \text{ km}}$$

9.

bus $t + 1$
       72 km/h →

car   80 km/h →
       t

$$80t \ = \ 72(t + 1)$$
$$80t \ = \ 72t + 72$$
$$8t \ = \ 72$$
$$t \ = \ \boxed{9 \text{ hours}}$$

11.

r $2r - 3$
360 miles
$t = 8$ hours

$$r(8) + (2r - 3)8 \ = \ 360$$
$$8r + 16r - 24 \ = \ 360$$
$$24r \ = \ 384$$
$$r \ = \ 16$$
$$2r - 3 \ = \ 32 - 3 = 29$$

The speeds of the trains are
$\boxed{16 \text{ mph and } 29 \text{ mph}}$.

13.

$$t \quad\quad 5-t$$
$$\text{50 mph} \quad \text{35 mph}$$
$$\longleftarrow\text{220 miles}\longrightarrow$$

$$\begin{aligned}
50t + 35(5-t) &= 220 \\
50t + 175 - 35t &= 220 \\
15t &= 45 \\
t &= 3 \\
5 - t &= 2
\end{aligned}$$

$\boxed{\text{3 hours at 50 mph, 2 hours at 35 mph}}$

15. Let x equal the amount at 9%.

$25000 - x$ equals the amount at 12%.

$$\begin{aligned}
0.09x + 0.12(25000 - x) &= 2550 \\
0.09x + 3000 - 0.12x &= 2550 \\
-0.03x &= -450 \\
x &= 15000 \\
25000 - x &= 10000
\end{aligned}$$

$\boxed{\$15000 \text{ at } 9\%, \$10000 \text{ at } 12\%}$

17. Let x equal the amount of 14%. $2x$ equals the amount at 12%.

$$\begin{aligned}
0.14x + 0.12(2x) &= 256.50 \\
0.14x + 0.24x &= 256.50 \\
0.38x &= 256.50 \\
x &= 675 \\
2x &= 1350
\end{aligned}$$

$\boxed{\$675 \text{ at } 14\%, \$1350 \text{ at } 12\%}$

19. Let x equal the amount at 10%. $2x + 300$ equals the amount at 13%.

$$\begin{aligned}
0.10x + 0.13(2x + 300) &= 338.52 \\
0.10x + 0.26x + 39 &= 338.52 \\
0.36x &= 299.52 \\
x &= 832 \\
2x + 300 &= 1664 + 300 = 1964
\end{aligned}$$

$\boxed{\$832 \text{ at } 10\%, \$1964 \text{ at } 13\%}$

21. Let x equal the amount at 9%. $6000 - x$ equals the amount at 6%.

$$\begin{aligned}
0.09x &= 0.06(6000 - x) \\
0.09x &= 360 - 0.06x \\
0.15x &= 360 \\
x &= 2400 \\
6000 - x &= 3600
\end{aligned}$$

$\boxed{\$2400 \text{ at } 9\%, \$3600 \text{ at } 6\%}$

23. Amount invested = $\$40000 - \$25000 = \$15000$.

Let x equal the amount at 7%. $15000 - x$ equals the amount at 6.5%.

$$\begin{aligned}
0.07x + 0.065(15000 - x) &= 1020 \\
0.07x + 975 - 0.065x &= 1020 \\
0.005x &= 45 \\
x &= 9000 \\
15000 - x &= 6000
\end{aligned}$$

$\boxed{\$9000 \text{ at } 7\%, \$6100 \text{ at } 6.5\%}$

25. Let x equal the amount at 9%. $2x$ equals the amount at 12%. $3x$ equals the amount at 15%.

$$\begin{aligned}
0.09x + 0.12(2x) + 0.15(3x) &= 585 \\
0.09x + 0.24x + 0.45x &= 585 \\
0.78x &= 585 \\
x &= 750 \\
2x &= 1500 \\
3x &= 2250
\end{aligned}$$

$\boxed{\$750 \text{ at } 9\%, \$1500 \text{ at } 12\%, \$2250 \text{ at } 15\%}$

27. Amount of acid at 30% + amount of acid at 12% = amount of acid at 20%.
   Let x equal the amount of acid at 30%. $50 - x$ equals the amount of acid at 12%.
$$
\begin{aligned}
0.30x + 0.12(50 - x) &= 0.20(50) \\
0.30x + 6 - 0.12x &= 10 \\
0.18x &= 4 \\
x &= 22\tfrac{2}{9} \\
50 - x &= 27\tfrac{7}{9}
\end{aligned}
$$
   $\boxed{22\tfrac{2}{9} \text{ liters at 30\%, } 27\tfrac{7}{9} \text{ liters at 12\%}}$

29. Let x equal the amount of alcohol at 15%. $x + 4$ equals the amount of alcohol at 17%.
$$
\begin{aligned}
0.15x + 0.20(4) &= 0.17(x + 4) \\
0.15x + 0.8 &= 0.17x + 0.68 \\
0.12 &= 0.02x \\
6 &= x \\
x + 4 &= 10
\end{aligned}
$$
   $\boxed{6 \text{ ounces at 15\%}}$

31. Let x equal the amount of pure water added. $x + 10$ equals the amount of milliliters at 10%.
$$
\begin{aligned}
0\%x + 0.15(50) &= 0.10(x + 50) \\
0 + 7.5 &= 0.10x + 5 \\
2.5 &= 0.10x \\
25 &= x
\end{aligned}
$$
   $\boxed{25 \text{ milliliters of pure water}}$

33. Let x equal the amount of oak chips at \$0.70 per pound. $30 + x$ equals the combined amount at \$1.00 per pound.
$$
\begin{aligned}
0.70x + 30(1.20) &= (30 + x)(\$1.00) \\
0.70x + 36 &= 30 + x \\
6 &= 0.30x \\
20 &= x
\end{aligned}
$$
   $\boxed{20 \text{ pounds of oak chips}}$

35. Let x equal the amount of tea at 60¢ per pound. $100 - x$ equals the amount of tea at 75¢ per pound.
$$
\begin{aligned}
0.60x + 0.75(100 - x) &= 0.72(100) \\
0.60x + 75 - 0.75x &= 72 \\
-0.15x &= -3 \\
x &= 20 \\
100 - x &= 80
\end{aligned}
$$
   $\boxed{20 \text{ pounds at 60¢ per pound, 80 pounds at 75¢ per pound}}$

37. Let x equal the cost per pound of nut mixture.
$$
\begin{aligned}
0.60(120) + 1.00(40) &= x(120 + 40) \\
72 + 40 &= 160x \\
112 &= 160x \\
0.70 &= x
\end{aligned}
$$
   $\boxed{\$0.70 \text{ per pound}}$ for the nut mixture

39. Let x equal the number of students at **the** second school. $x + 900$ equals the number of students in combined school.

$$
\begin{aligned}
0.40(900) + 0.75x &= 0.525(x + 900) \\
360 + 0.75x &= 0.525x + 472.50 \\
0.225x &= 112.50 \\
x &= 500
\end{aligned}
$$

$\boxed{500 \text{ students}}$ in the second school

41. Let t equal the number of hours to complete job.
Fractional part done by first person + fractional part done by second person = one whole job.

$$
\begin{aligned}
\frac{t}{55} + \frac{t}{66} &= 1 \\
(55)(66)\left[\frac{t}{55} + \frac{t}{66}\right] &= (55)(66) \cdot 1 \\
65t + 55t &= 3630 \\
121t &= 3630 \\
t &= 30
\end{aligned}
$$

Working together they can complete the job in $\boxed{30 \text{ hours}}$.

43. Let t equal the time (in minutes) together.

$$
\begin{aligned}
\frac{t}{15} + \frac{t}{10} &= 1 \\
30\left(\frac{t}{15} + \frac{t}{10}\right) &= 30(1) \\
2t + 3t &= 30 \\
5t &= 30 \\
t &= 6
\end{aligned}
$$

$\boxed{6 \text{ minutes}}$

45. Let t equal the time (in hours) together.

$$
\begin{aligned}
\frac{t}{15} - \frac{t}{20} &= 1 \\
60\left(\frac{t}{15} - \frac{t}{20}\right) &= 60(1) \\
4t - 3t &= 60 \\
t &= 60
\end{aligned}
$$

$\boxed{60 \text{ hours}}$

47.

	Principal	Interest	Amount at end of year
End of first year	6000	6000(0.07) = 420	6000 + 420 = 6420
End of second year	6420	6420(0.07) = 449.40	6420 + 449.40 = 6869.40

49.

$$
\begin{aligned}
8r - 3(2r) + 1(0.40r) &= 9.6 \\
8r - 6r + 0.40r &= 9.6 \\
2.4r &= 9.6 \\
r &= 4 \text{ miles per hour} \\
2r &= 8 \text{ miles per hour} \\
0.40r &= 1.6 \text{ miles per hour}
\end{aligned}
$$

$\boxed{4 \text{ mph for 8 hours, 8 mph for 3 hours, 1.6 mph for 1 hour}}$

51. D is true;
   A. The mixture will be between 7% and 12%.
   B. $\dfrac{600 \text{ miles}}{y \text{ miles per hour}} = \dfrac{600}{y}$ hours
   C. annual simple interest $= 0.11(25000 - x)$
   D. none of A–C in the text is true

Review Problems

54.
$$\begin{aligned} 6 - (x - 3) - x &= 4x \\ 6 - x + 3 - x &= 4x \\ 9 &= 6x \\ \frac{3}{2} &= x \end{aligned}$$

$$\boxed{\left\{ \frac{3}{2} \right\}}$$

55.
$$\begin{aligned} P &= h(a + b) \text{ for } b \\ P &= ha + hb \\ P - ha &= hb \\ \frac{P - ha}{h} &= b \end{aligned}$$

$$\boxed{b = \frac{P - ha}{h} \text{ or } \frac{P}{h} - a}$$

56.
$$\begin{aligned} 5 - 2(x - 8) &\le 4(2x - 1) \\ 5 - 2x + 16 &\le 8x - 4 \\ 21 - 2x &\le 8x - 4 \\ 25 &\le 10x \\ 10x &\ge 25 \\ x &\ge \frac{5}{2} \end{aligned}$$

$$\boxed{\left\{ x \mid x \ge \frac{5}{2} \right\}}$$

Chapter 3 Review Problems

Chapter 3 Review Problems, pp. 242-245

1. Let x equal the number.
$$\begin{aligned} 3x + 11 &= 50 \\ 3x &= 39 \\ x &= \boxed{13} \end{aligned}$$

2. Let x equal the regular price of TV set.
$$\begin{aligned} x - 0.30x &= 350 \\ 0.70x &= 350 \\ x &= 500 \end{aligned}$$
$$\boxed{\$500}$$

3. Let x equal the maximum price of dinner.
$$\begin{aligned} x + 0.25x &= 21.25 \\ 1.25x &= 21.25 \\ x &= 17 \end{aligned}$$
$$\boxed{\$17}$$

4. Let x equal the number.
$$\begin{aligned} 2x + 6 &= 5(x - 3) \\ 2x + 6 &= 5x - 15 \\ 21 &= 3x \\ \boxed{7} &= x \end{aligned}$$

5.   Let x equal the smaller number. $56 - x$ equals the larger number.

$$
\begin{aligned}
7x &= 3(56 - x) + 12 \\
7x &= 168 - 3x + 12 \\
10x &= 180 \\
x &= 18 \\
56 - 18 &= 38
\end{aligned}
$$

The numbers are $\boxed{18 \text{ and } 38}$.

6.   Let x equal the number of labor hours.

$$
\begin{aligned}
17 + 26x &= 199 \\
26x &= 182 \\
x &= 7
\end{aligned}
$$

$\boxed{7 \text{ hours}}$

7.   Let t equal the average temperature in Bismarck. $6t + 12$ equals the average temperature in Miami. $6t + 12 - 25 = 6t - 13 = $ average temperature in Atlanta.

$$
\begin{aligned}
t + (6t + 12) + (6t - 13) &= 122.5 \\
13t - 1 &= 122.5 \\
13t &= 123.5 \\
t &= 9.5 \\
6t + 12 &= 57 + 12 = 69 \\
6t - 13 &= 57 - 13 = 44
\end{aligned}
$$

$\boxed{\text{Bismarck, } 9.5\,^\circ\text{F}; \text{ Miami, } 69^\circ\text{F}; \text{ Atlanta, } 44^\circ\text{F}}$

8.   Let x equal the cost of the other desk. $4c$ equals the cost of **the** first desk.

$$
\begin{aligned}
4c + c &= 2075 \\
5c &= 2075 \\
c &= 415 \\
4c &= 1660
\end{aligned}
$$

The desks cost $\boxed{\$415 \text{ and } \$1660}$.

9.   Let x equal the number of nickels. $3x$ equals the number of quarters. $x + 5$ equals the number of dimes.

$$
\begin{aligned}
0.05x + 0.25(3x) + 0.10(x + 5) &= 3.20 \\
0.05x + 0.75x + 0.10x + 0.50 &= 3.20 \\
0.90x &= 2.70 \\
x &= 3 \\
3x &= 9 \\
x + 5 &= 8
\end{aligned}
$$

$\boxed{3 \text{ nickels, } 9 \text{ quarters, } 8 \text{ dimes}}$

10.   Let x equal the number of nickels. $15 - x$ equals the number of dimes.

$$
\begin{aligned}
0.05x + 0.10(15 - x) &= 1.35 \\
0.05x + 1.5 - 0.10x &= 1.35 \\
-0.05x &= -0.15 \\
x &= 3 \\
15 - x &= 12
\end{aligned}
$$

$\boxed{3 \text{ nickels, } 12 \text{ dimes}}$

11. Let x equal the number of $95 tickets sold. $x + 1000$ equals the number of $40 tickets sold. $2x - 1000$ equals the number of $60 tickets sold.

$$
\begin{aligned}
95x + 40(x + 1000) + 60(2x - 1000) &= 745000 \\
95x + 40x + 40000 + 120x - 60000 &= 745000 \\
255x - 20000 &= 745000 \\
255x &= 765000 \\
x &= 3000 \\
x + 1000 &= 4000 \\
2x - 1000 &= 5000
\end{aligned}
$$

$\boxed{3000 \text{ \$95 tickets; } 4000 \text{ \$40 tickets; } 5000 \text{ \$60 tickets}}$

12. Let x equal the regular hourly rate.

$$
\begin{aligned}
40x + 6(1.5x) + 4(2x) &= 342 \\
40x + 9x + 8x &= 342 \\
57x &= 342 \\
x &= 6
\end{aligned}
$$

regular hourly rate: $\boxed{\$6}$

13. $\dfrac{-6}{9} = \dfrac{-2}{3}$; $\dfrac{4}{-6} = \dfrac{-2}{3}$; To obtain the next term, multiply the previous term by $-\dfrac{2}{3}$.

$$
4\left(-\frac{2}{3}\right) = \boxed{-\frac{8}{3}}
$$

$$
-\frac{8}{3}\left(-\frac{2}{3}\right) = \frac{16}{9} \ \sqrt{}
$$

14.

$$
\frac{A^2}{B} = C
$$

$$
\frac{7^2}{5} = \frac{49}{5} = \frac{49}{5} \ \sqrt{}
$$

$$
\frac{8^2}{2} = \frac{64}{2} = 32 \ \sqrt{}
$$

$$
\boxed{C = \frac{A^2}{B}}
$$

15. Cost $= c + d(k - 12)$ cents $= \boxed{c + dk - 12d \text{ cents}}$.

16. Let x equal the total weight of the banana. $\dfrac{1}{8}x$ equals the weight of the banana peel. $\dfrac{7}{8}x$ equals the weight of the peeled banana.

$$
\begin{aligned}
x &= \frac{7}{8}x + \frac{7}{8} \\
\frac{1}{8}x &= \frac{7}{8} \\
x &= 7
\end{aligned}
$$

$\boxed{7 \text{ ounces}}$

17. Let x equal the number of bicycles. $13 - x$ equals the number of tricycles.

$$
\begin{aligned}
2x + 3(13 - x) &= 31 \\
2x + 39 - 3x &= 31 \\
-x &= -8 \\
x &= 8 \\
13 - x &= 5
\end{aligned}
$$

$\boxed{8 \text{ bicycles; } 5 \text{ tricycles}}$

18.

pennies		nickels		dimes		total
no.	amount	no.	amount	no.	amount	
		1	5	1	10	15
5	5			1	10	15
		3	15			15
5	5	2	10			15
10	10	1	5			15
15	15					15

6 ways

19. $\dfrac{6 \text{ inches}}{4 \text{ feet}} = \dfrac{6 \text{ inches}}{4(12 \text{ inches})} = \dfrac{6}{48} = \boxed{\dfrac{1}{8} \text{ or } 1:8}$

20. $\dfrac{10 \text{ centimeters}}{3 \text{ meters}} = \dfrac{10 \text{ cm}}{3(100 \text{ cm})} = \dfrac{10}{30} = \boxed{\dfrac{1}{30} \text{ or } 1:30}$

21. 40 people: $2(12) = 24$ men; $40 - 24 = 16$ women

$\dfrac{\text{women}}{\text{men}} = \dfrac{16}{24} = \boxed{\dfrac{2}{3} \text{ or } 2:3}$

22. Let $3x$ equal one number. $5x$ equals the other number.

$$\begin{aligned} 2(3x) &= 5x + 34 \\ 6x &= 5x + 34 \\ x &= 34 \\ 3x &= 102 \\ 5x &= 170 \end{aligned}$$

The numbers are $\boxed{102 \text{ and } 170}$.

23.
$$\begin{aligned} \frac{32}{x+8} &= \frac{12}{x-2} \\ 32(x-2) &= 12(x+8) \\ 32x - 64 &= 12x + 96 \\ 20x &= 160 \\ x &= 8 \end{aligned}$$

$\boxed{\{8\}}$

24.
$$\begin{aligned} \frac{3y+3}{3} &= \frac{7y-1}{5} \\ 5(3y+3) &= 3(7y-1) \\ 15y + 15 &= 21y - 3 \\ 18 &= 6y \\ 3 &= y \end{aligned}$$

$\boxed{\{3\}}$

25.
$$\begin{aligned} \frac{2\frac{1}{5}}{3\frac{1}{4}} &= \frac{x}{2\frac{1}{3}} \\ 3\frac{1}{4}x &= 2\frac{1}{5}\left(2\frac{1}{3}\right) \\ \frac{13}{4}x &= \frac{11}{5}\left(\frac{7}{3}\right) \\ x &= \frac{11}{5} \cdot \frac{7}{3} \div \frac{13}{4} \\ x &= \frac{11}{5} \cdot \frac{7}{3} \cdot \frac{4}{13} \\ x &= \frac{308}{195} \end{aligned}$$

$\boxed{\left\{\dfrac{308}{195}\right\}}$

26.
$$\begin{aligned} \frac{5 \text{ miles}}{7 \text{ feet}} &= \frac{x}{3 \text{ feet}} \\ 7x &= 15 \text{ miles} \\ x &= \frac{15}{7} \text{ miles} \\ x &= \boxed{2\frac{1}{7} \text{ miles}} \end{aligned}$$

27. $\dfrac{3 \text{ teachers}}{50 \text{ students}} = \dfrac{x}{5400 \text{ students}}$

$50x = 3(5400) \text{ teachers}$

$x = \boxed{324 \text{ teachers}}$

28. $\dfrac{32 \text{ tagged}}{82 \text{ caught}} = \dfrac{112 \text{ tagged}}{x}$

$32x = 82(112)$

$x = 287$

$\boxed{287 \text{ trout}}$ in lake.

29. $(5x + 31) + (7x - 13) = 180$

$12x + 18 = 180$

$12x = 162$

$\boxed{x = 13.5}$

30. $2x + 21 = 4x - 95$

$116 = 2x$

$58 = x$

$\boxed{x = 58}$

31. $2x = 170 - 3x$

$5x = 170$

$\boxed{x = 34}$

$(170 - 3x)° = (170 - 3 \cdot 34)° = \boxed{64°}$

$(2x)° = (2 \cdot 34)° = \boxed{68°}$

32. $(3x + 10) + (2x - 40) = 90$

$5x - 30 = 90$

$5x = 120$

$\boxed{x = 24}$

$(3x + 10)° = (3 \cdot 24 + 10)° = \boxed{82°}$

$(2x - 40)° = (2 \cdot 24 - 40)° = \boxed{8°}$

33. $x + 2x + 3x = 180$

$6x = 180$

$\boxed{x = 30}$

$2x = 60$

$3x = 90$

$\boxed{30°, 60°, 90°}$

34. $5x - 3 = 2x + 6$

$3x = 9$

$\boxed{x = 3}$

$2x + 6 = 2 \cdot 3 + 6 = 6 + 6 = \boxed{12}$

$5x - 3 = 5 \cdot 3 - 3 = 15 - 3 = \boxed{12}$

$9x + 2 = 9 \cdot 3 + 2 = 27 + 2 = \boxed{29}$

35. Let x equal the measure of the angle. $90 - x$ equals the measure of its complement.

$90 - x = 3x - 10$

$100 = 4x$

$25 = x$

$90 - x = 65$

$\boxed{\text{angle}, 25°; \text{complement}, 65°}$

36. Let x equal the measure of the angle. $180 - x$ equals the measure of the supplement.

$180 - x = 4x - 45$

$225 = 5x$

$45 = x$

$180 - x = 135$

$\boxed{\text{angle}, 45°; \text{supplement}, 135°}$

37. Let x equal the measure of the angle. $90 - x$ equals the measure of the complement.

$x = \dfrac{3}{4}(90 - x)$

$4x = 270 - 3x$

$7x = 270$

$x = \dfrac{270}{7} = 38\dfrac{4}{7}$

$90 - x = 51\dfrac{3}{7}$

$\boxed{\text{angle}, 38\dfrac{4}{7}°; \text{complement}, 51\dfrac{3}{7}°}$

38. $m \angle B = 7m \angle A + 11$
$m \angle C = 5m \angle A$

$$
\begin{aligned}
m \angle A + m \angle B + m \angle C &= 180 \\
m \angle A + (7m \angle A + 11) + (5m \angle A) &= 180 \\
13m \angle A + 11 &= 180 \\
13m \angle A &= 169 \\
m \angle A &= \boxed{13°} \\
m \angle B &= 7(13) + 11 = 91 + 11 = \boxed{102°} \\
m \angle C &= 5(13) = \boxed{65°}
\end{aligned}
$$

39.

$$
\begin{aligned}
180 - m \angle B &= \text{supplement of } \angle B \\
m \angle C &= (180 - m \angle B) - 4 = 176 - m \angle B \\
m \angle A &= \frac{1}{2}m \angle B - 30 \\
m \angle A + m \angle B + m \angle C &= 180 \\
\left(\frac{1}{2}m \angle B - 30\right) + m \angle B + (176 - m \angle B) &= 180 \\
\frac{1}{2}m \angle B + 146 &= 180 \\
\frac{1}{2}m \angle B &= 34 \\
m \angle B &= \boxed{68\ °} \\
m \angle A &= 34 - 30 = \boxed{4°} \\
m \angle C &= 176 - 68 = \boxed{108°}
\end{aligned}
$$

40. Let w equal the width. $2w - 1$ equals the length.

$$
\begin{aligned}
2w + 2(2w - 1) &= 16 \\
2w + 4w - 2 &= 16 \\
6w &= 18 \\
w &= 3 \\
2w - 1 &= 6 - 1 = 5
\end{aligned}
$$

$\boxed{3 \text{ meters} \times 5 \text{ meters}}$

41. Let w equal the width of the original rectangle. $3w$ equals the length of the original rectangle. $2w$ equals the width of the new rectangle. $3w - 4$ equals the length of the new rectangle.

$$
\begin{aligned}
2(w + 3w) &= 2(2w + 3w - 4) \\
2(4w) &= 2(5w - 4) \\
4w &= 5w - 4 \\
4 &= w \\
3w &= 12
\end{aligned}
$$

$\boxed{\text{width, 4 inches; length, 12 inches}}$

42. Let x equal the length of the shortest side. $4x$ equals the length of the longest side. $4x - 15$ equals the length of the third side.

$$
\begin{aligned}
x + 4x + (4x - 15) &= 488 \\
9x - 15 &= 488 \\
9x &= 503 \\
x &= 55\frac{8}{9} \\
4x &= 223\frac{5}{9} \\
4x - 15 &= 208\frac{5}{9}
\end{aligned}
$$

The lengths of the sides of the triangle are $\boxed{55\frac{8}{9}\text{ cm},\ 208\frac{5}{9}\text{ cm},\ 223\frac{5}{9}\text{ cm}}$.

43.

$$
\begin{aligned}
A &= \frac{1}{2}h(B + b) & A = 36 \text{ sq yd},\ B = 7 \text{ yd},\ h = 6 \text{ yd},\ b = ? \\
36 &= \frac{1}{2}6(7 + b) \\
36 &= 3(7 + b) \\
12 &= 7 + b \\
5 &= b
\end{aligned}
$$

$\boxed{5 \text{ yards}}$

44. Let $x, x + 2, x + 4$ represent the sides of the triangle.

$$
\begin{aligned}
x + (x + 2) + (x + 4) &= 87 \\
3x + 6 &= 87 \\
3x &= 81 \\
x &= 27 \\
x + 2 &= 29 \\
x + 4 &= 31
\end{aligned}
$$

The lengths of the sides of the triangle are $\boxed{27 \text{ yards, } 29 \text{ yards, and } 31 \text{ yards}}$.

45.

$$
\begin{aligned}
A &= \frac{1}{2}bh & A = 133 \text{ sq yd},\ b = 14 \text{ yd},\ h = ? \\
133 &= \frac{1}{2}(14)h \\
133 &= 7h \\
19 &= h
\end{aligned}
$$

$\boxed{19 \text{ yd}}$

46.

$$
\begin{aligned}
A &= LW & A = 216 \text{ sq yd},\ L = 9 \text{ ft} = \frac{9}{3} \text{ yd} = 3 \text{ yd},\ W = ? \\
216 &= 3W \\
72 &= W
\end{aligned}
$$

$\boxed{72 \text{ yd}}$

47. area of triangle + area outside triangle and inside trapezoid = area of trapezoid

$$
\begin{aligned}
\frac{1}{2}(1)(1) + 31.5 &= \frac{1}{2}x(6 + 10) \\
0.5 + 31.5 &= \frac{1}{2}x(16) \\
32 &= 8x \\
4 &= x
\end{aligned}
$$

$\boxed{4 \text{ feet}}$

48. $\dfrac{\pi 2^2}{\pi 8^2} = \dfrac{4}{x}$

$\quad\quad \dfrac{4}{64} = \dfrac{4}{x}$

$\quad\quad 4x = 4(64)$

$\quad\quad\; x = 64$

$\boxed{\$64}$

49. $\dfrac{1}{6}$ ft $= \dfrac{1}{6}$(12 in.) $= 2$ in.

$\quad \dfrac{1}{12}$ ft $= \dfrac{1}{12}$(12 in.) $= 1$ in.

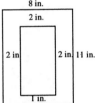

Writing area: width: $8 - (2 + 2) = 4$ in.

$\quad\quad\quad\quad\quad\quad\quad$ length: $11 - (2 + 1) = 8$ in.

$A = 4(8) = \boxed{32 \text{ square inches}}$

50. $\dfrac{\text{bull's-eye area}}{\text{total area}} = \dfrac{\pi 2^2}{\pi 12^2} = \dfrac{4}{144} = \boxed{\dfrac{1}{36} \text{ or } 1 : 36}$

51. $\quad 60t + 80t = 400$

$\quad\quad\quad\quad 140t = 400$

$\quad\quad\quad\quad\quad\; t = 2\dfrac{6}{7}$

$\boxed{2\dfrac{6}{7} \text{ hours}}$

52. Let r equal the rate of one plane. $r + 60$ equals the rate of the other plane.

$\quad\quad r(5) + (r + 60)(5) = 2800$

$\quad\quad\quad\; 5r + 5r + 300 = 2800$

$\quad\quad\quad\quad\quad\quad\quad 10r = 2500$

$\quad\quad\quad\quad\quad\quad\quad\quad r = 250$

$\quad\quad\quad\quad\quad\; r + 60 = 310$

$\boxed{250 \text{ mph, } 310 \text{ mph}}$

53.

$$t$$
$$40 \text{ mph}$$
$$45 \text{ mph}$$
$$t - \tfrac{1}{2}$$

since $30 \text{ min} = \dfrac{1}{2}\text{h}$

$$40t = 45\left(t - \frac{1}{2}\right)$$

$$40t = 45t - \frac{45}{2}$$

$$\frac{45}{2} = 5t$$

$$\frac{9}{2} = t$$

$$d = 40t = 40\left(\frac{9}{2}\right) = 180$$

$\boxed{180 \text{ miles}}$

54. Let x equal the amount at 8%. $1000 - x$ equals the amount at 10%.

$$0.08x + 0.10(1000 - x) = 94$$
$$0.08x + 100 - 0.10x = 94$$
$$-0.02x = -6$$
$$x = 300$$
$$1000 - x = 700$$

$\boxed{\$300 \text{ at } 8\%, \$700 \text{ at } 10\%}$

55. Let x equal the amount invested at 6%.

$$0.06x = 0.05(6000)$$
$$0.06x = 300$$
$$x = 5000$$

$\boxed{\$5000}$

56. Let x equal the amount invested at 8%. $2x + 100$ equals the amount invested at 9%.

$$0.08x + 0.09(2x + 100) = 1910$$
$$0.08x = 0.09(2x + 100) = 1910$$
$$0.08x + 0.18x + 9 = 1910$$
$$0.26x = 1901$$
$$x = 7311.54$$
$$2x + 100 = 14723.08$$

$\boxed{\$7311.54 \text{ at } 8\%; \$14723.08 \text{ at } 9\%}$

57. Let x equal the amount of 75% salt solution. $10 - x$ equals the amount of 50% salt solution.

$$0.75x + 0.50(10 - x) = 0.60(10)$$
$$0.75x + 5 - 0.5x = 6$$
$$0.25x = 1$$
$$x = 4$$
$$10 - x = 6$$

$\boxed{4 \text{ gallons of } 75\% \text{ salt solution, } 6 \text{ gallons of } 50\% \text{ salt solution}}$

58. Let x equal the amount of candy at 15¢ per pound. $30 - x$ equals the amount of candy at 25¢ per pound.

$$
\begin{aligned}
0.15x + 0.25(30 - x) &= 0.18(30) \\
0.15x + 7.5 - 0.25x &= 5.4 \\
-0.10x &= -2.1 \\
x &= 21 \\
30 - x &= 9
\end{aligned}
$$

$\boxed{\text{21 pounds at 15¢ per pound, 9 pounds at 25¢ per pound}}$

59. Let x equal the amount of Chinese tea at 57¢ per kg. $x + 14$ equals the amount of blend at 50¢ per kg.

$$
\begin{aligned}
0.57x + 0.48(14) &= 0.50(x + 14) \\
0.57x + 6.72 &= 0.5x + 7 \\
0.07x &= 0.28 \\
x &= 4
\end{aligned}
$$

$\boxed{\text{4 kg}}$ of Chinese tea at 57¢ per kg.

60.

$$
\begin{aligned}
\frac{t}{6} + \frac{t}{12} &= 1 \qquad t = \text{time (in hours) together} \\
12\left(\frac{t}{6} + \frac{t}{12}\right) &= 12(1) \\
2t + t &= 12 \\
3t &= 12 \\
t &= 4
\end{aligned}
$$

$\boxed{\text{4 hours}}$ to complete **the** job working together.

61. t equals the time in minutes together

$$
\begin{aligned}
\frac{t}{8} + \frac{t}{12} + \frac{t}{24} &= 1 \\
24\left(\frac{t}{8} + \frac{t}{12} + \frac{t}{24}\right) &= 24(1) \\
3t + 2t + t &= 24 \\
6t &= 24 \\
t &= 4
\end{aligned}
$$

$\boxed{\text{4 minutes}}$ together

62. x equals the total distance. $\frac{1}{3}x$ equals the distance by foot. $\frac{1}{6}x$ equals the distance by water. 10 miles equals the distance by boat.

$$
\begin{aligned}
x - \left(\frac{1}{3}x + \frac{1}{6}x\right) &= 10 \\
x - \frac{1}{3}x - \frac{1}{6}x &= 10 \\
6\left(x - \frac{1}{3}x - \frac{1}{6}x\right) &= 6(10) \\
6x - 2x - x &= 60 \\
3x &= 60 \\
x &= 20
\end{aligned}
$$

$\boxed{\text{20 miles}}$

63. Let x equal the number of hours worked by mechanic. $x - 2$ equals the number of hours worked by assistant.

$$
\begin{aligned}
6x + 4(x - 2) + 98 &= 190 \\
6x + 4x - 8 + 98 &= 190 \\
10x + 90 &= 190 \\
10x &= 100 \\
x &= 10 \\
x - 2 &= 8
\end{aligned}
$$

$\boxed{\text{mechanic, 10 hours; assistant, 8 hours}}$

Chapter 3 Cumulative Review

Cumulative Review, pp. 245-246

1. a. $2 + (4 + 9) = (2 + 4) + 9$
Associative Property of Addition
b. $5(7 + 3) = 5(3 + 7)$
Commutative Property of Addition
c. $6 + (-6) = 0$
Additive Inverse Property

2. when $x = 4$, $y = 3$,
$$\frac{2x^2 - y^2}{3x - 2} = \frac{2(4^2) - 3^2}{3(4) - 2} = \frac{2(16) - 9}{12 - 2} = \frac{32 - 9}{10} = \boxed{\frac{23}{10}}$$

3. $\dfrac{-9(3 - 6)}{(-12)(3) + (-3 - 5)(8 - 4)} = \dfrac{-9(-3)}{-36 + (-8)(4)} = \dfrac{27}{-36 - 32} = \dfrac{27}{-68} = \boxed{-\dfrac{27}{68}}$

4. temperature at 10 P.M. + change in temperature at 3 A.M. + change in temperature at 12 A.M.
$= -4°\text{F} - 11°\text{F} + 21°\text{F} = \boxed{6°\text{F}}$

5. C is true.
A. $12 - \left| -3 \right| = 12 - 3 = 9 \neq 15$
B. $4[5(-2) - 36] = 4(-46) = -184 \leq 138 \ not \geq 138$
C. $\dfrac{-14 - 2(-7)}{(-3)(-4) - (-2)} - \dfrac{-14 + 14}{12 + 2} = 0 \geq 0$; true
D. $\dfrac{26 - 7(-2)}{-3(3 - 6) - 1} = \dfrac{26 + 14}{-3(-3) - 1} = \dfrac{40}{10} = 4 < 5 \ not > 5$

6. a. Natural numbers: $\left\{ 8, \sqrt{25} \right\}$
b. Whole numbers: $\left\{ 0, 8, \sqrt{25} \right\}$
c. Integers: $\left\{ -3, 0, 8, \sqrt{25} \right\}$
d. Irrational numbers: $\left\{ \sqrt{29} \right\}$
e. Real numbers: $\left\{ -3, -\dfrac{1}{2}, \dfrac{1}{7}, 0, 8, 9.\overline{3}, \sqrt{25}, \sqrt{29} \right\}$

7. $7x - (3x + 5) = -4(2x - 1)$

$7x - 3x - 5 = -8x + 4$

$4x - 5 = -8x + 4$

$12x = 9$

$x = \dfrac{3}{4}$

$\boxed{\left\{\dfrac{3}{4}\right\}}$

8. $\dfrac{1}{5}y + \dfrac{2}{3}y = y + \dfrac{1}{15}$

$15\left(\dfrac{1}{5}y + \dfrac{2}{3}y\right) = 15\left(y + \dfrac{1}{15}\right)$

$3y + 10y = 15y + 1$

$13y = 15y + 1$

$-2y = 1$

$y = -\dfrac{1}{2}$

$\boxed{\left\{-\dfrac{1}{2}\right\}}$

9.   Let $x, x + 1, x + 2$ represent three consecutive integers.

$x + (x + 1) + (x + 2) = 2x + 13$

$3x + 3 = 2x + 13$

$x = 10$

$x + 1 = 11$

$x + 2 = 12$

The integers are $\boxed{10, 11, 12}$.

10. Let x equal the number.

$7x + 3 = 8x - 2$

$\boxed{5} = x$

11. Let x equal the price before reduction.

$x - 0.39x = 13.98$

$0.61x = 13.98$

$x \approx 22.92$

$\boxed{\$22.92}$

12. $3x - 7 = 4x - 7 - (7 + x)$

$3x - 7 = 4x - 7 - 7 - x$

$3x - 7 = 3x - 14$

$-7 \neq -14$

no solution; $\boxed{\varnothing}$

13. $-1 < 4y + 7 \leq 15$

$-8 < 4y \leq 8$

$-2 < y \leq 2$

$\boxed{\{y \mid -2 < y \leq 2\}}$

14. Let x equal the present age of the bronze coin. $x + 74$ equals the present age of the silver coin. $x - 30$ equals the age of bronze coin 30 years ago. $x + 44$ equals the age of the silver coin 30 years ago.

$x + 44 = 3(x - 30)$

$x + 44 = 3x - 90$

$134 = 2x$

$67 = x$

$x + 74 = 141$

$\boxed{\text{bronze coin, 67 years old; silver coin, 141 years old}}$

15. $4(2x - 1) - 3(x - 11) - 2(-4x - 5) = 8x - 4 - 3x + 33 + 8x + 10 = \boxed{13x + 39}$

16. $4x - 8(x + 1) \geq -2(5x - 4) - 10$

$4x - 8x - 8 \geq -10x + 8 - 10$

$-4x - 8 \geq -10x - 2$

$6x \geq 6$

$x \geq 1$

$\boxed{\{x \mid x \geq 1\}}$

17. $(4x + 7) + (11x - 22) = 180$

$15x - 15 = 180$

$15x = 195$

$\boxed{x = 13}$

$(4x + 7)^\circ = (4 \cdot 13 + 7)^\circ = (52 + 7)^\circ = \boxed{59^\circ}$

$(11x - 22)^\circ = (11 \cdot 13 - 22)^\circ = (143 - 22)^\circ = \boxed{121^\circ}$

18. $\dfrac{3y-2}{5} = \dfrac{4y}{1}$

$3y-2 = 20y$

$-2 = 17y$

$-\dfrac{2}{17} = y$

$$\boxed{\left\{-\dfrac{2}{17}\right\}}$$

19. $\dfrac{9}{135} = \dfrac{5}{x}$

$9x = 5(135)$

$x = 75$

$\boxed{\$75}$

20. Let x equal the side of the square. x equals the width of the rectangle. $2x + 9$ equals the length of the rectangle.

perimeter of rectangle $= 3$(perimeter of square)

$2x + 2(2x+9) = 3(4x)$

$2x + 4x + 18 = 12x$

$18 = 6x$

$3 = x$

$2x+9 = 6 + 9 = 15$

dimensions of the square: $\boxed{3 \text{ feet} \times 3 \text{ feet}}$

dimensions of the rectangle: $\boxed{3 \text{ feet} \times 15 \text{ feet}}$

21. Let x equal Stevenson's share. $x + 8$ equals Louis' share. $2(x+8) + 3 = 2x + 19$ equals Robert's share.

$x + (x+8) + (2x+19) = 119$

$4x + 27 = 119$

$4x = 92$

$x = 23$

$x + 8 = 31$

$2x + 19 = 46 + 19 = 65$

$\boxed{\text{Robert, 65 novels; Louis, 31 novels; Stevenson, 23 novels}}$

22. $3^2 + 4^2 = 5^2$

$10^2 + 11^2 + 12^2 = 13^2 + \boxed{14^2}$

$21^2 + 22^2 + 23^2 + \boxed{24^2} = \boxed{25^2} + \boxed{26^2} + \boxed{27^2}$

23. D is not possible; $2x > 23$ $9x < 865$

$x > \dfrac{23}{2}$ $x < 96\dfrac{1}{9}$

$x > 11.5$

$11.5 < x < 96\dfrac{1}{9}$

A. $x = 96$; possible

B. $x < 50$; possible

C. $x > 23$; possible

D. $x > 100$ *not* possible

24.
$$3x - 17 = 8443$$
$$3x = 8460$$
$$x = 2820$$
$$(\text{Gunyashev}) + (\text{gorilla}) = x - 5$$
Let G equal the amount of Gunyashev lifted.
$$G + (G + 775) = 2820 - 5$$
$$2G = 2815 - 775$$
$$2G = 2040$$
$$G = 1020$$

$\boxed{1020 \text{ pounds}}$

25. Let x equal the measure of the angle. $90 - x$ equals the measure of its complement.
$$x - (90 - x) = 16$$
$$2x - 90 = 16$$
$$2x = 106$$
$$x = 53$$

$\boxed{53°}$

26. Let x equal the number of sheets of paper.
$$2x + 4 \le 29$$
$$2x \le 25$$
$$x \le 12.5$$

$\boxed{12 \text{ sheets}}$

27. Let r equal the radius of the circle. $r + 2$ equals the radius of the larger circle.
$$2\pi(r + 2) = 16\pi$$
$$r + 2 = 8$$
$$r = 6$$

$\boxed{6 \text{ meters}}$

28. Let x equal the weekly salary.
$$0.15x = 14.95 + 2(15.99) + 2(31.99) + 3(1.89)$$
$$0.15x = 116.58$$
$$x = 777.20$$
weekly salary: $\boxed{\$777.20}$

29. $\boxed{12 \div 2 \div 3 = 2}$
$x = 12, y = 2, z = 3$

30.
$$\frac{803}{\frac{2}{3}(803)} = \frac{750}{x}$$
$$803x = 750\left(\frac{2}{3}\right)(803)$$
$$x = 500$$

$\boxed{\$500}$

Chapter 4   Linear Equations and Inequalities in Two Variables

Section 4.1 Linear Equations in Two Variables

Problem Set 4.1, pp. 253-256

1. $2x - y = 3$
$x = -1;\ -2 - y = 3$
$-y = 5$
$y = \boxed{-5}$

$(-1, -5)$

3. $2x - y = 3$
$y = -1:\ 2x - (-1) = 3$
$2x = 2$
$x = \boxed{1}$

$(1, -1)$

5. $2x - y = 3$
$x = 3:\ 2(3) - y = 3$
$-y = -3$
$y = \boxed{3}$

$(3, 3)$

7. $4x - 3y = 12$
$x = 6:\ 4(6) - 3y = 12$
$-3y = -12$
$y = \boxed{4}$

$(6, 4)$

9. $4x - 3y = 12$
$y = 0:\ 4x - 3(0) = 12$
$4x = 12$
$x = \boxed{3}$

$(3, 0)$

11. $4x - 3y = 12$
$x = \frac{1}{4}:\ 4\left(\frac{1}{4}\right) - 3y = 12$
$-3y = 1$
$y = \boxed{-\frac{11}{3}}$

$\left(\frac{1}{4}, -\frac{11}{3}\right)$

13. $4x - 3y = 12$
$x = -2:\ 4(-2) - 3y = 12$
$-3y = 20$
$y = \boxed{-\frac{20}{3}}$

$\left(-2, -\frac{20}{3}\right)$

15. $2x - y = 4$

x	0	3	$\frac{11}{4}$	0
y	-4	2	$\frac{3}{2}$	-4

$x = 0:\qquad 0 - y = 4;\ y = -4$
$x = 3:\qquad 6 - y = 4;\ y = 2$
$y = \frac{3}{2}:\qquad 2x - \frac{3}{2} = 4$
$\qquad\qquad 2x = \frac{11}{2};\ x = \frac{11}{4}$
$y = -4:\ 2x - (-4) = 4;\ 2x = 0;\ x = 0$

17. $y = 4x - 1$

x	1	$-\dfrac{3}{2}$	$-\dfrac{3}{4}$	2
y	3	-7	-4	7

$y = 3$: $3 = 4x - 1$
$\qquad\quad 4 = 4x, 1 = x$
$y = -7$: $-7 = 4x - 1$
$\qquad\qquad -6 = 4x; -\dfrac{3}{2} = x$
$x = -\dfrac{3}{4}$: $y = 4\left(-\dfrac{3}{4}\right) - 1 = -3 - 1 = -4$
$x = 2$: $y = 4(2) - 1 = 8 - 1 = 7$

19. $3x + 5y = 0$

x	-5	5	10	0
y	3	-3	-6	0

$y = 3$: $3x + 5(3) = 0$
$\qquad\qquad 3x = -15$
$\qquad\qquad\quad x = -5$
$x = 5$: $3(5) + 5y = 0$
$\qquad\qquad 5y = -15$
$\qquad\qquad\; y = -3$
$x = 10$: $3(10) + 5y = 0$
$\qquad\qquad\; 5y = -30$
$\qquad\qquad\;\; y = -6$
$y = 0$: $3x + 5(0) = 6$
$\qquad\qquad\; x = 0$

21. $x = 2$

x	2	2	2
y	-3	5	0

when $x = 2$, y can be any value
$y = -3$, $x = 2$
$y = 5$, $x = 2$
$y = 0$, $x = 2$

23. $y = 2x + 6$
$\boxed{(0, 6)}$: $6 = 0 + 6$
a solution $6 = 6$ True

$\boxed{(-3, 0)}$: $0 = 2(-3) + 6$
a solution $0 = 0$ True

$(2, -2)$: $-2 = 2(2) + 6$
not a solution $-2 = 10$ False

25. $3x + 5y = 15$
$\boxed{(-5, 6)}$: $3(-5) + 5(6) = 15$
a solution $-15 + 30 = 15$
$\qquad\qquad\qquad 15 = 15$ True

$(0, 5)$: $3(0) + 5(5) = 15$
not a solution $3(0) + 5(5) = 15$
$\qquad\qquad\qquad 25 = 15$ False

$\boxed{(0, -3)}$: $3(10) + 5(-3) = 15$
$\qquad\qquad\qquad 30 - 15 = 15$
a solution $15 = 15$ True

27. $x + 3y = 0$

$\boxed{(0, 0)}$: $0 + 3(0) = 0$

a solution $0 = 0$ True

$\left(1, \dfrac{1}{3}\right)$: $1 + 3\left(\dfrac{1}{3}\right) = 0$

not a solution $1 + 1 = 0$
                               $2 = 0$ False

$\boxed{\left(2, -\dfrac{2}{3}\right)}$: $2 + 3\left(-\dfrac{2}{3}\right) = 0$

a solution $2 - 2 = 0$
                               $0 = 0$ True

29. $x - 4 = 0$

$x = 4$

only ordered pairs where $x = 4$ are solutions

$\boxed{(4, 7)}$ is a solution

$(3, 4)$ not a solution $x \neq 4$

$(0, -4)$ not a solution $x \neq 4$

31.
$$
\begin{aligned}
2x + 2y &= 2x \\
x &= 5 \text{ yards} \\
10 + 2y &= 24 \\
2y &= 14 \\
y &= 7
\end{aligned}
$$
$\boxed{\text{width}, 7 \text{ yards}}$

33.
$$
\begin{aligned}
4x - 3y &= 15 \\
x = -3: \quad 4(-3) - 3y &= 15 \\
-12 - 3y &= 15 \\
-3y &= 27 \\
y &= \boxed{-9}
\end{aligned}
$$

35. $\boxed{4x - 3y = 9}$
$$
\begin{aligned}
x = -3: \quad 4(-3) - 3y &= 9 \\
-3y &= 21 \\
y &= \boxed{-7}
\end{aligned}
$$

37. $\boxed{x = 2y + 4}$
$$
\begin{aligned}
x = -16: \quad -16 &= 2y + 4 \\
-20 &= 2y \\
\boxed{-10} &= y
\end{aligned}
$$

39. a. $(x + 2y) + (x + 3y) + (7x + 2y) + (5x + 4y) = 360$

$\boxed{14x + 11y = 360}$

b. $x = 10$: $14(10) + 11y = 360$
                          $11y = 220$

$\boxed{y = 20}$

$$
\begin{aligned}
m\angle A &= (x + 2y)^\circ = (10 + 2 \cdot 20)^\circ = (10 + 40)^\circ = \boxed{50^\circ} \\
m\angle B &= (x + 3y)^\circ = (10 + 3 \cdot 20)^\circ = (10 + 60)^\circ = \boxed{70^\circ} \\
m\angle C &= (7x + 2y)^\circ = (7 \cdot 10 + 2 \cdot 20)^\circ = (70 + 40)^\circ = \boxed{110^\circ} \\
m\angle D &= (5x + 4y)^\circ = (5 \cdot 10 + 4 \cdot 20)^\circ = (50 + 80)^\circ = \boxed{130^\circ}
\end{aligned}
$$

41. a. $\boxed{8x + 6y = 14.50}$

b. $y = 0.75$: $8x + 6(0.75) = 14.50$
                     $8x + 4.5 = 14.50$
                     $8x = 10.00$
                     $x = 1.25$

cost of one pen: $\boxed{\$1.25}$

43. a. $\boxed{4x + 2y = 50}$

    b. $x = 10$: $4(10) + 2y \;=\; 50$

$$2y \;=\; 10$$
$$y \;=\; 5$$

$\boxed{\text{5 parrots}}$

45. a.   width of larger rectangle: $8 + y$

        height of larger rectangle: $5 + x + 5 = 10 + x$

        perimeter is 58 meters: $\boxed{2(8 + y) + 2(10 + x) = 58}$

        $16 + 2y + 20 + 2x = 58$

        $2x + 2y = 22$

        $\boxed{x + y = 11}$

    b. $x = 4.5$ meters: $4.5 + y \;=\; 11$

                    $y \;=\; \boxed{6.5 \text{ meters}}$

47. C is true;   A. $2y - 3x = -8$: $(2, 4)$: $8 - 6 = -8$ false

              B.   not necessarily

              C. $x + 3y = 30$

                 $(b, -5)$: $b - 15 = 30$

                          $b = 45$ true

              D.   only ordered pairs where $x = 5$ are solutions.

Review Problems

52. $\quad 11 - 3x \;>\; 7x - 19$

          $-10x \;>\; -30$

              $x \;<\; 3$

$\boxed{\{x \mid x < 3\}}$

53. Let x equal Ken's present age.

    $x + 3$ equals Bob's present age.

    $x + 10$ equals Ken's age in 10 years.

    $x + 13$ equals Bob's age in 10 years.

        $(x + 10) + (x + 13) \;=\; 33$

                  $2x \;=\; 10$

                    $x \;=\; 5$

            $x + 3 \;=\; 8$

Bob is $\boxed{\text{8 years old}}$.

54. Let x equal the number of dimes.

    $x + 12$ equals the number of nickels.

      $0.10x + 0.05(x + 2) \;=\; 1.75$

      $0.10x + 0.05x + 0.10 \;=\; 1.75$

                $0.15x \;=\; 1.65$

                  $x \;=\; 11$

$\boxed{\text{11 dimes}}$

Section 4.2 Graphing Linear Equations in Two Variables

Problem Set 4.2, pp. 266-269

1. (3, 4): I   **3.** (–4, 1): II   **5.** (–2, –5): III   **7.** (4, –3): IV

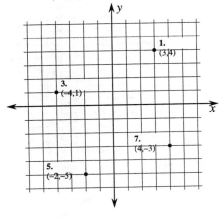

9–23.

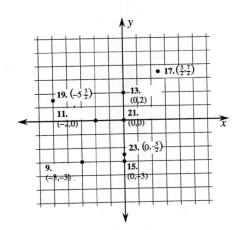

25. A (5, 2)   **27.** C (–6, 5)

29. E (–2, –3)   **31.** G (5, –3)

33. $x + y = 4$

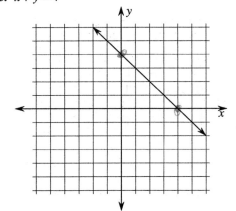

35. $x - 3y = 6$

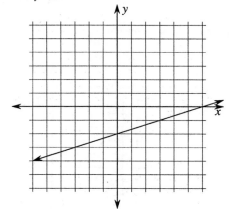

37. $2x = 5y - 10$

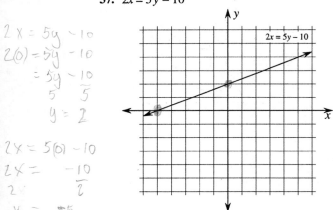

$2x = 5y - 10$
$2(0) = 5y - 10$
$\quad = 5y - 10$
$\quad \dfrac{5}{5} \quad \dfrac{10}{5}$
$\quad y = 2$

$2x = 5(0) - 10$
$2x = \quad -10$
$\dfrac{2}{2} \qquad \dfrac{-10}{2}$
$x = -5$

39. $2x + 3y = 12$

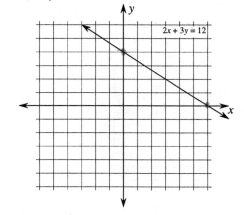

41. $5x + 3y = 15$

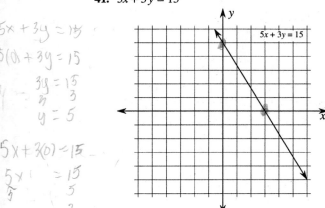

$5x + 3y = 15$
$5(0) + 3y = 15$
$\quad \dfrac{3y}{3} = \dfrac{15}{3}$
$\quad y = 5$

$5x + 3(0) = 15$
$\dfrac{5x}{5} = \dfrac{15}{5}$
$x = 3$

43. $5x - 6y = 30$

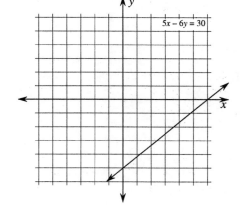

45. $3x + y = 0$

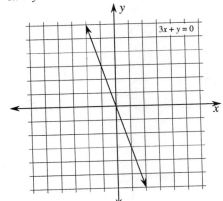

47. $y = 5x$

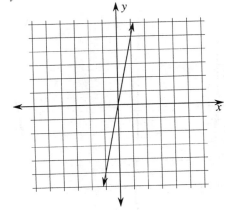

49. $y = -5x$

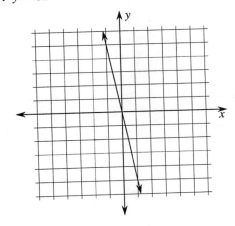

51. $y = \dfrac{x}{3}$

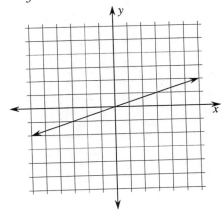

53. $y = 2x - 1$

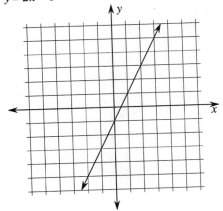

55. $y = -6$

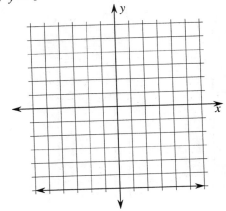

57. $x = 5$

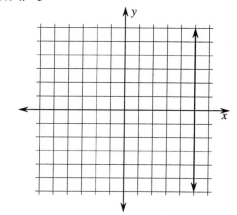

59. $x = 0$

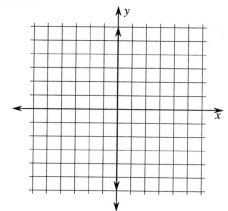

61. $y + 3 = 0$

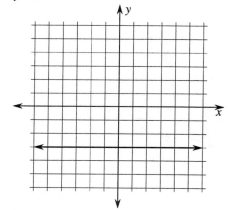

63. $3x = -15$

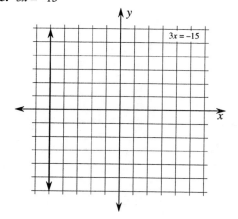

65. $5y = 20$

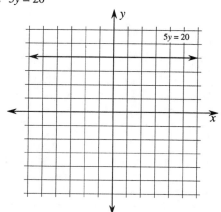

67. $y = x - 2$

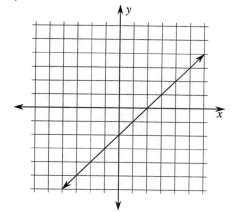

69. $x = 2y + 3$

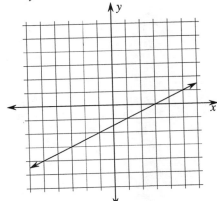

71. $x + 2y = 6$

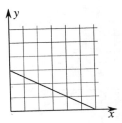

73. $3x - 2y = 6$

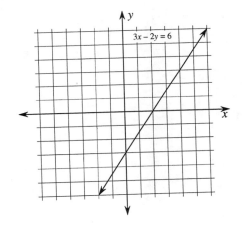

75. Answers will vary. Sample given.
$y = x - 1$

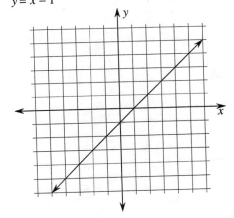

77. $T = \frac{1}{4} C + 37$

a.

C	0	4	8	12	16
T	37	38	39	40	41

b.

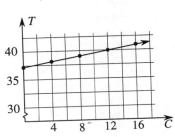

79. *x*-intercept: 2; *y*-intercept: −3

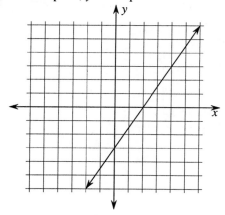

81. *y*-intercept: 2
when *x* changes by +2 units, *y* changes by +3 units
(0, 2) → (2, 5)

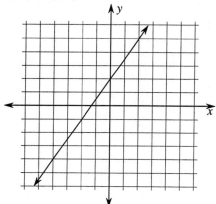

83. vertical line; *x*-intercept: −2

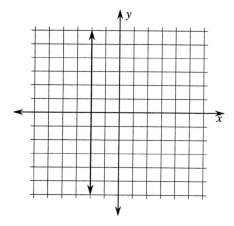

85. *x* coordinate is always zero
The line is the *y*-axis.

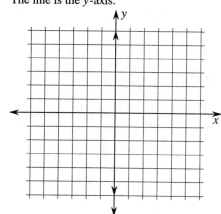

87. $y = 60{,}000 - 5000x$

   $0 \le x \le 12$

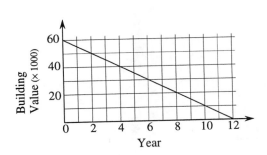

89. coordinate of fourth vertex: $\boxed{(5,\,2)}$ using
   the x-value from $(5, -4)$ and the y-value from $(-3, 2)$

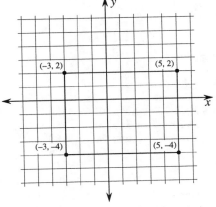

91. Area $= bh$

   $b \;=\; 4 - (-3) = 7$

   $h \;=\; 5 - (-5) = 10$

   $A \;=\; 7(10) = \boxed{70 \text{ square units}}$

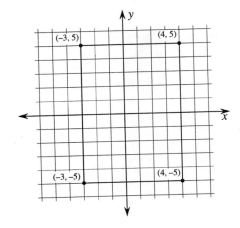

93. $A = \dfrac{1}{2}h(B + b)$ $h = 3,\; B = 5,\; b = 2$

   $A = \dfrac{1}{2}\,3(5 + 2)$

   $A = \dfrac{1}{2}\,3(7)$

   $A = \boxed{10\frac{1}{2} \text{ square units}}$

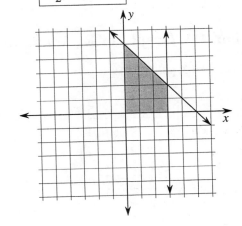

95. $\boxed{B}$ is true;   B. $x + 0y \;=\; 4$

   $x \;=\; 4$ is a vertical line

Review Problems

99. $3(y-2)+y = y-7$
$3y-6+y = y-7$
$3y = -1$
$y = -\dfrac{1}{3}$

$$\boxed{\left\{-\dfrac{1}{3}\right\}}$$

100. $x+2(x+12)+3(x+10) = 180$
$x+2x+24+3x+30 = 180$
$6x+54 = 180$
$6x = 126$
$x = 21$
$2(x+12) = 2(21+12)=2(33)=66$
$3(x+10) = 3(21+10)=3(31)=93$
The measure of the angles are $\boxed{21°,\ 66°,\ \text{and } 93°}$.

101. Let x equal the value of the lot.
$5x$ equals the value of the house.
$x+5x = 112{,}200$
$6x = 112{,}200$
$x = 18{,}700$
The lot is worth $\boxed{\$18{,}700}$.

Section 4.3 Graphs of Equations and Functions

Problem Set 4.3, pp. 276-279

1. $y = x^2 - x - 2$

x	-2	-1	0	1	2	3
y	4	0	-2	-2	0	4

$x=-2:\ y = (-2)^2 - (-2) - 2 = 4$
$x=-1:\ y = (-1)^2 - (-1) - 2 = 0$
$x=0:\ y = 0 - 0 - 2 = -2$
$x=1:\ y = 1 - 1 - 2 = -2$
$x=2:\ y = 2^2 - 2 - 2 = 0$
$x=3:\ y = 3^2 - 3 - 2 = 4$

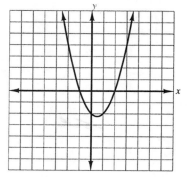

3. $y = x^2 + 2x + 1$

x	-3	-2	-1	0	1	2
y	4	1	0	1	4	9

$x = -3$: $(-3)^2 + 2(-3) + 1 \;=\; 4$
$x = -2$: $(-2)^2 + 2(-2) + 1 \;=\; 1$
$x = -1$: $(-1)^2 + 2(-1) + 1 \;=\; 0$
$x = 0$: $0 + 0 + 1 \;=\; 1$
$x = 1$: $1 + 2(1) + 1 \;=\; 4$
$x = 2$: $2^2 + 2(2) + 1 \;=\; 9$

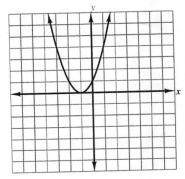

5. $y = -x^2 + x + 1$

x	-1	0	1	2	3	4
y	-1	1	1	-1	-5	-11

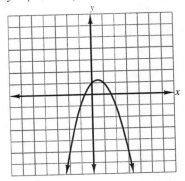

7. $f(x) = x^3 - 2x$

x	-2	-1	0	1	2
y	-4	1	0	-1	4

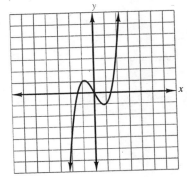

9. $f(x) = x^3 - 3x - 1$

x	-2	-1	0	1	2
y	-3	1	-1	-3	1

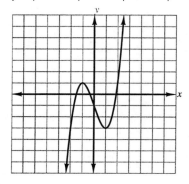

11. $N(x) = 5000\sqrt{100-x}$

x	0	19	36	51	64	75	84	91	100
y	50,000	45,000	40,000	35,000	30,000	25,000	20,000	15,000	0

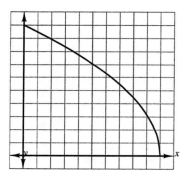

13. y is a function of x
Each vertical line that can be drawn intersects the graph only once.

15. y is not a function of x
A vertical line drawn through $x = 0$ intersects the graph three times.
Vertical lines drawn through other values of x also intersect the graph more than once.

17. y is not a function of x
A vertical line drawn through the three points to the right of the y-axis intersects all three points.

19. y is a function of x
Each vertical line that can be drawn intersects the graph only once.

21. a. For each value of x there is only one value of y. Thus, y is a function of x.
   b. day 15: 50 fruit flies
day 25: 200 fruit flies
   c. The number of flies increases with time, increasing slowly between 0 and 15 days, increasing more rapidly between 15 and 35 days, and increasing slowly for 35 days and beyond.

23. a. $N(2) = 80$ milligrams
   b. 40 milligrams are present in the bloodstream after 6 hours.
   c. $N(8) = 20$ milligrams
   d. after 3 hours; maximum: 100 milligrams
$N(3) = 100$ milligrams

25. $f(x) = 3x + 4$
$f(-1) = -3 + 4 = \boxed{1}$

27. $f(x) = 3x + 4$
$f(3) = 9 + 4 = \boxed{13}$

29. $f(x) = 3x + 4$
$f(a) = \boxed{3a + 4}$

31. $f(x) = 3x + 4$
$-17 = 3x + 4$
$-21 = 3x$
$\boxed{-7} = x$

33. $f(x) = 3x + 4$
$f(a + h) = 3(a + h) + 4$
$= \boxed{3a + 3h + 4}$

35. $g(x) = x^2 - 3x + 7$
$g(4) = 4^2 - 3(4) + 7$
$= 16 - 12 + 7$
$= \boxed{11}$

37. $g(x) = x^2 - 3x + 7$
$g(-1) = (-1)^2 - 3(-1) + 7$
$= 1 + 3 + 7$
$= \boxed{11}$

39. $g(x) = x^2 - 3x + 7$
$g(b) = \boxed{b^2 - 3b + 7}$

41. $g(x) = x^2 - 3x + 7$
$g(0) \cdot g(1) = (0 \cdot 0 + 7) \cdot (1 - 3 + 7)$
$= 7 \cdot 5 = \boxed{35}$

43. $f(x) = 5$
$f(4) = \boxed{5}$

45. $f(x) = 5$
$f(0) = \boxed{5}$

47. $g(x) = |x - 2|$
$g(5) = |5 - 2|$
$= \boxed{3}$

49. $g(x) = |x - 2|$
$g(0) = |-2|$
$= \boxed{2}$

51. $g(-3) = |-3 - 2|$
$= |-5|$
$= \boxed{5}$

53. $g(x) = |x - 2|$
$g(7) = |7 - 2|$
$= |5|$
$= \boxed{5}$

55. $\dfrac{g(-13)}{f(-13)} = \dfrac{|-13 - 2|}{5} = \dfrac{|-15|}{5} = \dfrac{15}{5} = \boxed{3}$

57. $14 = |x - 2|$
$x - 2 = 14 \quad \text{or} \quad -14$
$x = \boxed{16 \text{ or } -12}$

Review Problems

60. $150t + 250t = 800$
$$400t = 800$$
$$t = 2$$
$\boxed{2 \text{ hours}}$

61. Let x equal the number of dozen large eggs at $0.60.
$15 - x$ equals the number of dozen medium eggs at $0.48.
$$0.60x + 0.48(15 - x) = 8.00$$
$$0.60x + 7.2 - 0.48x = 8.00$$
$$0.12x = 0.80$$
$$x = 6\frac{2}{3}$$
$$15 - x = 8\frac{1}{3}$$

$\boxed{6\frac{2}{3} \text{ dozen large eggs at } 60¢, 8\frac{1}{3} \text{ dozen medium eggs at } 80¢}$

62. Let $x, x + 2, x + 4$ represent three even integers.
$$x + (x + 2) + (x + 4) = 144$$
$$3x = 138$$
$$x = 46$$
$$x + 2 = 48$$
$$x + 4 = 50$$
The integers are $\boxed{46, 48 \text{ and } 50}$.

Section 4.4 The Slope of a Line

Problem Set 4.4, pp. 288-292

For problems 1-19, use the formula for slope, $m = \dfrac{y_2 - y_1}{x_2 - x_1}$.

1. $(x_1, y_1) = (2, 6), (x_2, y_2) = (3, 5)$
$m = \dfrac{5 - 6}{3 - 2} = \dfrac{-1}{1} = \boxed{-1}$; line $\boxed{\text{falls}}$

3. $(4, 7), (8, 10)$
$m = \dfrac{10 - 7}{8 - 4} = \boxed{\dfrac{3}{4}}$; line $\boxed{\text{rises}}$

5. $(-2, 1), (2, 2)$
$m = \dfrac{2 - 1}{2 - (-2)} = \boxed{\dfrac{1}{4}}$; line $\boxed{\text{rises}}$

7. $(4, -2), (3, -2)$
$m = \dfrac{-2 - (-2)}{3 - 4} = \dfrac{0}{-1} = \boxed{0}$; line is $\boxed{\text{horizontal}}$

9. $(-2, 4), (-1, -1)$
$m = \dfrac{-1 - 4}{-1 - (-2)} = \dfrac{-5}{1} = \boxed{-5}$; line $\boxed{\text{falls}}$

11. $(5, 3), (5, -2)$
$m = \dfrac{-2 - 3}{5 - 5} = \dfrac{-5}{0} \boxed{\text{undefined}}$; line is $\boxed{\text{vertical}}$

13. $(5, -2), (1, 0)$
$m = \dfrac{0 - (-2)}{1 - 5} = \dfrac{2}{-4} = \boxed{-\dfrac{1}{2}}$; line $\boxed{\text{falls}}$

15. $(2, 0), (0, 8)$
$m = \dfrac{8 - 0}{0 - 2} = \dfrac{8}{-2} = \boxed{-4}$; line $\boxed{\text{falls}}$

17. $(5, 1), (-2, 1)$

$m = \dfrac{1-1}{-2-5} = \dfrac{0}{7} = \boxed{0}$; line is $\boxed{\text{horizontal}}$

19. $(-1, 2), (-1, 3)$

$m = \dfrac{3-2}{-1-(-1)} = \dfrac{1}{0}\ \boxed{\text{undefined}}$; line is $\boxed{\text{vertical}}$

21. $(1, 2), (3, 6)$

$m = \dfrac{6 \cdot 2}{3-1} = \dfrac{4}{2} = \boxed{2}$

23. $(3, 1), (6, -2)$

$m = \dfrac{-2-1}{6-3} = \dfrac{-3}{3} = \boxed{-1}$

25. $(-2, 2), (-2, -5)$

$m = \dfrac{-5-2}{-2-(-2)} = \dfrac{-7}{0} = \boxed{\text{undefined}}$

27. $(-4, -1), (-4, 5)$

$m = \dfrac{5-(-1)}{-4-(-4)} = \dfrac{6}{0} = \boxed{\text{undefined}}$

29. x-intercept: 6 $(6, 0)$

y-intercept: -2 $(0, -2)$

$m = \dfrac{-2-0}{0-6} = \dfrac{-2}{-6} = \boxed{\dfrac{1}{3}}$

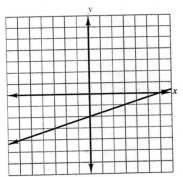

31. pitch $= \dfrac{\text{rise}}{\text{run}} = \dfrac{12-3}{21-6} = \dfrac{9}{15} = \boxed{\dfrac{3}{5}}$

33. pitch $= \dfrac{\text{rise}}{\text{run}}$: $\dfrac{4}{x} = \dfrac{1}{5}$

$\boxed{20} = x$

35. $(2, 1), (x, 4)$ $m = 3$

$3 = \dfrac{4-1}{x-2}$

$3(x-2) = 3$

$x-2 = 1$

$x = \boxed{3}$

37. $(2, 4), (-4, y)$ $m = -\dfrac{5}{2}$

$-\dfrac{5}{2} = \dfrac{y-4}{-4-2} = \dfrac{y-4}{-6}$

$-5(-6) = 2(y-4)$

$30 = 2y - 8$

$38 = 2y$

$\boxed{19} = y$

39. $(8, y), (4, 1)$ $m = -1$

$-1 = \dfrac{1-y}{4-8} = \dfrac{1-y}{-4}$

$-1(-4) = 1 - y$

$4 = 1 - y$

$y = \boxed{-3}$

41. $(4, -2), (3, y)$ $m = 0$

$$0 = \frac{y - (-2)}{3 - 4} = \frac{y + 2}{-1}$$

$$0 = y + 2$$

$$\boxed{-2} = y$$

43. L_1: $(0, 2), (3, 3)$ $m = \dfrac{1}{3}$

L_2: $m = 0$

L_3: $(-2, 0), (3, 1)$ $m = \dfrac{1}{5}$

45. y-intercept $= 4$; $m = 3$

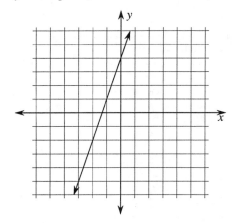

47. y-intercept $= -1$, $m = \dfrac{1}{2}$

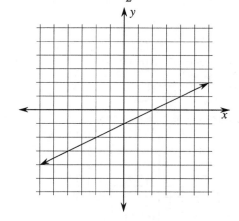

49. y-intercept $= 1$, $m = -\dfrac{1}{2}$

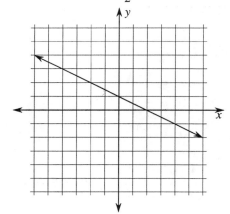

51. x-intercept $= 2$, $m = \dfrac{2}{3}$

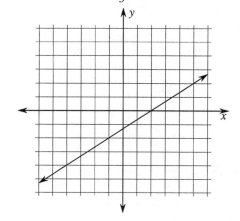

53. x-intercept $= 1$, $m = -\dfrac{3}{4}$

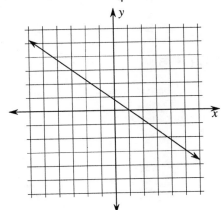

55. y-intercept $= -3$, $m = 0$

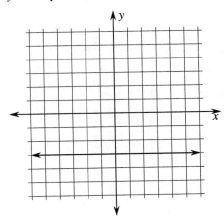

57. x-intercept $= 3$, slope is undefined

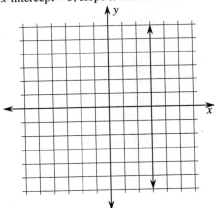

59. $(1, 3)$, no slope

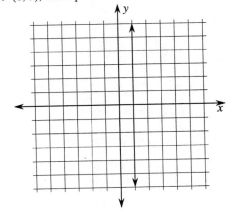

61. $(0, -3)$, $(-1, -5)$: $m = \dfrac{-5 - (-3)}{-1 - 0} = \dfrac{-5 + 3}{-1} = 2$

$(1, 3)$, $(-2, -3)$: $m = \dfrac{-3 - 3}{-2 - 1} = \dfrac{-6}{-3} = 2$

The slopes are the same.

Therefore, the lines are $\boxed{\text{parallel}}$.

63. $(4, 1)$, $(-2, -5)$: $m = \dfrac{-5 - 1}{-2 - 4} = \dfrac{-6}{-6} = 1$

$(1, 3)$, $(-5, 9)$: $m = \dfrac{9 - 3}{-5 - 1} = \dfrac{6}{-6} = -1$

$\boxed{\text{No}}$; the slopes are not equal

The lines are not parallel. They are perpendicular.

65. The slopes of the opposite sides are

$(-3, -3)$, $(2, -5)$ $\qquad m = \dfrac{-5 + 3}{2 + 3} = \dfrac{-2}{5}$

$(2, -5)$, $(5, -1)$ $\qquad m = \dfrac{-1 + 5}{5 - 2} = \dfrac{4}{3}$

$(5, -1)$, $(0, 1)$ $\qquad m = \dfrac{1 + 1}{0 - 5} = \dfrac{-2}{5}$

$(0, 1)$, $(-3, -3)$ $\qquad m = \dfrac{-3 - 1}{-3 - 0} = \dfrac{4}{3}$

The slopes of the opposite sides are equal. Since the opposite sides are parallel, the figure is a parallelogram.

67. a parallel line has the same slope: $(2, -3), (5, 2)$ $m = \dfrac{2-(-3)}{5-2} = \boxed{\dfrac{5}{3}}$

69. $(5, y), (1, 0)$ $m_1 = \dfrac{y-1}{5-1} = \dfrac{y-1}{4}$

$(2, 3), (-2, 1)$ $m_2 = \dfrac{1-3}{-2-2} = \dfrac{-2}{-4} = \dfrac{1}{2}$

The lines are parallel so $m_1 = m_2$

$$\dfrac{y-1}{4} = \dfrac{1}{2}$$
$$2y - 2 = 4$$
$$2y = 6$$
$$y = \boxed{3}$$

71. $(-4, 1), (4, 3)$ $m_1 = \dfrac{3-1}{4-(-4)} = \dfrac{2}{8} = \dfrac{1}{4}$

$(-4, 1), (-3, -3)$ $m_2 = \dfrac{-3-1}{-3-(-4)} = \dfrac{-4}{1} = -4$

Since the slopes are negative reciprocals $\dfrac{1}{4} = \dfrac{-1}{-4}$, the lines are perpendicular.

73. $(-4, -3), (5, 4)$ $m_1 = \dfrac{4-(-3)}{5-(-4)} = \dfrac{7}{9}$

$(-4, -3), (-3, -4)$ $m_2 = \dfrac{-4-(-3)}{-3-(-4)} = \dfrac{-1}{1} = -1$

$\boxed{\text{No}}$, the slopes are not equal or negative reciprocals. The lines are not perpendicular.

75. $(-5, 3), (-1, 4)$ $m = \dfrac{4-3}{-1+5} = \dfrac{1}{4}$

$(-1, 4), (1, -4)$ $m = \dfrac{-4-4}{1-(-1)} = \dfrac{-8}{2} = -4$

$(1, -4), (-5, 3)$ $m = \dfrac{3-(-4)}{-5-1} = \dfrac{7}{-6}$

Since $\dfrac{1}{4}(-4) = -1$, the lines ending at $(1, -4)$ are perpendicular, and the figure is a right triangle.

77. $(-1, y), (1, 0)$ $m_1 = \dfrac{y-0}{-1-1} = \dfrac{y}{-2}$

$(2, 3), (-2, 1)$ $m_2 = \dfrac{1-3}{-2-2} = \dfrac{-2}{-4} = \dfrac{1}{2}$

$$m_1 m_2 = -1$$
$$\dfrac{y}{-2}\left(\dfrac{1}{2}\right) = -1$$
$$y = \boxed{4}$$

79. $y = 2x + 4$

$m = \boxed{2}$

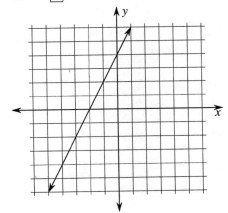

81. $y = -3x + 6$

$m = \boxed{-3}$

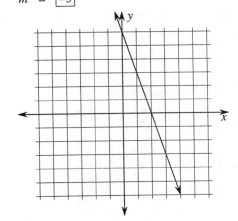

83. $(3, 1), (6, 3) \quad m_1 = \dfrac{3-1}{6 \cdot 3} = \dfrac{2}{3}$

$(6, 3), (9, 5) \quad m_2 = \dfrac{5-2}{9 \cdot 6} = \dfrac{2}{3}$

$m_1 = m_2 \qquad$ The lines are $\boxed{\text{collinear}}$.

85. $(0, -1), (4, -16) \quad m_1 = \dfrac{-16+1}{4-0} = \dfrac{-15}{4}$

$(4, -16), (-2, 7) \quad m_2 = \dfrac{7+16}{-2-4} = \dfrac{23}{-6}$

$m_1 \neq m_2 \qquad$ The lines are $\boxed{\text{not collinear}}$.

87. D is true.

89. $(5286, 4719), (-2754, 8243)$

$m = \dfrac{8243 - 4719}{-2754 - 5286} = \dfrac{3524}{-8040} \approx \boxed{-0.4383}$

Review Problems

93. $\qquad 0.05x = 36$

$\qquad\qquad x = \boxed{720}$

94. $\dfrac{45 \text{ liters acid}}{135 \text{ liters solution}} = \dfrac{x}{600 \text{ liter solution}}$

$135x = 45(600) \text{ liters acid}$

$x = \boxed{200 \text{ liters acid}}$

95. $\quad 1 + 4(3y + 3) = 2 - [2y - (y - 2)]$

$\qquad 1 + 12y + 12 = 2 - [2y - y + 2]$

$\qquad\quad 12y + 13 = 2 - [y + 2]$

$\qquad\qquad\quad 13 = 2 - y - 2$

$\qquad\qquad\quad 13 = -13y$

$\qquad\qquad\quad\; y = -1$

$\boxed{\{-1\}}$

Section 4.5 Equations of Lines

Problem Set 4.5, pp. 300-303

1. $y = 2x + 3$
$m = 2, (0, 3)$

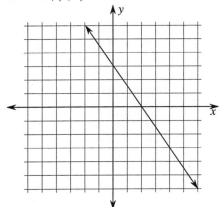

3. $y = -2x + 4$
$m = -2, (0, 4)$

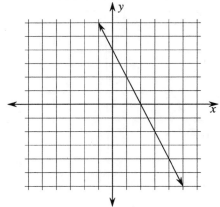

5. $y = \frac{1}{2}x + 3$
$m = \frac{1}{2}, (0, 3)$

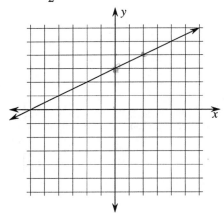

7. $y = \frac{2}{3}x - 4$
$m = \frac{2}{3}, (0, -4)$

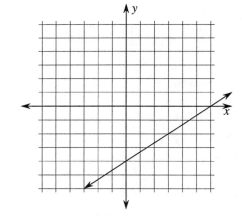

9.
$$y = -\frac{3}{4}x + 4$$
$$m = -\frac{3}{4}, (0, 4)$$

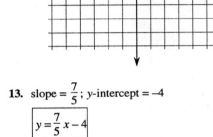

11. slope = 6; y-intercept = 5

$$\boxed{y = 6x + 5}$$

13. slope = $\frac{7}{5}$; y-intercept = -4

$$\boxed{y = \frac{7}{5}x - 4}$$

15. $y = 3x - 4$

slope = $\boxed{3}$; y-intercept = $\boxed{-4}$

17. $-5x + y = 7$
$$y = 5x + 7$$
slope = $\boxed{5}$; y-intercept = $\boxed{7}$

19. $8x + 4y = 8$
$$4y = -8x + 8$$
$$y = -2x + 2$$
slope = $\boxed{-2}$; y-intercept = $\boxed{2}$

21. $3x - 2y = 6$
$$2y = 3x - 6$$
$$y = \frac{3}{2}x - 3$$

slope = $\boxed{\frac{3}{2}}$; y-intercept = $\boxed{-3}$

23. $x - y = 0$
$$y = x$$
slope = $\boxed{1}$; y-intercept = $\boxed{0}$

25. $y = -2x + 5$
$$\boxed{2x + y = 5}$$

27. $y = 3x - 7$
$$\boxed{3x - y = 7}$$

29.
$$y = -\frac{1}{2}x + 3$$
$$\frac{1}{2}x + y = 3$$
$$\boxed{x + 2y = 6}$$

31.
$$y = \frac{1}{3}x + 2$$
$$\frac{1}{3}x - y = -2$$
$$\boxed{x - 3y = -6}$$

33.
$$y = \frac{3}{4}x + 4$$
$$\frac{3}{4}x - y = -4$$
$$\boxed{3x - 4y = -16}$$

35.
$$y = -\frac{2}{3}x - 1$$
$$\frac{2}{3}x + y = -1$$
$$\boxed{2x + 3y = -3}$$

37. slope = 2, passing through (3, 5)

point-slope form: $\boxed{y - 5 = 2(x - 3)}$

$$y - 5 \ = \ 2x - 6$$

slope-intercept form: $\boxed{y = 2x - 1}$

39. slope = 6, passing through (–2, 5)

$$y - 5 \ = \ 6[x - (-2)]$$

point-slope form: $\boxed{y - 5 = 6(x + 2)}$

$$y - 5 \ = \ 6x + 12$$

slope-intercept form: $\boxed{y = 6x + 17}$

41. slope = $\dfrac{1}{2}$, passing through the origin: (0, 0)

point-slope form: $\boxed{y - 0 = \dfrac{1}{2}(x - 0)}$

slope-intercept form: $\boxed{y = \dfrac{1}{2}x}$

43. slope = $-\dfrac{2}{3}$, passing through (6, –2)

point-slope form: $\boxed{y + 2 = -\dfrac{2}{3}(x - 6)}$

slope-intercept form: $\boxed{y = -\dfrac{2}{3}x + 2}$

45. passing through (1, 2) and (5, 10)

slope = $\dfrac{10 - 2}{5 - 1} = \dfrac{8}{4} = 2$

point-slope form: $\boxed{y - 2 = 2(x - 1)}$ or $\boxed{y - 10 = 2(x - 5)}$

$$y - 2 \ = \ 2x - 2$$

slope-intercept form: $\boxed{y = 2x}$

47. passing through (–3, 0) and (0, 3)

slope = $\dfrac{3 - 0}{0 + 3} = \dfrac{3}{3} = 1$

point-slope form: $\boxed{y - 0 = 1(x + 3)}$ or $\boxed{y - 3 = 1(x - 0)}$

slope-intercept form: $\boxed{y = x + 3}$

49. passing through (–3, 1) and (2, 4)

slope = $\dfrac{4 + 1}{2 + 3} = \dfrac{5}{5} = 1$

point-slope form: $\boxed{y + 1 = 1(x + 3)}$ or $\boxed{y - 4 = 1(x - 2)}$

slope-intercept form: $\boxed{y = x + 2}$

51. passing through (–3, –2) and (3, 6)

slope = $\dfrac{6 + 2}{3 + 3} = \dfrac{8}{6} = \dfrac{4}{3}$

point-slope form: $\boxed{y + 2 = \dfrac{4}{3}(x + 3)}$ or $\boxed{y - 6 = \dfrac{4}{3}(x - 3)}$

$$y + 2 \ = \ \dfrac{4}{3}x + 4$$

slope-intercept form: $\boxed{y = \dfrac{4}{3}x + 2}$

53. passing through (–3, –1) and (4, –1)

slope = $\dfrac{-1 + 1}{4 + 3} = \dfrac{0}{7} = 0$

point-slope form: $\boxed{y + 1 = 0(x + 3)}$ or $\boxed{y + 1 = 0(x - 4)}$

slope-intercept form: $\boxed{y = -1}$

55. passing through $(2, -1)$ and $\left(\dfrac{3}{4}, \dfrac{1}{3}\right)$

slope $= \dfrac{\dfrac{1}{3} + 1}{\dfrac{3}{4} - 2} = \dfrac{\dfrac{4}{3}}{-\dfrac{5}{4}} = -\dfrac{16}{15}$

point-slope form: $\boxed{y + 1 = -\dfrac{16}{15}(x - 2)}$ or $\boxed{y - \dfrac{1}{3} = -\dfrac{16}{15}\left(x - \dfrac{3}{4}\right)}$

$y + 1 \;=\; -\dfrac{16}{15}x + \dfrac{32}{15}$

slope-intercept form: $\boxed{y = -\dfrac{16}{15}x + \dfrac{17}{15}}$

57. passing through $(2, 4)$ with x-intercept $= -2$; $(-2, 0)$

slope $= \dfrac{0 - 4}{-2 - 2} = \dfrac{-4}{-4} = 1$

point-slope form: $\boxed{y - 4 = 1(x - 2)}$

slope-intercept form: $\boxed{y = x + 2}$

59. x-intercept $= -\dfrac{1}{2}$ and y-intercept $= 4$

$\left(-\dfrac{1}{2}, 0\right), (0, 4)$

slope $= \dfrac{4 - 0}{0 + \dfrac{1}{2}} = \dfrac{4}{\dfrac{1}{2}} = 8$

point-slope form: $\boxed{y - 0 = 8\left(x + \dfrac{1}{2}\right)}$ or $\boxed{y - 4 = 8(x - 0)}$

slope-intercpet form: $\boxed{y = 8x + 4}$

61. passing through $(6, 8)$ with y-intercept $= 12$: $(0, 12)$

slope $= \dfrac{12 - 8}{0 - 6} = \dfrac{4}{-6} = -\dfrac{2}{3}$

points-slope form: $\boxed{y - 8 = -\dfrac{2}{3}(x - 6)}$

$y - 8 \;=\; -\dfrac{2}{3}x + 4$

slope-intercept form: $\boxed{y = -\dfrac{2}{3}x + 12}$

63. passing through $(-1, 4)$ and parallel to the line whose equation is $y = 3x + 2$

$m = 3$

point-slope form: $\boxed{y - 4 = 3(x + 1)}$

slope-intercept form: $y - 4 = 3x + 3$

$\boxed{y = 3x + 7}$

standard form: $\boxed{3x - y = -7}$

65. passing through (–2, –6) and parallel to the line whose equation is $3x + y = 6$

$y = -3x + 6$ $m = -3$

point-slope form: $y - (-6) = -3[x - (-2)]$

$\boxed{y + 6 = -3(x + 2)}$

slope-intercept form: $y + 6 = -3x - 6$

$\boxed{y = -3x - 12}$

standard form: $\boxed{3x + y = -12}$

67. passing through (2, –6) and parallel to the line whose equation is $4x - 2y = 12$

$2y = 4x - 12$ $y = 2x - 6$ $m = 2$

point-slope form: $y - (-6) = 2(x - 2)$

$\boxed{y + 6 = 2(x - 2)}$

slope-intercept form: $y + 6 = 2x - 4$

$\boxed{y = 2x - 10}$

standard form: $\boxed{2x - y = 10}$

69. passing through (–1, 1) and perpendicular to a line with slope $\frac{1}{4}$

slope of line $= -\dfrac{1}{\frac{1}{4}} = -4$

point-slope form: $y - 1 = -4[x - (-1)]$

$\boxed{y - 1 = -4(x + 1)}$

slope-intercept form: $y - 1 = -4x - 4$

$\boxed{y = -4x - 3}$

standard form: $\boxed{4x + y = -3}$

71. x-intercept = 8 and perpendicular to a line with slope $= -\dfrac{1}{5}$

$m = -\dfrac{1}{-\frac{1}{5}} = 5; \ (8, 0)$

point-slope form: $\boxed{y - 0 = 5(x - 8)}$

slope-intercept form: $\boxed{y = 5x - 40}$

standard form: $\boxed{5x - y = 40}$

73. passing through (–2, 6) and perpendicular to a line whose equation is $y = \frac{1}{8}x + 3$

slope of line perpendicular $= \frac{1}{8}$, slope of line $= -\dfrac{1}{\frac{1}{8}} = -8$

point-slope form: $\boxed{y - 6 = -8(x + 2)}$

$y - 6 = -8x - 16$

slope-intercept form: $\boxed{y = -8x - 10}$

standard form: $\boxed{8x + y = -10}$

75. passing through (6, 4) and perpendicular to a line whose equation is $2x + y = 10$

$y = -2x + 10$ slope of line perpendicular $= -1$, slope of line $= \dfrac{-1}{-2} = \dfrac{1}{2}$

point-slope form: $\boxed{y - 4 = \dfrac{1}{2}(x - 6)}$

slope-intercept form: $y - 4 = \dfrac{1}{2}x - 3$

$$\boxed{y = \dfrac{1}{2}x + 1}$$

standard form: $\dfrac{1}{2}x - y = -1$

$$\boxed{x - 2y = -2}$$

77. Person 1 (10, 115), Person 2 (30, 125)

slope $= \dfrac{125 - 115}{30 - 10} = \dfrac{10}{20} = \dfrac{1}{2}$

$y - 115 \ = \ \dfrac{1}{2}(x - 10)$

$y - 115 \ = \ \dfrac{1}{2}x - 5$

$$\boxed{y = \dfrac{1}{2}x + 110}$$

80-year old man: $x = 80$

$y \ = \ \dfrac{1}{2}(80) + 110$

$y \ = \ 40 + 110$

$y \ = \ 150$

$\boxed{\text{blood pressure}, 150}$

79. Person 1 (8, 5), Person 2 (12, 7)

slope $= \dfrac{7 - 5}{12 - 8} = \dfrac{2}{4} = \dfrac{1}{2}$

$y - 5 \ = \ \dfrac{1}{2}(x - 8)$

$y - 5 \ = \ \dfrac{1}{2}x - 4$

$$\boxed{y = \dfrac{1}{2}x + 1}$$

14 hours in insolation: $x = 14$

$y \ = \ \dfrac{1}{2}(14) + 1$

$y \ = \ 7 + 1 = 8$

$\boxed{8 \text{ minutes}}$ to find the way through the maze

81. $y = 4x + 5$ and $y = 4x - \dfrac{1}{5}$

$m_1 = 4 \quad m_2 = 4$

The slopes are equal.

The lines are $\boxed{\text{parallel}}$.

83. $y = 5x + 2$ and $y = \dfrac{1}{5}x - 1$

$m_1 = 5 \qquad m_2 = \dfrac{1}{5}$

$m_1 m_2 = 5\left(\dfrac{1}{5}\right) = 1. \neq -1; \, m_1 \neq m_2$

The lines are $\boxed{\text{neither}}$ parallel nor perpendicular.

85. $y = 8x + 9$ and $y = -\dfrac{1}{8}x$

$m_1 = 8 \qquad m_2 = -\dfrac{1}{8}$

$m_1 m_2 = 8\left(-\dfrac{1}{8}\right) = -1$

The lines are $\boxed{\text{perpendicular}}$ since $m_1 m_2 = -1$.

87. $y = x + \dfrac{3}{2}$ and $y = -x + 5$

$m_1 = 1 \qquad m_2 = -1$

$m_1 m_2 = 1(-1) = -1$

The lines are $\boxed{\text{perpendicular}}$ since $m_1 m_2 = -1$.

89. $2x + 3y = 7$ and $y = -\dfrac{2}{3}x + 3$

$y = -\dfrac{2}{3}x + \dfrac{7}{3}$

$m_1 = -\dfrac{2}{3} \qquad m_2 = -\dfrac{2}{3} \qquad m_1 = m_2$

The lines are $\boxed{\text{parallel}}$ since the slopes are equal.

91. $x = 0$ and $y = 3$

m_1 is undefined $m_2 = 0$

The lines are $\boxed{\text{perpendicular}}$ since $x = 0$ is a vertical line and $y = 3$ is a horizontal line.

93. a. $(0, 32), (100, 212)$

$$\text{slope} = \frac{212 - 32}{100 - 0} = \frac{180}{100} = \frac{9}{5}$$

$$F - 32 = \frac{9}{5}(C - 0)$$

$$\boxed{F = \frac{9}{5} C + 32}$$

b. $C = 50°$

$$F = \frac{9}{5}(50) + 32$$

$$F = 90 + 32 = \boxed{122°}$$

95. $\boxed{B}$ is true;   **A.** point-slope form: $y - y_1 = m(x - x_1)$ *not* $mx + b$

   B. true; $y = x$ and $y = -x$
       $m_1 = 1$ and $m_2 = -1$
       The lines are perpendicular and pass through the origin.
   C. A line with no slope is a vertical line. A line with zero slope is a horizontal line. A vertical line and a horizontal line are perpendicular.
   D. $y = 5x$ has y-intercept of zero.

97. $\boxed{B}$ is true;   **A.** To increase the coefficient of x in $y = mx + b$ is to increase the *slope, not* the y-intercept.

   B. true; $\frac{x}{3} + \frac{y}{4} = 1$ has x-intercept of 3 and y-intercept of 4.
   C. $(0, 1), (4, 9)$
       $m = \frac{8}{4} = 2$ $y - 1 = 2(x - 0)$
                                $y = 2x + 1$
       $(3, 7):\ 7 = 6 + 1$
                  $(3, 7)$ is a point on the line
   D. Vertical lines have undefined slope and can be written in standard form, $x = b$.

Review Problems

101. $-3(7x + 12) - 5\ >\ 8(x - 7) + 15$
       $-21x - 36 - 5\ >\ 8x - 56 + 15$
           $-21x - 41\ >\ 8x - 41$
               $-29x\ >\ 0$
                   $x\ <\ 0$
           $\boxed{\{x \mid x < 0\}}$

102.   Let w equal the width.
       $w + 8$ equals the length.
$$
\begin{aligned}
2w + 2(w + 18) &= 292 \\
2w + 2w + 36 &= 292 \\
4w &= 256 \\
w &= 64 \\
w + 18 &= 82
\end{aligned}
$$
The length is $\boxed{82 \text{ meters}}$ and the width is $\boxed{64 \text{ meters}}$.

103. $40t + 50t\ =\ 360$
       $40t\ =\ 360$
         $t\ =\ 4$
   $\boxed{4 \text{ hours}}$

Section 4.6 Graphing Linear Inequalities in Two Variables

Problem Set 4.6, pp. 311-313

1. $x + y > 4$
 (2, 2): $2 + 2 \; > \; 4; 4 > 4$ false
 $\boxed{(3, 2)}$: $3 + 2 \; > \; 4; 5 > 4$ true   (3, 2) satisfies the inequality
 $\boxed{(-3, 8)}$: $-3 + 8 \; > \; 4; 5 > 4$ true   (−3, 8) satisfies the inequality

3. $2x + y \geq 5$
 $\boxed{(4, 0)}$: $8 + 0 \; \geq \; 5$ true   (4, 0) satisfies the inequality
 $\boxed{(1, 3)}$: $2 + 3 \; \geq \; 5$ true   (1, 3) satisfies the inequality
 (0, 0): $0 + 0 \; \geq \; 5$ false

5. $y \geq -2x + 4$
 $\boxed{(4, 0)}$: $0 \; \geq \; -8 + 4 = -4$ true   (4, 0) satisfies the inequality
 $\boxed{(1, 3)}$: $3 \; \geq \; -2 + 2 = 2$ true   (1, 3) satisfies the inequality
 (−2, −4): $-4 \; \geq \; 4 + 4 = 8$ false

7. $y > -2x + 1$
 $\boxed{(2, 3)}$: $3 \; > \; -4 + 1 = -3$ true   (2, 3) satisfies the inequality
 (0, 0): $0 \; > \; 0 + 1 = 1$ false
 $\boxed{(0, 5)}$: $5 \; > \; 0 + 1 = 1$ true   (0, 5) satisfies the inequality

9. $3x - 2y \leq 6$

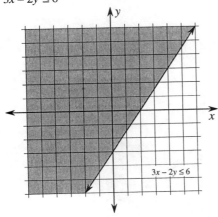

11. $4x + 3y > 12$

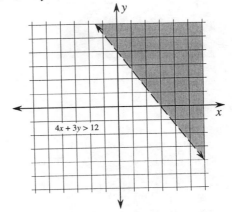

13. $5x - y < -10$

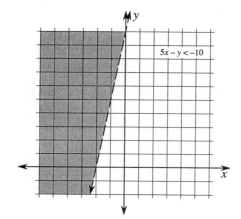

15. $2x - \frac{1}{2}y \geq 2$

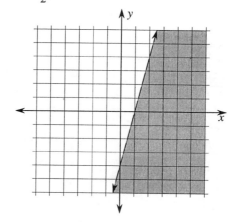

17. $x + y \leq 0$

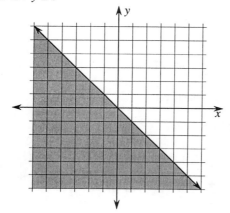

19. $y < 2x + 3$

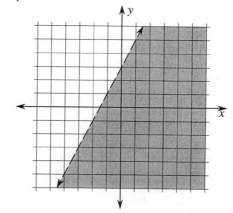

21. $y \geq 3x - 2$

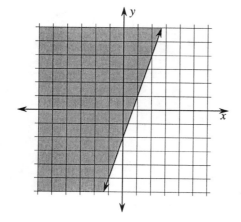

23. $y > \frac{1}{2}x + 2$

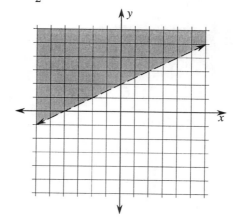

25. $y < \dfrac{3}{4}x - 3$

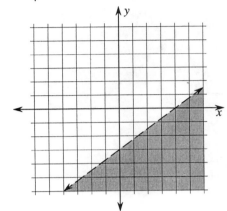

27. $y > 2x$

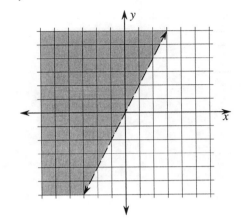

29. $y \le \dfrac{5}{4}x$

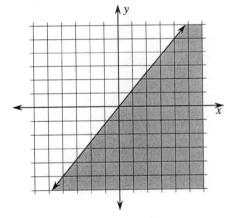

31. $y > -\dfrac{2}{3}x + 1$

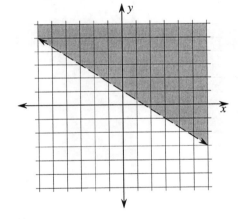

33. $x \ge 3$

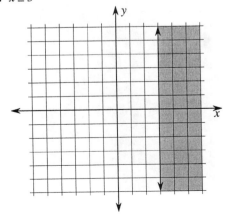

35. $x > -4$

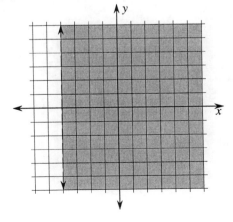

37. $y \le 2$

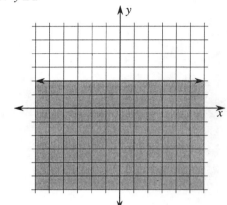

39. $y > -1$

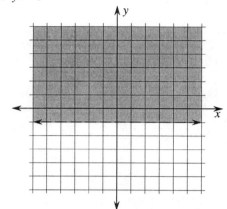

41. $x \ge 0$

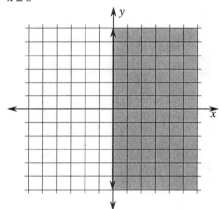

43. $x + y > 4$

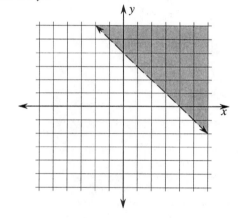

45. $3x - 5y \le 15$

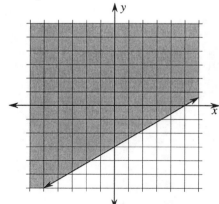

47. $y - 3x < 0$

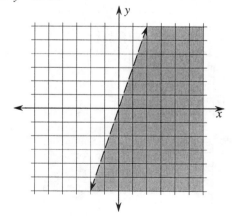

49. $2x + y \le 4$ $\boxed{f}$ **51.** $y < 1$ $\boxed{b}$

53. $y \ge 3x$ $\boxed{e}$ **55.** $\boxed{x \ge 4}$

57. $\boxed{y > 2x + 3}$ **59.** $\boxed{y \le -x + 4}$

61. a. $\boxed{75x}$ **63. a.** $\boxed{0.06x}$

    b. $\boxed{50y}$ **b.** $\boxed{0.08y}$

    c. $\boxed{75x + 50y > 300}$ **c.** $\boxed{0.06x + 0.08y \ge 1.2}$

    d.   **d.**

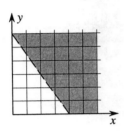   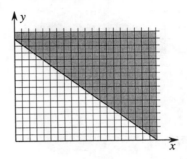

65. $x \ge 0$ $y \le 2$

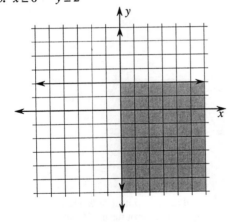

67. $\boxed{D}$ is true;   A. $(0, 3)$: $y > 2x - 3$
                          $-3 > 0 - 3$ false
             B.   The graph is above the boundary line $x = y + 1$, *not* below.
             C. $y \ge 4x$, a solid line is used, *not* a dashed line
             D.   true; The graph of $x < 4$ is the half-plane to the left of $x = 4$.

69. $3x - 8 \geq 5y + 7$

$\boxed{3x - 5y \geq 15}$

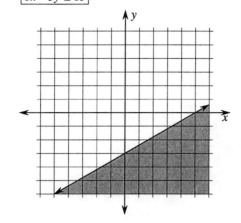

71. $y \leq -3x + 4$

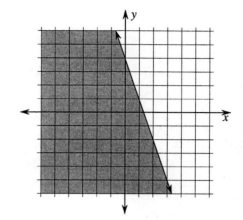

73. $y \geq \dfrac{1}{2}x + 4$

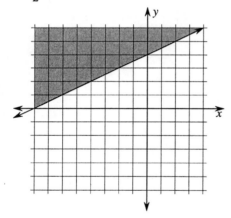

Review Problems

80. $3x - [2x - (3 - x)]$
$= 3x - (2x - 3 + x)$
$= 3x - (3x - 3)$
$= 3x - 3x + 3$
$= \boxed{3}$

81. Let x equal the amount of 50% alcohol solution.
$10 - x$ equals the amount of 75% alcohol solution.
$$0.50x + 0.75(10 - x) = 0.60(10)$$
$$0.50x + 7.5 - 0.75x = 6$$
$$-0.25x = -1.5$$
$$x = 6$$
$$10 - x = 4$$
$\boxed{\text{6 liters of 50\% alcohol, 4 liters of \textbf{75\% alcohol}}}$

82. $\dfrac{t}{10} + \dfrac{t}{15} + \dfrac{t}{12} = 1$ $t =$ time (in minutes) together

$$60\left(\dfrac{t}{10} + \dfrac{t}{15} + \dfrac{t}{12}\right) = 60(1)$$

$$6t + 4t + 5t = 60$$

$$15t = 60$$

$$t = 4$$

$\boxed{\text{4 minutes together}}$

Chapter 4 Review Problems

Chapter 4 Review Problems pp. 315-319

1. $3x - 2y = 12$

$x = 0$: $0 - 2y = 12$ $y = \boxed{-6}$ $(0, -6)$

$y = 0$: $3x - 0 = 12$ $x = \boxed{4}$ $(4, 0)$

$x = 2$: $6 - 2y = 12, -2y = 6, y = \boxed{-3}$ $(2, -3)$

$x = -4$: $-12 - 2y = 12, -2y = 24, y = \boxed{-12}$ $(-4, -12)$

$y = 6$: $3x - 12 = 12, 3x = 24, x = \boxed{8}$ $(8, 6)$

$y = -\dfrac{3}{2}$: $3x + 3 = 12, 3x = 9, x = \boxed{3}$ $\left(3, -\dfrac{3}{2}\right)$

2. $3x + 5y = 15$

$x = 0$: $0 + 5y = 15, y = \boxed{3}$

$y = 0$: $3x + 0 = 15, x = \boxed{5}$

$x = -5$: $-15 + 5y = 15, 5y = 30, y = \boxed{6}$

$y = -6$: $3x - 30 = 15, 3x = 45, x = \boxed{15}$

$x = 10$: $30 + 5y = 15, 5y = -15, y = \boxed{-3}$

x	0	5	-5	15	10
y	3	0	6	-6	-3

3. $y = \boxed{6}$ for all values of x.

x	-7	$\dfrac{2}{13}$	0	3000
y	6	6	6	6

4. $3x - y = 12$

$\boxed{(0, -12)}$: $0 + 12 = 12$ true

$(0, 4)$: $0 - 4 = 12$ false

$(-1, 15)$: $-3 - 15 = 12$ false

$\boxed{(-2, -18)}$: $-6 + 18 = 12$ true

5. $\boxed{5x + 2y = 25}$

$x = -3$: $-15 + 2y = 25$

$2y = 40$

$y = \boxed{20}$

6. $\boxed{y = 3x + 6}$

$y = -18$: $-18 = 3x + 6$

$\; -24 = 3x$

$\;\boxed{-8} = x$

7. III   **8.** IV   **9.** I   **10.** II

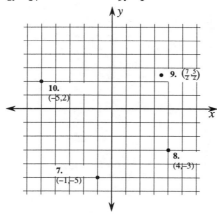

11. A (5, 6)   **12.** B (–2, 0)

13. C (–5, 2)   **14.** D (–4, 2)

15. E (0, –5)   **16.** F (3, –1)

17. $2x + y = 4$

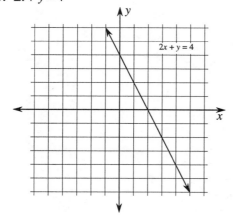

18. $3x - 2y = 12$

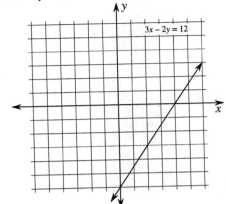

19. $3x = 6 - 2y$

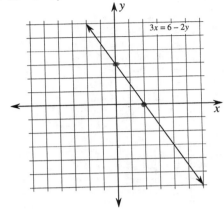

20. $3x - y = 0$

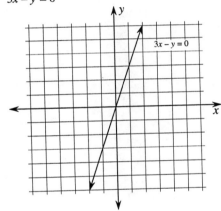

21. $x = 3$

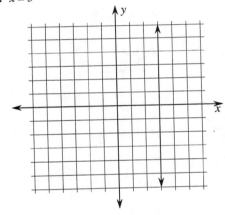

22. $2y = -10$

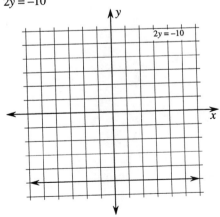

23. $x = 2y - 4$

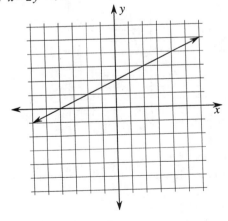

24. $y = x^2 - 3x - 4$

$x = -2$: $y = 4 + 6 - 4 = 6$
$x = -1$: $y = 1 + 3 - 4 = 0$
$x = 0$: $y = 0 - 0 - 4 = -4$
$x = 1$: $y = 1 - 3 - 4 = -6$
$x = 2$: $y = 4 - 6 - 4 = -6$
$x = 3$: $y = 9 - 9 - 4 = -4$
$x = 4$: $y = 16 - 12 - 4 = 0$
$x = 5$: $y = 25 - 15 - 4 = 6$

x	−2	−1	0	1	2	3	4	5
y	6	0	−4	−6	−6	−4	0	6

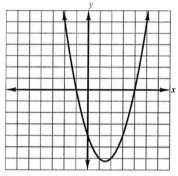

25. $y = x^3 - x$

$x = -2$: $y = -8 + 2 = -6$
$x = -1$: $y = -1 + 1 = 0$
$x = 0$: $y = 0 - 0 = 0$
$x = 1$: $y = 1 - 1 = 0$
$x = 2$: $y = 8 - 2 = 6$

x	−2	−1	0	1	2
y	−6	0	0	0	6

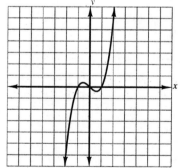

26. y is not a function

A vertical line drawn through $x = 0$ intersects the graph 2 times.

27. y is a function of x

Each vertical line that can be drawn intersects the graph only once.

28. y is a $\boxed{\text{function}}$ of x
Each vertical line that can be drawn intersects the graph only once.

29. y is $\boxed{\text{not a function of } x}$
A vertical line drawn through $x = 1$ intersects the graph 2 times.

30. y is $\boxed{\text{not a function}}$ of x
A vertical line drawn through $x = -5$ intersects the graph an infinite number of times.

31. y is a $\boxed{\text{function}}$ of x
Each vertical line that can be drawn intersects the graph only once.

32. y is $\boxed{\text{not a function}}$ of x
A vertical line drawn through $x = 2$ intersects the graph 2 times.

33. y is a $\boxed{\text{function}}$ of x
Each vertical line that can be drawn intersects the graph only once.

34. a. time: $\boxed{\text{5 P.M.}}$; minimum temperature: $\boxed{-4°\text{F}}$
 b. time: $\boxed{\text{8 P.M.}}$; maximum temperature: $\boxed{16°\text{F}}$
 c. x-intercepts: $\boxed{4 \text{ and } 6}$
 The Fahrenhiet temperature is $0°$ at 4 P.M. and 6 P.M.
 d. y-intercept: $\boxed{12}$
 At 12 noon, the Fahrenheit temperature is $12°$.
 e. For each value of time, there is only one value of temperature.
 f. 7 P.M.: $4°\text{F}$; 8 P.M.: $16°\text{F}$
 percent increase $= \dfrac{\text{amount of increase}}{\text{original amount}} \times 100\% = \dfrac{16-4}{4} \times 100\% = \dfrac{12}{4} \times 100\% = \boxed{300\%}$

35. $f(x) = 5x - 3$
 $f(-2) = 5(-2) - 3 = -10 - 3 = \boxed{-13}$

36. $f(x) = 4x^2 - x + 5$
 $f(3) = 4(-3)^2 - (-3) + 5$
 $= 4(9) + 3 + 5$
 $= 36 + 8 = \boxed{44}$

37. $f(x) = 17$
 $f(5) = \boxed{17}$

38. $(3, 2), (5, 1)$
 $m = \dfrac{1-2}{5-3} = \boxed{-\dfrac{1}{2}}$, line $\boxed{\text{falls}}$

39. $(-1, -2), (-3, 4)$
 $m = \dfrac{-4+2}{-3+1} = \dfrac{-2}{-2} = \boxed{1}$; line $\boxed{\text{rises}}$

40. $\left(-3, \dfrac{1}{4}\right), \left(6, \dfrac{1}{4}\right)$
 $m = \dfrac{\frac{1}{4} - \frac{1}{4}}{6+3} = \dfrac{0}{9} = \boxed{0}$; line is $\boxed{\text{horizontal}}$

41. $(-2, 5), (-2, 10)$

$m = \dfrac{10 - 5}{-2 + 2} = \dfrac{5}{0}$ is $\boxed{\text{undefined}}$; line is $\boxed{\text{vertical}}$

42. x-intercept = 2, $(2, 0)$

y-intercept = 4, $(0, 4)$

$m = \dfrac{4 - 0}{0 - 2} = \dfrac{4}{-2} = \boxed{-2}$

43. $(3, 5), (-2, y) \quad m = -4$

$-4 = \dfrac{y - 5}{-2 - 3} = \dfrac{y - 5}{-5}$

$20 = y - 5$

$\boxed{25} = y$

44. $(2, K), (5, 3K) \quad m = \dfrac{1}{2}$

$\dfrac{1}{2} = \dfrac{3K - K}{5 - 2} = \dfrac{2K}{3}$

$3 = 4K$

$\boxed{\dfrac{3}{4}} = K$

45. $(6, -1), (8, -2)$

$m_1 = \dfrac{-2 + 1}{8 - 6} = \dfrac{-1}{2}$

$(-4, 2), (6, -3)$

$m_2 = \dfrac{-3 - 2}{6 + 4} = -\dfrac{5}{10} = -\dfrac{1}{2}$

$m_1 = m_2$

The lines are parallel since the slopes are equal.

46. $(1, 2), (4, 7)$

$m_1 = \dfrac{7 - 2}{4 - 1} = \dfrac{5}{3}$

$(3, 3), (8, 0)$

$m_2 = \dfrac{0 - 3}{8 - 3} = -\dfrac{3}{5}$

$m_1 m_2 = \dfrac{5}{3}\left(-\dfrac{3}{5}\right) = -1$

The lines are perpendicular since $m_1 m_2 = -1$.

47. $y = 3x + 2$

$m = 3, b = 2$

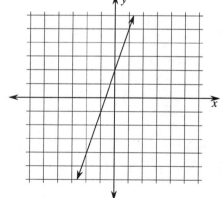

48. $y = \dfrac{2}{3}x - 1$

$m = \dfrac{2}{3}, b = -1$

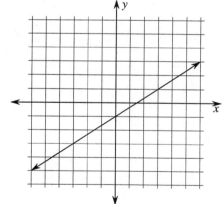

49. $y = -\dfrac{1}{4}x + 3$

$m = -\dfrac{1}{4}, \ b = 3$

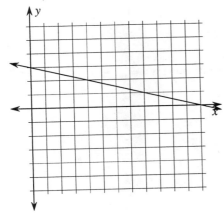

50. slope = 5, y-intercept = -8

$\boxed{y = 5x - 8}$

51. $\begin{aligned} 4x + 8y &= 16 \\ 8y &= -4x + 16 \\ y &= -\frac{1}{2}x + 2 \end{aligned}$

slope = $\boxed{m = -\dfrac{1}{2}}$

y-intercept = $\boxed{b = 2}$

52. $y = -\dfrac{1}{3}x + 5$

$\dfrac{1}{3}x + y = 5$

$\boxed{x + 3y = 15}$

53. slope = 6 passing through $(-4, 7)$

point-slope: $\boxed{y - 7 = 6(x + 4)}$

slope-intercept: $y - 7 = 6x + 24$

$\boxed{y = 6x + 31}$

standard: $\boxed{6x - y = -31}$

54. passing through $(3, 4)$ and $(2, 1)$

$m = \dfrac{1 - 4}{2 - 3} = \dfrac{-3}{-1} = 3$

point-slope: $\boxed{y - 4 = 3(x - 3)}$ or $\boxed{y - 1 = 3(x - 2)}$

slope-intercept: $y - 4 = 3x - 9$

$\boxed{y = 3x - 5}$

standard: $\boxed{3x - y = 5}$

55. passing through $(-2, -3)$ and $(4, -1)$

$m = \dfrac{-1 + 3}{4 + 2} = \dfrac{2}{6} = \dfrac{1}{3}$

point-slope: $\boxed{y + 3 = \dfrac{1}{3}(x + 2)}$ or $\boxed{y + 1 = \dfrac{1}{3}(x - 4)}$

slope-intercept: $y + 3 = \dfrac{1}{3}x + \dfrac{2}{3}$

$\boxed{y = \dfrac{1}{3}x - \dfrac{7}{3}}$

standard: $\dfrac{1}{3}x - y = \dfrac{7}{3}$

$\boxed{x - 3y = 7}$

56. passing through (–4, 6) and parallel to the line whose equation is $y = -9x + 2$

$m = -9$

point-slope: $\boxed{y - 6 = -9(x + 4)}$

slope-intercept: $y - 6 = -9x - 36$

$\boxed{y = -9x - 30}$

standard: $\boxed{9x + y = -30}$

57. passing through (1, –5) and perpendicular to the line whose equation is $y = \dfrac{1}{4}x - 3$

slope of line perpendicular $= \dfrac{1}{4}$

slope of line $= -\dfrac{1}{\frac{1}{4}} = -4$

point-slope: $\boxed{y + 5 = -4(x - 1)}$

slope-intercept: $y + 5 = -4x + 4$

$\boxed{y = -4x - 1}$

standard: $\boxed{4x + y = -1}$

58. having x-intercept $= -4$ and parallel to the line whose equation is $5x + y = 11$

$y = -5x + 11 \qquad m = -5 \qquad (-4, 0)$

point-slope: $\boxed{y - 0 = -5(x + 4)}$

slope-intercept: $\boxed{y = -5x - 20}$

standard: $\boxed{5x + y = -20}$

59. having a y-intercept $= 3$ and perpendicular to the line whose equation is $4x - 8y = 16$

$8y = 4x - 16 \qquad\qquad y\text{-intercept} = 3: (0, 3)$

$y = \dfrac{1}{2}x - 2$

slope of line perpendicular $= \dfrac{1}{2}$

slope of line $= -\dfrac{1}{\frac{1}{2}} = -2$

point-slope: $\boxed{y - 3 = -2(x - 0)}$

slope-intercept: $\boxed{y = -2x + 3}$

standard: $\boxed{2x + y = 3}$

60. Person 1: (31, 61)

Person 2: (38, 75)

$m = \dfrac{75 - 61}{38 - 31} = \dfrac{14}{7} = 2$

$y - 61 = 2(x - 31)$ or $y - 75 = 2(x - 38)$

$y - 61 = 2x - 62$

$\boxed{y = 2x - 1}$

2-year-old child whose height is 36 inches: $x = 36$

$y = 2(36) - 1 = 71$ adult height: $\boxed{71 \text{ inches}}$

61. $3x - 4y > 7$

$(0, 0)$:	$0 - 0$	> 7	false
$(-2, -1)$:	$-6 + 4$	> 7	false
$(-2, -5)$:	$-6 + 20$	> 7	true
$(-3, 4)$:	$-9 - 16$	> 7	false
$(3, -6)$:	$9 + 24$	> 7	true

$(-2, -5)$ satisfies inequality

$(3, -6)$ satisfies inequality

62. $x - 2y > 6$

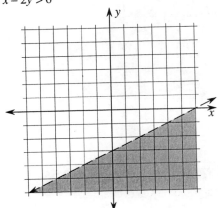

63. $4x - 6y \leq 12$

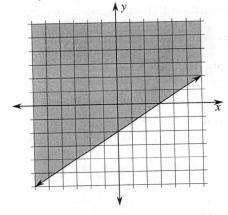

64. $x + 2y \leq 0$

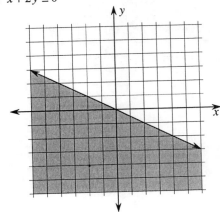

65. $y > 3x + 2$

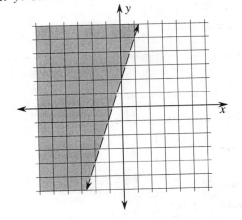

66. $y \leq \frac{1}{3}x + 2$

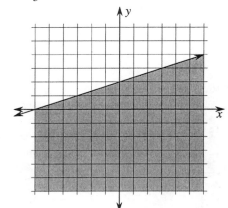

67. $y < -\frac{1}{2}x$

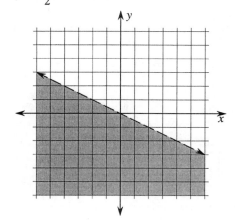

68. $x < 4$

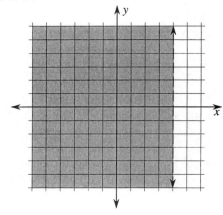

69. $y \geq -2$

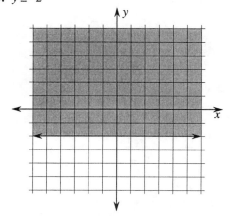

70. $x - 2y \leq 4$

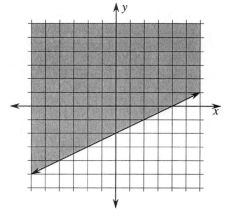

71. a. $3x$
  b. $5y$
  c. $3x + 5y \le 15$
  d.

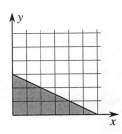

Cumulative Review Problems (Chapters 1-4)

Cumulative Review, pp. 319-320

1. $\dfrac{5(-3) - 3(-4)}{5(-10) + 2}$

$= \dfrac{-15 + 12}{-50 + 2} = \dfrac{-3}{-48} = \boxed{\dfrac{1}{16}}$

2. $3(y + 1) + 11 = 16 + 5y$

$\quad 3y + 3 + 11 = 16 + 5y$

$\quad\quad 3y + 14 = 16 + 5y$

$\quad\quad\quad -2 = 2y$

$\quad\quad\quad -1 = y$

$\boxed{\{-1\}}$

3. $\dfrac{1}{4}y + \dfrac{2}{3}y = \dfrac{1}{6}$

$12\left(\dfrac{1}{4}y + \dfrac{2}{3}y\right) = 12\left(\dfrac{1}{6}\right)$

$\quad 3y + 8y = 2$

$\quad\quad 11y = 2$

$\quad\quad\quad y = \dfrac{2}{11}$

$\boxed{\left\{\dfrac{2}{11}\right\}}$

4. $-7(2y + 1) > 4(3 - y) + 1$

$\quad -14y - 7 > 12 - 4y + 1$

$\quad\quad -10y > 20$

$\quad\quad\quad y < -2$

$\boxed{\{y \mid y < -2\}}$

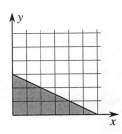

5. Let x equal the price before reduction.
  $x - 0.35x = 185.25$
  $\quad 0.65x = 185.25$
  $\quad\quad x = 285$

$\boxed{\$285}$

6. Let x equal the number of years ago when the rug was twice as old as the dresser.
  $35 - x = 2(25 - x)$
  $35 - x = 50 - 2x$
  $\quad\quad x = 15$

$\boxed{\text{15 years ago}}$

7. $-8 < 3y - 5 \le 7$
  $-3 < 3y \le 12$
  $-1 < y \le 4$
  $\boxed{\{y \mid -1 < y \le 4\}}$

8. Let x equal the number of 4-room apartments.
$2x$ equals the number of 5-room apartments.
$2x + 200$ equals the number of 6-room apartments.

$$\begin{aligned} x + 2x + (2x + 200) &= 1550 \\ 5x &= 1350 \\ x &= 270 \\ 2x &= 540 \\ 2x + 200 &= 740 \end{aligned}$$

270 four-room apartments, 540 five-room apartments, 740 six-room apartments

9.
$$\begin{aligned} 3x + 8 &= 7x - 24 \\ 32 &= 4x \\ 8 &= x \end{aligned}$$

$(3x + 8)° = (3 \cdot 8 + 8)° = (24 + 8)° = \boxed{32°}$

$(7x - 24)° = (7 \cdot 8 - 24)° = (56 - 24)° = \boxed{32°}$

10. Let x equal the number of hours plumber worked.
$$\begin{aligned} 18 + 35x &= 228 \\ 35x &= 210 \\ x &= 6 \end{aligned}$$

$\boxed{6 \text{ hours}}$

11. Let x equal the running speed of a cheetah.
$x - 29$ equals the running speed of Robert Hayes.
$$\begin{aligned} 3x + 5 &= 242 \\ 3x &= 237 \end{aligned}$$
Solution: $x = 79$
sum of average running speeds exceeds solution by 4:
$$\begin{aligned} x + (x - 29) &= 79 + 4 \\ 2x &= 112 \\ x &= 56 \\ x - 29 &= 27 \end{aligned}$$

Robert Haye's running speed was $\boxed{27 \text{ miles per hour}}$.

12.
$$\begin{aligned} \frac{3y + 5}{y - 2} &= \frac{1}{4} \\ 4(3y + 5) &= y - 2 \\ 12y + 20 &= y - 2 \\ 11y &= -22 \\ y &= -2 \end{aligned}$$

$\boxed{\{-2\}}$

13. Let x equal the measure of the angle.

$90 - x$ equals the measure of its complement.
$180 - x$ equals the measure of its supplement.
$$\begin{aligned} (90 - x) + (180 - x) &= 114 \\ 270 - 2x &= 114 \\ 156 &= 2x \\ 78 &= x \end{aligned}$$
The angle measures $\boxed{78°}$.

14.
$$\begin{aligned} f(x) &= x^2 - 5x - 2 \\ f(-1) &= (-1)^2 - 5(-1) - 2 = 1 + 5 - 2 = 4 \\ f(-2) &= (-2)^2 - 5(-2) - 2 = 4 + 10 - 2 = 12 \\ f(-1) - f(-2) &= 4 - 12 = \boxed{-8} \end{aligned}$$

15. $y = -3x + 2$
   $m = -3,\ b = 2$

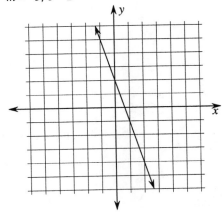

16. $\dfrac{3}{\$0.87} = \dfrac{5}{x}$

$3x = 5(\$0.87)$

$x = \boxed{\$1.45}$

17. Let x equal the length of each side of the square.
   $x + 6$ equals the length of each side of the equilateral triangle.
   perimeter of square = perimeter of triangle.

$$4x = 3(x + 6)$$
$$4x = 3x + 18$$
$$x = 18$$
$$x + 6 = 24$$

length of each side of the triangle: $\boxed{24 \text{ decimeters}}$

18. $0.76z + 0.80(11 - z) = 0.45(20)$
   $0.76z + 8.8 - 0.80z = 9$
   $-0.04z = 0.2$

$x = -5.0$

$\boxed{\{-5.0\}}$

19.
$$P = 2L + 2W$$
$$2L + 2W = P$$
$$2L = P - 2W$$

$$\boxed{L = \dfrac{P - 2W}{2} \text{ or } \dfrac{P}{2} - W}$$

20. $(1, 3),\ (3, 5)$

$m = \dfrac{5 - 3}{3 - 1} = \dfrac{2}{2} = 1$

point-slope: $\boxed{y - 3 = 1(x - 1) \text{ or } y - 5 = 1(x - 3)}$

slope-intercept: $y - 3 = x - 1$

$\boxed{y = x + 2}$

standard: $\boxed{x - y = -2}$

21. $3x - 4y > 12$

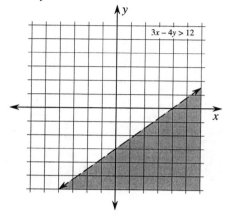

22. $\begin{aligned} 8y - 2(3 - 4y) &= 4(4y - 1) - 2 \\ 8y - 6 + 8y &= 16y - 4 - 2 \\ 16y - 6 &= 16y - 6 \\ -6 &= -6 \end{aligned}$

true for all real numbers

$\boxed{\{x \mid x \in R\}}$

23. Let x equal the number of nickels.
$x + 6$ equals the number of dimes.
$2x$ equals the number of quarters.

$\begin{aligned} 0.05x + 0.10(x + 6) + 0.25(2x) &= 10.35 \\ 0.05x + 0.10x + 0.60 + 0.50x &= 10.35 \\ 0.65x &= 9.75 \\ x &= 15 \\ x + 6 &= 21 \\ 2x &= 30 \end{aligned}$

$\boxed{15 \text{ nickels}, 21 \text{ dimes}, 30 \text{quarters}}$

24. $\dfrac{12 \cdot 2}{6} = 4$ $\boxed{x = 12, y = 2, z = 6}$

25. $\begin{aligned} A &= \frac{1}{2}bh \\ h &= 20 \text{ yards}, A = 150 \text{ square yards} \\ 150 &= \frac{1}{2}b\,(20) \\ 150 &= 10b \\ 15 &= b \end{aligned}$

base, $\boxed{15 \text{ yards}}$

26. $y = x^2 + 2x + 2$

$x = -3$: $y = (-3)^2 + 2(-3) + 2 = 9 - 6 + 2 = 5$
$x = -2$: $y = (-2)^2 + 2(-2) + 2 = 4 - 4 + 2 = 2$
$x = -1$: $y = (-1)^2 + 2(-1) + 2 = 1 - 2 + 2 = 1$
$x = 0$: $y = 0 + 0 + 2 = 2$
$x = 1$: $y = 1^2 + 2(1) + 2 = 1 + 2 + 2 = 5$
$x = 2$: $y = 2^2 + 2(2) + 2 = 4 + 4 + 2 = 10$

x	-3	-2	-1	0	1	2
y	5	2	1	2	5	10

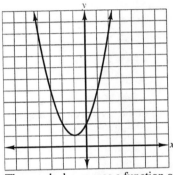

The graph shows y as a function of x since every x-value has only one y-value.

27. width of box: $12 - 2x$
length of box: $16 - 2x$

$16 - 2x = 3(12 - 2x)$
$16 - 2x = 36 - 6x$
$4x = 20$
$x = 5$

5 feet $\times$ 5 feet square

28.
$40t + 60t = 350$
$100t = 350$
$t = 3.5$

3.5 hours

29. Let x equal the amount invested at 12%.

$10,000 + x$ equals the amount invested at 14%

$0.12x + 0.14(10,000 + x) = 6210$
$0.12x + 1400 + 0.14x = 6210$
$0.26x = 4810$
$x = 18,500$

$10,000 + x = 28,500$

$18,500 at 12%; $28,500 at 14%

30. t equals the time (in hours) together

$\dfrac{t}{3} + \dfrac{t}{5} = 1$

$15\left(\dfrac{t}{3} + \dfrac{t}{5}\right) = 15(1)$
$5t + 3t = 15$
$8t = 15$
$t = \dfrac{15}{8} = 1\dfrac{7}{8}$

time together: $1\dfrac{7}{8}$ hours

Chapter 5   Systems of Linear Equations and Inequalities

Section 5.1 Solving Systems of Linear Equations by Graphing

Problem Set 5.1, pp. 329-332

1. (2, 3): $x + 3y = 11$ $x - 5y = -13$
    $2 + 3(3) = 11$ $2 - 5(3) = -13$
    $2 + 9 = 11$ True $2 - 15 = -13$
                            $-13 = -13$ True

    Since the ordered pair (2, 3) satisfies both equations, it is a $\boxed{\text{solution}}$ of the given system of equations.

3. (–3, –1): $5x - 11y = -4$ $6x - 8y = -10$
    $5(-3) - 11(-1) = -4$ $6(-3) - 8(-1) = -10$
    $-15 + 11 = -4$ $-18 + 8 = -10$
    $-4 = -4$ True $-10 = -10$ True

    (–3, –1) is the $\boxed{\text{solution}}$ of the given system.

5. (2, 5): $2x + 3y = 17$ $x + 4y = 16$
    $2(2) + 3(5) = 17$ $2 + 4(5) = 16$
    $4 + 15 = 17$ $2 + 20 = 16$
    $19 = 17$ False $22 = 16$ False

    Since (2, 5) fails to satisfy both equations, it is $\boxed{\text{not a solution}}$ of the given system.

7. $\left(\frac{1}{3}, 1\right)$: $6x - 9y = -7$ $9x + 5y = 8$

    $6\left(\frac{1}{3}\right) - 9(1) = -7$ $9\left(\frac{1}{3}\right) + 5(1) = 8$

    $2 - 9 = -7$ $3 + 5 = 8$
    $-7 = -7$ True $8 = 8$ True

    $\left(\frac{1}{3}, 1\right)$ is the $\boxed{\text{solution}}$ of the given system.

9. (8, 5): $5x - 4y = 20$ $3y = 2x + 1$
    $5(8) - 4(5) = 20$ $3(5) = 2(8) + 1$
    $40 - 20 = 20$ $15 = 16 + 1$
    $20 = 20$ True $15 = 17$ False

    (8, 5) fails to satisfy both equations; it is $\boxed{\text{not a solution}}$ of the given system.

11. (0, 5): $\frac{3}{5}x + \frac{2}{5}y = 2$ $y = 5$

    $\frac{3}{5}(0) + \frac{2}{5}(5) = 2$ $5 = 5$ True

    $0 + 2 = 2$
    $2 = 2$ True

    (0, 5) is the $\boxed{\text{solution}}$ of the given system.

13. $x + y = 6$
$x - y = 2$
solution: $\boxed{\{(4, 2)\}}$

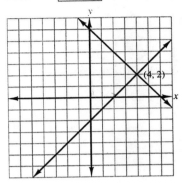

15. $x + y = 1$
$y - x = 3$
solution: $\boxed{\{(-1, 2)\}}$

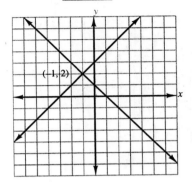

17. $3x + y = 3$
$6x + 2y = 12$
solution: $\boxed{\varnothing; \text{ inconsistent}}$

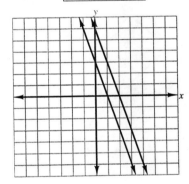

19. $2x - 3y = 6$
$4x + 3y = 12$
solution: $\boxed{\{(3, 0)\}}$

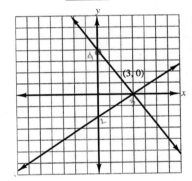

21. $x + y = 5$
$-x - y = -6$
solution: $\boxed{\varnothing; \text{ inconsistent}}$

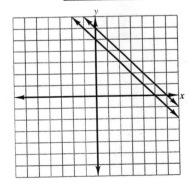

23. $x - y = 2$
$3x - 3y = -6$
solution: $\boxed{\varnothing; \text{ inconsistent}}$

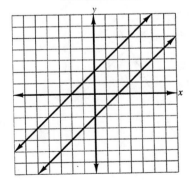

25. $4x + y = 4$
  $3x - y = 3$
  solution: $\boxed{\{(1, 0)\}}$

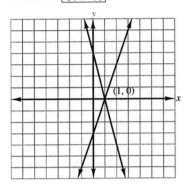

27. $x + y = 4$
  $x = -2$
  solution: $\boxed{\{(-2, 6)\}}$

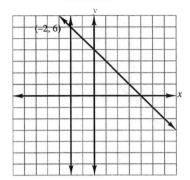

29. $x = -3$
  $y = 5$
  solution: $\boxed{\{(-3, 5)\}}$

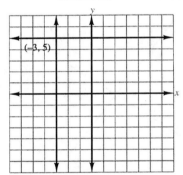

31. $y = x + 5$
  $y = -x + 3$
  solution: $\boxed{\{(-1, 4)\}}$

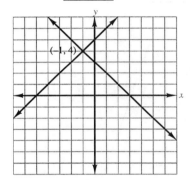

33. $y = 2x$
  $y = -x + 6$
  solution: $\boxed{\{(2, 4)\}}$

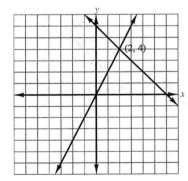

35. $y = 3x - 4$
  $y = -2x + 1$
  solution: $\boxed{\{(1, -1)\}}$

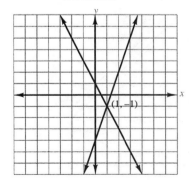

37. $y = 2x - 1$
$\ y = 2x + 1$
solution: $\boxed{\varnothing;\ \text{inconsistent}}$

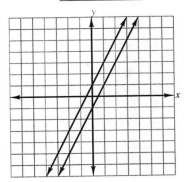

39. $x - y = 0$
$\ \ 2x = 2y$
solution: $\boxed{\{(x, y)\ \mid\ x - y = 0\};\ \text{dependent}}$

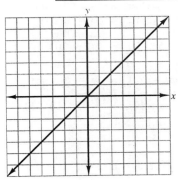

41. $y = 2x - 1$
$\ x - 2y = -4$
solution: $\boxed{\{(2, 3)\}}$

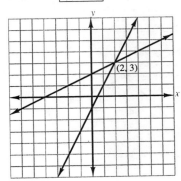

43. $y = \dfrac{1}{2}x - 1$
$\ x - y = -1$
solution: $\boxed{\{(-4, -3)\}}$

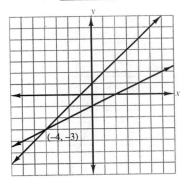

45. $\ x = 2$
$\ x = -1$
solution: $\boxed{\varnothing;\ \text{inconsistent}}$

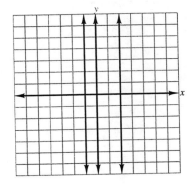

47. $\boxed{\text{B}}$ is true.

49. a. The artist must sell 4 pieces to break even. The graphs intersect at $x = 4$.

   b. $x > 4$

   c.   Revenue $-$ Cost $= 9x - (20 + 4x)$
                         Profit $= 9x - 20 - 4x$
                                      $= 5x - 20$
             $x = 2$: $5x - 20 = 5(2) - 20 = 10 - 20 = -10$
             artist's loss: $10

   d.   Revenue $-$ Cost $= 9x - (20 + 4x)$
                                      $= 5x - 20$
             $x = 10$; $= 5(10) - 20 = 30$
             artist's profit: $30

51. a. $40
   b.   supply exceeds demand

53. $y = 2x + 2$
        $y = -2x + 6$
        solution: $\{(1, 4)\}$

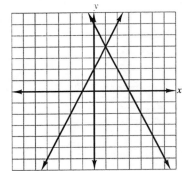

55. $x + 2y = 2$
        $x - 2y = 2$
        solution: $\{(2, 0)\}$

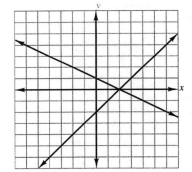

Review Problems

61. $3(y - 4) - (y + 7) - 2(3y - 6)$
         $= 3y - 12 - y - 7 - 6y + 12$
         $= \boxed{-4y - 7}$

62. $6y - 2(y + 4) - 2y = -4(y - 1)$
         $6y - 2y - 8 - 2y = -4y + 4$
                  $2y - 8 = -4y + 4$
                       $6y = 12$
                        $y = 2$
                      $\{2\}$

63. $\dfrac{5 \text{ people}}{13 \text{ kilograms}} = \dfrac{700{,}000 \text{ people}}{x}$
              $5x = 13(700{,}000) \text{ kilograms}$
               $x = \boxed{1{,}820{,}000 \text{ kilograms}}$

Section 5.2 Solving Systems of Linear Equations by the Addition (Elimination) Method

Problem Set 5.2, pp. 339-341

1.
$$
\begin{aligned}
x + y &= 1 \\
\underline{x - y} &\underline{= 3} \\
2x &= 4 \qquad \text{add to eliminate } y \\
x &= 2 \\
2 + y &= 1 \qquad \text{substitute } x = 2 \text{ into either equation and solve for } y \\
y &= -1
\end{aligned}
$$
$$\boxed{\{(2, -1)\}}$$

3.
$$
\begin{aligned}
2x + 3y &= 6 \\
\underline{2x - 3y} &\underline{= 6} \\
4x &= 12 \\
x &= 3 \\
2(3) + 3y &= 6 \\
3y &= 0 \\
y &= 0
\end{aligned}
$$
$$\boxed{\{(3, 0)\}}$$

5.
$$
\begin{aligned}
x + 2y &= 7 \\
\underline{-x + 3y} &\underline{= 18} \\
5y &= 25 \\
y &= 5 \\
x + 2(5) &= 7 \\
x + 10 &= 7 \\
x &= -3
\end{aligned}
$$
$$\boxed{\{(-3, 5)\}}$$

7.
$$
\begin{aligned}
5x - y &= 9 \\
\underline{-5x + 2y} &\underline{= -8} \\
y &= 1 \\
5x - 1 &= 9 \\
5x &= 10 \\
x &= 2
\end{aligned}
$$
$$\boxed{\{(2, 1)\}}$$

9.
$$
\begin{array}{ll}
\begin{aligned}
x + 2y &= 2 \\
\underline{-4x + 3y} &\underline{= 25}
\end{aligned}
&
\begin{array}{l}
\xrightarrow{(\times 4)} \\
\xrightarrow{\text{(no change)}}
\end{array}
\begin{aligned}
4x + 8y &= 8 \\
\underline{-4x + 3y} &\underline{= 25} \\
11y &= 33 \\
y &= 3 \\
x + 2(3) &= 2 \\
x + 6 &= 2 \\
x &= -4
\end{aligned}
\end{array}
$$
$$\boxed{\{(-4, 3)\}}$$

11.
$$
\begin{array}{ll}
\begin{aligned}
2x - 7y &= 2 \\
\underline{3x + y} &\underline{= -20}
\end{aligned}
&
\begin{array}{l}
\xrightarrow{\text{(no change)}} \\
\xrightarrow{(\times 7)}
\end{array}
\begin{aligned}
2x - 7y &= 2 \\
\underline{21x + 7y} &\underline{= -140} \\
23x &= -138 \\
x &= -6 \\
3(-6) + y &= -20 \\
-18 + y &= -20 \\
y &= -2
\end{aligned}
\end{array}
$$
$$\boxed{\{(-6, -2)\}}$$

13.
$$
\begin{array}{ll}
\begin{aligned}
x + 5y &= -1 \\
\underline{2x + 7y} &\underline{= 1}
\end{aligned}
&
\begin{array}{l}
\xrightarrow{(\times -2)} \\
\xrightarrow{\text{(no change)}}
\end{array}
\begin{aligned}
-2x - 10y &= 2 \\
\underline{2x + 7y} &\underline{= 1} \\
-3y &= 3 \\
y &= -1 \\
x + 5(-1) &= -1 \\
x - 5 &= -1 \\
x &= 4
\end{aligned}
\end{array}
$$
$$\boxed{\{(4, -1)\}}$$

15. $\begin{aligned} 4x + 3y &= 15 \\ \underline{2x - 5y} &= \underline{1} \end{aligned}$ $\begin{array}{c} \text{(no change)} \\ \xrightarrow{\hspace{0.8cm}} \\ \xrightarrow[(\times -2)]{} \end{array}$ $\begin{aligned} 4x + 3y &= 15 \\ \underline{-4x + 10y} &= \underline{-2} \\ 13y &= 13 \\ y &= 1 \\ 2x - 5(1) &= 1 \\ 2x &= 6 \\ x &= 3 \end{aligned}$

$$\boxed{\{(3, 1)\}}$$

17. $\begin{aligned} 3x - y &= 1 \\ \underline{3x - y} &= \underline{2} \end{aligned}$ $\begin{array}{c} \xrightarrow{(\times -1)} \\ \text{(no change)} \\ \xrightarrow{\hspace{0.8cm}} \end{array}$ $\begin{aligned} -3x + y &= -1 \\ \underline{3x - y} &= \underline{2} \\ 0 &= 1 \quad \text{False} \end{aligned}$

no solution to this $\boxed{\text{inconsistent}}$ system.

$$\boxed{\varnothing}$$

19. $\begin{aligned} 3x - 4y &= 11 \\ \underline{2x + 3y} &= \underline{-4} \end{aligned}$ $\begin{array}{c} \xrightarrow{(\times 3)} \\ \xrightarrow{(\times 4)} \end{array}$ $\begin{aligned} 9x - 12y &= 33 \\ \underline{8x + 12y} &= \underline{-16} \\ 17x &= 17 \\ x &= 1 \\ 2(1) + 3y &= -4 \\ 3y &= -6 \\ y &= -2 \end{aligned}$

$$\boxed{\{(1, -2)\}}$$

21. $\begin{aligned} 3x + 2y &= -1 \\ \underline{-2x + 7y} &= \underline{9} \end{aligned}$ $\begin{array}{c} \xrightarrow{(\times 2)} \\ \xrightarrow{(\times 3)} \end{array}$ $\begin{aligned} 6x + 4y &= -2 \\ \underline{-6x + 21y} &= \underline{27} \\ 25y &= 25 \\ y &= 1 \\ 3x + 2(1) &= -1 \\ 3x &= -3 \\ x &= -1 \end{aligned}$

$$\boxed{\{(-1, 1)\}}$$

23. $\begin{aligned} x + 3y &= 2 \\ \underline{3x + 9y} &= \underline{6} \end{aligned}$ $\begin{array}{c} \xrightarrow{(\times -3)} \\ \text{(no change)} \\ \xrightarrow{\hspace{0.8cm}} \end{array}$ $\begin{aligned} -3x - 9y &= -6 \\ \underline{3x + 9y} &= \underline{6} \\ 0 &= 0 \quad \text{True} \end{aligned}$

The system has infinitely many solutions. The equations are $\boxed{\text{dependent}}$.
The solution set consists of all ordered pairs that satisfy either equation.

$$\boxed{\{(x, y) \mid x + 3y = 2\}}$$

25. $3x = 2y + 7$
$5x = 2y + 13$

Rewrite: $\begin{aligned} 3x - 2y &= 7 \\ 5x - 2y &= 13 \end{aligned}$ $\begin{array}{c} \xrightarrow{(\times -1)} \\ \text{(no change)} \\ \xrightarrow{\hspace{0.8cm}} \end{array}$ $\begin{aligned} -3x + 2y &= -7 \\ \underline{5x - 2y} &= \underline{13} \\ 2x &= 6 \\ x &= 3 \\ 3(3) &= 2y + 7 \\ 2 &= 2y \\ 1 &= y \end{aligned}$

$$\boxed{\{(3, 1)\}}$$

27. $\begin{aligned} 2x &= 3y - 4 \\ -6x + 12y &= 6 \end{aligned}$ $\xrightarrow[\text{(+ 3)}]{\text{(rewrite)}}$ $\begin{aligned} 2x - 3y &= -4 \\ \underline{-2x + 4y} &= \underline{2} \\ y &= -2 \\ 2x &= 3(-2) - 4 \\ 2x &= -10 \\ x &= -5 \end{aligned}$

$$\boxed{\{(-5, -2)\}}$$

29. $\begin{aligned} 7x - 3y &= 4 \\ -14x + 6y &= -7 \end{aligned}$ $\xrightarrow[\text{(no change)}]{\text{(×2)}}$ $\begin{aligned} 14x - 6y &= 8 \\ \underline{-14x + 6y} &= \underline{-7} \\ 0 &= 1 \quad \text{False} \end{aligned}$

no solution; The system is $\boxed{\text{inconsistent}}$.

$$\boxed{\varnothing}$$

31. $\begin{aligned} 2x - y &= 3 \\ 4x + 4y &= -1 \end{aligned}$ $\xrightarrow[\text{(no change)}]{\text{(×4)}}$ $\begin{aligned} 8x - 4y &= 12 \\ \underline{4x + 4y} &= \underline{-1} \\ 12x &= 11 \\ x &= \frac{11}{12} \end{aligned}$

$$2\left(\frac{11}{12}\right) - y = 3$$

$$y = \frac{22}{12} - \frac{36}{12}$$

$$y = -\frac{14}{12} = -\frac{7}{6}$$

$$\boxed{\left\{ \left(\frac{11}{12}, -\frac{7}{6}\right) \right\}}$$

33. $\begin{aligned} 4x &= 5 + 2y \\ 2x + 3y &= 4 \end{aligned}$ $\xrightarrow[\text{(×-2)}]{\text{(rewrite)}}$ $\begin{aligned} 4x - 2y &= 5 \\ \underline{-4x - 6y} &= \underline{-8} \\ -8x &= -3 \\ y &= \frac{3}{8} \end{aligned}$

$$4x = 5 + 2\left(\frac{3}{8}\right)$$

$$4x = 5 + \frac{3}{4} = \frac{23}{4}$$

$$x = \frac{23}{16}$$

$$\boxed{\left\{ \left(\frac{23}{16}, \frac{3}{8}\right) \right\}}$$

35. $\begin{aligned} 4x - 8y &= 36 \\ \underline{3x - 6y} &= \underline{27} \end{aligned}$ $\xrightarrow[\text{(÷ 3)}]{\text{(÷ -4)}}$ $\begin{aligned} -x + 2y &= -9 \\ \underline{x - 2y} &= \underline{9} \\ 0 &= 0 \quad \text{True} \end{aligned}$

The system has infinitely many solutions.

The equations are $\boxed{\text{dependent}}$.

$$\boxed{\{(x, y) \mid x - 2y = 9\}}$$

37. $2x + 4y = 5$ $\xrightarrow[\text{(× −2/3)}]{\text{(no change)}}$ $2x + 4y = 5$
 $\underline{3x + 6y = 6}$ $\underline{-2x - 4y = -4}$
 $0 = 1$ False

No solution; The system is $\boxed{\text{inconsistent}}$.

$\boxed{\varnothing}$

39. $5x + y = 2$ $\xrightarrow[\text{(× −1)}]{\text{(no change)}}$ $5x + y = 2$
 $\underline{3x + y = 1}$ $\underline{-3x - y = -1}$
 $2x = 1$
 $x = \dfrac{1}{2}$

 $3\left(\dfrac{1}{2}\right) + y = 1$

 $y = -\dfrac{1}{2}$

$$\boxed{\left\{\left(\dfrac{1}{2}, -\dfrac{1}{2}\right)\right\}}$$

41. $2x + 2y = -2 - 4y$
 $3x + y = 7y + 27$
 Rewrite: $2x + 6y = -2$
 $\underline{3x - 6y = 27}$
 $5x = 25$
 $x = 5$
 $2(5) + 2y = -2 - 4y$
 $2(5) + 6y = -2$
 $6y = -12$
 $y = -2$

$\boxed{\{(5, -2)\}}$

43. $3(x - 3) = 2y - 1$
 $4(2x - y) = 18$
 Rewrite: $3x - 9 = 2y - 1$ $\xrightarrow[\text{(÷ −2)}]{\text{(rewrite)}}$ $3x - 2y = 8$
 $8x - 4y = 18$ $\underline{-4x + 2y = -9}$
 $-x = -1$
 $x = 1$
 $3(1 - 3) = 2y - 1$
 $-6 = 2y - 1$
 $-5 = 2y$
 $-\dfrac{5}{2} = y$

$$\boxed{\left\{\left(1, -\dfrac{5}{2}\right)\right\}}$$

45. $3(x + 3) = 5 - 2(y - 4)$ $\xrightarrow{\text{(rewrite)}}$ $3x + 9 = 5 - 2y + 8$
 $3(x - 1) + 2(y + 2) = 0$ $\xrightarrow{\text{(rewrite)}}$ $3x - 3 + 2y + 4 = 0$
 Rewrite: $3x + 2y = 4$ $\xrightarrow[\text{(no change)}]{\text{(× −1)}}$ $-3x - 2y = -4$
 $3x + 2y = -1$ $\underline{3x + 2y = -1}$
 $0 = -5$

No solution; the system is $\boxed{\text{inconsistent}}$.

$\boxed{\varnothing}$

47.

$$
\begin{array}{rcl}
0.2x + 0.3y & = & 0 \\
0.3x + 0.4y & = & 0.1
\end{array}
\quad
\begin{array}{l}
\xrightarrow{\times -40} \\
\xrightarrow{\times 30}
\end{array}
$$

$$
\begin{array}{rcl}
-8x - 12y & = & 0 \\
9x + 12 & = & 3 \\ \hline
x & = & 3 \\
0.2(3) + 0.3y & = & 0 \\
0.3y & = & -0.6 \\
y & = & -2
\end{array}
$$

$$\boxed{\{(3, -2)\}}$$

49.

$$
\begin{array}{rcl}
9x & = & 5y + 7 \\
21x & = & 10 - 20y
\end{array}
$$

$$
\text{Rewrite:}\quad
\begin{array}{rcl}
9x - 5y & = & 7 \\
21x + 20y & = & 10
\end{array}
\qquad
\begin{array}{l}
(\times 4) \to \\
(\text{no change}) \to
\end{array}
\qquad
\begin{array}{rcl}
36x - 20y & = & 28 \\
21x + 20y & = & 10 \\ \hline
57x & = & 38 \\
x & = & \dfrac{2}{3}
\end{array}
$$

$$
\begin{array}{rcl}
9\left(\dfrac{2}{3}\right) & = & 5y + 7 \\
-1 & = & 5y \\
y & = & -\dfrac{1}{5}
\end{array}
$$

$$\boxed{\left\{\dfrac{2}{3}, -\dfrac{1}{5}\right\}}$$

51.

$$
\begin{array}{rcl}
x + y & = & 11 \\
\dfrac{1}{5}x + \dfrac{1}{7}y & = & 1
\end{array}
\qquad
\begin{array}{l}
(\times -5) \to \\
(\times 35) \to
\end{array}
\qquad
\begin{array}{rcl}
-5x - 5y & = & -55 \\
7x + 5y & = & 35 \\ \hline
2x & = & -20 \\
x & = & -10 \\
-10 + y & = & 11 \\
y & = & 21
\end{array}
$$

$$\boxed{\{(-10, 21)\}}$$

53.

$$
\begin{array}{rcl}
\dfrac{3}{5}x + \dfrac{4}{5}y & = & 1 \\
\dfrac{1}{4}x - \dfrac{3}{8}y & = & -1
\end{array}
\qquad
\begin{array}{l}
(\times 5) \to \\
(\times 8) \to \\
(\times 3) \to \\
(\times 4) \to
\end{array}
\qquad
\begin{array}{rcl}
3x + 4y & = & 5 \\
2x - 3y & = & -8 \\
9x + 12y & = & 15 \\
8x - 12x & = & -32 \\ \hline
17x & = & -17 \\
x & = & -1 \\
x & = & 1
\end{array}
$$

$$
\begin{array}{rcl}
\dfrac{3}{5}(1) + \dfrac{4}{5}y & = & 1 \\
\dfrac{4}{5}y & = & \dfrac{8}{5} \\
y & = & 2
\end{array}
$$

$$\boxed{\{-1, 2\}}$$

55. $\dfrac{4}{5}x - y = -1$ $(\times 5) \rightarrow$ $4x - 5y = -5$

$\dfrac{2}{5}x + y = 1$ $(\times 5) \rightarrow$ $\underline{2x + 5y = 5}$

$6x = 0$

$x = 0$

$\dfrac{2}{5}(0) + y = 1$

$y = 1$

$\boxed{\{(0, 1)\}}$

57. $\boxed{C}$ is true.

59. $1.25x - 1.5y = -1.75$ $(\times -2) \rightarrow$ $-2.5x + 3y = 3.5$
$2.5x - 1.75y = -1$ $\underline{2.5x - 1.75y = -1}$

$1.25y = 2.5$

$y = 2$

$1.25x - 1.5(2) = -1.75$

$x = 1$

$\boxed{\{(1, 2)\}}$

Review Problems

63. Answers may vary. Samples given.
  a. $12 + 3 - 6 = 9$
  b. $12 \div 6 + 3 = 5$
  c. $(2)(12) \div 6 = 4$

64. Let x equal the measure of the angle.
$90 - x$ equals the measure of its complement.
$180 - x$ equals the measure of its supplement.
$$(90 - x) + (180 - x) = 196$$
$$270 - 2x = 196$$
$$-2x = -74$$
$$x = 37$$
The angle measures $\boxed{37°}$.

65. $6(y - 5) - 9y < -4y - 5(2y - 5)$
$6y - 30 - 9y < -4y - 10y + 25$
$-3y - 30 < -14y + 25$
$11y < 55$
$y < 5$
$\boxed{\{y \mid y < 5\}}$

Section 5.3 Solving System of Linear Equations by the Substitution Method

Problem Set 5.3, pp. 347-348

1. $x + y = 4$
$ y = 3x$
Substitute $3x$ for y
$ x + (3x) = 4$
$ 4x = 4$
$ x = 1$
$ y = 3(1) = 3$
$\boxed{\{(1,\ 3)\}}$

3. $x + 3y = 8$ $\qquad$ (substitute for y) $\rightarrow$ $\qquad$ $x + 3(2x - 9) = 8$
$ y = 2x - 9$ $x + 6x - 27 = 8$
 $7x = 35$
 $x = 5$
 $y = 2(5) - 9 = 10 - 9 = 1$
 $\boxed{\{(5,\ 1)\}}$

5. $ x = 9 - 2y$
$x + 2y = 13$ $\qquad$ (substitute for x) $\rightarrow$ $\qquad$ $(9 - 2y) + 2y = 13$
 $9 = 13$ False
 no solution;
 The system is $\boxed{\text{inconsistent}}$.
 $\boxed{\varnothing}$

7. $2(x - 1) - y = -3$ $\quad$ (substitute for y) $\rightarrow$ $\quad$ $2(x - 1) - (2x + 3) = -3$
$ y = 2x + 3$ $2x - 2 - 2x - 3 = -3$
 $-5 = -3$ False
 no solution;
 The system is $\boxed{\text{inconsistent}}$.
 $\boxed{\varnothing}$

9. $x + 3y = 5$ $\qquad$ (solve for x) $\rightarrow$ $x = 5 - 3y$
$ 4x + 5y = 13$ $\quad$ (substitute for x) $\rightarrow$ $\quad$ $4(5 - 3y) + 5y = 13$
 $20 - 12y + 5y = 13$ $\cdot 2$
 $-7y = -7$
 $y = 1$
 $x = 5 - 3(1) = 5 - 3 = 2$
 $\boxed{\{(2,\ 1)\}}$

11. $2x - y = -5$ $\qquad$ (solve for y) $\rightarrow$ $y = 2x + 5$
$ x + 5y = 14$ $\quad$ (substitute for y) $\rightarrow$ $\quad$ $x + 5(2x + 5) = 14$
 $x + 10x + 25 = 14$
 $11x = -11$
 $x = -1$
 $y = 2(-1) + 5 = 3$
 $\boxed{\{(-1,\ 3)\}}$

13.
$$21x - 35 = 7y$$
$$y = 3x - 5$$
(substitute for y) →
$$21x - 35 = 7(3x - 5)$$
$$21x - 35 = 21x - 35$$
$$-35 = -35$$
$$0 = 0 \quad \text{True}$$

The system has infinitely many solutions.

The equations are $\boxed{\text{dependent}}$.

$$\boxed{\{(x, y) \mid y = 3x - 5\}}$$

15.
$$x - y = 11 \quad \text{(solve for } y) \rightarrow \quad x = y + 11$$
$$x - 6y = -9 \quad \text{(substitute for } x) \rightarrow \quad (y + 11) - 6y = -9$$
$$-5y = -20$$
$$y = 4$$
$$x = 4 + 11 = 15$$
$$\boxed{\{(15, 4)\}}$$

17.
$$2x - y = 3 \quad \text{(solve for } y) \rightarrow \quad y = 2x - 3$$
$$5x - 2y = 10 \qquad 5x - 2(2x - 3) = 10$$
$$5x - 4x + 6 = 10$$
$$x = 4$$
$$y = 2(4) - 3 = 8 - 3 = 5$$
$$\boxed{\{(4, 5)\}}$$

19.
$$x + 8y = 6 \quad \text{(solve for } x) \rightarrow \quad x = -8y + 6$$
$$2x + 4y = -3 \qquad 2(-8y + 6) + 4y = -3$$
$$-16y + 12 + 4y = -3$$
$$-12y = -15$$
$$y = \frac{5}{4}$$
$$x = -8\left(\frac{5}{4}\right) + 6 = -10 + 6 = -4$$
$$\boxed{\left\{\left(-4, \frac{5}{4}\right)\right\}}$$

21.
$$x = 4y - 2$$
$$x = 6y + 8 \quad \text{(substitute for } x) \rightarrow \quad 4y - 2 = 6y + 8$$
$$-10 = 2y$$
$$-5 = y$$
$$x = 4(-5) - 2 = -20 - 2 = -22$$
$$\boxed{\{(-22, -5)\}}$$

23.
$$y = 2x - 8$$
$$y = 3x - 13 \quad \text{(substitute for } y) \rightarrow \quad 2x - 8 = 3x - 13$$
$$5 = x$$
$$y = 2(5) - 8$$
$$y = 2$$
$$\boxed{\{(5, 2)\}}$$

25.
$$5x + 2y = 0$$
$$x - 3y = 0 \quad \text{(solve for } x) \rightarrow \quad x = 3y$$
$$\text{(substitute for } x) \rightarrow \quad 5(3y) + 2y = 0$$
$$17y = 0$$
$$y = 0$$
$$x = 3(0) = 0$$
$$\boxed{\{(0, 0)\}}$$

27. $\begin{aligned} 6x + 2y &= 7 \\ y &= 2 - 3x \end{aligned}$ (substitute for y) $\rightarrow$ $\begin{aligned} 6x + 2(2 - 3x) &= 7 \\ 6x + 4 - 6x &= 7 \\ 4 &= 7 \quad \text{False} \end{aligned}$

no solution; the system is $\boxed{\text{inconsistent}}$.

$\boxed{\varnothing}$

29. $\begin{aligned} 2x + 5y &= -4 \\ 3x - y &= 11 \end{aligned}$ (solve for y) $\rightarrow$ $\begin{aligned} y &= 3x - 11 \end{aligned}$

(substitute for y) $\rightarrow$ $\begin{aligned} 2x + 5(3x - 11) &= -4 \\ 2x + 15x - 55 &= -4 \\ 17x &= 51 \\ x &= 3 \\ y = 3(3) - 11 &= -2 \end{aligned}$

$\boxed{\{(3, -2)\}}$

31. $\begin{aligned} \frac{x}{2} + \frac{y}{2} &= -1 \\[2mm] \frac{x}{3} - \frac{y}{2} &= -4 \end{aligned}$

$\begin{aligned} (\times 2) \rightarrow \quad x + y &= -2 \\ (\times 6) \rightarrow \quad 2x - 3y &= -24 \end{aligned}$ (solve for y) $\rightarrow$ $\begin{aligned} y &= -x - 2 \end{aligned}$

(substitute for x) $\rightarrow$ $\begin{aligned} 2x - 3(-x - 2) &= -24 \\ 2x + 3x + 6 &= -24 \\ 5x &= -30 \\ x &= -6 \\ y = -(-6) - 2 &= 4 \end{aligned}$

$\boxed{\{(-6, 4)\}}$

33. $\begin{aligned} \frac{x}{4} + \frac{y}{2} &= \frac{7}{8} \\[2mm] x &= 5 - 3y \end{aligned}$ (substitute for x) $\rightarrow$ $\begin{aligned} \frac{(5 - 3y)}{4} + \frac{y}{2} &= \frac{7}{8} \end{aligned}$

$(\times 8) \rightarrow$ $\begin{aligned} 2(5 - 3y) + 4y &= 7 \\ 10 - 6y + 4y &= 7 \\ -2y &= -3 \\ y &= \frac{3}{2} \end{aligned}$

$$x = 5 - 3\left(\frac{3}{2}\right) = 5 - \frac{9}{2} = \frac{1}{2}$$

$$\boxed{\left\{ \left(\frac{1}{2}, \frac{3}{2} \right) \right\}}$$

35. $\begin{aligned} 2x + 3y &= 2 \\ \underline{x - 3y} &= \underline{-6} \\ \text{(add)} \quad 3x &= -4 \\ x &= -\frac{4}{3} \end{aligned}$

$$-\frac{4}{3} - 3y = -6$$

$$-3y = -6 + \frac{4}{3} = -\frac{14}{3}$$

$$y = \frac{14}{9}$$

$$\boxed{\left\{ \left(-\frac{4}{3}, \frac{14}{9} \right) \right\}}$$

37. $\begin{aligned} x + y &= 1 \\ \underline{3x - y} &= \underline{3} \\ \text{(add)} \quad 4x &= 4 \\ x &= 1 \end{aligned}$

$$1 + y = 1$$

$$y = 0$$

$$\boxed{\{(1, 0)\}}$$

39. $\begin{aligned} 3x + 2y &= -3 \\ 2x - 5y &= 17 \end{aligned}$ $\begin{aligned} (\times 5) &\rightarrow \\ (\times 2) &\rightarrow \end{aligned}$ $\begin{aligned} 15x + 10y &= -15 \\ \underline{4x - 10y} &= \underline{\ \ 34} \\ 19x &= 19 \\ x &= 1 \\ 2(1) - 5y &= 17 \\ -5y &= 15 \\ y &= -3 \end{aligned}$

$$\boxed{\{(1, -3)\}}$$

41. $\begin{aligned} 3x - 2y &= 6 \\ y &= 3 \end{aligned}$ (substitute for y) $\rightarrow$ $\begin{aligned} 3x - 2(3) &= 6 \\ 3x &= 12 \\ x &= 4 \end{aligned}$

$$\boxed{\{(4, 3)\}}$$

43. $\begin{aligned} 3x + 7y &= -10 \\ x + 2 &= 0 \end{aligned}$
(solve for x) $\rightarrow$
(substitute for x) $\rightarrow$
$\begin{aligned} x &= -2 \\ 3(-2) + 7y &= -10 \\ -6 + 7y &= -10 \\ 7y &= -4 \\ y &= -\dfrac{4}{7} \end{aligned}$

$$\boxed{\left\{ \left(-2, -\dfrac{4}{7} \right) \right\}}$$

45. $\begin{aligned} 3x - 2y &= 8 \\ x &= -2y \end{aligned}$ (substitute for x) $\rightarrow$ $\begin{aligned} 3(-2y) - 2y &= 8 \\ -6y - 2y &= 8 \\ -8y &= 8 \\ y &= -1 \\ x = -2(-1) &= 2 \end{aligned}$

$$\boxed{\{(2, -1)\}}$$

47. $\begin{aligned} 4x + y &= -12 \\ -3x - y &= 10 \\ x &= -2 \\ 4(-2) + y &= -12 \\ y &= -4 \end{aligned}$

$$\boxed{\{(-2, -4)\}}$$

49. $\begin{aligned} 3(1 - 2x) - 2(3y + 4) &= 1 \\ 3(x - 1) - 2y &= -5 \end{aligned}$ $\rightarrow$

Simplify: $\begin{aligned} -6x - 6y &= 6 \\ 3x - 2y &= -2 \end{aligned}$ $\begin{aligned} (\div -3) &\rightarrow \\ (\text{no change}) &\rightarrow \end{aligned}$

$\begin{aligned} 3 - 6x - 6y - 8 &= 1 \\ 3x - 3 - 2y &= -5 \\ 2x + 2y &= -2 \\ \underline{3x - 2y} &= \underline{-2} \\ 5x &= -4 \\ x &= -\dfrac{4}{5} \end{aligned}$

$$y = -1 - x = -1 - \left(-\dfrac{4}{5} \right) = -\dfrac{1}{5}$$

$$\boxed{\left\{ \left(-\dfrac{4}{5}, -\dfrac{1}{5} \right) \right\}}$$

51.
$$
\begin{aligned}
y &= 3x - 1 \\
-12x + 4y &= -3
\end{aligned}
$$

(substitute for y) $\rightarrow$
$$
\begin{aligned}
-12x + 4(3x - 1) &= -3 \\
-12x + 12x - 4 &= -3 \\
-4 &= -3 \quad \text{False}
\end{aligned}
$$

no solution;

The system is $\boxed{\text{inconsistent}}$.

$\boxed{\varnothing}$

53.
$$
\begin{array}{ll}
3x - 4y = 19 & (\times -7) \rightarrow \\
7x + 18y = 17 & (\times 3) \rightarrow
\end{array}
\quad
\begin{aligned}
-21x + 28y &= -133 \\
\underline{21x + 54y} &= \underline{51} \\
82y &= -82 \\
y &= -1 \\
3x - 4(-1) &= 19 \\
3x &= 15 \\
x &= 5
\end{aligned}
$$

$$\boxed{\{(5, -1)\}}$$

55.
$$
\begin{array}{llll}
\dfrac{x}{5} - \dfrac{y}{4} = \dfrac{1}{10} & (\times 20) \rightarrow & 4x - 5y = 2 & (\times -2) \rightarrow \\[2ex]
\dfrac{x}{2} - \dfrac{y}{3} = \dfrac{5}{6} & (\times 6) \rightarrow & 3x - 2y = 5 & (\times 5) \rightarrow
\end{array}
$$

$$
\begin{aligned}
-8x + 10y &= -4 \\
\underline{15x - 10y} &= \underline{25} \\
7x &= 21 \\
x &= 3 \\
\dfrac{3}{2} - \dfrac{y}{3} &= \dfrac{5}{6} \\
\dfrac{-y}{3} &= \dfrac{5}{6} - \dfrac{3}{2} = \dfrac{-4}{6} \\
y &= 2
\end{aligned}
$$

$$\boxed{\{(3, 2)\}}$$

57.
$$
\begin{array}{ll}
5(x + 1) = 7(y + 1) - 7 & \text{(Simplify)} \rightarrow \\
6(x + 1) + 5 = 5(y + 1) & \rightarrow
\end{array}
$$

$$
\begin{aligned}
5x + 5 &= 7y + 7 - 7 \\
6x + 6 + 5 &= 5y + 5
\end{aligned}
$$

$$
\begin{array}{ll}
\text{Simplify}: \quad 5x - 7y = -5 & (\times -5) \rightarrow \\
6x - 5y = -6 & (\times 7) \rightarrow
\end{array}
$$

$$
\begin{aligned}
-25x + 35y &= 25 \\
\underline{42x - 35y} &= \underline{-42} \\
17x &= -17 \\
x &= -1 \\
5(-1) - 7y &= -5 \\
-7y &= 0 \\
y &= 0
\end{aligned}
$$

$$\boxed{\{(-1, 0)\}}$$

59.
$$
\begin{array}{ll}
3(x + 1) - 2y = 2y & \text{(simplify)} \rightarrow \\
4(x - 2) + 5y = 19
\end{array}
$$

$$
\begin{aligned}
3x + 3 - 2y &= 2y \\
4x - 8 + 5y &= 19
\end{aligned}
$$

$$
\begin{array}{ll}
\text{Simplify}: \quad 3x - 4y = -3 & (\times 5) \rightarrow \\
4x + 5y = 27 & (\times 4) \rightarrow
\end{array}
$$

$$
\begin{aligned}
15x - 20y &= -15 \\
\underline{16x + 20y} &= \underline{108} \\
31x &= 93 \\
x &= 3 \\
4(3) + 5y &= 27 \\
5y &= 15 \\
y &= 3
\end{aligned}
$$

$$\boxed{\{(3, 3)\}}$$

61. $8y = 15 - 4x$
         $x + 2y = 4$ (solve for x) $\rightarrow$ $x = 4 - 2y$
                        (substitute for x) $\rightarrow$ $8y = 15 - 4(4 - 2y)$
                                                      $8y = 15 - 16 + 8y$
                                                      $0 = -1$ False

no solution; **the** system is $\boxed{\text{inconsistent}}$.

$\boxed{\varnothing}$

63. $\dfrac{x}{2} - \dfrac{y}{4} = 1$

   $\dfrac{x}{3} + y = 3$ (solve for y) $\rightarrow$ $y = 3 - \dfrac{x}{3}$

                    (substitute for y) $\rightarrow$ $\dfrac{x}{2} - \dfrac{1}{4}\left(3 - \dfrac{x}{3}\right) = 1$

                    ($\times 12$) $6x - 3\left(3 - \dfrac{x}{3}\right) = 12$

                                    $6x - 9 + x = 12$
                                    $7x = 21$
                                    $x = 3$

                             $y = 3 - \dfrac{3}{3} = 3 - 1 = 2$

                             $\boxed{\{(3, 2)\}}$

65. $x + 1.5y = 0$ (solve for x) $\rightarrow$ $x = -1.5y$
        $4x + 5y = -2.2$ (substitute for x) $\rightarrow$ $4(-1.5y) + 5y = -2.2$
                                              $-6y + 5y = -2.2$
                                                  $-y = -2.2$
                                                   $y = 2.2$
                                $x = -1.5(2.2) = -3.3$
                                $\boxed{\{(-3.3, 2.2)\}}$

67. $\boxed{D}$ is true;   D. $x = 2y - 2$
                     $2(2y - 2) - 2 = 1$
                     $4y - 4 - 2y = 1$
                     $2y - 4 = 1$ True

69. $x = 3 - y - z$
        $2x + y - z = -6$ (substitute for x) $\rightarrow$ $2(3 - y - z) + y - z = -6$
        $3x - y + z = 11$ (substitute for x) $\rightarrow$ $3(3 - y - z) - y + z = 11$
     Simplify: $6 - 2y - 2z + y - z = -6$ $\rightarrow$ $-y - 3z = -12$
              $9 - 3y - 3z + z = 11$ $\rightarrow$ $-4y - 2z = 2$
        ($\times -2$) $\rightarrow$ $2y + 6z = 24$
        ($\div 2$) $\rightarrow$ $\underline{-2y - z = 1}$
                          $5z = 25$
                          $z = 5$
                          $y = -3z + 12 = -15 + 12 = -3$
                          $x = 3 - y - z = 3 + 3 - 5 = 1$
                     $\boxed{x = 1, y = -3, z = 5}$

Review Problems

72. $2x - 3y < 6$

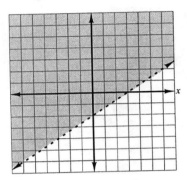

73. passing through $(-1, 6)$ and parallel to a line whose equation is $y = -4x + 3$

$m = -4$

point-slope: $\boxed{y - 6 = -4(x + 1)}$

$y - 6 = -4x - 4$

slope-intercept: $\boxed{y = -4x + 2}$

74. Let x equal the amount invested at 7.5%.

$x + 800$ equals the amount invested at 9%.

$$
\begin{aligned}
0.075x + 0.09(x + 800) &= 270 \\
0.075x + 0.09x + 72 &= 270 \\
0.165x &= 198 \\
x &= 1200 \\
x + 800 &= 2000
\end{aligned}
$$

$\boxed{\$1200 \text{ at } 7.5\%; \$2000 \text{ at } 9\%}$

Section 5.4 Problem Solving Using Systems of Equations

Problem Set 5.4, pp. 358-361

1. Let x equal the first number and y equal the second number.

$$
\begin{array}{llll}
2x - y &= 3 & \rightarrow & 4x - 2y &= 6 \\
3x + 2y &= 8 & & \underline{3x + 2y} &= 8 \\
& & & 7x &= 14 \\
& & & x &= 2 \\
& & & y &= 2x - 3 = 2(2) - 3 = 1
\end{array}
$$

The numbers are $\boxed{2 \text{ and } 1}$.

3. Let x equal the first number and y equal the second number.

$$
\begin{array}{ll}
x + y &= 3 \\
\underline{x - y} &= 7 \\
2x &= 10 \\
x &= 5 \\
y &= 3 - x = 3 - 5 = -2
\end{array}
$$

The numbers are $\boxed{5 \text{ and } -2}$.

5. Let x equal the first number and y equal the second number.

$$\begin{array}{rclcrclcrcl}
2x &=& y+3 & \to & 2x-y &=& 3 & \to & -6x+3y &=& -9 \\
5x-3y &=& 5 & & 5x-3y &=& 5 & & \underline{5x-3y} &=& \underline{5} \\
& & & & & & & & -x &=& -4 \\
& & & & & & & & x &=& 4 \\
& & & & & & & & 8 &=& y+3 \\
& & & & & & & & 5 &=& y
\end{array}$$

The numbers are $\boxed{4 \text{ and } 5}$.

7. Let t equal the tens' digit and u equal the units' digit.

$$\begin{array}{rclcrclcrcl}
2t+3u &=& 11 & \to & 2t+3u &=& 11 & \to & 2t+3u &=& 11 \\
10t+u &=& 21+5t & & 5t+u &=& 21 & & \underline{-15t-3u} &=& \underline{-63} \\
& & & & & & & & -13t &=& -52 \\
& & & & & & & & t &=& 4 \\
& & & & & & & & u = 21-5t &=& 21-20=1
\end{array}$$

The number is $\boxed{41}$.

9. Let t equal the tens' digit and u equal the units' digit.

$$\begin{array}{rclcrclcrcl}
t+2u &=& 11 & \to & t+2u &=& 11 & \to & t+2u &=& 11 \\
10t+u &=& 20+11u & & 10t-10u &=& 20 & & \underline{2t-2u} &=& \underline{4} \\
& & & & & & & & 3t &=& 15 \\
& & & & & & & & t &=& 5 \\
& & & & & & & & u = t-2 = 5-2 &=& 3
\end{array}$$

The number is $\boxed{53}$.

11. Let t equal the tens' digit and u equals the units' digit.

$$\begin{array}{rcl}
t &>& u \\
t+u &=& 12 \\
t-u &=& \frac{1}{2}(12) = 6 \\
\hline
2t &=& 18 \\
t &=& 9 \\
u &=& 12-9=3
\end{array}$$

The number is $\boxed{93}$.

13. $(2x+y)+(12x+7y)=180$

$\underline{\quad 2x+y \;=\; 7y-x \quad}$

Simplify: $14x+8y = 180$ Simplify $\to$ $\qquad 7x+4y = 90$

$\qquad\qquad\quad 3x-6y = 0$ (Solve for y) $\to$ $\qquad y = \frac{1}{2}x$

$\qquad$ (Substitute for y) $\to$ $7x+4\left(\frac{1}{2}x\right) = 90$

$\qquad\qquad\qquad\qquad\qquad 7x+2x = 90$

$\qquad\qquad\qquad\qquad\qquad 9x = 90$

$\qquad\qquad\qquad\qquad\qquad x = \boxed{10}$

$\qquad\qquad\qquad\qquad\boxed{y} = \frac{1}{2}x = \boxed{5}$

$(2x+y)^\circ = (2\cdot10+5)^\circ = \boxed{25^\circ}$

$(12x+7y)^\circ = (12\cdot10+7\cdot5)^\circ = (120+35)^\circ = \boxed{155^\circ}$

$(7y-x)^\circ = (7\cdot5-10)^\circ = \boxed{25^\circ}$

15. $m\angle A + m\angle B = 180$

$\underline{m\angle A = m\angle C}$

$(4x - 2y + 4) + (12x + 6y + 12) = 180$

$\qquad\qquad 4x - 2y + 4 \;=\; 6x - 24$

(Simplify) $\rightarrow \qquad 16x + 4y \;=\; 164 \qquad \rightarrow \qquad 4x + y \;=\; 41$

$\qquad\qquad\qquad -2x - 2y \;=\; -28 \qquad\qquad\qquad \underline{-x - y \;=\; -14}$

$\qquad\qquad\qquad\qquad\qquad\qquad\qquad\qquad\qquad\qquad 3x \;=\; 27$

$\qquad\qquad\qquad\qquad\qquad\qquad\qquad\qquad\qquad\qquad\; x \;=\; \boxed{9}$

$\qquad\qquad\qquad\qquad\qquad\qquad\qquad\qquad\quad \boxed{y} \;=\; 14 - x = 14 - 9 = \boxed{5}$

$m\angle A = (4x - 2y + 4)° = (4 \cdot 9 - 2 \cdot 5 + 4)° = \boxed{30°}$

$m\angle B = (12x + 6y + 12)° = (12 \cdot 9 + 6 \cdot 5 + 12)° = \boxed{150°}$

$m\angle C = (6x - 24)° = (6 \cdot 9 - 24)° = \boxed{30°}$

17. $\qquad 2L + 2W \;=\; P \qquad P = 20$ meters

$\rightarrow \qquad L + 4W \;=\; 19$

$\qquad\qquad \underline{-L - W \;=\; -10}$

$\qquad\qquad\quad 3W \;=\; 9$

$\qquad\qquad\quad\; W \;=\; 3 \qquad$ width: 3 meters

$\qquad\qquad\quad\; L \;=\; 10 - W = 10 - 3 = 7 \qquad$ length: 7 meters

$\boxed{3 \text{ meters} \times 7 \text{ meters}}$

19. length of large rectangle $= y + 8$

width of the large rectangle $= x + 10$

$2(x + 10) + 2(y + 8) = 58$

$y + x + y = 17.5$

(Simplify) $\rightarrow \qquad 2x + 2y \;=\; 22 \qquad \rightarrow \qquad 2x + 2y \;=\; 22$

$\qquad\qquad\qquad\quad x + 2y \;=\; 17.5 \qquad\qquad\qquad \underline{-x - 2y \;=\; -17.5}$

$\qquad\qquad\qquad\qquad\qquad\qquad\qquad\qquad\qquad\qquad\;\; x \;=\; 4.5$

$\qquad\qquad\qquad\qquad\qquad\qquad\qquad\qquad\qquad\; y \;=\; 11 - x = 11 - 4.5 = 6.5$

$\boxed{x = 4.5,\, y = 6.5}$

21. Let x equal the cost of each sweater and y equal the cost of each shirt.

$\qquad x + 3y \;=\; 42 \qquad \rightarrow \qquad -3x - 9y \;=\; -126$

$\qquad 3x + 2y \;=\; 56 \qquad\qquad\qquad \underline{3x + 2y \;=\; \quad 56}$

$\qquad\qquad\qquad\qquad\qquad\qquad\qquad -7y \;=\; -70$

$\qquad\qquad\qquad\qquad\qquad\qquad\qquad\quad y \;=\; 10$

$\qquad\qquad\qquad\qquad\qquad\qquad\qquad\quad x \;=\; -3y + 42 = -30 + 42 = 12$

$\boxed{\text{cost of one sweater, \$12; cost of one shirt, \$10}}$

23. Let A equal the number of ounces of health bar A.

Let B equal the number of ounces of health bar B.

Ascorbic acid: $15A + 10B \;=\; 45$

Niacin: $\qquad\quad \underline{2A + 4B \;=\; \;14}$

$\qquad (\div 5) \rightarrow \quad 3A + 2B \;=\; 9$

$\qquad (\div -2) \rightarrow \;\; \underline{-A - 2B \;=\; -7}$

$\qquad\qquad\qquad\qquad\quad 2A \;=\; 2$

$\qquad\qquad\qquad\qquad\quad\;\; A \;=\; 1$

$\qquad\qquad\qquad\qquad\quad 2B \;=\; 7 - A = 6$

$\qquad\qquad\qquad\qquad\quad\;\; B \;=\; 3$

$\boxed{1 \text{ ounce of } A;\; 3 \text{ ounces of } B}$

25. $(-1, 10), (2, 1)$

$$
\begin{aligned}
y &= mx + b \\
10 &= -m + b \\
1 &= 2m + b
\end{aligned}
\qquad \rightarrow \qquad
\begin{aligned}
10 &= -m + b \\
\underline{-1} &= \underline{-2m - b} \\
9 &= -3m \\
-3 &= m \\
b &= 10 + m = 10 - 3 = 7
\end{aligned}
$$

$\boxed{y = -3m + 7}$ $\qquad$ $\boxed{m = -3,\, b = 7}$

27. $(1, 4), (-1, 8)$

$$
\begin{aligned}
Ax + By &= 12 \\
A + 4B &= 12 \\
\underline{-A + 8B} &= \underline{12} \\
12B &= 24 \\
B &= 2 \\
A &= 12 - 4B = 12 - 8 = 4
\end{aligned}
$$

$\boxed{4x + 2y = 12}$ $\qquad$ $\boxed{A = 4,\, B = 2}$

29. Let x equal the number of shoes sold.
C equals the cost.
R equals the revenue.

$\boxed{C = 20x + 120{,}000}$

$\boxed{R = 50x}$

break-even point:
$$
\begin{aligned}
R &= C \\
50x &= 20x + 120{,}000 \\
30x &= 120{,}000 \\
x &= 4000;\ 4000 \text{ pairs of shoes sold} \\
R &= 50(4000) = 200{,}000
\end{aligned}
$$

$\boxed{(4000,\ \$200{,}000)}$

31. $\boxed{\text{first car: } y = 40{,}000 + 4000x}$

$\boxed{\text{second car: } y = 36{,}000 + 4800x}$

$$
\begin{aligned}
40{,}000 + 4000x &= 36{,}000 + 4800x \\
4000 &= 800x \\
5 &= x
\end{aligned}
$$

$\boxed{\text{5 years}}$

33. Let x equal the speed of plane in still air.
y equals the speed of wind.

	R	$\times$	T	$= D$
Trip with the wind	$x + y$		5 hours	3000 km
Trip against the wind	$x - y$		6 hours	3000 km

$$
\begin{aligned}
5(x + y) &= 3000 \\
6(x - y) &= 3000
\end{aligned}
\qquad
\begin{aligned}
\rightarrow \\
\rightarrow
\end{aligned}
\qquad
\begin{aligned}
x + y &= 600 \\
\underline{x - y} &= \underline{500} \\
2x &= 1100 \\
x &= 550 \\
y &= 50
\end{aligned}
$$

$\boxed{\text{speed of plane in still air, 550 kilometers per hour; speed of wind, 50 kilometers per hour}}$

35. Let x equal the rate of rowing in still water.
y equals the rate of current.

	R	$\times$	T	$= D$
Trip with the current	$x+y$		2 hours	16 km
Trip against the current	$x-y$		2 hours	8 km

$$
\begin{aligned}
2(x+y) &= 16 \\
2(x-y) &= 8
\end{aligned}
\quad
\begin{aligned}
\rightarrow \\
\rightarrow
\end{aligned}
\quad
\begin{aligned}
x+y &= 8 \\
x-y &= 4 \\
\hline
2x &= 12 \\
x &= 6 \\
y &= 2
\end{aligned}
$$

rate of rowing in still water; 6 miles per hour; rate of current, 2 miles per hour

37. Let x equal the rate of swimmer in still water.
y equals the rate of current.

	R	$\times$	T	$= D$
Trip with the current	$x+y$		2 hours	10 miles
Trip against the current	$x-y$		4(2 hours)	10 miles

$$
\begin{aligned}
2(x+y) &= 10 \\
8(x-y) &= 10
\end{aligned}
\quad
\begin{aligned}
\rightarrow \\
\rightarrow
\end{aligned}
\quad
\begin{aligned}
x+y &= 5 \\
x-y &= \frac{5}{4} \\
\hline
2x &= \frac{25}{4} \\
x &= \frac{25}{8} = 3\frac{1}{8} \\
y &= 5 - \frac{25}{8} = \frac{15}{8} = 1\frac{7}{8}
\end{aligned}
$$

The rate *or* speed of the current is $1\frac{7}{8}$ miles per hour .

39. Let x equal the speed of plane in still air.
y equals the speed of wind.

	R	$\times$	T	$= D$
Trip with no wind	x		$1\frac{1}{2}$ hours	300 miles
Trip against the wind	$x-y$		2 hours	300 miles

$$
\begin{aligned}
\frac{3}{2}x &= 300 \\
x &= 200 \\
2(x-y) &= 300 \\
x-y &= 150 \\
y &= x - 150 = 200 - 150 = 50
\end{aligned}
$$

rate or speed of wind: 50 miles per hour

41. Let x equal the number of tricycles.
  y equals the number of bicycles.

$$
\begin{array}{rclcrcl}
3x + 2y & = & 132 & \rightarrow & 3x + 2y & = & 132 \\
2x + 2y & = & 112 & \rightarrow & \underline{-2x - 2y} & = & \underline{-112} \\
& & & & x & = & 20 \\
& & & & y & = & 56 - x = 36
\end{array}
$$

20 tricycles, 36 bicycles

Review Problems

43. $(-1, 0)$: $m = \dfrac{2}{3}$

$$y - 0 = \frac{2}{3}(x + 1)$$

$$y = \frac{2}{3}x + \frac{2}{3}$$

$(14, -2)$: $m = \dfrac{-2}{3}$

$$y + 2 = \frac{-2}{3}(x - 14)$$

$$y + 2 = \frac{-2}{3}x + \frac{28}{3}$$

$$y = \frac{-2}{3}x + \frac{22}{3}$$

The intersection of the two equations gives the coordinates of the boat in distress.

$$\frac{2}{3}x + \frac{2}{3} = -\frac{2}{3}x + \frac{22}{3}$$

$$\frac{4x}{3} = \frac{20}{3}$$

$$x = 5$$

$$y = \frac{2}{3}x + \frac{2}{3} = \frac{10}{3} + \frac{2}{3} = \frac{12}{3} = 4$$

$(5, 4)$

45.

	beginning	*A*	*B*
number of people in upstairs apartment	x	$x - 1$	$x + 1$
number of people in downstairs apartment	y	$y + 1$	$y - 1$
equation		$x - 1 = y + 1$	$x + 1 = 2(y - 1)$

A: one upstairs moves downstairs
B: one downstairs moves upstairs

$$
\begin{array}{rclcrcl}
x - 1 & = & y + 1 & \rightarrow & -x + y & = & -2 \\
x + 1 & = & 2y - 2 & \rightarrow & \underline{x - 2y} & = & \underline{-3} \\
& & & & -y & = & -5 \\
& & & & y & = & 5 \\
& & & & x & = & y + 2 = 7
\end{array}
$$

7 people upstairs; 5 people downstairs

48. $4x - 2y > 8$

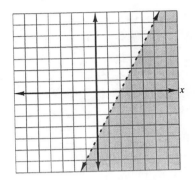

49. Let x equal the number of twenties.
$32 + x$ equals the number of tens.

$$20x + 10(32 + x) = 800$$
$$20x + 320 + 10x = 800$$
$$30x = 480$$
$$x = 16$$
$$32 + x = 48$$

16 $20 bills

50.
$$f(x) = x^2 - x - 2$$
$$f(-1) = (-1)^2 + 1 - 2$$
$$= 1 + 1 - 2 = \boxed{0}$$

Section 5.5 Solving Systems of Inequalities

Problem Set 5.5, pp. 365-366

1. $x + y \leq 4$
$x - y \leq 1$

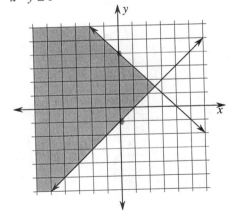

3. $2x - 4y \leq 8$
$x + y \geq -1$

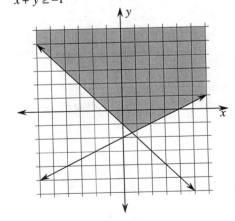

5. $x + 3y \leq 6$
   $x - 2y \leq 4$

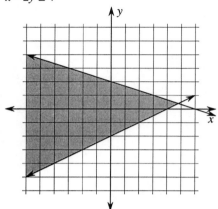

7. $x - 4y \leq 4$
   $x \geq 2y$

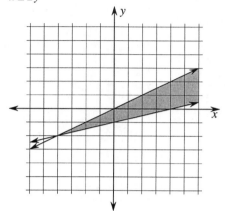

9. $2x + y \leq 4$
   $x + 2 > y$

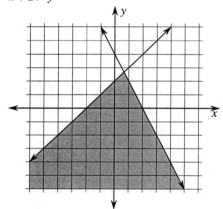

11. $y \leq 2x + 2$
   $y \geq 2x + 1$

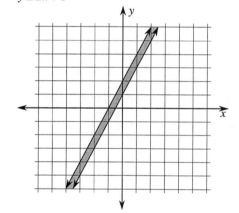

13. $y > 2x - 3$
   $y < 2x + 1$

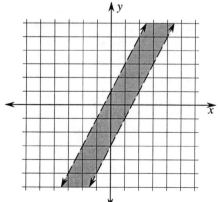

15. $x - 2y > 4$
   $2x + y \geq 6$

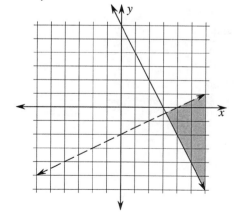

17. $x \geq 3$
$y \geq 3$

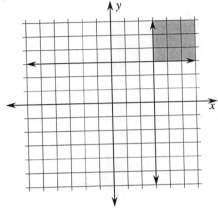

19. $x \geq 2$
$y < 3$

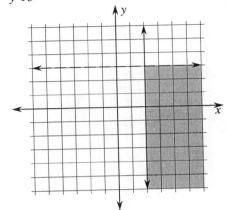

21. $x + y < 1$
$x + y > 4$
no solution

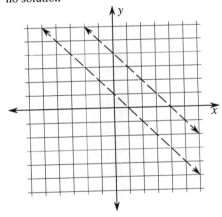

23. $x > 0$
$y \leq 0$

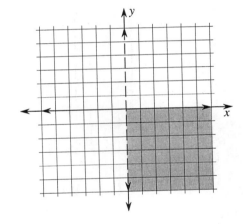

25. $2x + y \geq 6$
$y \leq -2x - 4$
no solution

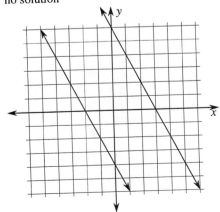

27. $y \geq 2x + 1$
$y \leq 5$

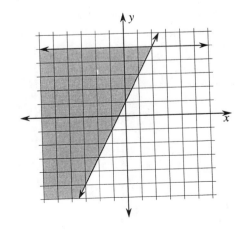

29. $x + y \le 5$
$x \ge 0$
$y \ge 0$

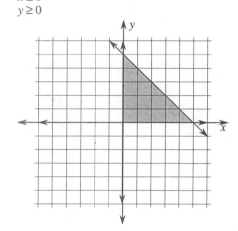

31. $4x - 3y > 12$
$x \ge 0$
$y \le 0$

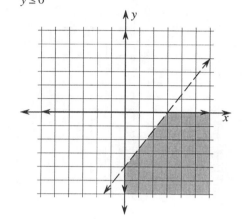

33. $0 \le x \le 3$
$0 \le y \le 3$

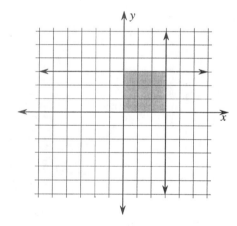

35. $x - y \le 4$
$x + 2y \le 4$
$x \ge 0$

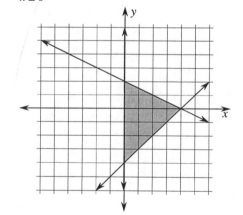

37. a. $0 \le y \le 4000$
$0 \le x + y \le 10,000$

b.

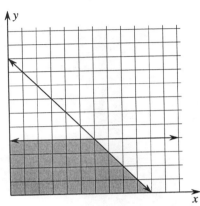

c. Answers may vary. Samples given.
(0, 4000), (1000, 2000), (2000, 3000), (3000, 4000), (5000, 4000), (10,000,0)

39. a. $x + y \ge 20$
$5x + 10y \ge 200$

b.

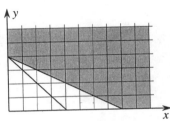

c. Answers may vary. Samples given.
(40, 0), (0, 20), (20, 10), (10, 15)

41. B is true.
(0, 6): $x + y > 5$ $0 + 6 > 5$ true
$x - y < 0$ $0 - 6 < 0$ true

Review Problems

44. $y = x^2 - 1$

x	-2	-1	0	1	2
y	3	0	-1	0	3

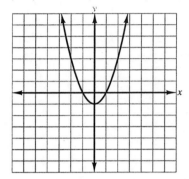

45. $(-5, -2), (-1, 6)$

$m = \dfrac{6 + 2}{-1 + 5} = \dfrac{8}{4} = 2$

point-slope: $\boxed{y + 2 = 2(x + 5) \text{ or } y - 6 = 2(x + 1)}$

slope-intercept: $y + 2 = 2x + 10$

$\boxed{y = 2x + 8}$

standard: $\boxed{2x - y = -8}$

46. $-5 + [(-11 + 3) - (-1 - 9)]$

$= -5 + (-8 + 10)$

$= -5 + 2$

$= \boxed{-3}$

Chapter 5 Review Problems

Chapter 5 Review Problems pp. 368-369

1. $(1, -5)$:

$4x - y$	$=$	9			
$4 + 5$	$=$	9			
9	$=$	9	True		

$2x + 3y = -13$
$2 - 15 = -13$
$-13 = -13$ True

$(1, -5)$ is the $\boxed{\text{solution}}$ of the given system.

2. $(-5, 2)$:

$2x + 3y = -4$
$-10 + 6 = -4$
$-4 = -4$ True

$x - 4y = -10$
$-5 - 8 = -10$
$-13 = -10$ False

$(-5, 2)$ fails to satisfy both equations. It is $\boxed{\text{not a solution}}$ of the given system.

3. $x + y = 6$
$x - y = 6$
solution: $\boxed{\{(6, 0)\}}$

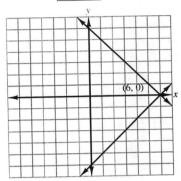

4. $2x - 3y = 12$
$-2x + y = -8$
solution: $\boxed{\{(3, -2)\}}$

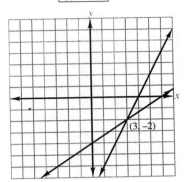

5. $y = \dfrac{1}{2}x$
$y = 2x - 3$
solution: $\boxed{\{(2, 1)\}}$

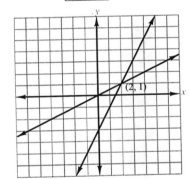

6. $3x + 2y = 6$
$3x - 2y = 6$
solution: $\boxed{\{(2, 0)\}}$

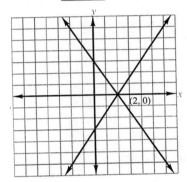

7. $y = 4x$
$y = 4x - 2$
solution: $\boxed{\text{Ø; inconsistent}}$

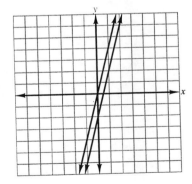

8. $2x - 4y = 8$
$x = 2y + 4$
solution: $\boxed{\{(x, y) \mid 2x - 4y = 8\} \text{ dependent}}$

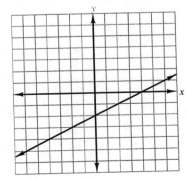

9. $x - y = 4$
  $x = -2$
  solution: $\boxed{\{(-2, -6)\}}$

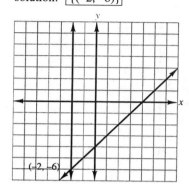

10. $x = -3$
  $y = 6$
  solution: $\boxed{\{(-3, 6)\}}$

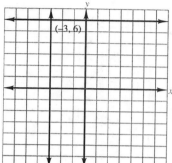

11.

$$\begin{aligned} x + y &= 6 \\ 2x + y &= 8 \end{aligned} \quad\rightarrow\quad \begin{aligned} x + y &= 6 \\ \underline{-2x - y} &= \underline{-8} \\ -x &= -2 \\ x &= 2 \\ y &= 4 \end{aligned}$$

$$\boxed{\{(2, 4)\}}$$

12.

$$\begin{aligned} 3x - 4y &= 1 \\ 12x - y &= -11 \end{aligned} \quad\rightarrow\quad \begin{aligned} -3x + 4y &= -1 \\ 48x - 4y &= -44 \\ 45x &= -45 \\ x &= -1 \\ y = 12x + 11 &= -1 \end{aligned}$$

$$\boxed{\{(-1, -1)\}}$$

13.

$$\begin{aligned} 3x - 7y &= 13 \\ 6x + 5y &= 7 \end{aligned} \quad\rightarrow\quad \begin{aligned} -6x + 14y &= -26 \\ 6x + 5y &= 7 \\ 19y &= -19 \\ y &= -1 \\ 3x + 7 &= 13 \\ 3x &= 6 \\ x &= 2 \end{aligned}$$

$$\boxed{\{(2, -1)\}}$$

14.

$$\begin{aligned} 8x - 4y &= 16 \\ 4x + 5y &= 22 \end{aligned} \quad\rightarrow\quad \begin{aligned} -8x + 4y &= -16 \\ 8x + 10y &= 44 \\ 14y &= 28 \\ y &= 2 \\ 4x + 10 &= 22 \\ 4x &= 12 \\ x &= 3 \end{aligned}$$

$$\boxed{\{(3, 2)\}}$$

15. $\begin{aligned} 5x - 2y &= 8 \\ 3x - 5y &= 1 \end{aligned}$ $\quad \rightarrow \quad$ $\begin{aligned} 25x - 10y &= 40 \\ \underline{-6x + 10y} &= \underline{-2} \\ 19x &= 38 \\ x &= 2 \\ 10 - 2y &= 8 \\ -2y &= -2 \\ y &= 1 \end{aligned}$

$$\boxed{\{(2, 1)\}}$$

16. $\begin{aligned} x &= 2y \\ 2x + 6y &= 5 \end{aligned}$ $\quad \rightarrow \quad$ $\begin{aligned} -2x + 4y &= 0 \\ \underline{2x + 6y} &= \underline{5} \\ 10y &= 5 \\ y &= \frac{1}{2} \\ x &= 2\left(\frac{1}{2}\right) = 1 \end{aligned}$

$$\boxed{\left\{\left(1, \frac{1}{2}\right)\right\}}$$

17. $\begin{aligned} 4(x + 3) &= 3y + 7 \\ 2(y - 5) &= x + 5 \end{aligned}$ $\quad \rightarrow \quad$ $\begin{aligned} 4x + 12 &= 3y + 7 \\ 2y - 10 &= x + 5 \end{aligned}$

$\rightarrow \begin{aligned} 4x - 3y &= -5 \\ -x + 2y &= 15 \end{aligned}$ $\quad \rightarrow \quad$ $\begin{aligned} 4x - 3y &= -5 \\ \underline{-4x + 8y} &= \underline{60} \\ 5y &= 55 \\ y &= 11 \end{aligned}$

$x = 2y - 15 = 22 - 15 = 7$

$$\boxed{\{(7, 11)\}}$$

18. $\begin{aligned} 2x + y &= 5 \\ 2x + y &= 7 \end{aligned}$ $\quad \rightarrow \quad$ $\begin{aligned} 2x + y &= 5 \\ \underline{-2x - y} &= \underline{-7} \\ 0 &= -2 \ \text{False} \end{aligned}$

no solution; The system is $\boxed{\text{inconsistent}}$.

$$\boxed{\varnothing}$$

19. $\begin{aligned} 3x - 4y &= -1 \\ -6x + 8y &= 2 \end{aligned}$ $\quad \rightarrow \quad$ $\begin{aligned} 3x - 4y &= -1 \\ \underline{-3x + 4y} &= \underline{1} \\ 0 &= 0 \ \text{True} \end{aligned}$

The system has infinitely many solutions.
The equations are $\boxed{\text{dependent}}$.

$$\boxed{\{(x, y) \mid 3x - 4y = -1\}}$$

20. $\begin{aligned} 2x + 7y &= 0 \\ 7x + 2y &= 0 \end{aligned}$ $\quad \rightarrow \quad$ $\begin{aligned} -4x - 14y &= 0 \\ \underline{49x + 14y} &= \underline{0} \\ 45x &= 0 \\ x &= 0 \\ y &= 0 \end{aligned}$

$$\boxed{\{(0, 0)\}}$$

21. $\frac{1}{2}x + \frac{1}{4}y = \frac{1}{2}$ $\rightarrow$ $2x + y = 2$ $\rightarrow$ $-10x - 5y = -10$

$\frac{1}{5}x + \frac{1}{8}y = 0$ $\qquad\qquad$ $8x + 5y = 0$ $\qquad\qquad$ $\underline{8x + 5y = 0}$

$$-2x = -10$$
$$x = 5$$
$$y = 2 - 2(5) = -8$$
$$\boxed{\{(5, -8)\}}$$

22. $6x + 4y = -3$ $\rightarrow$ $-12x - 8y = 6$

$12x - 10y = -15$ $\qquad$ $\underline{12x - 10y = -15}$

$$-18y = -9$$
$$y = \frac{1}{2}$$
$$6x + 4\left(\frac{1}{2}\right) = -3$$
$$6x = -5$$
$$x = -\frac{5}{6}$$
$$\boxed{\left\{\left(-\frac{5}{6}, \frac{1}{2}\right)\right\}}$$

23. $x = -3y$

$3y + x = -1$ (substitute for x) $\rightarrow$ $3y + (-3y) = -1$

$$0 = -1 \text{ False}$$

no solution; The system is $\boxed{\text{inconsistent}}$.

$\boxed{\varnothing}$

24. $x + y = 3$ (solve for y) $\rightarrow$ $y = 3 - x$

$3x + 2y = 9$ (substitute for y) $\rightarrow$ $3x + 2(3 - x) = 9$

$$3x + 6 - 2x = 9$$
$$x = 3$$
$$y = 3 - 3 = 0$$
$$\boxed{\{(3, 0)\}}$$

25. $x + 3y = -4$ (solve for x) $\rightarrow$ $x = -3y - 4$

$3x + 2y = 3$ (substitute for x) $\rightarrow$ $3(-3y - 4) + 2y = 3$

$$-9y - 12 + 2y = 3$$
$$-7y = 15$$
$$y = -\frac{15}{7}$$
$$x = -3\left(-\frac{15}{7}\right) - 4 = \frac{45}{7} - \frac{28}{7} = \frac{17}{7}$$
$$\boxed{\left\{\left(\frac{17}{7}, -\frac{15}{7}\right)\right\}}$$

26.　$\begin{aligned} y + 1 &= 3x \\ 8x - 1 &= 4y \end{aligned}$　(solve for y) →　(substitute for y) →　$\begin{aligned} y &= 3x - 1 \\ 8x - 1 &= 4(3x - 1) \\ 8x - 1 &= 12x - 4 \\ 3 &= 4x \\ \frac{3}{4} &= x \end{aligned}$

$$y = 3\left(\frac{3}{4}\right) - 1 = \frac{5}{4}$$

$$\boxed{\left\{\left(\frac{3}{4}, \frac{5}{4}\right)\right\}}$$

27.　$\begin{aligned} 3x - 2y &= -4 \\ x &= -2 \end{aligned}$　(substitute for x) →　$\begin{aligned} -6 - 2y &= -4 \\ -2y &= 2 \\ y &= -1,\, x = -2 \end{aligned}$

$$\boxed{\{(-2, -1)\}}$$

28.　$\begin{aligned} y &= 39 - 3x \\ y &= 2x - 61 \end{aligned}$　(substitute for y) →　$\begin{aligned} 39 - 3x &= 2x - 61 \\ -5x &= -100 \\ x &= 20 \\ y &= 39 - 60 = -21 \end{aligned}$

$$\boxed{\{(20, -21)\}}$$

29.　$\begin{aligned} 3x + 4y &= 6 \\ y - 6x &= 6 \end{aligned}$　(solve for y) →　$\begin{aligned} y &= 6x + 6 \\ 3x + 4(6x + 6) &= 6 \\ 3x + 24x + 24 &= 6 \end{aligned}$

(simplify) →　$\begin{aligned} 27x &= -18 \\ x &= -\frac{2}{3} \\ y &= 6\left(-\frac{2}{3}\right) + 6 = -4 + 6 = 2 \end{aligned}$

$$\boxed{\left\{\left(-\frac{2}{3}, 2\right)\right\}}$$

30.　$\begin{aligned} 2x - y &= 4 \\ x &= y + 1 \end{aligned}$　(substitute for x) →　$\begin{aligned} 2(y + 1) - y &= 4 \\ y &= 2 \\ x &= 2 + 1 = 3 \end{aligned}$

$$\boxed{\{(3, 2)\}}$$

31.　$\begin{aligned} 4x + y &= 5 \\ 12x &= 15 - 3y \end{aligned}$　(solve for y) →　(substitute for y) →　$\begin{aligned} y &= 5 - 4x \\ 12x &= 15 - 3(5 - 4x) \\ 12x &= 15 - 15 + 12x \\ 0 &= 0 \ \text{True} \end{aligned}$

The system has infinitely many solutions.

The equations are $\boxed{\text{dependent}}$.

$$\boxed{\{(x, y) \mid 4x + y = 5\}}$$

32.
$$4x - y = -3 \quad \text{(substitute for } y) \rightarrow \quad 4x - 4x = -3$$
$$y = 4x \qquad\qquad\qquad\qquad\qquad 0 = -3 \;\; \text{False}$$

no solution; The system is $\boxed{\text{inconsistent}}$.

$\boxed{\varnothing}$

33.
$$2x + 7y = -4$$
$$3x - y = 17 \qquad \text{(solve for } y) \rightarrow \qquad\qquad\qquad y = 3x - 17$$
$$\text{(substitute for } y) \rightarrow \quad 2x + 7(3x - 17) = -4$$
$$2x + 21x - 119 = -4$$
$$23x = 115$$
$$x = 5$$
$$y = 15 - 17 = -2$$

$\boxed{\{(5, -2)\}}$

34.
$$\frac{1}{4}x + 2y = \frac{1}{2} \qquad \rightarrow \qquad x + 8y = 2$$
$$\frac{1}{2}x + \frac{1}{3}y = -\frac{8}{3} \qquad\qquad 3x + 2y = -16$$

$$\text{(solve for } x \text{ in first equation)} \rightarrow \qquad x = 2 - 8y$$
$$\text{(substitute for } x \text{ in second equation)} \rightarrow \quad 3(-8y + 2) + 2y = -16$$
$$-24y + 6 + 2y = -16$$
$$-22y = -22$$
$$y = 1$$
$$x = -8 + 2 = -6$$

$\boxed{\{(-6, 1)\}}$

35. Let x equal the first number and y equal the second number.
$$3x + 8y = -1 \qquad \rightarrow \qquad 3x + 8y = -1$$
$$x - 2y = -5 \qquad\qquad\qquad \underline{4x - 8y = -20}$$
$$7x = -21$$
$$x = -3$$
$$-3 - 2y = -5$$
$$-2y = -2$$
$$y = 1$$

The numbers are $\boxed{-3 \text{ and } 1}$.

36. Let t equal the tens' digit and u equals the units' digit.
$$3t + 4u = 43 \qquad \rightarrow \qquad 3t + 4u = 43 \qquad \rightarrow \qquad 3t + 4u = 43$$
$$10t + u = 15 + 6u \qquad\quad 10t - 5u = 15 \qquad\qquad\quad 2t - u = 3$$
$$\rightarrow \qquad 3t + 4u = 43$$
$$\underline{8t - 4u = 12}$$
$$11t = 55$$
$$t = 5$$
$$u = 2t - 3 = 7$$

The number is $\boxed{57}$.

37. Let w equal the width.
l equals the length.
$$2w + 2l = 24 \qquad\qquad -2w - 2l = -24$$
$$2(2w) + 3(2l) = 62 \qquad\quad \underline{2w + 3l = 31}$$
$$l = 7$$
$$w = 12 - 7 = 5$$

$\boxed{\text{length, 7 yards; width, 5 yards}}$

38. $m\angle A = m\angle B \rightarrow 3y + 20 = 4x - 30$

$(3y + 20) + (4x - 30) + (x + 5y + 10) = 180$

$$\begin{array}{rclcrcl}
\rightarrow \quad -4x + 3y & = & -50 & \rightarrow & -20x + 15y & = & -250 \\
5x + 8y & = & 180 & & \underline{20x + 32y} & = & \underline{720} \\
& & & & 47y & = & 470 \\
& & & & y & = & 10 \\
& & & & -4x + 30 & = & -50 \\
& & & & -4x & = & -80 \\
& & & & x & = & 20
\end{array}$$

$m\angle A = (3y + 20)^\circ = \boxed{50^\circ}$

$m\angle B = (4x - 30)^\circ = \boxed{50^\circ}$

$m\angle C = (x + 5y + 10)^\circ = \boxed{80^\circ}$

39.
$$\begin{array}{rclcrcl}
10y + 5 & = & 3x + 10 & \rightarrow & 3x - 10y & = & -5 \\
8x + 5 + 10y + 5 & = & 180 & & \underline{8x + 10y} & = & \underline{170} \\
& & & & 11x & = & 165 \\
& & & & x & = & \boxed{15} \\
& & & & 10y & = & 3x + 5 = 50 \\
& & & & y & = & \boxed{5}
\end{array}$$

$(8x + 5)^\circ = (8 \cdot 15 + 5)^\circ = \boxed{125^\circ}$

$(10y + 5)^\circ = (10 \cdot 5 + 5)^\circ = \boxed{55^\circ}$

$(3x + 10)^\circ = (3 \cdot 5 + 10) = \boxed{55^\circ}$

40. Let x equal the cost of one pen.
y equals the cost of one pad.

$$\begin{array}{rclcrcl}
8x + 6y & = & 3.90 & \rightarrow & -8x - 6y & = & -3.90 \\
3x + 2y & = & 1.40 & & \underline{9x + 6y} & = & \underline{4.20} \\
& & & & x & = & 0.30 \\
& & & & 2y & = & 1.40 - 0.90 = 0.50 \\
& & & & y & = & 0.25
\end{array}$$

$\boxed{\text{cost of one pen, } \$0.30}$

41. $y = mx + b$

$(-4, 8):\quad 8 = -4m + b$

$(2, 5):\quad 5 = 2m + b$

$$\begin{array}{rcl}
\rightarrow \quad 8 & = & -4m + b \\
\underline{10} & = & \underline{4m + 2b} \\
18 & = & 3b \\
6 & = & b \\
2m & = & 5 - 6 = -1 \\
m & = & -\dfrac{1}{2}
\end{array}$$

$y = -\dfrac{1}{2}x + 6 \qquad \boxed{m = -\dfrac{1}{2}, b = 6}$

42. cost: $\boxed{C = 50x + 80{,}000}$

revenue: $\boxed{R = 250x}$

break-even point:
$$\begin{array}{rcl}
R & = & C \\
250x & = & 50x + 80{,}000 \\
200x & = & 80{,}000 \\
x & = & \boxed{400} \\
R & = & 250x = \$100{,}000 \\
(400, & & \$100{,}000)
\end{array}$$

43. Let x equal the speed of the plane in still air.
   y equals the speed of the wind.

$$
\begin{aligned}
6(x+y) &= 1080 \\
3(x-y) &= 360
\end{aligned}
\qquad \rightarrow \qquad
\begin{aligned}
x+y &= 180 \\
\underline{x-y} &= \underline{120} \\
2x &= 300 \\
x &= 150 \\
y &= 30
\end{aligned}
$$

speed of plane in still air, 150 mph; speed of wind, 30 mph

44. Let x equal the speed of boat in still water
   y equals the speed of current.

$$
\begin{aligned}
1(x+y) &= 12 \\
1.5(x-y) &= 12
\end{aligned}
\qquad
\begin{aligned}
\rightarrow \\
\rightarrow
\end{aligned}
\qquad
\begin{aligned}
x+y &= 12 \\
\underline{x-y} &= \underline{8} \\
2x &= 20 \\
x &= 10 \\
y &= 2
\end{aligned}
$$

speed of boat in still water, 10 mph; speed of current, 2 mph

45. Let x equal the present age of Lestat.
   y equals the present age of Louis.
   $x - 2 =$ age of Lestat two years ago
   $y - 2 =$ age of Louis two years ago
   $x + 4 =$ age of Lestat four years from now
   $y - 1 =$ age of Louis one year ago

$$
\begin{aligned}
y-2 &= 2(x-2) \\
x+4 &= y-1
\end{aligned}
\qquad \rightarrow \qquad
\begin{aligned}
-2x+y &= -2 \\
\underline{x-y} &= \underline{-5} \\
-x &= -7 \\
x &= 7 \\
y &= 12
\end{aligned}
$$

Lestat, 7 years old; Louis, 12 years old

46. Let x equal the number of apples Bud has.
   y equals the number of apples Lou has.

$$
\begin{aligned}
x-1 &= y+1 \\
2(y-1) &= x+1
\end{aligned}
\qquad \rightarrow \qquad
\begin{aligned}
x-y &= 2 \\
\underline{-x+2y} &= \underline{3} \\
y &= 5 \\
x &= 7
\end{aligned}
$$

Bud, 7 apples; Lou, 5 apples

47. $2x + y < 6$
    $y - 2x < 6$

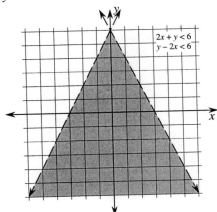

48. $2x + 3y \le 6$
    $y > 3x$

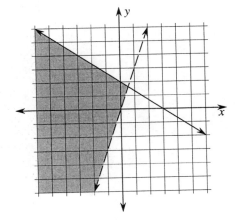

49. $y < 2x - 2$
    $x > 3$

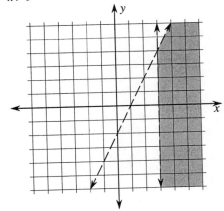

50. $y \ge 5x - 4$
    $y \le 5x + 1$

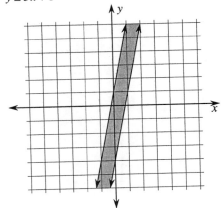

51. $x < 6$
    $y \ge -1$

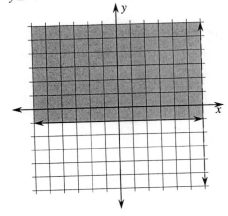

52. $2x + 3y \ge 6$
    $3x - y \le 3$

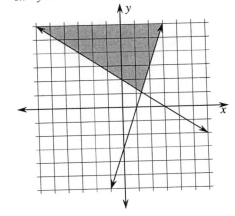

53. a. $2x + 2y \leq 2600$
   $x > 800$

  b.

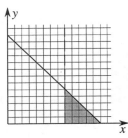

  c. Answers will vary. Samples given.
      (900, 400), (1000, 300), (1100, 200)

Cumulative Review Problems (Chapters 1-5)

Cumulative Review, pp. 369-370

1. $6(3y - 2) - (y - 14) - 2(8y + 7)$
   $= 18y - 12 - y + 14 - 16y - 14$
   $= \boxed{y - 12}$

2. $-14 - [18 - (6 - 10)]$
   $= -14 - (18 + 4)$
   $= -14 - 22$
   $= \boxed{-36}$

3. Let $x, x + 2, x + 4$ represent three consecutive odd integers.
$$
\begin{aligned}
x + (x + 4) &= 11 + (x + 2) \\
2x + 4 &= x + 13 \\
x &= 9 \\
x + 2 &= 11 \\
x + 4 &= 13
\end{aligned}
$$
  The integers are $\boxed{9, 11, \text{ and } 13}$.

4. $\begin{aligned} 3y + 2y - 7(y + 1) &= -3(y + 4) \\ -2y - 7 &= -3y - 12 \\ y &= -5 \end{aligned}$

  $\boxed{\{-5\}}$

5. Let x equal the number.
$$
\begin{aligned}
6(x + 5) &\leq 72 \\
x + 5 &\leq 12 \\
x &\leq 7
\end{aligned}
$$
  $\boxed{\text{The number is at most } 7}$.

6. $A = p + prt$

  $A - p = prt$

  $\boxed{\dfrac{A - p}{pr} = t}$ or $\boxed{t = \dfrac{A}{pr} - \dfrac{1}{r}}$

7. $\dfrac{y + 7}{3} = \dfrac{y - 1}{4}$

  $4(y + 7) = 3(y - 1)$

  $4y + 28 = 3y - 3$

  $y = -31$

  $\boxed{\{-31\}}$

8. $a > 2$: 3, 5, 7, 9
$b < 8$: 1, 3, 5, 7
c, d even: 2, 4, 6, 8
$a < c < d$
$c < d < b$
$a < c < d < b$
$3 < 4 < 6 < 7$
$\boxed{d = 6}$

9. Let x equal the length of each side of the square.
$2x - 10$ equals the length of each side of the equilateral triangle.
perimeter of triangle = perimeter of square
$$
\begin{aligned}
3(2x - 10) &= 4x \\
6x - 30 &= 4x \\
2x &= 30 \\
x &= 15 \\
2x - 10 &= 30 - 10 = 20
\end{aligned}
$$
length of each side of triangle: $\boxed{20 \text{ centimeters}}$

10.
$$
\begin{aligned}
t5 + 37 + 51 + 4u &= 161 \\
t5 + 4u &= 161 - 88 \\
t5 + 4u &= 73 \\
t + 4 &= 7 - 1 \qquad \boxed{t = 2} \\
5 + u &= 13 \qquad \boxed{u = 8}
\end{aligned}
$$

11. $6x - 3y = 12$

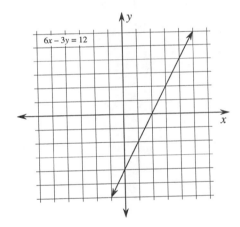

12. $y = \dfrac{1}{2}x - 2$

$m = \dfrac{1}{2},\ b = -2$

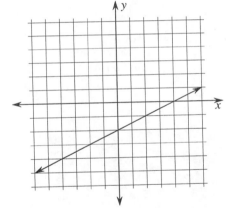

13. $y \geq 3x - 1$

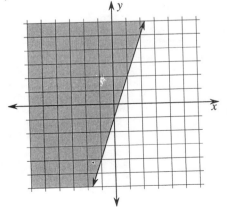

14. $3x - 4y = 8$ → $15x - 20y = 40$
         $4x + 5y = -10$ $\underline{16x + 20y = -40}$
                                              $31x = 0$
                                              $x = 0$
                                              $-4y = 8 - 0$
                                              $y = -2$
                                              $\boxed{\{(0, -2)\}}$

15. passing through $(2, -3)$ and parallel to the line whose equation is

$x + 2y = 4$ → $y = -\dfrac{1}{2}x + 2$

$m = -\dfrac{1}{2}$

point-slope: $\boxed{y + 3 = -\dfrac{1}{2}(x - 2)}$

slope-intercept: $y + 3 = -\dfrac{1}{2}x - 2$

$\boxed{y = -\dfrac{1}{2}x - 5}$

standard: $\dfrac{1}{2}x + y = -5$

$\boxed{x + 2y = -4}$

16. $f(x) = -x^2 - 3x + 4$
         $f(-2) = -(-2)^2 - 3(-2) + 4$
              $= -4 + 6 + 4 = \boxed{6}$

17. $2x - y < 0$

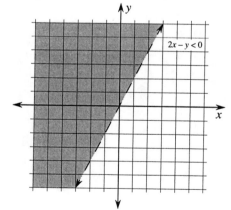

18. Let x equal the number of nickels.
$x + 4$ equals the number of dimes
$3(x + 4)$ = number of quarters

$0.05x + 0.10(x + 4) + 0.25(3)(x + 4) = 9.70$
$0.05x + 0.10x + 0.40 + 0.75x + 3.00 = 9.70$
$0.90x = 6.30$
$x = 7$
$x + 4 = 11$
$3(x + 4) = 33$

$\boxed{\text{7 nickels, 11 dimes, 33 quarters}}$

19. $3(y + 1) \le 5(2y - 4) + 2$

$3y + 3 \le 10y - 20 + 2$

$-7y \le -21$

$y \ge 3$

$\boxed{\{y \mid y \ge 3\}}$

20. $4 - \dfrac{1}{3}z \le 6 + \dfrac{2}{3}z$

$-z \le 2$

$z \ge -2$

$\boxed{\{z \mid z \ge -2\}}$

21. Let x equal the amount invested at 9%.

$17,000 - x$ equals the amount invested at 11%.

$$0.09x + 0.11(17,000 - x) = 1670$$
$$0.09x + 1870 - 0.11x = 1670$$
$$0.09x + 1870 - 0.11x = 1670$$
$$-0.02x = -200$$
$$x = 10,000$$
$$17,000 - x = 7000$$

$\boxed{\$10,000 \text{ at } 9\%, \$7000 \text{ at } 11\%}$

22. y is $\boxed{\text{not a function}}$ of x; for some values of x there are 2 values of y.

23. $\dfrac{1}{4}x + \dfrac{1}{4}y = 1 \quad \rightarrow \quad x + y = 4 \quad \rightarrow \quad x + y = 4$

$\dfrac{1}{2}x + \dfrac{1}{2}y = -1 \qquad x + y = -2 \qquad \underline{-x - y = 2}$

$0 = 6 \;\; \text{False}$

no solution; The system is $\boxed{\text{inconsistent}}$.

$\boxed{\varnothing}$

24. Let x equal the number.

$2x + 15 = 9 + x$

$x = \boxed{-6}$

25. $y = -x^2 + 4x - 3$

x	-1	0	1	2	3	4
y	-8	-3	0	1	0	-3

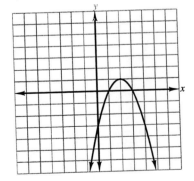

26. $y < -3$

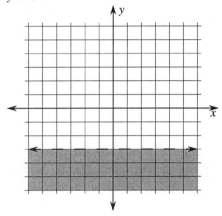

27.
$$0.24y - 0.38(2 + y) = -0.34(y + 4)$$
$$0.24y - 0.76 - 0.38y = -0.34y - 1.36$$
$$0.20y = -0.60$$
$$y = -3$$

$\boxed{\{-3\}}$

28. Let x equal the amount of 25% salt solution.
$x + 20$ equals the amount of 20% salt solution.
$$0.25x + 0.12(20) = 0.20(x + 20)$$
$$0.25x + 2.4 = 0.20x + 4$$
$$0.05x = 1.6$$
$$x = 32$$

$\boxed{32 \text{ liters}}$ of 25% salt solution

29. Let t equal the time walked.
$\frac{1}{2} - t$ equals the time ran
$$3t = 6\left(\frac{1}{2} - t\right)$$
$$3t = 3 - 6t$$
$$9t = 3$$
$$t = \frac{1}{3}\text{hr}$$

distance walked $= 3\left(\frac{1}{3}\right) = \boxed{1 \text{ mile}}$

30. Let B equal the number of votes for candidate B.
$A = B - 30$ equals the number of votes for candidate A.
$$A + B + 40 = 200$$
$$A + B = 160$$
$$A - B = -30$$
$$2A = 130$$
$$A = 65$$
$$B = 95$$

$\boxed{\text{candidate A, 65 votes; candidate B, 95 votes}}$

31. $\dfrac{432}{-1.045^8} \approx \boxed{-304}$

32. $3.8(50 - 2.3^4) \div 6.1 \approx \boxed{13.71}$

Chapter 6   Exponents and Polynomials

Section 6.1 Adding and Subtracting Polynomials

Problem Set 6.1, pp. 379-381

1. $3x + 7$ is a $\boxed{\text{binomial}}$ of degree $\boxed{1}$.

3. -9 is a $\boxed{\text{monomial}}$ of degree $\boxed{\text{zero}}$.

5. $x^3 - 2x$ is a $\boxed{\text{binomial}}$ of degree $\boxed{3}$.

7. $x^2 - 3x + 4$ is a $\boxed{\text{trinomial}}$ of degree $\boxed{2}$.

9. $3y^{17}$ is a $\boxed{\text{monomial}}$ of degree $\boxed{17}$.

11. $7y^2 - 9y^4 + 5$ is a $\boxed{\text{trinomial}}$ of degree $\boxed{4}$.

13. $4x - 10x$ is a $\boxed{\text{binomial}}$ of degree $\boxed{1}$.

15. $5x - 10x^2 = \boxed{-10x^2 + 5x}$; degree $\boxed{2}$

17. $3x + 4x^5 - 3x^2 - 2 = \boxed{4x^5 - 3x^2 + 3x - 2}$; degree $\boxed{5}$

19. $3 - 3y^4 = \boxed{-3y^4 + 3}$; degree $\boxed{4}$

21. $13 = \boxed{13}$; degree $\boxed{0}$

23. A trinomial is $\boxed{\text{always}}$ polynomial.

25. A monomial is $\boxed{\text{never}}$ trinomial.

27. A binomial $\boxed{\text{sometimes}}$ has a degree of 3.

29. $\boxed{D}$ is true
A. The coefficient of x is -5 not 5.
B. The degree is 3 *not* 2.
C. $\dfrac{1}{5x^2} + \dfrac{1}{3x}$ is *not* a polynomial .
D. $\begin{aligned} (2x^2 - 8x + 6) - (x^2 - 3x + 5) &= x^2 - 5x + 1 \\ 2x^2 - 8x + 6 - x^2 + 3x - 5 &= x^2 - 5x + 1 \\ x^2 - 5x + 1 &= x^2 - 5x + 1 \\ 0 &= 0 \quad \text{True} \end{aligned}$

31. $\begin{aligned} f(x) &= 60x^2 - x^3 \\ f(10) &= 60(10)^2 - 10^3 = 6000 - 1000 = \boxed{5000} \end{aligned}$
$\boxed{\text{5000 people are sick with the flu after 10 days.}}$

33. $x^2 - x + 41 = 8^2 - 8 + 41 = 64 - 8 + 41 = \boxed{97}$

35. $\begin{aligned} f(x) &= 0.002x^2 - 0.21x + 15 \\ f(80) &= 0.002(80)^2 - 0.21(80) + 15 \\ &= 12.8 - 16.8 + 15 = \boxed{11} \end{aligned}$
$\boxed{\text{The cost of operating an automobile at 80 kilometers per hour is 11 cents per kilometer.}}$

37. $(5x^2 - 3x) + (2x^2 - x)$
$= (5x^2 + 2x^2) + (-3x - x)$
$= \boxed{7x^2 - 4x}$

39. $(3x^2 - 7x + 10) + (x^2 + 6x + 8)$
$= (3x^2 + x^2) + (-7x + 6x) + (10 + 8)$
$= \boxed{4x^2 - x + 18}$

41. $(4y^3 + 7y - 5) + (10y^2 - 6y + 3)$
$= 4y^3 + 10y^2 + (7y - 6y) + (-5 + 3)$
$= \boxed{4y^3 + 10y^2 + y - 2}$

43. $(2x^2 - 6x + 7) + (3x^3 - 3x)$
$= 3x^3 + 2x^2 + (-6x - 3x) + 7$
$= \boxed{3x^3 + 2x^2 - 9x + 7}$

45. $(4y^2 + 8y + 11) + (-2y^3 + 5y + 2)$
$= -2y^3 + 4y^2 + (8y + 5y) + (11 + 2)$
$= \boxed{-2y^3 + 4y^2 + 13y + 13}$

47. $(-2y^6 + 3y^4 - y^2) + (-y^6 + 5y^4 + 2y^2)$
$= (-2y^6 - y^6) + (3y^4 + 5y^4) + (-y^2 + 2y^2)$
$= \boxed{-3y^6 + 8y^4 + y^2}$

49. $(3x^2 - 2x) - (5x^2 - 6x)$
$= 3x^2 - 2x - 5x^2 + 6x$
$= \boxed{-2x^2 + 4x}$

51. $(x^2 - 5x + 3) - (x^2 - 6x - 8)$
$= x^2 - 5x + 3 - x^2 + 6x + 8$
$= \boxed{x + 11}$

53. $(y - 2) - (7y - 9)$
$= y - 2 - 7y + 9$
$= \boxed{-6y + 7}$

57. $(2n^2 - n^7 - 6) - (2n^3 - n^7 - 8)$
$= 2n^2 - n^7 - 6 - 2n^3 + n^7 + 8$
$= \boxed{-2n^3 + 2n^2 + 2}$

59. $(y^5 - y^3) - (y^4 - y^2)$
$= y^5 - y^3 - y^4 + y^2)$
$= \boxed{y^5 - y^4 - y^3 + y^2}$

61. $(-3x^6 + 3x^4 - x^2) - (-x^6 + 2x^4 + 2x^2)$
$= -3x^6 + 3x^4 - x^2 + x^6 - 2x^4 - 2x^2$
$= \boxed{-2x^6 + x^4 - 3x^2}$

63. $(9 - 4y + y^2) - (3 - 5y - y^2)$
$= 9 - 4y + y^2 - 3 + 5y + y^2$
$= \boxed{2y^2 + y + 6}$

65. $(2x^2 + 3x + 7) - (-6x^2 - 7x + 3) + (5x^2 - 4x + 8)$
$= 2x^2 + 3x + 7 + 6x^2 + 7x - 3 + 5x^2 - 4x + 8$
$= \boxed{13x^2 + 6x + 12}$

67. $(-4y^2 - 5y + 7) + (3y^2 + 6y - 1) + (2y^2 - 3y + 2)$
$= -4y^2 - 5y + 7 + 3y^2 + 6y - 1 + 2y^2 - 3y + 2$
$= \boxed{y^2 - 2y + 8}$

69. $\left(\dfrac{5}{9}y^3 + \dfrac{1}{3}y^2 - 4y + 3\right) - \left(-\dfrac{2}{3}y^3 + y^2 + \dfrac{1}{2}y - 2\right)$

$= \dfrac{5}{9}y^3 + \dfrac{1}{3}y^2 - 4y + 3 + \dfrac{2}{3}y^3 - y^2 - \dfrac{1}{2}y + 2$

$= \left(\dfrac{5}{9}y^3 + \dfrac{6}{9}y^3\right) + \left(\dfrac{1}{3}y^2 - y^2\right) + \left(-4y - \dfrac{1}{2}y\right) + (3 + 2)$

$= \boxed{\dfrac{11}{9}y^3 - \dfrac{2}{3}y^2 - \dfrac{9}{2}y + 5}$

71. Add:
$\begin{array}{r} 5y^3 - 7y^2 \\ 6y^3 + 4y^2 \\ \hline \boxed{11y^3 - 3y^2} \end{array}$

73. Add:
$\begin{array}{r} y^3 + 5y^2 - 7y - 3 \\ -2y^3 + 3y^2 + 4y - 11 \\ \hline \boxed{-y^3 + 8y^2 - 3y - 14} \end{array}$

75. Add:
$\begin{array}{r} 4x^3 - 6x^2 + 5x - 7 \\ -9x^3 \qquad -4x + 3 \\ \hline \boxed{-5x^3 - 6x^2 + x - 4} \end{array}$

77. Add:
$\begin{array}{r} 7x^4 - 3x^3 + x^2 \\ x^3 - x^2 + 4x - 2 \\ \hline \boxed{7x^4 - 2x^2 + 4x - 2} \end{array}$

79. Add:
$$7x^2 - 9x + 3$$
$$4x^2 + 11x - 2$$
$$\underline{-3x^2 + 5x - 6}$$
$$\boxed{8x^2 + 7x - 5}$$

81. Subtract:
$$7x + 1 \quad \rightarrow \quad 7x + 1$$
$$\underline{-(3x - 5)} \qquad \underline{-3x + 5}$$
$$\qquad\qquad \boxed{4x + 6}$$

83. Subtract:
$$7x^2 - 3 \quad \rightarrow \quad 7x^2 - 3$$
$$\underline{-(-3x^2 + 4)} \qquad \underline{3x^2 - 4}$$
$$\qquad\qquad \boxed{10x^2 - 7}$$

85. Subtract:
$$7y^2 - 5y + 2 \quad \rightarrow \quad 7y^2 - 5y + 2$$
$$\underline{-(11y^2 + 2y - 3)} \qquad \underline{-11y^2 - 2y + 3}$$
$$\qquad\qquad\qquad \boxed{-4y^2 - 7y + 5}$$

87. Subtract:
$$7x^3 + 5x^2 - 3 \quad \rightarrow \quad 7x^3 + 5x^2 - 3$$
$$\underline{-(-2x^3 - 6x^2 + 5)} \qquad \underline{2x^3 + 6x^2 - 5}$$
$$\qquad\qquad\qquad \boxed{9x^3 + 11x^2 - 8}$$

89. Subtract:
$$5y^3 + 6y^2 - 3y + 10 \quad \rightarrow \quad 5y^3 + 6y^3 - 3y + 10$$
$$\underline{-(6y^3 - 2y^2 - 4y - 4)} \qquad \underline{-6y^3 + 2y^2 + 4y + 4}$$
$$\qquad\qquad\qquad\qquad \boxed{-y^3 + 8y^2 + y + 14}$$

91. Subtract:
$$7x^4 - 3x^3 + 2x^2 \quad \rightarrow \quad 7x^4 - 3x^3 + 2x^2$$
$$\underline{-(-x^3 - x^2 + x - 2)} \qquad \underline{x^3 + x^2 - x + 2}$$
$$\qquad\qquad\qquad \boxed{7x^4 - 2x^3 + 3x^2 - x + 2}$$

93. Subtract:
$$4y^3 - \frac{1}{2}y^2 + \frac{3}{8}y + 1 \quad \rightarrow \quad 4y^3 - \frac{1}{2}y^2 + \frac{3}{8}y + 1$$
$$-\left(\frac{9}{2}y^3 + \frac{1}{4}y^2 - y + \frac{3}{4}\right) \qquad \underline{-\frac{9}{2}y^3 - \frac{1}{2}y^2 + y - \frac{3}{4}}$$
$$\qquad\qquad\qquad \boxed{-\frac{1}{2}y^3 - y^3 + \frac{11}{8}y + \frac{1}{4}}$$

95.
$$3x^2 + x - 2$$
$$4x - 3x + 2$$
$$\underline{3x^2 - 2x - 5}$$
$$\boxed{10x^2 - 2x - 5}$$

97.
$$11x^2 - 20x + 10 \quad \rightarrow \quad 11x^2 - 20x + 10$$
$$\underline{-(-9x^2 + 24x - 30)} \qquad \underline{9x^2 - 24x + 30}$$
$$\qquad\qquad\qquad \boxed{20x^2 - 44x + 40}$$

99.
$$4x^3 + x^2$$
$$\underline{+(-x^3 + \qquad 7x - 3)}$$
sum: $\quad 3x^3 + x^2 + 7x - 3 \quad \rightarrow \quad 3x^3 + x^2 + 7x - 3$
$$\underline{-(x^3 - 2x^2 \qquad + 2)} \qquad \underline{-x^3 + 2x^2 \qquad - 2}$$
$$\qquad\qquad\qquad\qquad \boxed{2x^3 + 3x^2 + 7x - 5}$$

101.
$$f(x) = x^3 - x^2 - x - 1$$
$$f(-1) = (-1)^3 - (-1)^2 - (-1) - 1$$
$$= -1 - 1 + 1 - 1 = \boxed{-2}$$

103. $C - R$
$$= (t^3 - 3t^2 + 5t) - \left(\frac{1}{3}t^3 - t^2 + t\right)$$
$$= t^3 - 3t^2 + 5t - \frac{1}{3}t^3 + t^2 - t$$
$$= \boxed{\frac{2}{3}t^3 - 2t^2 + 4t}$$

105.

$$-3x^2 + x$$
$$\underline{+ \quad\quad (?)} \quad \rightarrow$$
sum: $\ x^2 + 7x$

$\rightarrow$

$$x^2 + 7x$$
$$\underline{-(-3x^2 + x)}$$

$\rightarrow$

$$x^2 + 7x$$
$$\underline{3x^2 - x}$$
$$\boxed{4x^2 + 6x}$$

107.

$$x^2 - 3x$$
$$\underline{-(\quad ? \quad)} \quad \rightarrow$$
difference: $\ 3x^2 - 4x$

$$x^2 - 3x$$
$$\underline{-(3x^2 - 4x)} \quad \rightarrow$$

$$x^2 - 3x$$
$$\underline{-3x^2 + 4x}$$
$$\boxed{-2x^2 + x}$$

Review Problems

111.

$$\frac{1}{3}y + \frac{1}{4}y = \frac{7}{4}y - \frac{5}{2}$$

$$\frac{1}{3}y - \frac{6}{4}y = -\frac{5}{2}$$

$$\frac{1}{3}y - \frac{3}{2}y = -\frac{5}{2}$$

$$\frac{-7}{6}y = -\frac{5}{2}$$

$$y = \left(-\frac{6}{7}\right)\left(-\frac{5}{2}\right) = \frac{15}{7}$$

$$\boxed{\left\{\frac{15}{7}\right\}}$$

112. $(6, 1), (-2, -5)$

$$m = \frac{-5 - 1}{-2 - 6} = \frac{-6}{-8} = \frac{3}{4}$$

$$y - 1 = \frac{3}{4}(x - 6)$$

$$4y - 4 = 3x - 18$$

$$\boxed{3x - 4y = 14}$$

113. Let x equal the length of the side of the square.
$x + 2.5$ equals the length of the side of the equilateral triangle.

$$\text{perimeter of square} = \text{perimeter of triangle}$$

$$4x = 3(x + 2.5)$$
$$4x = 3x + 7.5$$
$$x = 7.5$$
$$x + 2.5 = 10$$

length of each side of the triangle: $\boxed{10 \text{ centimeters}}$

Section 6.2 Multiplying Polynomials

Problem Set 6.2, pp. 390-393

1. $x^3 \cdot x = x^{3+7} = \boxed{x^{10}}$

3. $(2x^2)(4x^3) = 8x^{2+3} = \boxed{8x^5}$

5. $(-7y)(3y^7) = -21y^{1+7} = \boxed{-21y^8}$

7. $(-2x^3)(-3x^2) = 6x^{3+2} = \boxed{6x^5}$

9. $x^3 \cdot x^2 \cdot x = x^{3+2+1} = \boxed{x^6}$

11. $(x^3y^2)(x^4y) = x^{3+4}y^{2+1} = \boxed{x^7y^3}$

13. $(2x^2y)(-4x^4y) = x^{2+4}y^{1+1} = \boxed{-8x^6y^2}$

15. $(x^3)^4 = x^{3 \cdot 4} = \boxed{x^{12}}$

17. $(5x)^2 = 5^2x^2 = \boxed{25x^2}$

19. $(-2y)^3 = (-2)^3y^3 = \boxed{-8y^3}$

21. $(2x^2)^2 = 2^2(x^2)^2 = 4x^{2 \cdot 2} = \boxed{4x^4}$

23. $(4y^2)^3 = 4^3(y^2)^3 = 64y^{2 \cdot 3} = \boxed{64y^6}$

25. $(-3y^4)^3 = (-3)^3y^{4 \cdot 3} = \boxed{-27y^{12}}$

27. $(-x^2)^2 = (-1)^2\,x^{2 \cdot 2} = \boxed{x^4}$

29. $(-2^2)^3 = (-1)^3\,2^{2 \cdot 3} = -2^6 = \boxed{-64}$

31. $(x^3y^2)^2 = x^{3 \cdot 2}y^{2 \cdot 2} = \boxed{x^6y^4}$

33. $(3x^2y)^2 = 3^2x^{2 \cdot 2}y^2 = \boxed{9x^4y^2}$

35. $(-2ab^3)^4 = (-2)^4a^4b^{3 \cdot 4} = \boxed{16a^4b^{12}}$

37. $(-4x^2y^3)^3 = (-4)^3x^{2 \cdot 3}y^{3 \cdot 3} = \boxed{-64x^6y^9}$

39. $x^3(3x^2)^3 = x^3(3^3x^6) = 27x^{3+6} = \boxed{27x^9}$

41. $(-2y)(2y^3)^2 = (-2y)(4y^6) = \boxed{-8y^7}$

43. $(xy^2)(x^2y)^3 = xy^2x^{2 \cdot 3}y^3 = xx^6y^{2+3} = \boxed{x^7y^5}$

45. $(xy^2)^2(xy)^2 = x^2y^4x^2y^2 = \boxed{x^4y^6}$

47. $(-2a)(-2a^3b)^3 = (-2a)(-8a^9b^3) = \boxed{16a^{10}b^3}$

49. $(-2x)(-3x^2y)^2 = (-2x)(9x^4y^2) = \boxed{-18x^5y^2}$

51. $(xy^2)(-2x^2y)^3 = (xy^2)(-8x^6y^3) = \boxed{-8x^7y^5}$

53. $(-2x^3)(3x^2y)^3 = (-2x^3)(27x^6y^3) = \boxed{-54x^9y^3}$

55. $(-3xy)^2(-2xy)^3 = 9x^2y^2(-8x^3y^3) = \boxed{-72x^5y^5}$

57. $(4x)(2x^2) + (4x^2)(3x) = 8x^3 + 12x^3 = \boxed{20x^3}$

59. $x(x-3) = x \cdot x - 3x = \boxed{x^2 - 3x}$

61. $-x(x+4) = \boxed{-x^2 - 4x}$

63. $4x^2(x-2) = 4x^2(x) - 4x^2(2) = \boxed{4x^3 - 8x^2}$

65. $2x^2(x^2 + 3x) = 2x^2(x^2) + 2x^2(3x) = \boxed{2x^4 + 6x^3}$

67. $-5x^2(x^2 - x) = -5x^2(x^2) - 5x^2(-x) = \boxed{-5x^4 + 5x^3}$

69. $-y^3(3y^2 - 5) = -y^3(3y^2) - y^3(-5) = \boxed{-3y^5 + 5y^3}$

71. $3x(6x^2 - 5x) = 3x(6x^2) - 3x(5x) = \boxed{18x^3 - 15x^2}$

73. $(4x-3)5x = 4x(5x) - 3(5x) = \boxed{20x^2 - 15x}$

75. $(3x^3 - 4x^2)(-2x) = 3x^3(-2x) - 4x^2(-2x) = \boxed{-6x^4 + 8x^3}$

77. $x(3x^3 - 2x + 5) = x(3x^3) - x(2x) + x(5) = \boxed{3x^4 - 2x^2 + 5x}$

79. $-y(-3y^2 - 2y - 4) = -y(-3y^2) - y(-2y) - y(-4) = \boxed{3y^3 + 2y^2 + 4y}$

81. $x^2(3x^4 - 5x - 3) = x^2(3x^4) + x^2(-5x) + x^2(-3) = \boxed{3x^6 - 5x^3 - 3x^2}$

83. $2x^2(3x^2 - 4x + 7) = 2x^2(3x^2) + 2x^2(-4x) + 2x^2(7) = \boxed{6x^4 - 8x^3 + 14x^2}$

85. $(x^2 + 5x - 3)(-2x) = x^3(-2x) + 5x(-2x) - 3(-2x) = \boxed{-2x^3 - 10x^2 + 6x}$

87. $-3x^2(-4x^2 + x - 5) = -3x^2(-4x^2) - 3x^2(x) - 3x^2(-5) = \boxed{12x^4 - 3x^3 + 15x^2}$

89. $3y^3(5y^6 - 2y^4 - 9) = 3y^3(5y^6) + 3y^3(-2y^4) + 3y^3(-9) = \boxed{15y^9 - 6y^7 - 27y^3}$

91. $-5x^4(10x^5 - x^3 - x) = -5x^4(10x^5) - 5x^4(-x^3) - 5x^4(-x) = \boxed{-50x^9 + 5x^7 + 5x^5}$

93. $(2x^3 + 5x^2 - 7x + 4)(-3x) = (2x^3)(-3x) + (5x^2)(-3x) - 7x(-3x) + 4(-3x) = \boxed{-6x^4 - 15x^3 + 21x^2 - 12x}$

95. $(x + 3)(x + 5) = x(x + 5) + 3(x + 5) = x^2 + 5x + 3x + 15 = \boxed{x^2 + 8x + 15}$

97. $(2x - 1)(x + 4) = 2x(x + 4) - 1(x + 4) = 2x^2 + 8x - x - 4 = \boxed{2x^2 + 7x - 4}$

99. $(3y - 2)(5y - 4) = 3y(5y - 4) - 2(5y - 4) = 15y^2 - 12y - 10y + 8 = \boxed{15y^2 - 22y + 8}$

101. $(2x + 3)(2x - 3) = 2x(2x - 3) + 3(2x - 3) = 4x^2 - 6x + 6x - 9 = \boxed{4x^2 - 9}$

103. $\begin{aligned}(y + 1)(y^2 + 2y + 3) &= y(y^2 + 2y + 3) + 1(y^2 + 2y + 3) \\ &= y^3 + 2y^2 + 3y + y^2 + 2y + 3 = \boxed{y^3 + 3y^2 + 5y + 3}\end{aligned}$

105. $\begin{aligned}(y - 3)(y^2 - 3y + 4) &= y(y^2 - 3y + 4) - 3(y^2 - 3y + 4) \\ &= y^3 - 3y^2 + 4y - 3y^2 + 9y - 12 = \boxed{y^3 - 6y^2 + 13y - 12}\end{aligned}$

107. $\begin{aligned}(2a - 3)(a^2 - 3a + 5) &= 2a(a^2 - 3a + 5) - 3(a^2 - 3a + 5) \\ &= 2a^3 - 6a^2 + 10a - 3a^2 + 9a - 15 \\ &= \boxed{2a^3 - 9a^2 + 19a - 15}\end{aligned}$

109. $\begin{aligned}(z - 4)(-2z^2 - 3z + 2) &= z(-2z^2 - 3z + 2) - 4(-2z^2 - 3z + 2) \\ &= -2z^3 - 3z^2 + 2z + 8z^2 + 12z - 8 \\ &= \boxed{-2z^3 + 5z^2 + 14z - 8}\end{aligned}$

111. $\begin{aligned}(2y - 5)(-2y^2 + 4y - 3) &= 2y(-2y^2 + 4y - 3) - 5(-2y^2 + 4y - 3) \\ &= -4y^3 + 8y^2 - 6y + 10y^2 - 20y + 15 \\ &= \boxed{-4y^3 + 18y^2 - 26y + 15}\end{aligned}$

113. $\begin{aligned}(y^2 + 5)(y - 6) &= y^2(y - 6) + 5(y - 6) \\ &= \boxed{y^3 - 6y^2 + 5y - 30}\end{aligned}$

115. $\begin{aligned}(y^3 - 3y + 2)(y - 4) &= (y^3 - 3y + 2)(y) + (y^3 - 3y + 2)(-4) \\ &= y^4 - 3y^2 + 2y - 4y^3 + 12y - 8 \\ &= \boxed{y^4 - 4y^3 - 3y^2 + 14y - 8}\end{aligned}$

117. $\begin{aligned}(x + 4)^2 &= (x + 4)(x + 4) \\ &= x(x + 4) + 4(x + 4) \\ &= x^2 + 4x + 4x + 16 = \boxed{x^2 + 8x + 16}\end{aligned}$

119. $\begin{aligned}(4x-3)^2 &= (4x-3)(4x-3)\\ &= 4x(4x-3)-3(4x-3)\\ &= 16x^2-12x-12x+9 = \boxed{16x^2-24x+9}\end{aligned}$

121. $\begin{aligned}(2y^4+3y^3)^2 &= (2y^4+3y^3)(2y^4+3y^3)\\ &= 2y^4(2y^4+3y^3)+3y^3(2y^4+3y^3)\\ &= 4y^8+6y^7+6y^7+9y^6 = \boxed{4y^8+12y^7+9y^6}\end{aligned}$

123. $\begin{aligned}(2z^3+3z)(5z^2+3) &= 2z^3(5z^2+3)+3z(5z^2+3)\\ &= 10z^5+6z^3+15z^3+9z\\ &= \boxed{10z^5+21z^3+9z}\end{aligned}$

125. $\begin{aligned}(3y^3-y)(2y^2-6) &= 3y^3(2y^2-6)-y(2y^2-6)\\ &= 6y^5-18y^3-2y^3+6y\\ &= \boxed{6y^5-20y^3+6y}\end{aligned}$

127. $\begin{aligned}(x+2)^3 &= (x+2)(x+2)(x+2)\\ &= (x+2)[x(x+2)+2(x+2)]\\ &= (x+2)(x^2+2x+2x+4)\\ &= (x+2)(x^2+4x+4)\\ &= x(x^2+4x+4)+2(x^2+4x+4)\\ &= x^3+4x^2+4x+2x^2+8x+8\\ &= \boxed{x^3+6x^2+12x+8}\end{aligned}$

129. $\begin{aligned}(2z-1)^3 &= (2z-1)(2z-1)(2z-1)\\ &= (2z-1)[2z(2z-1)-1(2z-1)]\\ &= (2z-1)(4z^2-2z-2z+1)\\ &= (2z-1)(4z^2-4z+1)\\ &= 2z(4z^2-4z+1)-1(4z^2-4z+1)\\ &= 8z^3-8z^2+2z-4z^2+4z-1\\ &= \boxed{8z^3-12z^2+6z-1}\end{aligned}$

131. $\begin{aligned}4x(x+3)+(x+3)(x-1) &= (4x^2+12x)+x(x-1)+3(x-1)\\ &= 4x^2+12x+x^2-x+3x-3\\ &= \boxed{5x^2+14x-3}\end{aligned}$

133. $(2x-3)(x+2)+(3x+1)(x-3)$
$= 2x(x+2)-3(x+2)+3x(x-3)+1(x-3)$
$= 2x^2+4x-3x-6+3x^2-9x+x-3$
$= \boxed{5x^2-7x-9}$

135. $(z-1)(2z+3)(2z+1)$
$= (z-1)[2z(2z+1)+3(2z+1)]$
$= (z-1)(4z^2+2z+6z+3)$
$= (z-1)(4z^2+8z+3)$
$= z(4z^2+8z+3)-1(4z^2+8z+3)$
$= 4z^3+8z^2+3z-4z^2-8z-3$
$= \boxed{4z^3+4z^2-5z-3}$

137.
$$\begin{array}{r}2x^3+x^2+2x+3\\ x+4\\ \hline 8x^3+4x^2+8x+12\\ 2x^4+x^3-2x^2+3x\\ \hline \boxed{2x^4+9x^3+6x^2+11x+12}\end{array}$$

139.

$$x^2 - 3x + 9$$
$$2x - 3$$
$$\overline{}$$
$$-3x^2 + 9x - 27$$
$$\underline{2x^3 - 6x^2 + 18x}$$
$$\boxed{2x^3 - 9x^2 + 27x - 27}$$

141.

$$4z^3 - 2z^2 + 5z - 4$$
$$3z - 2$$
$$\overline{}$$
$$-8z^3 + 4z^2 - 10z + 8$$
$$\underline{12z^4 - 6z^3 + 15z^2 - 12z}$$
$$\boxed{12z^4 - 14z^3 + 19z^2 - 22z + 8}$$

143.

$$2y^5 - 3y^3 + y^2 - 2y + 3$$
$$2y - 1$$
$$\overline{}$$
$$-2y^5 + 3y^3 - y^2 + 2y - 3$$
$$\underline{4y^6 - 6y^4 + 2y^3 - 4y^2 + 6y}$$
$$\boxed{4y^6 - 2y^5 - 6y^4 + 5y^3 - 5y^2 + 8y - 3}$$

145.

$$2x^4 - 5x^3 - x^2 - 3$$
$$-3x - 4$$
$$\overline{}$$
$$-8x^4 + 20x^3 + 4x^2 + 12$$
$$\underline{-6x^5 + 15x^4 + 3x^3 + 9x}$$
$$\boxed{-6x^5 + 7x^4 + 23x^3 + 4x^2 + 9x + 12}$$

147.

$$3x^2 + x + 1$$
$$\underline{x^2 + 2x + 3}$$
$$9x^2 + 3x + 3$$
$$6x^3 + 2x^2 + 2x$$
$$\underline{3x^4 + x^3 + x^2}$$
$$\boxed{3x^4 + 7x^3 + 12x^2 + 5x + 3}$$

149.

$$2y^2 + y - 4$$
$$y^2 - 5y - 3$$
$$\overline{}$$
$$-6y^2 - 3y + 12$$
$$-10y^3 - 5y^2 + 20y$$
$$\underline{2y^4 + y^3 - 4y^2}$$
$$\boxed{2y^4 - 9y^3 - 15y^2 + 17y + 12}$$

151.

$$3n^3 - 2n^2 - 5n - 1$$
$$4n^2 - 3n - 2$$
$$\overline{}$$
$$-6n^3 + 4n^2 + 10n + 2$$
$$-9n^4 + 6n^3 + 15n^2 + 3n$$
$$\underline{12n^5 - 8n^4 - 20n^3 - 4n^2}$$
$$\boxed{12n^5 - 17n^4 - 20n^3 + 15n^2 + 13n + 2}$$

153.   Area $= (x + 3)(x + 2) \quad = \; x^2 + 2x + 3x + 6$

$ = \; \boxed{x^2 + 5x + 6}$

155.   Area $= x^2 + 3x + 2x + 6$

length $= \boxed{x + 3}$

width $= \boxed{x + 2}$

157.   Area $= (x + 3)(x + 5)$

$(x + 3)(x + 5) \quad = \; x^2 + 3x + 5x + 15$

$ = \; \boxed{x^2 + 8x + 15}$

159.   area of border $=$ area of large square $-$ area of small gray square

$ = \; (x + 4)(x + 4) - x(x)$

$ = \; x^2 + 4x + 4x + 16 - x^2$

$ = \; \boxed{8x + 16}$

161.   Let x equal Annie's present age.

$5x$ equals Warbuck's present age.

$x + 6$ equals Annie's age 6 years from now.

$5x + 6$ equals Warbuck's age 6 years from now.

$(x + 6)(5x + 6) \quad = \; 5x^2 + 6x + 30x + 36$

$ = \; \boxed{5x^2 + 36x + 36}$

163.   Let x equal one number.

$8 - x$ equals other number.

$x(8 - x) = \boxed{8x - x^2}$

165. Let x equal the length of box.
$x + 1$ equals the width of box.
$x + 2$ equals the height of box.

$$
\begin{aligned}
\text{Volume} \ &= \ x(x+1)(x+2) \\
&= \ x(x^2 + 2x + x + 2) \\
&= \ x(x^2 + 3x + 2) \\
&= \ \boxed{x^3 + 3x^2 + 2x}
\end{aligned}
$$

167. **a.** $(x-1)(x+1) = x^2 + x - x - 1 = \boxed{x^2 - 1}$

   b. $(x-1)(x^2 + x + 1) = x^3 + x^2 + x - x^2 - x - 1 = \boxed{x^3 - 1}$

   c. $(x-1)(x^3 + x^2 + x + 1) = x^4 + x^3 + x^2 + x - x^3 - x^2 - x - 1 = \boxed{x^4 - 1}$

   d. $(x-1)(x^4 + x^3 + x^2 + x + 1) = \boxed{x^5 - 1}$

169. $(\ \boxed{}\ )\left(-\dfrac{1}{3} xy^4\right) = -6x^3 y^{12}$

$$
\begin{aligned}
(18x^2 y^8)\left(-\tfrac{1}{3} xy^4\right) \ &= \ -6x^3 y^{12} \\
\boxed{} \ &= \ \boxed{18x^2 y^8}
\end{aligned}
$$

171. $\boxed{\text{B}}$ is true:
$$
\begin{aligned}
(5y^2)^3(2y-1) \ &= \ 125y^6(2y-1) \\
&= \ \boxed{250y^7 - 125y^6}
\end{aligned}
$$

Review Problems

176.
$$
\begin{array}{rl}
3x + 5y &= 9 \\
4x + 3y &= 1
\end{array}
\quad \rightarrow \quad
\begin{array}{rl}
-9x - 15y &= -27 \\
\underline{20x + 15y} &= \underline{\ 5\ } \\
11x &= -22 \\
x &= -2 \\
-8 + 3y &= 1 \\
y &= 9 \\
y &= 3
\end{array}
$$
$$\boxed{\{-2, 3\}}$$

177. $5x - 4y \geq -20$

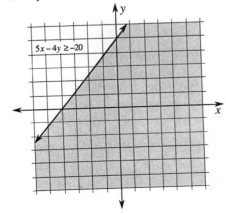

178. Let x equal the numerator.

$3x - 2$ equals the denominator.

$$x + (3x - 2) = 79$$
$$4x = 81$$
$$x = \frac{81}{4}$$
$$3x - 2 = \frac{243 - 8}{4} = \frac{235}{4}$$
$$\frac{\frac{81}{4}}{\frac{235}{4}} = \boxed{\frac{81/4}{235/4}} \text{ or } \boxed{\frac{81}{235}}$$

Section 6.3 Multiplying Binomials; Special Binomial Products

Problem Set 6.3, pp. 399-400

1. $(x + 3)(x + 5) = x^2 + 5x + 3x + 15 = \boxed{x^2 + 8x + 15}$

3. $(y - 5)(y + 3) = y^2 + 3y - 5y - 15 = \boxed{y^2 - 2y - 15}$

5. $(2b - 1)(b + 2) = 2b^2 + 4b - b - 2 = \boxed{2b^2 + 3b - 2}$

7. $(2x - 3)(x + 1) = 2x^2 + 2x - 3x - 3 = \boxed{2x^2 - x - 3}$

9. $(2y - 3)(5y + 3) = 10y^2 + 6y - 15y - 9 = \boxed{10y^2 - 9y - 9}$

11. $(3y - 7)(4y - 5) = 12y^2 - 15y - 28y + 35 = \boxed{12y^2 - 43y + 35}$

13. $(x^2 - 5)(x^2 - 3) = x^4 - 3x^2 - 5x^2 + 15 = \boxed{x^4 - 8x^2 + 15}$

15. $(3y^3 + 2)(y^3 + 4) = 3y^6 + 12y^3 + 2y^3 + 8 = \boxed{3y^6 + 14y^3 + 8}$

17. $(3y^6 - 5)(2y^6 - 2) = 6y^{12} - 6y^6 - 10y^6 + 10 = \boxed{6y^{12} - 16y^6 + 10}$

19. $(x^2 - 3)(x + 2) = \boxed{x^3 + 2x^2 - 3x - 6}$

21. $(4 + 5y)(5 - 4y) = 20 - 16y + 25y - 20y^2 = \boxed{20 + 9y - 20y^2}$

23. $(-3 + 2y)(4 + y) = -12 - 3y + 8y + 2y^2 = \boxed{-12 + 5y + 2y^2}$

25. $(-3 + r)(-5 - 2r) = 15 + 6r - 5r - 2r^2 = \boxed{15 + r - 2r^2}$

27. $(x + 3y)(x + y) = x^2 + xy + 3xy + 3y^2 = \boxed{x^2 + 4xy + 3y^2}$

29. $(9r + 4s)(2r - 3s) = 18r^2 - 27rs + 8rs - 12s^2 = \boxed{18r^2 - 19rs - 12s^2}$

31. $(4m + 9n)(-2m + 5n) = -8m^2 + 20mn - 18mn + 45n^2 = \boxed{-8m^2 + 2mn + 45n^2}$

33. $\left(x + \dfrac{2}{3}\right)\left(x - \dfrac{1}{2}\right) = x^2 - \dfrac{1}{2}x + \dfrac{2}{3}x - \dfrac{1}{3} = \boxed{x^2 + \dfrac{1}{6}x - \dfrac{1}{3}}$

35. $\begin{aligned} 2x(3x - 1)(2x + 5) &= 2x(6x^2 + 15x - 2x - 5) \\ &= 2x(6x^2 + 13x - 5) = \boxed{12x^3 + 26x^2 - 10x} \end{aligned}$

37. $\begin{aligned} (x - 3)(x - 2)(x + 1) &= (x - 3)(x^2 - x - 2) \\ &= x^3 - x^2 - 2x - 3x^2 + 3x + 6 \\ &= \boxed{x^3 - 4x^2 + x + 6} \end{aligned}$

39. $(x + 2)^2 = x^2 + 2(2x) + 2^2 = \boxed{x^2 + 4x + 4}$

41. $(y - 3)^2 = y^2 - 2(3y) + 9 = \boxed{y^2 - 6y + 9}$

43. $(x + 3y)^2 = x^2 + 2(3xy) + 9y^2 = \boxed{x^2 + 6xy + 9y^2}$

45. $(5x + 2y)^2 = 25x^2 + 2(5x)(2y) + 4y^2 = \boxed{25x^2 + 20xy + 4y^2}$

47. $(x - 5y)^2 = \boxed{x^2 - 10xy + 25y^2}$

49. $(3r - 2s)^2 = \boxed{9r^2 - 12rs + 4s^2}$

51. $(2x^2 + 3)^2 = \boxed{4x^4 + 12x^2 + 9}$

53. $(4x^2 - 1)^2 = \boxed{16x^4 - 8x^2 + 1}$

55. $(x^2 + y^2)^2 = \boxed{x^4 + 2x^2y^2 + y^4}$

57. $(2x^2 + 3y^3)^2 = \boxed{4x^4 + 12x^2y^3 + 9y^6}$

59. $(x^3 - 2y^2)^2 = \boxed{x^6 - 4x^3y^2 + 4y^4}$

61. $\left(2x + \dfrac{1}{2}\right)^2 = 4x^2 + 2(2x)\left(\dfrac{1}{2}\right) + \dfrac{1}{4} = \boxed{4x^2 + 2x + \dfrac{1}{4}}$

63. $\left(4y - \dfrac{1}{4}\right)^2 = 16y^2 - 2(4y)\left(\dfrac{1}{4}\right) + \dfrac{1}{16} = \boxed{16y^2 - 2y + \dfrac{1}{16}}$

65. $\begin{aligned} (x + 4)^3 = (x + 4)(x + 4)^2 &= (x + 4)(x^2 + 8x + 16) \\ &= x^3 + 8x^2 + 16x + 4x^2 + 32x + 6y = \boxed{x^3 + 12x^2 + 48x + 64} \end{aligned}$

67. $\begin{aligned} (2r - 1)^3 &= (2r - 1)(2r - 1)^2 = (2r - 1)(4r^2 - 4r + 1) \\ &= 8r^3 - 8r^2 + 2r - 4r^2 + 4r - 1 = \boxed{8r^2 - 12r^2 + 6r - 1} \end{aligned}$

69. $\begin{aligned} (x + 3)^2 - (x + 1)^2 &= x^2 + 6x + 9 - (x^2 + 2x + 1) \\ &= x^2 + 6x + 9 - x^2 - 2x - 1 \\ &= \boxed{4x + 8} \end{aligned}$

71. $\begin{aligned} (2r - 4)^2 - (r - 1)^2 &= 4r^2 - 16r + 16 - (r^2 - 2r + 1) \\ &= 4r^2 - 16r + 16 - r^2 + 2r - 1 \\ &= \boxed{3r^2 - 14r + 15} \end{aligned}$

73. $(5 - 3y)^2 = \boxed{25 - 30y + 9y^2}$

75. $(x + 3)(x - 3) = x^2 - 3^2 = \boxed{x^2 - 9}$

77. $(3x + 2)(3x - 2) = (3x)^2 - 2^2 = \boxed{9x^2 - 4}$

79. $(3r - 4)(3r + 4) = (3r)^2 - 4^2 = \boxed{9r^2 - 16}$

81. $(3 + r)(3 - r) = 3^2 - r^2 = \boxed{9 - r^2}$

83. $(5 - 7x)(5 + 7x) = 5^2 - (7x)^2 = \boxed{25 - 49x^2}$

85. $\left(2x + \dfrac{1}{2}\right)\left(2x - \dfrac{1}{2}\right) = (2x)^2 - \left(\dfrac{1}{2}\right)^2 = \boxed{4x^2 - \dfrac{1}{4}}$

87. $(3r + 4s)(3r - 4s) = (3r)^2 - (4s)^2 = \boxed{9r^2 - 16s^2}$

89. $(3c - 2d)(3c + 2d) = (3c)^2 - (2d)^2 = \boxed{9c^2 - 4d^2}$

91. $(y^2 + 1)(y^2 - 1) = \boxed{y^4 - 1}$

93. $(r^3 + 2)(r^3 - 2) = \boxed{r^6 - 4}$

95. $(1 - y^4)(1 + y^4) = \boxed{1 - y^8}$

97. $(7x^2 + 10y)(7x^2 - 10y) = (7x)^2 - (10y)^2 = \boxed{49x^4 - 100y^2}$

99. $(3x + 4)(3x - 4) - (2x + 3)(2x - 3)$
$= (9x^2 - 16) - (4x^2 - 9) = 9x^2 - 16 - 4x^2 + 9 = \boxed{5x^2 - 7}$

101. $(x + 4)(x - 4) + (x + 4)^2$
$= (x^2 - 16) + (x^2 + 8x + 16)$
$= \boxed{2x^2 + 8x}$

103. $(3x - 5)^2 - (2x + 3)(2x - 3)$
$= (9x^2 - 30x + 25) - (4x^2 - 9)$
$= 9x^2 - 30x + 25 - 4x^2 + 9$
$= \boxed{5x^2 - 30x + 34}$

105. $(x + 1)^2 - (x + 2)^2 + (x + 3)^2$
$= (x^2 + 2x + 1) - (x^2 + 4x + 4) + (x^2 + 6x + 9)$
$= x^2 + 2x + 1 - x^2 - 4x - 4 + x^2 + 6x + 9$
$= \boxed{x^2 + 4x + 6}$

107. $A = (x + 1)^2 = \boxed{x^2 + 2x + 1}$

109. $A = (2x - 3)(2x + 3) = \boxed{4x^2 - 9}$

111. Let x equal an integer.
$x + 1$ equals consecutive integer.
$x^2 + (x + 1)^2 = x^2 + x^2 + 2x + 1 = \boxed{2x^2 + 2x + 1}$

113. area of the region inside the right triangle and outside the square = area of triangle − area of square
$= \dfrac{1}{2}(2x - 3)(2x + 3) - (x + 1)^2$
$= \dfrac{1}{2}(4x^2 - 9) - (x^2 + 2x + 1)$
$= 2x^2 - \dfrac{9}{2} - x^2 - 2x - 1$
$= \boxed{x^2 - 2x - \dfrac{11}{2}}$

115. area of new garden $= (x + 4)(x - 4)$
$= \boxed{x^2 - 16}$ square feet
difference in area $= x^2 - 16 - x^2$
$= \boxed{-16}$

117. number of desks
$= (5d + 3)(4d - 2)$
$= \boxed{20d^2 + 2d - 6}$

119. $\boxed{A}$ is true; $(40 + 1)(40 - 1) = (40)^2 - 1^2 = 1600 - 1 = 1599.$

121. $(20,000 + 79)(0.537x - 3.814)$
$= \boxed{10,740x^2 - 76,237.577x - 301.306}$

Review Problems

126. $y = -\dfrac{1}{2}x + 3$

$m = -\dfrac{1}{2}, \ b = 3$

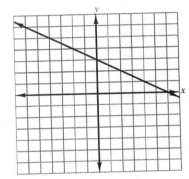

127. $\dfrac{y - 10}{7} = \dfrac{y - 5}{6}$

$6y - 60 = 7y - 35$

$-25 = y$

$\boxed{\{-25\}}$

128. Let x equal the number.

$-x$ equals its opposite

$x = 2(-x) - 18$

$x = -2x - 18$

$3x = -18$

$x = \boxed{-6}$

Section 6.4 Problem Solving

Problem Set 6.4, pp. 405-407

1. $(x - 5)(x - 2) = x^2 - 4$

$x^2 - 7x + 10 = x^2 - 4$

$-7x = -14$

$x = 2$

$\boxed{\{2\}}$

3. $(y - 8)^2 = (y + 4)(y - 4)$

$y^2 - 16y + 64 = y^2 - 16$

$-16y = -80$

$y = 5$

$\boxed{\{5\}}$

5. $(x + 4)(x - 4) = x(x + 2)$

$x^2 - 16 = x^2 + 2x$

$-16 = 2x$

$-8 = x$

$\boxed{\{-8\}}$

7. $(4y - 7)(y - 5) + (2y + 1)(3 - 2y) = -8$

$4y^2 - 27y + 35 + 6y + 3 - 4y^2 - 2y = -8$

$-23y = -46$

$y = 2$

$\boxed{\{2\}}$

9.
$$\begin{aligned}
(4z+5)(z-8) &= z+4z(z-2) \\
4z^2 - 27z - 40 &= z + 4z^2 - 8z \\
-20z &= 40 \\
z &= -2
\end{aligned}$$

$$\boxed{\{-2\}}$$

11.
$$\begin{aligned}
(3y+8)(9-2y) &= 1 - 2(y-8)(3y-4) \\
27y + 72 - 6y^2 - 16y &= 1 - 2(3y^2 - 28y + 32) \\
-6y^2 + 11y + 72 &= 1 - 6y^2 + 56y - 64 \\
-45y &= -135 \\
y &= 3
\end{aligned}$$

$$\boxed{\{3\}}$$

13.
$$\begin{aligned}
(y+5)^2 - (y+4)^2 &= 5 \\
y^2 + 10y + 25 - (y^2 + 8y + 16) &= 5 \\
2y + 9 &= 5 \\
2y &= -4 \\
y &= -2
\end{aligned}$$

$$\boxed{\{-2\}}$$

15. Let x equal the number.
$$\begin{aligned}
(x-3)(x+4) &= x^2 - 5 \\
x^2 + x - 12 &= x^2 - 5 \\
x &= 7
\end{aligned}$$

The number is $\boxed{7}$.

17. Let x and $x+1$ represent two consecutive positive integers.
$$\begin{aligned}
(x+1)^2 - x^2 &= 15 \\
x^2 + 2x + 1 - x^2 &= 15 \\
2x &= 14 \\
x &= 7
\end{aligned}$$

The numbers are $\boxed{7 \text{ and } 8}$.

19. Let x and $x+2$ represent two consecutive odd positive integers.
$$\begin{aligned}
(x+2)^2 - x^2 &= 40 \\
x^2 + 4x + 4 - x^2 &= 40 \\
4x &= 36 \\
x &= 9 \\
x+2 &= 11
\end{aligned}$$

The numbers are $\boxed{9 \text{ and } 11}$.

21. Let x and $x+1$ represent two consecutive integers.
$$\begin{aligned}
x(x+1) &= (x+1)^2 - 28 \\
x^2 + x &= x^2 + 2x + 1 - 28 \\
-x &= -27 \\
x &= 27 \\
x+1 &= 28
\end{aligned}$$

The numbers are $\boxed{27 \text{ and } 28}$.

23. Let $x, x+1,$ and $x+2$ represent three consecutive integers.
$$\begin{aligned}
x^2 + (x+1)^2 + (x+2)^2 &= 3(x+2)^2 - 25 \\
x^2 + x^2 + 2x + 1 + x^2 + 4x + 4 &= 3(x^2 + 4x + 4) - 25 \\
3x^2 + 6x + 5 &= 3x^2 + 12x - 13 \\
-6x &= -18 \\
x &= 3 \\
x+1 &= 4 \\
x+2 &= 5
\end{aligned}$$

The numbers are $\boxed{3, 4, \text{ and } 5}$.

25. Let x equal the smaller number.
$x + 3$ equals the larger number.
$$\begin{aligned}(x+3)^2 - x^2 &= 1 + 7x \\ x^2 + 6x + 9 - x^2 &= 1 + 7x \\ -x &= -8 \\ x &= 8 \\ x + 3 &= 11\end{aligned}$$
The numbers are $\boxed{8 \text{ and } 11}$.

27. Let x equal the smaller number.
$x + 2$ equals the larger number.
$$\begin{aligned}x(x+2) &= 24 + x^2 \\ x^2 + 2x &= 24 + x^2 \\ 2x &= 24 \\ x &= 12 \\ x + 2 &= 14\end{aligned}$$
The numbers are $\boxed{12 \text{ and } 14}$.

29. Let x equal the length of side of original square.
$x + 3$ equals the length of new rectangle.
$x - 2$ equals the width of new rectangle.
$$\begin{aligned}(x+3)(x-2) &= x^2 - 1 \\ x^2 + x - 6 &= x^2 - 1 \\ x &= 5\end{aligned}$$
The dimensions of the original square are
$\boxed{5 \text{ yards} \times 5 \text{ yards}}$.

31. Let x equal the length of side of original square.
$x + 6$ equals the length of new rectangle.
$x - 4$ equals the width of new rectangle.
$$\begin{aligned}(x+6)(x-4) &= x^2 \\ x^2 + 2x - 24 &= x^2 \\ 2x &= 24 \\ x &= 12\end{aligned}$$
The dimensions of the original square are $\boxed{12 \text{ meters} \times 12 \text{ meters}}$.

33. Let x equal the length of side of square picture.
$x + 2$ equals the length of side of picture plus frame.
$$\begin{aligned}(x+2)^2 - x^2 &= 48 \\ x^2 + 4x + 4 - x^2 &= 48 \\ 4x &= 44 \\ x &= 11\end{aligned}$$
The dimensions of the picture are $\boxed{11 \text{ inches} \times 11 \text{ inches}}$.

35. Let x equal the width of painting.
$2x - 8$ equals the length of painting.
$$\begin{aligned}(x+8)(2x-8+8) - x(2x-8) &= 816 \\ (x+8)2x - x(2x-8) &= 816 \\ 2x^2 + 16x - 2x^2 + 8x &= 816 \\ 24x &= 816 \\ x &= 34 \\ 2x - 8 &= 68 - 8 = 60\end{aligned}$$
The dimensions of the painting are $\boxed{34 \text{ inches} \times 60 \text{ inches}}$.

37. Let x equal the width of rectangle.

$x + 4$ equals the length of rectangle.

$x + 1$ equals the length of side of square.

$$
\begin{aligned}
x(x + 4) - (x + 1)^2 &= 5 \\
x^2 + 4x - x^2 - 2x - 1 &= 5 \\
2x &= 6 \\
x &= 3 \\
x + 4 &= 7 \\
x + 1 &= 4
\end{aligned}
$$

The dimensions of the rectangle are $\boxed{3 \text{ meters} \times 7 \text{ meters}}$.

The dimensions of the square are $\boxed{4 \text{ meters} \times 4 \text{ meters}}$.

39. Take a number: x

Add 2: $x + 2$

Square the sum: $(x + 2)^2$

Add 25 to the result: $(x + 2)^2 + 25$

Subtract the product of the original number times four more than the original number:

$(x + 2)^2 + 25 - x(x + 4)$

Add 6 to the difference: $(x + 2)^2 + 25 - x(x + 4) + 6$

Using polynomial operations:

$$
\begin{aligned}
(x + 2)^2 + 25 - x(x + 4) + 6 &= 35 \\
x^2 + 4x + 4 + 25 - x^2 - 4x + 6 &= 35 \\
35 &= 35
\end{aligned}
$$

The answer is 35, regardless of what number is originally chosen.

41.
$$
\begin{aligned}
\frac{x + 5}{x + 9} &= \frac{x + 10}{x + 4} \\
(x + 5)(x + 4) &= (x + 9)(x + 10) \\
x^2 + 9x + 20 &= x^2 + 19x + 90 \\
-10x &= 70 \\
x &= -7
\end{aligned}
$$

$\boxed{\{-7\}}$

43.
$$
\begin{aligned}
\frac{2(x + 1)}{x - 2} &= \frac{2x + 3}{x - 1} \\
2(x + 1)(x - 1) &= (x - 2)(2x + 3) \\
2(x^2 - 1) &= 2x^2 - x - 6 \\
2x^2 - 2 &= 2x^2 - x - 6 \\
x &= -4
\end{aligned}
$$

$\boxed{\{-4\}}$

45. Let $3x$ equal the first multiple of 3.

$3x + 3$ equals the next consecutive multiple by 3.

$$
\begin{aligned}
(3x + 3)^2 - (3x)^2 &= 45 \\
9x^2 + 18x + 9 - 9x^2 &= 45 \\
18x &= 36 \\
x &= 2 \\
3x &= 6 \\
3x + 3 &= 19
\end{aligned}
$$

The numbers are $\boxed{6 \text{ and } 9}$.

47. Let x equal your age.

y equals change in cents in your pocket.

$100x$ equals the first two digits given age in years.

$$
\begin{aligned}
50(2x + 5) - 365 + y + 115 &= 100x + y \\
100x + 250 - 365 + y + 115 &= 100x + y \\
100x + y &= 100x + y
\end{aligned}
$$

Review Problems

48. $\dfrac{16(-1-4)}{12(-5)+(6-9)(-3-1)}$

$=\dfrac{16(-5)}{-60+(-3)(-4)}$

$=\dfrac{-80}{-60+12}$

$=\dfrac{-80}{-48}=\boxed{\dfrac{5}{3}}$

49.
$$6(5-x)-3 \; > \; -3(x-7)-x$$
$$30-6x-3 \; > \; -3x+21-x$$
$$-2x \; > \; -6$$
$$x \; < \; 3$$

$\boxed{\{x \mid x<3\}}$

50. $2x-y=6$
$x+2y=-2$
solution: $(2,-2)$

Section 6.5 Integral Exponents and Dividing Polynomials

Problem Set 6.5, pp. 415-417

1. $\dfrac{x^5}{x^2}=x^{5-2}=\boxed{x^3}$

3. $\dfrac{z^{13}}{z^5}=z^{13-5}=\boxed{z^8}$

5. $\dfrac{30y^{10}}{10y^5}=3y^{10-5}=\boxed{3y^5}$

7. $\dfrac{-8x^{22}}{4x^2}=-2x^{22-2}=\boxed{-2x^{20}}$

9. $\dfrac{-9a^8}{18a^5}=-\dfrac{1}{2}a^{8-5}=\boxed{\dfrac{-a^3}{2}}$

11. $\dfrac{7x^{17}}{5x^5}=\dfrac{7x^{17-5}}{5}=\boxed{\dfrac{7}{5}x^{12}}$

13. $\dfrac{x^5y^3}{x^2y}=x^{5-2}y^{3-1}=\boxed{x^3y^2}$

15. $\dfrac{-20a^{11}b^9}{10a^6b^3}=-2a^{11-6}b^{9-3}=\boxed{-2a^5b^6}$

17. $\dfrac{-20x^5y^8}{60x^2y^4}=-\dfrac{1}{3}x^{5-2}y^{8-4}=\boxed{\dfrac{-x^3y^4}{3}}$

19. $30b^{10}\div b^6=30b^{10-6}=\boxed{30b^4}$

21. $-4x^2y^4 \div -2xy^3 = 2x^{2-1}y^{4-3} = \boxed{2xy}$

23. $\dfrac{4^5 x^8 y^6}{4^2 xy^2} = 4^{5-2}x^{8-1}y^{6-2} = 4^3x^7y^4 = \boxed{64x^7y^4}$

25. $\dfrac{18x^5 + 6x^4 + 9x^3}{3x^2} = \dfrac{18x^5}{3x^2} + \dfrac{6x^4}{3x^2} + \dfrac{9x^3}{3x^2} = \boxed{6x^3 + 2x^2 + 3x}$

27. $\dfrac{12x^4 - 8x^3 + 40x^2}{4x} = \dfrac{12x^4}{4x} - \dfrac{8x^3}{4x} + \dfrac{40x^2}{4x} = \boxed{3x^3 - 2x^2 + 10x}$

29. $(4x^2 - 6x) \div x = \dfrac{4x^2}{x} - \dfrac{6x}{x} = \boxed{4x - 6}$

31. $\dfrac{30z^3 + 10z^2}{-5z} = \dfrac{30z^3}{-5z} + \dfrac{10z^2}{-5z} = \boxed{-6z^2 - 2z}$

33. $\dfrac{8x^3 + 3x^2 - 2x}{2x} = \dfrac{8x^3}{2x} + \dfrac{3x^2}{2x} - \dfrac{2x}{2x} = \boxed{4x^2 + \dfrac{3}{2}x - 1}$

35. $\dfrac{25x^7 - 15x^5 - 5x^4}{5x^3} = \dfrac{25x^7}{5x^3} - \dfrac{15x^5}{5x^3} - \dfrac{5x^4}{5x^3} = \boxed{5x^4 - 3x^2 - x}$

37. $\dfrac{18x^7 - 9x^6 + 20x^3 - 10x^4}{-2x^4} = \dfrac{18x^7}{-2x^4} - \dfrac{9x^6}{-2x^4} + \dfrac{20x^5}{-2x^4} - \dfrac{10x^4}{-2x^4} = \boxed{-9x^3 + \dfrac{9}{2}x^2 - 10x + 5}$

39. $7^0 + 4^0 = 1 + 1 = \boxed{2}$

41. $-5^0 + (-5)^0 = -1 + 1 = \boxed{0}$

43. $5^{-3} = \dfrac{1}{5^3} = \boxed{\dfrac{1}{125}}$

45. $(-4)^{-1} = \dfrac{1}{-4} = \boxed{-\dfrac{1}{4}}$

47. $2^{-1} + 3^{-1} = \dfrac{1}{2} + \dfrac{1}{3} = \boxed{\dfrac{5}{6}}$

49. $4^{-1} - 5^{-1} = \dfrac{1}{4} - \dfrac{1}{5} = \boxed{\dfrac{1}{20}}$

51. $2^{-3} + 4^{-2} = \dfrac{1}{2^3} + \dfrac{1}{4^2} = \dfrac{1}{8} + \dfrac{1}{16} = \boxed{\dfrac{3}{16}}$

53. $\dfrac{1}{3^{-2}} = 3^2 = \boxed{9}$

55. $\dfrac{1}{4^{-3}} = 4^3 = \boxed{64}$

57. $\left(\dfrac{1}{4}\right)^{-2} = \dfrac{1}{\left(\dfrac{1}{4}\right)^2} = 4^2 = \boxed{16}$

59. $\left(\dfrac{2}{3}\right)^{-2} = \left(\dfrac{3}{2}\right)^2 = \boxed{\dfrac{9}{4}}$

61. $\dfrac{2^{-3}}{5^{-2}} = \dfrac{5^2}{2^3} = \boxed{\dfrac{25}{8}}$

63. $-4^{-2} = -\dfrac{1}{4^2} = \boxed{-\dfrac{1}{16}}$

65. $(-4)^{-2} = \dfrac{1}{(-4)^2} = \boxed{\dfrac{1}{16}}$

67. $\dfrac{1}{(-2)^{-4}} = \dfrac{1}{\dfrac{1}{(-2)^4}} = (-2)^4 = \boxed{16}$

69. $\dfrac{x^3}{x^9} = \dfrac{1}{x^{9-3}} = \boxed{\dfrac{1}{x^6}}$

71. $\dfrac{z^5}{z^{13}} = \dfrac{1}{z^{13-5}} = \boxed{\dfrac{1}{z^8}}$

73. $\dfrac{30y^5}{10y^{10}} = \dfrac{3}{y^{10-5}} = \boxed{\dfrac{3}{y^5}}$

75. $\dfrac{-8x^3}{2x^7} = \dfrac{-4}{x^{7-4}} = \boxed{\dfrac{-4}{x^3}}$

77. $\dfrac{-9x^5}{27a^4} = \boxed{\dfrac{-1}{3a^3}}$

79. $\dfrac{7w^5}{5w^{13}} = \boxed{\dfrac{7}{5w^8}}$

81. $\dfrac{x^2 y}{x^5 y^3} = \dfrac{1}{x^{5-2}y^{3-1}} = \boxed{\dfrac{1}{x^3 y^2}}$

83. $\dfrac{m^6 n^2}{m^2 n^6} = \dfrac{m^{6-2}}{n^{6-2}} = \boxed{\dfrac{m^4}{n^4}}$

85. $\dfrac{40a^6 b^4}{-10a^{12}b^4} = -\dfrac{4b^{4-4}}{a^{12-6}} = \dfrac{-4b^0}{a^6} = \boxed{\dfrac{-4}{a^6}}$

87. $\dfrac{-40x^7 y^5}{-5x^{11}y^3} = \dfrac{8y^{5-3}}{x^{11-7}} = \boxed{\dfrac{8y^2}{x^4}}$

89. $\dfrac{-2x^3 y}{6x^2 y^5} = \dfrac{-x^{3-2}}{3y^{5-1}} = \boxed{\dfrac{-x}{3y^4}}$

91. $\dfrac{-3a^2 b^{11}}{7a^4 b^4} = \dfrac{-3b^{11-4}}{7a^{4-2}} = \boxed{\dfrac{-3b^7}{7a^2}}$

93. $\dfrac{-20xy^4 z^{10}}{40x^3 y^2 z^{13}} = \dfrac{-y^{4-2}}{2x^{3-1}z^{13-10}} = \boxed{\dfrac{-y^2}{2x^2 z^3}}$

95. $\dfrac{6x^4 - 8x^3 + 20x}{2x^2} = \dfrac{6x^4}{2x^2} - \dfrac{8x^3}{2x^2} + \dfrac{20x}{2x^2} = \boxed{3x^2 - 4x + \dfrac{10}{x}}$

97. $\dfrac{8y^4 - 20y^3 - 10y^2 + 8y - 6}{2y} = \dfrac{8y^4}{2y} - \dfrac{20y^3}{2y} - \dfrac{10y^2}{2y} + \dfrac{8y}{2y} - \dfrac{6}{2y} = \boxed{4y^3 - 10y^2 - 5y + 4 - \dfrac{3}{y}}$

99. $\dfrac{x^6 - x^4 + 2x^3 - 5x^2 + 9x}{x^3} = \dfrac{x^6}{x^3} - \dfrac{x^4}{x^3} + \dfrac{2x^3}{x^3} - \dfrac{5x^2}{x^3} + \dfrac{9x}{x^3}$

$\qquad = \boxed{x^3 - x + 2 - \dfrac{5}{x} + \dfrac{9}{x^2}}$

101. $\dfrac{8x^8 - 12x^4 - 16x^3 + 20x}{4x^4} = \dfrac{8x^8}{4x^4} - \dfrac{12x^4}{4x^4} - \dfrac{16x^3}{4x^4} + \dfrac{20x}{4x^4}$

$\qquad = \boxed{2x^4 - 3 - \dfrac{4}{x} + \dfrac{5}{x^3}}$

103. $\dfrac{18x^5 - 24x^3 + 30x^2 + 12x + 24}{6x^3} = \dfrac{18x^5}{6x^3} - \dfrac{24x^3}{6x^3} + \dfrac{30x^2}{6x^3} + \dfrac{12x}{6x^3} + \dfrac{24}{6x^3}$

$\qquad = \boxed{3x^2 - 4 + \dfrac{5}{x} + \dfrac{2}{x^2} + \dfrac{4}{x^3}}$

105. $\dfrac{12x^7 - 6x^6 - 15x^4 + 3x^2 - 1}{-3x^5} = \dfrac{12x^7}{-3x^5} - \dfrac{6x^6}{-3x^5} - \dfrac{15x^4}{-3x^5} + \dfrac{3x^2}{-3x^5} - \dfrac{1}{-3x^5}$

$\qquad = \boxed{-4x^2 + 2x + \dfrac{5}{x} - \dfrac{1}{x^3} + \dfrac{1}{3x^5}}$

107. $(2.4x^4 - 3.6x^2) \div 1.2x^5 = \dfrac{2.4x^4}{1.2x^5} - \dfrac{3.6x^2}{1.2x^5} = \boxed{\dfrac{2}{x} - \dfrac{3}{x^3}}$

109. $\dfrac{2y^3(4y+2) - 3y^2(2y-4)}{2y^2}$

$= \dfrac{8y^4 + 4y^3 - 6y^3 + 12y^2}{2y^2}$

$= \dfrac{8y^4 - 2y^3 + 12y^2}{2y^2} = \dfrac{8y^4}{2y^2} - \dfrac{2y^3}{2y^2} + \dfrac{12y^2}{2y^2} = \boxed{4y^2 - y + 6}$

111.
$$\frac{(y+2)^2 + (y-2)^2}{2y} = \frac{y^2 + 4y + 4 + y^2 - 4y + 4}{2y}$$
$$= \frac{2y^2 + 8}{2y} = \frac{2y^2}{2y} + \frac{8}{2y} = \boxed{y + \frac{4}{y}}$$

113.
$$\frac{(y+5)^2 + (y+5)(y-5)}{2y} = \frac{y^2 + 10y + 25 + y^2 - 25}{2y}$$
$$= \frac{2y^2 + 10y}{2y} = \frac{2y^2}{2y} + \frac{10y}{2y}$$
$$= \boxed{y + 5}$$

115. Let y equal the monomial.
$$y(2x^5) = -6x^{12}$$
$$y = \frac{-6x^{12}}{2x^5} = \boxed{-3x^7}$$

117. Let y equal the trinomial.
$$4x(y) = 20x^4 + 8x^2 + 12x$$
$$y = \frac{20x^4 + 8x^2 + 12x}{4x}$$
$$= \boxed{5x^3 + 2x + 3}$$

119. Let y equal the polynomial.
$$\frac{(y)}{4x^2} = 10x - 22$$
$$y = 4x^2(10x - 22)$$
$$y = \boxed{40x^3 - 88x^2}$$

121. Let w equal the width.
$$w = \frac{\text{area}}{\text{length}} = \frac{x^5 + 3x^4 - x^3}{x^2} = \boxed{x^3 + 3x^2 - x}$$

123. Dean Wormer: $4^{-2} + 2^{-4}$
Bluto: $1 - (4^{-2} + 2^{-4})$
fractional part of cake Bluto devoured:
$$\frac{1 - 4^{-2} - 2^{-4}}{1} = 1 - \frac{1}{4^2} - \frac{1}{2^4} = 1 - \frac{1}{16} - \frac{1}{16} = 1 - \frac{1}{8} = \boxed{\frac{7}{8}}$$

125. $\boxed{B}$ is true; $\frac{4y^2 - 6y}{2y^3} = \frac{4y^2}{2y^3} - \frac{6y}{2y^3} = \frac{2}{y} - \frac{3}{y^2}$ True

127. $\boxed{B}$ is true; $(12x^2 + 16x^5) \div 2x^4$
$$= \frac{12x^2}{2x^4} + \frac{16x^5}{2x^4} = \frac{6}{x^2} + 8x = 8x + \frac{6}{x^2}$$ True

Review Problems

130.

	10	−5	1	Score
1	3	0	0	30
2	2	1	0	15
3	2	0	1	21
4	1	2	0	0
5	1	0	2	12
6	1	1	1	6
7	0	3	0	−15
8	0	2	1	−9
9	0	1	2	−3
10	0	0	3	3

$\boxed{10}$ possible different scores

131. Let x equal the number.

$$6x + 4 \;<\; 2x$$
$$4x \;<\; -4$$
$$x \;<\; -1$$
$$\{x \mid x < -1\}$$

$\boxed{\text{all real numbers less than } -1}$

132. Let x equal the number of dimes.

$x + 2$ equals the number of nickels.

$(x + 2) + 3 = x + 5$ equals the number of quarters.

$$0.10x + 0.05(x + 2) + 0.25(x + 5) \;=\; 3.35$$
$$0.10x + 0.50x + 0.10 + 0.25x + 1.25 \;=\; 3.35$$
$$0.40x \;=\; 2.00$$
$$x \;=\; 5$$
$$x + 2 \;=\; 7$$
$$x + 5 \;=\; 10$$

$\boxed{7 \text{ nickels}, 5 \text{ dimes}, 10 \text{ quarters}}$

Section 6.6 Dividing Polynomials by Binomials

Problem Set 6.6, pp. 424-425

1. $\dfrac{x^2 - 5x + 6}{x - 3} = \boxed{x - 2}$

$$
\begin{array}{r}
x - 2 \\
x - 3 \;\overline{\big)\; x^2 - 5x + 6} \\
\underline{x^2 - 3x} \\
-2x + 6 \\
\underline{-2x + 6} \\
\varnothing
\end{array}
$$

3. $\dfrac{2y^2 + 5y + 2}{y + 2} = \boxed{2y + 1}$

$$
\begin{array}{r}
2y + 1 \\
y + 2 \;\overline{\big)\; 2y^2 + 5y + 2} \\
\underline{2y^2 + 4y} \\
y + 2 \\
\underline{y + 2} \\
\varnothing
\end{array}
$$

5. $\dfrac{x^3 - 6x^2 + 7x - 2}{x - 1} = \boxed{x^2 - 5x + 2}$

$$
\begin{array}{r}
x^2 - 5x + 2 \\
x - 1 \;\overline{\big)\; x^3 - 6x^2 + 7x - 2} \\
\underline{x^3 - x^2} \\
-5x^2 + 7x \\
\underline{-5x^2 + 5x} \\
2x - 2 \\
\underline{2x - 2} \\
\varnothing
\end{array}
$$

7. $\dfrac{12y^2 - 20y + 3}{2y - 3} = \boxed{6y - 1}$

$$
\begin{array}{r}
6y - 1 \\
2y - 3 \;\overline{\big)\; 12y^2 - 20y + 3} \\
\underline{12y^2 - 18y} \\
-2y + 3 \\
\underline{-2y + 3} \\
\varnothing
\end{array}
$$

9. $\dfrac{4a^2 + 4a - 3}{2a - 1} = \boxed{2a + 3}$

$$
\begin{array}{r}
2a + 3 \\
2a - 1 \;\overline{\big)\; 4a^2 + 4a - 3} \\
\underline{4a^2 - 2a} \\
6a - 3 \\
\underline{6z - 3} \\
\varnothing
\end{array}
$$

11. $\dfrac{3y - y^2 + 2y^3 + 2}{2y + 1}$

$= \dfrac{2y^3 - y^2 + 3y + 2}{2y + 1} = \boxed{y^2 - y + 2}$

$$
\begin{array}{r}
y^2 - y + 2 \\
2y + 1 \;\overline{\big)\; 2y^3 - y^2 + 3y + 2} \\
\underline{2y^3 + y^2} \\
-2y^2 + 3y \\
\underline{-2y^2 - y} \\
4y + 2 \\
\underline{4y + 2} \\
\varnothing
\end{array}
$$

13. $\dfrac{x^2 - 5x + 8}{x - 3} = \boxed{x - 2 + \dfrac{2}{x - 3}}$

$$
\begin{array}{r}
x - 2 \\
x - 3 \overline{\smash{\big)}\, x^2 - 5x + 8} \\
\underline{x^2 - 3x} \\
-2x + 8 \\
\underline{-2x + 6} \\
2
\end{array}
$$

15. $\dfrac{5y + 10 + y^2}{y + 2} = \dfrac{y^2 + 5y + 10}{y + 2} = \boxed{y + 3 + \dfrac{4}{y + 2}}$

$$
\begin{array}{r}
y + 3 \\
y + 2 \overline{\smash{\big)}\, y^2 + 5y + 10} \\
\underline{y^2 + 2y} \\
3y + 10 \\
\underline{3y + 6} \\
4
\end{array}
$$

17. $\dfrac{2x^2 - 9x + 8}{2x + 3} = \boxed{x - 6 + \dfrac{26}{2x + 3}}$

$$
\begin{array}{r}
x - 6 \\
2x + 3 \overline{\smash{\big)}\, 2x^2 - 9x + 8} \\
\underline{2x^2 + 3x} \\
-12x + 8 \\
\underline{-12x - 18} \\
26
\end{array}
$$

19. $\dfrac{x^3 + 4x - 3}{x - 2} = \boxed{x^2 + 2x + 8 + \dfrac{13}{x - 2}}$

$$
\begin{array}{r}
x^2 + 2x + 8 \\
x - 2 \overline{\smash{\big)}\, x^3 + 0x^2 + 4x - 3} \\
\underline{x^3 - 2x^2} \\
2x^2 + 4x \\
\underline{2x^2 - 4x} \\
8x - 3 \\
\underline{8x - 16} \\
13
\end{array}
$$

21. $\dfrac{4y^3 + 8y^2 + 5y + 9}{2y + 3} = \boxed{2y^2 + y + 1 + \dfrac{6}{2y + 3}}$

$$
\begin{array}{r}
2y^2 + y + 1 \\
2y + 3 \overline{\smash{\big)}\, 4y^3 + 8y^2 + 5y + 9} \\
\underline{4y^3 + 6y^2} \\
2y^2 + 5y \\
\underline{2y^2 + 3y} \\
2y + 9 \\
\underline{2y + 3} \\
6
\end{array}
$$

23. $\dfrac{6y^3 - 5y^2 + 5}{3y + 2} = \boxed{2y^2 - 3y + 2 + \dfrac{1}{3y + 2}}$

$$
\begin{array}{r}
2y^2 - 3y + 2 \\
3y + 2 \overline{\smash{\big)}\, 6y^3 - 5y^2 + 0y + 5} \\
\underline{6y^3 + 4y^2} \\
-9y^2 + 0y \\
\underline{-9y^2 - 6y} \\
6y + 5 \\
\underline{6y + 4} \\
1
\end{array}
$$

25. $\dfrac{8x^3 - 1}{2x - 1} = \boxed{4x^2 + 2x + 1}$

$$
\begin{array}{r}
4x^2 + 2x + 1 \\
2x - 1 \overline{\smash{\big)}\, 8x^3 + 0x^2 + 0x - 1} \\
\underline{8x^3 - 4x^2} \\
4x^2 + 0x \\
\underline{4x^2 - 2x} \\
2x - 1 \\
\underline{2x - 1} \\
\varnothing
\end{array}
$$

27. $\dfrac{81 - 12y^3 + 54y^2 + y^4 - 108y}{y - 3}$

$= \dfrac{y^4 - 12y^3 + 54y^2 - 108y + 81}{y - 3}$

$= \boxed{y^3 - 9y^2 + 27y - 27}$

$$
\begin{array}{r}
y^3 - 9y^2 + 27y - 27 \\
y - 3 \enclose{longdiv}{y^4 - 12y^3 + 54y^2 - 108y + 81} \\
\underline{y^4 - 3y^3} \\
-9y^3 + 54y^2 \\
\underline{-9y^3 + 27y^2} \\
27y^2 - 108y \\
\underline{27y^2 - 81y} \\
-27y + 81 \\
\underline{-27y + 81} \\
0
\end{array}
$$

29. $\dfrac{4y^2 + 6y}{2y - 1} = \boxed{2y + 4 + \dfrac{4}{2y - 1}}$

$$
\begin{array}{r}
2y + 4 \\
2y - 1 \enclose{longdiv}{4y^2 + 6y + 0} \\
\underline{4y^2 - 2y} \\
8y + 0 \\
\underline{8y - 4} \\
4
\end{array}
$$

31. $\dfrac{y^4 - 2y^2 + 5}{y - 1} = \boxed{y^3 + y^2 - y - 1 + \dfrac{4}{y - 1}}$

$$
\begin{array}{r}
y^3 + y^2 - y - 1 \\
y - 1 \enclose{longdiv}{y^4 + 0y^3 - 2y^2 + 0y + 5} \\
\underline{y^4 - y^3} \\
y^3 - 2y^2 \\
\underline{y^3 - y^2} \\
-y^2 + 0y \\
\underline{-y^2 + 5} \\
-y + 5 \\
\underline{-y + 1} \\
4
\end{array}
$$

33. $\dfrac{x^2 - x - 3x^3 + 4x^4}{4x^2 + 3}$

$= \dfrac{4x^4 - 3x^3 + x^2 - x}{4x^2 + 3} = \boxed{x^2 - \dfrac{3}{4}x - \dfrac{1}{2} + \dfrac{\frac{5}{4}x + \frac{3}{2}}{4x^2 + 3}}$

$$
\begin{array}{r}
x^2 - 3/4x - 1/2 \\
4x^2 + 3 \enclose{longdiv}{4x^4 - 3x^3 + x^2 - x + 0} \\
\underline{4x^4 + 3x^3} \\
-3x^3 - 2x^2 - x \\
\underline{-3x^3 - 9/4x} \\
-2x^2 + 5/4x + 0 \\
\underline{-2x^2 - 3/2} \\
\dfrac{5}{4}x + \dfrac{3}{2}
\end{array}
$$

35. $\dfrac{y^3 + y}{y + 3} = \boxed{y^2 - 3y + 10 - \dfrac{30}{y + 3}}$

$$
\begin{array}{r}
y^2 - 3y + 10 \\
y + 3 \enclose{longdiv}{y^3 + 0y^2 + y + 0} \\
\underline{y^3 + 3y^2} \\
-3y^2 + y \\
\underline{-3y^2 - 9y} \\
10y + 0 \\
\underline{10y + 30} \\
-30
\end{array}
$$

37. $\dfrac{a^6 - 8}{a^2 - 2} = \boxed{a^4 + 2a^2 + 4}$

$$
\begin{array}{r}
a^4 + 2a^2 + 4 \\
a^2 - 2 \overline{\smash{\big)}\, a^6 + 0a^4 + 0a^2 - 8} \\
\underline{a^6 - 2a^4} \\
2a^4 + 0a^2 \\
\underline{2a^4 - 4a^2} \\
4a^2 - 8 \\
\underline{4a^2 - 8} \\
0
\end{array}
$$

39. Let y equal the trinomial.

$$
\begin{aligned}
\frac{y}{x+7} &= 2x - 5 \\
y &= (2x - 5)(x + 7) \\
y &= \boxed{2x^2 + 9x - 35}
\end{aligned}
$$

41. width $= \dfrac{\text{area}}{\text{length}} = \dfrac{2x^2 + 5x - 3}{2x - 1} = \boxed{x + 3 \text{ inches}}$

$$
\begin{array}{r}
x + 3 \\
2x - 1 \overline{\smash{\big)}\, 2x^2 + 5x - 3} \\
\underline{2x^2 - x} \\
6x - 3 \\
\underline{6x - 3} \\
0
\end{array}
$$

43. Let y equal the polynomial.

$$
\begin{aligned}
\frac{y}{2x+4} &= x - 3 + \frac{17}{2x+4} \\
y &= (2x + 4)\left(x - 3 + \frac{17}{2x+4}\right) \\
y &= (2x + 4)(x - 3) + (2x + 4)\left(\frac{17}{2x+4}\right) \\
&= 2x^2 - 2x - 12 + 17 \\
&= \boxed{2x^2 - 2x + 5}
\end{aligned}
$$

45. $\dfrac{7x^2 + x + k}{x - 1} = 7x + 8$, remainder $= 0$ when $k = \boxed{-8}$

$$
\begin{array}{r}
7x + 8 \\
x - 1 \overline{\smash{\big)}\, 7x^2 + x + k} \\
\underline{7x^2 - 7x} \\
8x + k \\
\underline{8x - 8} \\
k + 8 \quad k + 8 = 0,\ k = -8
\end{array}
$$

47. $\dfrac{2x(4x-3)-2x^2-7x+7}{3x-2}$

$= \dfrac{8x^2-6x-2x^2-7x+7}{3x-2}$

$= \dfrac{6x^2-13x+7}{3x-2}$

$= \boxed{2x-3+\dfrac{1}{3x-2}}$

$$
\begin{array}{r}
2x-3 \\
3x-2 \ \overline{)\ 6x^2-13x+7} \\
\underline{6x^2-4x} \\
-9x+7 \\
\underline{-9x+6} \\
1
\end{array}
$$

49. $\boxed{B}$ is true.

Review Problems

53.

$$
\begin{aligned}
x &= 3-5y & \rightarrow && x+5y &= 3 \\
\underline{x-2y} &= 10 && & \underline{-x+2y} &= -10 \\
&&&& 7y &= -7 \\
&&&& y &= -1 \\
&&&& x &= 3-5y = 3+5 = 8
\end{aligned}
$$

$$\boxed{\{(8,-1)\}}$$

54.

$$
\begin{aligned}
52t+58t &= 385 \\
110t &= 385 \\
t &= 3.5
\end{aligned}
$$

$$\boxed{3.5 \text{ hours}}$$

55. Let x the equal amount of acid.
$x+30$ equals the amount of 40% acid mixture.

$$
\begin{aligned}
x+0.15(30) &= 0.40(x+30) \\
x+4.5 &= 0.40x+12 \\
0.60x &= 7.5 \\
x &= 12.5
\end{aligned}
$$

$$\boxed{12.5 \text{ liters}}$$

Section 6.7 Exponents and Scientific Notation

Problem Set 6.7, pp. 435-438

1. $x^{-8}\cdot x^3 = x^{-8+3} = x^{-5} = \boxed{\dfrac{1}{x^5}}$

3. $(4x^{-5})(2x^2) = 8x^{-5+2} = 8x^{-3} = \boxed{\dfrac{8}{x^3}}$

5. $(-6x^4)\left(\dfrac{1}{2}x^{-9}\right) = -3x^{4-9} = -3x^{-5} = \boxed{\dfrac{-3}{x^5}}$

7. $(xy^3)(xy^{-5}) = x^{1+1}y^{3-5} = x^2y^{-2} = \boxed{\dfrac{x^2}{y^2}}$

9. $(-8x^2y^{-4})(3x^{-6}y^4) = -24x^{2-6}y^{-4+4} = -24x^{-4} = \boxed{-\dfrac{24}{x^4}}$

11. $\left(\dfrac{1}{2}x^{-3}y^{-2}\right)(4x^{-2}y^{-5}) = 2x^{-3-2}y^{-2-5} = 2x^{-5}y^{-7} = \boxed{\dfrac{2}{x^5y^7}}$

13. $\dfrac{y^6}{y^{10}} = y^{6-10} = y^{-4} = \boxed{\dfrac{1}{y^4}}$

15. $\dfrac{y^{-6}}{y^{10}} = y^{-6-10} = y^{-16} = \boxed{\dfrac{1}{y^{16}}}$

17. $\dfrac{y^6}{y^{-10}} = y^{6+10} = \boxed{y^{16}}$

19. $\dfrac{z^3}{(z^4)^2} = \dfrac{z^3}{z^8} = z^{3-8} = z^{-5} = \boxed{\dfrac{1}{z^5}}$

21. $\dfrac{z^{-3}}{(z^4)^2} = \dfrac{z^{-3}}{z^8} = z^{-3-8} = z^{-11} = \boxed{\dfrac{1}{z^{11}}}$

23. $\dfrac{(4x^3)^2}{x^8} = \dfrac{16x^6}{x^8} = 16x^{6-8} = 16x^{-2} = \boxed{\dfrac{16}{x^2}}$

25. $\dfrac{(6a^4)^3}{a^{-5}} = \dfrac{6^3 a^{12}}{a^{-5}} = 216a^{12+5} = \boxed{216a^{17}}$

27. $\left(\dfrac{a^3}{b^{-5}}\right)^2 = \dfrac{a^6}{b^{-10}} = \boxed{a^6 b^{10}}$

29. $\left(\dfrac{2a^7}{b^{-4}}\right)^3 = \dfrac{8a^{21}}{b^{-12}} = \boxed{8a^{21}b^{12}}$

31. $\left(\dfrac{y^4}{y^2}\right)^{-3} = (y^{4-2})^{-3} = (y^2)^{-3} = y^{-6} = \boxed{\dfrac{1}{y^6}}$

33. $\left(\dfrac{4x^5}{2x^2}\right)^{-4} = (2x^{5-2})^{-4} = 2^{-4}(x^3)^{-4} = 2^{-4}x^{-12} = \boxed{\dfrac{1}{16x^{12}}}$

35. $(-2z^{-1})^{-2} = (-2)^{-2}z^{(-1)(-2)} = \dfrac{z^2}{4}$

37. $\left(\dfrac{2a^{-2}}{b^3}\right)^{-3} = \dfrac{2^{-3}a^{(-2)(-3)}}{b^{3(-3)}} = \dfrac{a^6}{2^3 b^{-9}} = \boxed{\dfrac{a^6 b^9}{8}}$

39. $\dfrac{25a^5 b^3}{75a^2 b^7} = \dfrac{1}{3}\,a^{5-2}b^{3-7} = \dfrac{1}{3}\,a^3 b^{-4} = \boxed{\dfrac{a^3}{3b^4}}$

41. $\dfrac{2x^5 \cdot 3x^7}{15x^6} = \dfrac{2x^{5+7-6}}{5} = \boxed{\dfrac{2x^6}{5}}$

43. $\left(\dfrac{x^6 y^8}{x^2 y^9}\right)^{-3} = (x^{6-2}y^{8-9})^{-3} = (x^4 y^{-1})^{-3} = x^{4(-3)}y^{-1(-3)} = x^{-12}y^3 = \boxed{\dfrac{y^3}{x^{12}}}$

45. $(x^3)^5 x^{-7} = x^{15}x^{-7} = x^{15-7} = \boxed{x^8}$

47. $(2y^3)^4 y^{-6} = 2^4 y^{12}y^{-6} = 16y^{12-6} = \boxed{16y^6}$

49. $(-2y^4)^3 y^{-13} = -8y^{12}y^{-13} = -8y^{12-13} = -8y^{-1} = \boxed{-\dfrac{8}{y}}$

51. $(x^3)^4 (x^{-2})^7 = x^{12}x^{-14} = x^{12-14} = x^{-2} = \boxed{\dfrac{1}{x^2}}$

53. $\dfrac{(y^3)^4}{(y^2)^7} = \dfrac{y^{12}}{y^{14}} = y^{12-14} = y^{-2} = \boxed{\dfrac{1}{y^2}}$

55. $\dfrac{(y^3)^4}{(y^{-2})^7} = \dfrac{y^{12}}{y^{-14}} = y^{12+14} = \boxed{y^{26}}$

57. $\dfrac{3a(a^2)^5}{6a^3 (a^4)^2} = \dfrac{a(a^{10})}{2a^3 a^8} = \dfrac{a^{1+10}}{2a^{3+8}} = \dfrac{a^{11}}{2a^{11}} = \dfrac{a^{11-11}}{2} = \dfrac{a^0}{2} = \boxed{\dfrac{1}{2}}$

59. $(-3y^4)(-5y^{-7}) = 15y^{4-7} = 15y^{-3} = \boxed{\dfrac{15}{y^3}}$

61. $(-3y^{-3})(-5y^{-5}) = 15y^{-3-5} = 15y^{-8} = \boxed{\dfrac{15}{y^8}}$

63. $(-4ab^5)(3ab^{-8}) = -12a^{1+1}b^{5-8} = -12a^2 b^{-3} = \boxed{\dfrac{-12a^2}{b^3}}$

65. $(x^{-2}y^{-3})^{-4} = x^{(-2)(-4)}y^{(-3)(-4)} = \boxed{x^8 y^{12}}$

67. $\dfrac{(x^{-2}y)^{-3}}{(x^2 y^{-1})^3} = \dfrac{x^{(-2)(-3)}y^{-3}}{x^{2(3)}y^{(-1)(3)}} = \dfrac{x^6 y^{-3}}{x^6 y^{-3}} = x^{6-6}y^{-3+3} = x^0 y^0 = \boxed{1}$

69. $\dfrac{x^0 x^3}{x^{-3}x^4} = x^{0+3+3-4} = \boxed{x^2}$

71. $\left(\dfrac{18x^2 y^3}{3x^{-1}y^{-9}}\right)^0 = \boxed{1}$

73. $2.7 \times 10^2 =$ $\boxed{270}$ (move decimal point 2 places right)

75. $9.12 \times 10^5 =$ $\boxed{912,000}$ (move right 5)

77. $3.4 \times 10^0 =$ $\boxed{3.4}$

79. $7.9 \times 10^{-1} =$ $\boxed{0.79}$ (move left 1)

81. $2.15 \times 10^{-2} =$ $\boxed{0.0215}$ (move left 2)

83. $7.86 \times 10^{-4} =$ $\boxed{0.000786}$ (move left 4)

85. $9.29 \times 10^7 =$ $\boxed{92,900,000}$

87. $4 \times 10^{-5} =$ $\boxed{0.00004}$

89. $32,400 =$ $\boxed{3.24 \times 10^4}$

91. $220,000,000 =$ $\boxed{2.2 \times 10^8}$

93. $713 =$ $\boxed{7.13 \times 10^2}$

95. $6751 =$ $\boxed{6.751 \times 10^3}$

97. $0.0027 =$ $\boxed{2.7 \times 10^{-3}}$

99. $0.000\ 020\ 2 =$ $\boxed{2.02 \times 10^{-5}}$

101. $0.005 =$ $\boxed{5 \times 10^{-3}}$

103. $3.141\ 59 =$ $\boxed{3.14159 \times 10^0}$

105. $650,000 =$ $\boxed{6.5 \times 10^5}$

107. $9230 =$ $\boxed{9.230 \times 10^3}$

109. $0.000\ 000\ 000\ 000\ 000\ 000\ 531 =$ $\boxed{5.31 \times 10^{-19}}$

111. $0.000\ 007\ 5 =$ $\boxed{7.5 \times 10^{-6}}$

113. $(2 \times 10^3)(3 \times 10^2) = 6 \times 10^{3+2} =$ $\boxed{6 \times 10^5}$ $=$ $\boxed{600,000}$

115. $(2 \times 10^5)(8 \times 10^3) = 16 \times 10^{5+3} - 1.6 \times 10 \times 10^8 =$ $\boxed{1.6 \times 10^9}$ $=$ $\boxed{1,600,000,000}$

117. $\dfrac{12 \times 10^6}{4 \times 10^2} = 3 \times 10^{6-2} = 3 \times 10^4 =$ $\boxed{30,000}$

119. $\dfrac{15 \times 10^4}{5 \times 10^{-2}} = 3 \times 10^{4+2} = 3 \times 10^6 =$ $\boxed{3,000,000}$

121. $\dfrac{15 \times 10^{-4}}{5 \times 10^2} = 3 \times 10^{-4-2} = 3 \times 10^{-6} =$ $\boxed{0.000\ 003}$

123. $\dfrac{180 \times 10^6}{2 \times 10^3} = 90 \times 10^{6-3} = 90 \times 10^3 =$ $\boxed{90,000}$

125. $\dfrac{3 \times 10^4}{12 \times 10^{-3}} = 0.25 \times 10^{4+3} = 0.25 \times 10^7 =$ $\boxed{2,500,000}$

127. $(5 \times 10^2)^3 = 5^3 \times 10^{2(3)} = 125 \times 10^6 =$ $\boxed{125,000,000}$

129. $(3 \times 10^{-2})^4 = 3^4 \times 10^{-2(4)} = 81 \times 10^{-8} =$ $\boxed{0.000\ 000\ 81}$

131. $(4 \times 10^6)^{-1} = 4^{-1} \times 10^{6(-1)} = 0.25 \times 10^{-6} =$ $\boxed{0.000\ 000\ 25}$

133. $(3,000,000)(0.002) = (3 \times 10^6)(2 \times 10^{-3})$
$= 6 \times 10^{6-3} =$ $\boxed{6 \times 10^3}$

135. $(2300)(2,000,000) = (2.3 \times 10^3)(2 \times 10^6) = 4.6 \times 10^{3+6} =$ $\boxed{4.6 \times 10^9}$

137. $(0.000\ 05)(4,000,000,000,000) = (5 \times 10^{-5})(4 \times 10^{12}) = 20 \times 10^{-5+12} = 2 \times 10 \times 10^7 =$ $\boxed{2 \times 10^8}$

139. $(200)^3(0.000\ 01) \ = \ (2 \times 10^2)^3(1.0 \times 10^{-5}) = 2^3 \times 10^6 \times 10^{-5} = \boxed{8 \times 10^1}$

141. $\begin{aligned} (200)^4(0.005)^3 \ &= \ (2 \times 10^2)^4(5 \times 10^{-3})^3 \\ &= \ (2^4 \times 10^{2(4)})(5^3 \times 10^{-3(3)}) \\ &= \ (16 \times 10^8)(125 \times 10^{-9}) \\ &= \ (16)(125) \times 10^{8-9} \\ &= \ 2000 \times 10^{-1} \\ &= \ 2 \times 10^3 \times 10^{-1} \\ &= \ 2 \times 10^{3-1} = \boxed{2 \times 10^2} \end{aligned}$

143. $(2 \times 10^9)(4 \times 10^{-5}) = 8 \times 10^{9-5} = 8 \times 10^4 = \boxed{80{,}000}$

145. $(3 \times 10^{-1})(2 \times 10^{-3}) = 6 \times 10^{-1-3} = 6 \times 10^{-4} = \boxed{0.0006}$

147. $(5 \times 10^2)(3 \times 10^3)(2 \times 10^{-2})$
$= (5)(3)(2) \times 10^{2+3-2} = 30 \times 10^3 = 3 \times 10 \times 10^3 = 3 \times 10^4$
$= \boxed{30{,}000}$

149. $\dfrac{8 \times 10^5}{2 \times 10^{-1}} = 4 \times 10^{5+1} = 4 \times 10^6 = \boxed{4{,}000{,}000}$

151. $\dfrac{16 \times 10^{-3}}{8 \times 10^{-2}} = 2 \times 10^{-3+2} = 2 \times 10^{-1} = \boxed{0.2}$

153. $\dfrac{2.8}{2} \times 10^{6-2} = 1.4 \times 10^4 = \boxed{14{,}000}$

155. $(5 \times 10^3)(1.2 \times 10^{-4}) \div (2.4 \times 10^2)$
$= \dfrac{(5)(1.2) \times 10^{3-4}}{(2.4) \times 10^2} = 2.5 \times 10^{-1-2} = 2.5 \times 10^{-3}$
$= \boxed{0.0025}$

157. $\begin{aligned} \dfrac{(1.6 \times 10^4)(7.2 \times 10^{-3})}{(3.6 \times 10^8)(4 \times 10^{-3})} \ &= \ \dfrac{(1.6)}{(4)}\dfrac{(7.2)}{(3.6)} \times \dfrac{10^{4-3}}{10^{8-3}} \\ &= \ (0.4)(2) \times \dfrac{10^1}{10^5} \ = \ 0.8 \times 10^{1-5} \\ &= \ 0.8 \times 10^{-4} \\ &= \ \boxed{0.00008} \end{aligned}$

159. Mercury: $(0.4)(9.3 \times 10^7 \text{ miles}) = \boxed{3.72 \times 10^7 \text{ miles}}$

Venus: $(0.7)(9.3 \times 10^7 \text{ miles}) = \boxed{6.51 \times 10^7 \text{ miles}}$

Earth: $(1.0)(9.3 \times 10^7 \text{ miles}) = \boxed{9.3 \times 10^7 \text{ miles}}$

Mars: $(1.5)(9.3 \times 10^7 \text{ miles}) = 13.95 \times 10^7 \text{ miles} = 1.395 \times 10 \times 10^7 \text{ miles} = \boxed{1.395 \times 10^8 \text{ miles}}$

Jupiter: $(5.2)(9.3 \times 10^7 \text{ miles}) = 48.36 \times 10^7 \text{ miles} = 4.836 \times 10 \times 10^7 \text{ miles} = \boxed{4.836 \times 10^8 \text{ miles}}$

Saturn: $(9.6)(9.3 \times 10^7 \text{ miles}) = 89.28 \times 10^7 \text{ miles} = 8.928 \times 10 \times 10^7 \text{ miles} = \boxed{8.928 \times 10^8 \text{ miles}}$

Uranus: $(19.2)(9.3 \times 10^7 \text{ miles}) = 178.56 \times 10^7 \text{ miles} = 1.7856 \times 10^2 \times 10^7 \text{ miles} = \boxed{1.7856 \times 10^9 \text{ miles}}$

Neptune: $(30.1)(9.3 \times 10^7 \text{ miles}) = 279.93 \times 10^7 \text{ miles} = 2.7993 \times 10^2 \times 10^7 \text{ miles} = \boxed{2.7993 \times 10^9 \text{ miles}}$

Pluto: $\begin{aligned} (39.3)(9.3 \times 10^7 \text{ miles}) \ &= \ 365.49 \times 10^7 \text{ miles} \\ &= \ 3.6549 \times 10^2 \times 10^7 \text{ miles} \\ &= \ \boxed{3.6549 \times 10^9 \text{ miles}} \end{aligned}$

161. area $= (4 \times 10^3 \text{ miles})(2 \times 10^2 \text{ meters})$

$= 8 \times 10^{3+2}$ square meters

$= \boxed{8 \times 10^5 \text{ square meters}}$

163. area $= \frac{1}{2}(4.6 \times 10^3 \text{ yards})(4.2 \times 10^{-3} \text{ yards})$

$= \frac{1}{2}(4.6)(4.2) \times 10^{3-3}$ square yards

$= 9.66 \times 10^0$ square yards

$= \boxed{9.66 \text{ square yards}}$

165. $T = \frac{D}{R}$ $D = 9.14 \times 10^7$ miles

$R = 1.86 \times 10^5$ miles per second

$T \approx \frac{9.14 \times 10^7 \text{ miles}}{1.86 \times 10^5 \text{ miles per second}} \approx 4.914 \times 10^{7-5}$ seconds

$\approx 4.914 \times 10^2$ seconds

$\approx \boxed{491 \text{ seconds}}$

167. $(3.2 \times 10^{-3})(3 \times 10^{-5}) = 9.6 \times 10^n$

$9.6 \times 10^{-3-5} = 9.6 \times 10^n$

$9.6 \times 10^{-8} = 9.6 \times 10^n$

$n = -8$

$\boxed{\{-8\}}$

169. $\boxed{B}$ is true; $5^{-2} > 2^{-5}$

$\frac{1}{5^2} > \frac{1}{2^5}$

$\frac{1}{25} > \frac{1}{32}$

$\frac{32}{(25)(32)} > \frac{25}{(25)(32)}$ True

171. $\boxed{A}$ is true; $\frac{x^3}{x^7} = x^{3-7} = x^{-4} = \frac{1}{x^4}$.

True, for any nonzero real number x.

173. $\boxed{D}$ is true; $\frac{8 \times 10^{-9}}{4 \times 10^{-5}} = 2 \times 10^{-9-(-5)} = 2 \times 10^{-4}$.

175. $x = (((2^2)^2)^2)^2 = ((2^{2 \cdot 2})^2)^2 = ((2^{4 \cdot 2})^2 = 2^{8 \cdot 2} = 2^{16}$
$y = 2^{2^{2^2}} = ((2^2)^2)^2 = (2^{2 \cdot 2})^2 = 2^{4 \cdot 2} = 2^8$
$\frac{x}{y} = \frac{2^{16}}{2^8} = 2^{16-8} = \boxed{2^8}$

177. $\frac{7.483 \times 10^{-7}}{5.2 \times 10^2} = \boxed{1.439 \times 10^{-9}}$

Review Problems

178. Let x equal the number of minutes to inspect a model B.
$x + 3$ equals the number of minutes to inspect a model A.
Inspecting $19B$s and $32A$s takes 5 hours $= 300$ minutes:

$19x + 32(x + 3) = 300$
$19x + 32x + 96 = 300$
$51x = 204$
$x = 4$

It takes $\boxed{4 \text{ minutes to inspect a } B}$ and $\boxed{7 \text{ minutes to inspect an } A}$.

179. $A = p + prt$

$A - p = prt$

$$\boxed{\dfrac{A - p}{pr} = t \text{ or } t = \dfrac{A}{pr} - \dfrac{1}{r}}$$

180. $\dfrac{9 \text{ melons}}{\$11.25} = \dfrac{16 \text{ melons}}{x}$

$9x = 180$

$x = 20$

16 melons cost $\boxed{\$20}$.

Chapter 6   Review Problems

Chapter 6 Review Problems pp. 440-442

1. $7x^4 + 9x$ is a $\boxed{\text{binomial}}$ of degree $\boxed{4}$.

2. $3x + 5x^2 - 2$ is a $\boxed{\text{trinomial}}$ of degree $\boxed{2}$.

3. $16x$ is a $\boxed{\text{monomial}}$ of degree $\boxed{1}$.

4. $\begin{aligned} f(x) &= 2x^2 + 24x + 100 \\ f(3) &= 2(3^2) + 24(3) + 100 \\ &= 2(9) + 72 + 100 \\ &= 18 + 72 + 100 \\ &= 90 + 100 \\ &= \boxed{190} \end{aligned}$

The population of a slowly growing bacterial colony after 3 hours is 190.

5. $(-6x^2 + 7x^2 - 9x + 3) + (14x^3 + 3x^2 - 11x - 7)$
$= (-6x^3 + 14x^3) + (7x^2 + 3x^2) + (-9x - 11x) + 3 - 7)$
$= \boxed{8x^3 + 10x^2 - 20x - 4}$

6. $(-7a^2 + 4 + 9a^3) + (-13 - 8a^3 + 3a^2)$
$= (9a^3 - 8a^3) + (-7a^2 + 3a^2) + (4 - 13)$
$= \boxed{a^3 - 4a^2 - 9}$

7. $(5y^2 - y - 8) - (-6y^2 + 3y - 4)$
$= (5y^2 - y - 8 + 6y^2 - 3y + 4)$
$= (5y^2 + 6y^2) + (-y - 3y) + (-8 + 4)$
$= \boxed{11y^2 - 4y - 4}$

8. $(13x^4 - 8x^3 + 2x^2) - (5x^4 - 3x^2 + 2x^2 - 6)$
$= 13x^4 - 8x^3 + 2x^2 - 5x^4 + 3x^2 - 2x^2 + 6$
$= (13x^4 - 5x^4) + (-8x^3 + 3x^3) + (2x^2 - 2x^2) + 6$
$= \boxed{8x^4 - 5x^3 + 6}$

9. $(-13x^4 - 6x^2 + 5) - (x^4 + 7x^2 - 11x)$
$= -13x^4 - 6x^2 + 5x - x^4 - 7x^2 + 11x$
$= (-13x^4 - x^4) + (-6x^2 - 7x^2) + (11x + 5x)$
$= \boxed{-14x^4 - 13x^2 + 16x}$

10. Add: $\begin{aligned} 7y^4 - 6y^3 + 4y^2 - 4y \\ y^3 - y^2 + 3y - 4 \\ \hline \boxed{7y^4 - 5y^3 + 3y^2 - y - 4} \end{aligned}$

11. Subtract:

$\begin{array}{c} 7x^2 - 9x + 2 \\ 4x^2 - 2x - 7 \\ \hline \end{array}$ $\rightarrow$ $\begin{array}{c} 7x^2 - 9x + 2 \\ -(4x^2 - 2x - 7) \\ \hline \end{array}$ $\rightarrow$ $\begin{array}{c} 7x^2 - 9x + 2 \\ -4x^2 + 2x + 7 \\ \hline \boxed{3x^2 - 7x + 9} \end{array}$

12. Subtract:

$\begin{array}{c} 5x^3 - 6x^2 - 9x + 14 \\ -5x^3 + 3x^2 - 11x + 3 \\ \hline \end{array}$ $\rightarrow$ $\begin{array}{c} 5x^3 - 6x^2 - 9x + 14 \\ -(-5x^3 + 3x^2 - 11x + 3) \\ \hline \end{array}$ $\rightarrow$ $\begin{array}{c} 5x^3 - 6x^2 - 9x + 14 \\ 5x^3 - 3x^2 + 11x - 3 \\ \hline \boxed{10x^3 - 9x^2 + 2x + 11} \end{array}$

13. $7x(3x - 9) = 7x(3x) - (7x)(9) = \boxed{21x^2 - 63x}$

14. $-5x^3(4x^2 - 11x) = \boxed{-20x^5 + 55x^4}$

15. $3y^2(-7y^2 + 3y - 6) = 3y^2(-7y^2) + 3y^2(3y) + 3y^2(-6)$
$= \boxed{-21y^4 + 9y^3 - 18y^2}$

16. $-2y^5(8y^3 - 4y^2 - 10y + 6) = -2y^5(8y^3) - 2y^5(-4y^2) - 2y^5(-10y) - 2y^5(6)$
$\qquad = \boxed{-16y^8 + 8y^7 + 20y^6 - 12y^5}$

17. $(x + 3)(x^2 - 5x + 2)$
$= x(x^2 - 5x + 2) + 3(x^2 - 5x + 2)$
$= x(x^2) + x(-5x) + x(2) + 3(x^2) + 3(-5x) + 3(2)$
$= x^3 - 5x^2 + 2x + 3x^2 - 15x + 6$
$= x^3 - 5x^2 + 3x^2 + 2x - 15x + 6$
$= \boxed{x^3 - 2x^2 - 13x + 6}$

18. $(3y - 2)(4y^2 + 3y - 5)$
$= 3y(4y^2 + 3y - 5) - 2(4y^2 + 3y - 5)$
$= 3y(4y^2) + 3y(3y) + 3y(-5) - 2(4y^2) - 2(3y) - 2(-5)$
$= 12y^3 + 9y^2 - 15y - 8y^2 - 6y + 10$
$= 12y^3 + 9y^2 - 8y^2 - 15y - 6y + 10$
$= \boxed{12y^3 + y^2 - 21y + 10}$

19. $(2y^2 + 3y)(y^3 - y^2 + 4)$
$= 2y^2(y^3 - y^2 + 4) + 3y(y^3 - y^2 + 4)$
$= 2y^2(y^3) + 2y^2(-y^2) + 2y^2(4) + 3y(y^3) + 3y(-y^2) + 3y(4)$
$= 2y^5 - 2y^4 + 8y^2 + 3y^4 - 3y^3 + 12y$
$= 2y^5 - 2y^4 + 3y^4 - 3y^3 + 8y^2 + 12y$
$= \boxed{2y^5 + y^4 - 3y^3 + 8y^2 + 12y}$

20. $(x - 6)(x + 2) = x^2 + 2x - 6x - 12 = \boxed{x^2 - 4x - 12}$

21. $(3y - 5)(2y + 1) = 6y^2 + 3y - 10y - 5 = \boxed{6y^2 - 7y - 5}$

22. $(4x^3 - 2x^2)(x^2 - 3) = 4x^3(x^2) - 3(4x^3) - 2x^2(x^2) + 2x^2(3)$
$\qquad = 4x^5 - 12x^3 - 2x^4 + 6x^2$
$\qquad = \boxed{4x^5 - 2x^4 - 12x^3 + 6x^2}$

23.
$$
\begin{array}{r}
y^2 - 4y + 7 \\
3y - 5 \\
\hline
-5y^2 + 20y - 35 \\
3y^3 - 12y^2 + 21y \qquad\quad \\
\hline
\boxed{3y^3 - 17y^2 + 41y - 35}
\end{array}
$$

24.
$$
\begin{array}{r}
4x^3 - 2x^2 - 6x - 1 \\
3x^2 - 2x + 3 \\
\hline
12x^3 - 6x^2 - 18x - 3 \\
-8x^4 + 4x^3 + 12x^2 + 2x \qquad\quad \\
12x^5 - 6x^4 - 18x^3 - 3x^2 \qquad\qquad\quad \\
\hline
\boxed{12x^5 - 14x^4 - 2x^3 + 3x^2 - 16x - 3}
\end{array}
$$

25. $(3x^3 - 2)(x^3 + 4) = 3x^6 + 12x^3 - 2x^3 - 8 = \boxed{3x^6 + 10x^3 - 8}$

26. $(x + 3)^2 = x^2 + 2x(3) + 3^2 = \boxed{x^2 + 6x + 9}$

27. $(3y - 4)^2 = (3y)^2 - 2(3y)(4) + 4^2 = \boxed{9y^2 - 24y + 16}$

28. $(2x + 5y)^2 = (2x)^2 + 2(2x)(5y) + (5y)^2$
$\qquad = \boxed{4x^2 + 20xy + 25y^2}$

29. $(7x - 2y)^2 = (7x)^2 - 2(7x)(2y) + (2y)^2$
$\qquad = \boxed{49x^2 - 28xy + 4y^2}$

30. $(4x + 5)(4x - 5) = (4x)^2 - 5^2 = \boxed{16x^2 - 25}$

31. $(2z + 9)(2z - 9) = (2z)^2 - 9^2 = \boxed{4z^2 - 81}$

32. $(5r + 6s)(5r - 6s) = (5r)^2 - (6s)^2 = \boxed{25r^2 - 36s^2}$

33. $(5x + 7)(5x - 7) - (3x + 4)(3x - 4)$
$= (25x^2 - 49) - (9x^2 - 16) = 25x^2 - 49 - 9x^2 + 16$
$= \boxed{16x^2 - 33}$

34. $(2x + 3)^2 - (4x - 2)^2$

$= (4x^2 + 12x + 9) - (16x^2 - 16x + 4)$

$= 4x^2 + 12x + 9 - 16x^2 + 16x - 4$

$= 4x^2 - 16x^2 + 12x + 16x + 9 - 4$

$= \boxed{-12x^2 + 28x + 5}$

35. $\dfrac{-15y^8}{3y^2} = -5y^{8-2} = \boxed{-5y^6}$

36. $\dfrac{-25x^{11}y^{20}}{1000x^4y^{11}} = \dfrac{-1}{4}x^{11-4}y^{20-11} = \boxed{\dfrac{-x^7y^9}{4}}$

37. $-16x^3y^4 \div (-2xy^2) = \dfrac{-16x^3y^4}{-2xy^2} = 8x^{3-1}y^{4-2} = \boxed{8x^2y^2}$

38. $\dfrac{18y^4 - 12y^2 + 36y}{6y}$

$= \dfrac{18y^4}{6y} - \dfrac{12y^2}{6y} + \dfrac{36y}{6y}$

$= 3y^{4-1} - 2y^{2-1} + 6y^{1-1}$

$= 3y^3 - 2y + 6y^0$

$= \boxed{3y^3 - 2y + 6}$

39. $(30x^8 - 25y^7 + 3x^6 - 40x^5) \div (-5x^5)$

$= \dfrac{30x^8}{-5x^5} - \dfrac{-25x^7}{-5x^5} + \dfrac{3x^6}{-5x^5} - \dfrac{40x^5}{-5x^5}$

$= -6x^{8-5} + 5x^{7-5} - \dfrac{3}{5}x^{6-5} + 8x^{5-5}$

$= \boxed{-6x^3 + 5x^2 - \dfrac{3}{5}x + 8}$

40. $\dfrac{2z^3 - 6z^2 + 5z}{2z^2}$

$= \dfrac{2z^3}{2z^2} - \dfrac{6z^2}{2z^2} + \dfrac{5z}{2z^2}$

$= z^{3-2} - 3z^{2-2} + \dfrac{5}{2}z^{1-2}$

$= \boxed{z - 3 + \dfrac{5}{2z}}$

41. $\dfrac{20x^7 - 8x^6 - 16x^4 + 12x^2 - 2}{4x^5}$

$= \dfrac{20x^7}{4x^5} - \dfrac{8x^6}{4x^5} - \dfrac{16x^4}{4x^5} + \dfrac{12x^2}{4x^5} - \dfrac{2}{4x^5}$

$= 5x^{7-5} - 2x^{6-5} - 4x^{4-5} + 3x^{2-5} - \dfrac{1}{2x^5}$

$= \boxed{5x^2 - 2x - \dfrac{4}{x} + \dfrac{3}{x^3} - \dfrac{1}{2x^5}}$

42. $\dfrac{2x^2 + 3x - 14}{x - 2} = \boxed{2x + 7}$

$$
\begin{array}{r}
2x + 7 \\
x - 2 \overline{\smash{\big)}\, 2x^2 + 3x - 14} \\
\underline{2x^2 - 4x} \\
7x - 14 \\
\underline{7x - 14} \\
\varnothing
\end{array}
$$

43. $\dfrac{2y^3 - 5y^2 + 7y + 5}{2y + 1} = \boxed{y^2 - 3y + 5}$

$$
\begin{array}{r}
y^2 - 3y + 5 \\
2y + 1 \overline{\smash{\big)}\, 2y^3 - 5y^2 + 7y + 5} \\
\underline{2y^3 + y^2} \\
-6y^2 + 7y \\
\underline{-6y^2 - 3y} \\
10y + 5 \\
\underline{10y + 5} \\
\varnothing
\end{array}
$$

44. $\dfrac{z^3 - 2z^2 - 33z - 7}{z - 7} = \boxed{z^2 + 5z + 2 + \dfrac{7}{z - 7}}$

$$
\begin{array}{r}
z^2 + 5z + 2 \\
z - 7 \overline{\smash{\big)}\, z^3 - 2z^2 - 33z - 7} \\
\underline{z^3 - 7z^2} \\
5z^2 - 33z \\
\underline{5z^2 - 35z} \\
2z - 7 \\
\underline{2z - 14} \\
7
\end{array}
$$

45. $\dfrac{9x^4 - 12x^3 + 24x^2 - 28x - 3}{3x^2 + 7} = \boxed{3x^2 - 4x + 1 - \dfrac{10}{3x^2 + 7}}$

$$
\begin{array}{r}
3x^2 - 4x + 1 \\
3x^2 + 7 \overline{\smash{\big)}\, 9x^4 - 12x^3 + 24x^2 - 28x - 3} \\
\underline{9x^4 \qquad\quad + 21x^2} \\
-12x^3 + 3x^2 - 28x \\
\underline{-12x^3 \qquad\quad - 28x} \\
3x^2 \qquad\quad - 3 \\
\underline{3x^2 \qquad\quad + 7} \\
-10
\end{array}
$$

46. $(3y^6)(-2y^4) = -6y^{6+4} = \boxed{-6y^{10}}$

47. $(-3x^2y^7)(-8xy^5) = 24x^{2+1}y^{7+5} = \boxed{24x^3y^{12}}$

48. $(3x^3)^4 = 3^4x^{3(4)} = \boxed{81x^{12}}$

49. $4(2y^5)^3 = 4(2^3y^{5(3)}) = 4(8)y^{15} = \boxed{32y^{15}}$

50. $(2x)(4x)^2 + 15x^3 = (2x)(16x^2) + 15x^3$
$= 32x^3 + 15x^3 = \boxed{47x^3}$

51. $(2xy^4)^3 = 2^3 x^3 y^{4(3)} = \boxed{8x^3 y^{12}}$

52. $(xy^2)(x^3 y)^3 = (xy^2)(x^{3(3)} y^3)$
$= xy^2 x^9 y^3 = x^{1+9} \cdot y^{2+3} = \boxed{x^{10} y^5}$

53. $\dfrac{x^3}{x^9} = x^{3-9} = x^{-6} = \boxed{\dfrac{1}{x^6}}$

54. $\dfrac{30y^6}{5y^8} = 6y^{6-8} = 6y^{-2} = \boxed{\dfrac{6}{y^2}}$

55. $\dfrac{-40x^2 y}{120x^4 y^3} = -\dfrac{1}{3} x^{2-4} y^{1-3} = -\dfrac{1}{3} x^{-2} y^{-2} = \boxed{\dfrac{-1}{3x^2 y^2}}$

56. $(5y^{-7})(6y^2) = 30y^{-7+2} = 30y^{-5} = \boxed{\dfrac{30}{y^5}}$

57. $\dfrac{x^4 \cdot x^{-2}}{x^{-6}} = x^{4-2+6} = \boxed{x^8}$

58. $\dfrac{xy^{-3}}{x^4 y^2} = x^{1-4} y^{-3-2} = x^{-3} y^{-5} = \boxed{\dfrac{1}{x^3 y^5}}$

59. $\left(\dfrac{x^4}{y^{-3}}\right)^5 = \dfrac{x^{4(5)}}{y^{-3(5)}} = \dfrac{x^{20}}{y^{-15}} = \boxed{x^{20} y^{15}}$

60. $\dfrac{(3y^3)^4}{y^{10}} = \dfrac{3^4 y^{3(4)}}{y^{10}} = 81y^{12-10} = \boxed{81y^2}$

61. $\dfrac{y^{-7}}{(y^4)^3} = \dfrac{y^{-7}}{y^{12}} = y^{-7-12} = y^{-19} = \boxed{\dfrac{1}{y^{19}}}$

62. $\left(\dfrac{x^7}{x^4}\right)^{-4} = (x^{7-4})^{-4} = x^{3(-4)} = x^{-12} = \boxed{\dfrac{1}{x^{12}}}$

63. $\left(\dfrac{x^8 z^{10}}{x^6 z^{12}}\right)^{-2} = (x^{8-6} z^{10-12})^{-2} = (x^2 z^{-2})^{-2} = (x^2)^{-2}(z^{-2})^{-2} = x^{-4} z^4 = \boxed{\dfrac{z^4}{x^4}}$

64. $\dfrac{(y^3)^4 y^{-3}}{(y^{-2})^4} = \dfrac{y^{3(4)} y^{-3}}{y^{(-2)(4)}} = \dfrac{y^{12} y^{-3}}{y^{-8}} = y^{12-3+8} = \boxed{y^{17}}$

65. $(-4xy^5)\left(-\dfrac{1}{2} xy^{-8}\right) = (-4)\left(-\dfrac{1}{2}\right) x^{1+1} y^{5-8} = 2x^2 y^{-3} = \boxed{\dfrac{2x^2}{y^3}}$

66. $2.3 \times 10^4 = \boxed{23{,}000}$ (move decimal point to right 4 places)

67. $1.76 \times 10^{-3} = \boxed{0.00176}$ (move left 3 places)

68. $9.84 \times 10^{-1} = \boxed{0.984}$ (move left 1 place)

69. $7^{-2} = \dfrac{1}{7^2} = \boxed{\dfrac{1}{49}}$

70. $2^{-1} + 4^{-1} = \dfrac{1}{2} + \dfrac{1}{4} = \boxed{\dfrac{3}{4}}$

71. $(2^3)^{-2} = 2^{3(-2)} = 2^{-6} = \dfrac{1}{2^6} = \boxed{\dfrac{1}{64}}$

72. $\dfrac{5^{-5}}{5^{-3}} = \dfrac{1}{5^2} = 5^{-5-(-3)} = 5^{-2} = \dfrac{1}{5^2} = \boxed{\dfrac{1}{25}}$

73. $73{,}900{,}000 = \boxed{7.39 \times 10^7}$

74. $0.000\ 089\ 4 = \boxed{8.94 \times 10^{-5}}$

75. $0.000\ 972\ 5 = \boxed{9.725 \times 10^{-4}}$

76. $0.38 = \boxed{3.8 \times 10^{-1}}$

77. $8.639 = \boxed{8.639 \times 10^0}$

78. $37{,}000 = \boxed{3.7 \times 10^4}$

79. $(6 \times 10^{-3})(1.5 \times 10^6) = 6(1.5) \times 10^{-3+6} = \boxed{9 \times 10^3 = 9000}$

80. $\dfrac{2 \times 10^2}{4 \times 10^{-3}} = \dfrac{10^{2+3}}{2} = 0.5 \times 10^5 = 0.5 \times 10 \times 10^4 = \boxed{5.0 \times 10^4}$

$\qquad\qquad\qquad\qquad\qquad\qquad\qquad\qquad = \boxed{50,000}$

81. $(4 \times 10^{-2})^2 = 4^2 \times 10^{-2(3)} = 16 \times 10^{-4} \;\; = \;\; 1.6 \times 10 \times 10^{-4}$

$\qquad\qquad\qquad\qquad\qquad\qquad\qquad\qquad = \;\; 1.6 \times 10^{1-4} \quad = \quad \boxed{1.6 \times 10^{-3}}$

$\qquad\qquad\qquad\qquad\qquad\qquad\qquad\qquad\qquad\qquad = \quad \boxed{0.0016}$

82. $(3,000,000)(0.000\ 02)$
$= (3 \times 10^6)(2 \times 10^{-5})$
$= 6 \times 10^{6-5} = \boxed{6 \times 10}$

83. $(20,000)^3(0.000\ 6)$
$= (2 \times 10^4)^3(6 \times 10^{-4})$
$= (2^3 \times 10^{4(3)})(6 \times 10^{-4})$
$= 8(6) \times 10^{12-4}$
$= 48 \times 10^8$
$= \boxed{4.8 \times 10^9}$

84. $\dfrac{96,000}{(12,000)(0.000\ 04)}$

$= \dfrac{9.6 \times 10^4}{(1.2 \times 10^4)(4 \times 10^{-5})}$

$= \dfrac{8}{4} \times 10^{4-4+5}$

$= \boxed{2.0 \times 10^5}$

85.
$$\begin{aligned}(x+7)(x-7) &= x(x+2) - 1\\ x^2 - 49 &= x^2 + 2x - 1\\ -48 &= 2x\\ -24 &= x\end{aligned}$$
$$\boxed{\{-24\}}$$

86.
$$\begin{aligned}(3y-1)(2y-3) &= 6y^2 - (4 - 3y)\\ 6y^2 - 9y - 2y + 3 &= 6y^2 - 4 + 3y\\ -11y + 3 &= -4 + 3y\\ -14y &= -7\\ y &= \tfrac{1}{2}\end{aligned}$$
$$\boxed{\left\{\tfrac{1}{2}\right\}}$$

87.
$$\begin{aligned}(2z+3)^2 - 5(z+2) &= (4z-2)(z+1)\\ 4z^2 + 12z + 9 - 5z - 10 &= 4z^2 + 4z - 2z - 2\\ 7z - 1 &= 2z - 2\\ 5z &= -1\\ z &= \tfrac{-1}{5}\end{aligned}$$
$$\boxed{\left\{-\tfrac{1}{5}\right\}}$$

88. Let x equal the number.
$$\begin{aligned}(x-5)(x+6) &= x^2 - 21\\ x^2 + x - 30 &= x^2 - 21\\ x &= 9\end{aligned}$$
The number is $\boxed{9}$.

89. Let $x, x+2, x+4$ represent three consecutive odd integers.
$$\begin{aligned}x^2 + (x+2)^2 + (x+4)^2 &= 3(x+4)^2 - 40\\ x^2 + x^2 + 4x + 4 + x^2 + 8x + 16 &= 3(x^2 + 8x + 16) - 40\\ 3x^2 + 12x + 20 &= 3x^2 + 24x + 48 - 40\\ -12x &= -12\\ x &= 1\\ x+2 &= 3\\ x+4 &= 5\end{aligned}$$
The integers are $\boxed{1,\ 3 \text{ and } 5}$.

90. Let x equal the length of side of square.
$x + 2$ equals the length of new rectangle.
$x - 3$ equals the width of new rectangle.

$$
\begin{aligned}
(x + 2)(x - 3) &= x^2 - 14 \\
x^2 - x - 6 &= x^2 - 14 \\
-x &= -8 \\
x &= 8
\end{aligned}
$$

dimensions of original square: $\boxed{8 \text{ meters} \times 8 \text{ meters}}$

91. Let x equal the length of side of square picture.
$x + 2 + 2 = x + 4$ equals length of side of picture plus frame.

$$
\begin{aligned}
(x + 4)^2 - x^2 &= 40 \\
x^2 + 8x + 16 - x^2 &= 40 \\
8x &= 24 \\
x &= 3
\end{aligned}
$$

dimensions of painting: $\boxed{3 \text{ inches} \times 3 \text{ inches}}$

92. Let x equal the width of rectangle.
$x + 3$ equals the length of rectangle.
$x + 2$ equals the length of side of square.
area of square − area of rectangle = 5 square centimeters

$$
\begin{aligned}
(x + 2)^2 - x(x + 3) &= 5 \\
(x^2 + 4x + 4) - (x^2 + 3x) &= 5 \\
x^2 + 4x + 4 - x^2 - 3x &= 5 \\
x &= 1 \qquad \text{width of rectangle: } 1 \text{ cm} \\
x + 3 &= 4 \qquad \text{length of rectangle: } 4 \text{ cm} \\
x + 2 &= 3 \qquad \text{length of side of square: } 3 \text{ cm}
\end{aligned}
$$

$\boxed{\text{dimensions of rectangle}, 1 \text{ cm} \times 4 \text{ cm}; \text{dimensions of square}, 3 \text{ cm} \times 3 \text{ cm}}$

Cumulative Review Problems (Chapters 1-6)

Cumulative Review, pp. 442-443

1.
$$
\begin{aligned}
3y(y - 1) - (3y + 1)(y - 2) &= 2y + 2 \\
3y^2 - 3y - (3y^2 - 6y + y - 2) &= 2y + 2 \\
3y^2 - 3y - (3y^2 - 5y - 2) &= 2y + 2 \\
3y^2 - 3y - 3y^2 + 5y + 2 &= 2y + 2 \\
2y + 2 &= 2y + 2 \\
2 &= 2 \quad \text{True for all real numbers; infinitely many solutions.}
\end{aligned}
$$

$\boxed{\{y \mid y \in R\}}$

2.
$$
\begin{aligned}
(8x + 35) + (4x + 13) &= 180 \\
12x + 48 &= 180 \\
12x &= 132 \\
x &= \boxed{11}
\end{aligned}
$$

$(8x + 35)° = (8 \cdot 11 + 35)° = (88 + 35)° = \boxed{123°}$

$(4x + 13)° = (4 \cdot 11 + 13)° = (44 + 13)° = \boxed{57°}$

3. $$3 - \frac{1}{4}x \;\leq\; 2 + \frac{3}{8}x$$

$$-\frac{1}{4}x - \frac{3}{8}x \;\leq\; 2 - 3$$

$$-\frac{5}{8}x \;\leq\; -1$$

$$x \;\geq\; \frac{-8}{5}\,(-1)$$

$$x \;\geq\; \frac{8}{5}$$

$$\boxed{\left\{\, x \mid x \geq \frac{8}{5} \right\}}$$

4. $5x - 2y = -10$

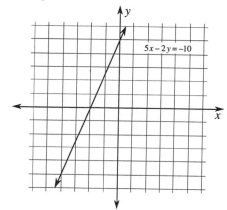

5. $y \geq -\dfrac{2}{5}x + 2$

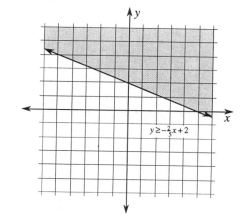

6. $\begin{aligned} 3x - 6y &= 1 \\ x &= 2y + 3 \end{aligned}$ $\rightarrow$ $\begin{aligned} 3(2y + 3) - 6y &= 1 \\ 6y + 9 - 6y &= 1 \\ 9 &= 1 \quad \text{False} \end{aligned}$

no solution; **the** system is $\boxed{\text{inconsistent}}$; $\boxed{\varnothing}$

7. Let $x, x + 2, x + 4$ represent three consecutive even integers.

$$\begin{aligned} x + 3(x + 4) &= (x + 2) + 64 \\ x + 3x + 12 &= x + 66 \\ 4x + 12 &= x + 66 \\ 3x &= 54 \\ x &= 18 \\ x + 2 &= 20 \\ x + 4 &= 22 \end{aligned}$$

The integers are $\boxed{18,\ 20 \text{ and } 22}$.

8. $x + y = -1$
    $-2x + y = 5$
    solution: $\boxed{\{(-2,\ 1)\}}$

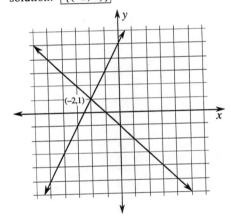

9. Let x equal the first number, and y equal the second number.

$$\begin{array}{ccccc} x + 2y = 8 & \rightarrow & x + 2y = 8 & \rightarrow & x + 2y = 8 \\ y + 2x = 25 & & 2x + y = 25 & & -4x - 2y = -50 \\ & & & & \overline{-3x = -42} \\ & & & & x = 14 \\ & & & & y + 2(14) = 25 \\ & & & & y + 28 = 25 \\ & & & & y = -3 \end{array}$$

The numbers are $\boxed{14 \text{ and } -3}$.

10. $(2x - 3)(4x - 1) - (5x - 2)^2$
    $= (8x^2 - 2x - 12x + 3) - (25x^2 - 20x + 4)$
    $= 8x^2 - 14x + 3 - 25x^2 + 20x - 4$
    $= \boxed{-17x^2 + 6x - 1}$

11. $f(x) = (x - 4)^2$
    $f(-2) = (-2 - 4)^2 = (-6)^2 = \boxed{36}$

12. passing through (–1, 3) and perpendicular to the line whose equation is $2x + y = 4$:
slope of line perpendicular: $y = -2x + 4$, $m = -2$

slope of line: $\dfrac{-1}{-2} = \dfrac{1}{2}$

point-slope: $y - 3 = \dfrac{1}{2}(x - (-1))$

$$\boxed{y - 3 = \dfrac{1}{2}(x + 1)}$$

slope-intercept: $y - 3 = \dfrac{1}{2}x + \dfrac{1}{2}$

$$\boxed{y = \dfrac{1}{2}x + \dfrac{7}{2}}$$

standard: $\dfrac{1}{2}x - y = -\dfrac{7}{2}$

$$\boxed{x - 2y = -7}$$

13. (–3, –5), (–2, 4)

slope $= \dfrac{4 - (-5)}{-2 - (-3)} = \dfrac{4 + 5}{-2 + 3} = \dfrac{9}{1} = \boxed{9}$

14. Let t equal the tens' digit and u equal the units' digit.

$2t + 4u$	=	24	(÷2) →		$t + 2u$	=	12
$10t + u$	=	$5u$	(simplify) →		$10t$	=	$4u$

(simplify) → $5t = 2u$
(substitute for $2u$) → $t + 5t = 12$
$6t = 12$
$t = 2$
$2u = 5t$
$2u = 10$
$u = 5$

The number is $\boxed{25}$.

15. $\dfrac{3y^4 + 2y^3 - 2y^2 - 2y - 2}{y^2 - 1} = \boxed{3y^2 + 2y + 1 - \dfrac{1}{y^2 - 1}}$

$$
\begin{array}{r}
3y^2 + 2y + 1 \\
y^2 - 1 \overline{)\,3y^4 + 2y^3 - 2y^2 - 2y - 2} \\
\underline{3y^4 \quad\quad - 3y^2} \\
2y^3 + y^2 - 2y \\
\underline{2y^3 \quad - 2y} \\
y^2 \quad - 2 \\
\underline{y^2 \quad - 1} \\
-1
\end{array}
$$

16.
$$\dfrac{1}{3}x - \dfrac{1}{4}x = -\dfrac{1}{12}x + \dfrac{1}{2}$$
$$12\left(\dfrac{1}{3}x - \dfrac{1}{4}x\right) = 12\left(-\dfrac{1}{12}x + \dfrac{1}{2}\right)$$
$$4x - 3x = -x + 6$$
$$x = -x + 6$$
$$2x = 6$$
$$x = 3$$

$\boxed{\{3\}}$

17.

$$9x^5 - 3x^3 + 2x - 7 \quad \rightarrow \quad \begin{array}{l} 9x^5 - 3x^3 + 2x - 7 \\ -9x^5 - 3x^3 + 7x + 9 \end{array}$$
$$-(9x^5 + 3x^3 - 7x - 9)$$

$$\boxed{-6x^3 + 9x + 2}$$

18. Let t equal the time (in hours).

$$\begin{aligned} 13t + 11t &= 72 \\ 24t &= 72 \\ t &= 3 \end{aligned}$$

They will meet in $\boxed{3 \text{ hours}}$.

19. Let w equals the width.
l equals the length.

$$\begin{aligned} 2w + 2l &= 16 \\ 1(2w) + 4(2l) &= 46 \end{aligned}$$

$(\div 2) \rightarrow$

$(\div 2 \text{ and simplify}) \rightarrow$

$$\begin{aligned} -w - l &= -8 \\ w + 4l &= 23 \\ \hline 3l &= 15 \\ l &= 5 \\ 2w + 2(5) &= 16 \\ 2w &= 6 \\ w &= 3 \end{aligned}$$

length, 5 yards; width, 3 yards
dimensions: $\boxed{3 \text{ yards} \times 5 \text{ yards}}$

20. $(5y^2 + y)(4y^2 - 3y - 2)$

$= 5y^2(4y^2 - 3y - 2) + y(4y^2 - 3y - 2)$

$= 5y^2(4y^2) + 5y^2(-3y) + (5y^2)(-2) + y(4y^2) + y(-3y) + y(-2)$

$= 20y^4 - 15y^3 - 10y^2 + 4y^3 - 3y^2 - 2y$

$= \boxed{20y^4 - 11y^3 - 13y^2 - 2y}$

21.

$$\begin{aligned} \frac{6x - 5}{x} &= \frac{11}{5} \\ 30x - 25 &= 11x \\ 19x &= 25 \\ x &= \frac{25}{19} \end{aligned}$$

$$\boxed{\left\{ \frac{25}{19} \right\}}$$

22. Let x equal the number of 20-cent stamps.
$2x + 1$ equals the number of 35-cent stamps.

$$\begin{aligned} 0.20(x) + (0.35)(2x + 1) &= 9.35 \\ 0.2x + 0.7x + 0.35 &= 9.35 \\ 0.9x &= 9.00 \\ x &= 10 \\ 2x + 1 &= 2(10) + 1 = 21 \end{aligned}$$

$\boxed{10 \text{ 20-cent stamps}, 21 \text{ 35-cent stamps}}$

23.

$$\begin{aligned} 3x + 2y &= 10 \\ 4x - 3y &= -15 \end{aligned}$$

$(\times 3) \rightarrow$

$(\times 2) \rightarrow$

$$\begin{aligned} 9x + 6y &= 30 \\ 8x - 6y &= -30 \\ \hline 17x &= 0 \\ x &= 0 \\ 3(0) + 2y &= 10 \\ 2y &= 10 \\ y &= 5 \end{aligned}$$

$\boxed{\{(0, 5)\}}$

24. $2x + 5y \leq 10$
$\quad\;\; x - y \geq 4$

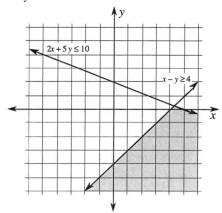

25. $\dfrac{(2.1 \times 10^{-3})(4.8 \times 10^{4})}{(1.6 \times 10^{-6})(7 \times 10^{6})}$

$= \left(\dfrac{2.1}{7}\right)\left(\dfrac{4.8}{1.6}\right) \times \dfrac{10^{-3+4}}{10^{-6+6}}$

$= (0.3)(3) \times \dfrac{10^{1}}{10^{0}}$

$= 0.9 \times 10^{1}$

$= \boxed{9}$

26. $x < -2$

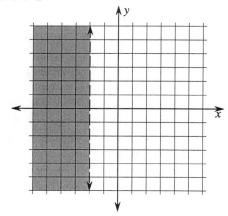

27. $\dfrac{(8-10)^{3} - (-4)^{2}}{2 + 8(2) \div 4}$

$= \dfrac{(-2)^{3} - 16}{2 + 16 \div 4}$

$= \dfrac{-8 - 16}{2 + 4}$

$= \dfrac{-24}{6} = \boxed{-4}$

28. $12 \div 2 - 3 = 3$

$\boxed{x = 12,\, y = 2,\, z = 3}$

29. Let x equal the speed of boat in still water.
y equals the speed of current.

	R	$\times$	T	$= D$
with the current	$x + y$		1 hour	16 miles
against the current	$x - y$		8 hours	16 miles

$$
\begin{array}{llll}
1(x + y) &= 16 & \rightarrow & x + y = 16 \\
8(x - y) &= 16 & (\div 8) \rightarrow & \underline{x - y = 2} \\
& & & 2x = 18 \\
& & & x = 9 \\
& & & 9 + y = 16 \\
& & & y = 7
\end{array}
$$

speed of boat in still water, 9 mph; speed of current, 7 mph

30.
$$
\begin{array}{rll}
5(2y - 6) + 4y &= 3(4y - 8) + 2y \\
10y - 30 + 4y &= 12y - 24 + 2y \\
14y - 30 &= 14y - 24 \\
-30 &= -24 \quad \text{False}
\end{array}
$$

no solution; the system is inconsistent ; $\varnothing$.

Chapter 7   Factoring Polynomials

Section 7.1 Factoring Polynomials with Common Factors

Problem Set 7.1, pp. 454-456

1. $12 = 2 \cdot 2 \cdot 3 = \boxed{2^2 \cdot 3}$

3. $76 = 2 \cdot 2 \cdot 19 = \boxed{2^2 \cdot 19}$

5. $40 = 2 \cdot 2 \cdot 2 \cdot 5 = \boxed{2^3 \cdot 5}$

7. $98 = 2 \cdot 7 \cdot 7 = \boxed{2 \cdot 7^2}$

9. $345 = 3 \cdot 5 \cdot 23 = \boxed{3 \cdot 5 \cdot 23}$

11. $8 = 2^3$
$28 = 2^2 \cdot 7$
GCF $(8, 28) = 2^2 = \boxed{4}$

13. $36 = 2^2 \cdot 3^2$
$84 = 2^2 \cdot 3 \cdot 7$
GCF $(36, 84) = 2^2 \cdot 3 = \boxed{12}$

15. $40 = 2^3 \cdot 5$
$48 = 2^4 \cdot 3$
$64 = 2^6$
GCF $(40, 48, 64) = 2^3 = \boxed{8}$

17. $76 = 2^2 \cdot 19$
$88 = 2^3 \cdot 11$
$100 = 2^2 \cdot 5^2$
GCF $(76, 88, 100) = 2^2 = \boxed{4}$

19. $49 = 7^2$
$56 = 2^3 \cdot 7$
$87 = 3 \cdot 29$
GCF $(49, 56, 87) = \boxed{1}$

21. $6x = 3 \cdot 2x$
$10x^3 = 5x^2 \cdot 2x$
GCF $(6x, 10x^3) = \boxed{2x}$

23. $6xy^2 = xy \cdot 6y$
$5x^2y = xy \cdot 5x$
GCF $(6xy^2, 5x^2y) = \boxed{xy}$

25. $24x^2yz = 24xy \cdot xz$
$48xy^2 = 24xy \cdot 2y$
GCF $(24x^2yz, 48xy^2) = \boxed{24xy}$

27. $12y^3 = 2y \cdot 6y^2$
$6y^2 = 2y \cdot 3y$
$4y = 2y \cdot 2$
GCF $(12y^3, 6y^2, 4y) = \boxed{2y}$

29. $30x^3y = 6xy \cdot 5x^2$
$18x^2y = 6xy \cdot 3x$
$54xy^3 = 6xy \cdot 9y^2$
GCF $(30x^3y, 18x^2y^2, 54xy^3) = \boxed{6xy}$

31. $14y^2 = 1 \cdot 14y^2$
$1 = 1$
$7y^4 = 1 \cdot 7y^4$
GCF $(14y^2, 1, 7y^4) = \boxed{1}$

33. $5x + 5 = 5(x) + 5(1) = \boxed{5(x + 1)}$

35. $3z - 3 = 3(z) - 3(1) = \boxed{3(z - 1)}$

37. $8x + 16 = 8(x) + 8(2) = \boxed{8(x + 2)}$

39. $25x - 10 = 5(5x) - 5(2) = \boxed{5(5x - 2)}$

41. $y^2 + y = y(y) + y(1) = \boxed{y(y + 1)}$

43. $18x^2 - 24 = 6(3x^2) - 6(4) = \boxed{6(3x^2 - 4)}$

45. $25y^2 - 13y = y(25y) - y(13) = \boxed{y(25y - 13)}$

47. $36x^3 + 24x^2 = 12x^2(3x) + 12x^2(2) = \boxed{12x^2(3x + 2)}$

49. $27y^6 + 9y^4 = 9y^4(3y^2) + 9y^4(1) = \boxed{9y^4(3y^2 + 1)}$

51. $8x^2 - 4x^4 = 4x^2(2) - 4x^2(x^2) = \boxed{4x^2(2 - x^2)}$

53. $12x^2 - 13y^3 = (1)(12x^2) - (1)(13y^3) = \boxed{1(12x^2 - 13y^3)}$

55. $12y^2 + 16y - 8 = 4(3y^2) + 4(4y) + 4(-2) = \boxed{4(3y^2 + 4y - 2)}$

57. $100 + 75y - 50y^2 = 25(4) + 25(3y) + 25(-2y^2) = \boxed{25(4 + 3y - 2y^2)}$

59. $9y^4 + 18y^3 + 6y^2 = 3y^2(3y^2) + 3y^2(6y) + 3y^2(2) = \boxed{3y^2(3y^2 + 6y + 2)}$

61. $100y^5 - 50y^3 + 100y^2 = 50y^2(2y^3) + 50y^2(-y) + 50y^2(2) = \boxed{50y^2(2y^3 - y + 2)}$

63. $10x - 20x^2 + 5x^3 = 5x(2) + 5x(-4x) + 5x(x^2) = \boxed{5x(2 - 4x + x^2)}$

65. $-2y^2 - 3y^3 + 6y^5 = y^2(-2) + y^2(-3y) + y^2(6y^3) = \boxed{y^2(-2 - 3y + 6y^3)}$

67. $2x^2y - 5x^2y^2 + 7xy^2 = xy(2x) + xy(-5xy) + xy(7y) = \boxed{xy(2x - 5xy + 7y)}$

69. $3x^2y^2 - 9xy^2 + 15y^2 = 3y^2(x^2) + 3y^2(-3x) + 3y^2(5) = \boxed{3y^2(x^2 - 3x + 5)}$

71. $15x^2y^2 - 6xy^2 + 9x^2y = 3xy(5xy) + 3xy(-2y) + 3xy(3x) = \boxed{3xy(5xy - 2y + 3x)}$

73. $24x^3y^3z^3 + 30x^2y^2z + 18x^2yz^2 = 6x^2yz(4xy^2z^2) + 6x^2yz(5y) + 6x^2yz(3z) = \boxed{6x^2yz(4xy^2z^2 + 5y + z)}$

75. $-2x^2 + 8x - 10$
$= \boxed{2(-x^2 + 4x - 5)}$
$= \boxed{-2(x^2 - 4x + 5)}$

77. $3a - 3b$
$= \boxed{3(a - b)}$
$= \boxed{-3(-a + b)}$

79. $-4x + 12x^2$
$= \boxed{4x(-1 + 3x)}$
$= \boxed{-4x(1 - 3x)}$

81. $-y^3 + 7y^2$
$= \boxed{y^2(-y + 7)}$
$= \boxed{-y^2(y - 7)}$

83. $y^2 + y$
$= \boxed{y(y + 1)}$
$= \boxed{-y(-y - 1)}$

85. $3 - x$
$= \boxed{1(3 - x)}$
$= \boxed{-1(-3 + x)}$

87. $x(a + b) + 7(a + b) = \boxed{(a + b)(x + 7)}$

89. $x(c + 3) - y(c + 3) = \boxed{(c + 3)(x - y)}$

91. $x(x - 2y) + z(x - 2y) = \boxed{(x - 2y)(x + z)}$

93. $4y(5y - 9) + (5y - 9) = \boxed{(5y - 9)(4y + 1)}$

95. $a(x - 2) - (x - 2) = \boxed{(x - 2)(a - 1)}$

97. $xy + 4x + 3y + 12 = x(y + 4) + 3(y + 4) = \boxed{(y + 4)(x + 3)}$

99. $xy + 8x + 2y + 16 = x(y + 8) + 2(y + 8) = \boxed{(y + 8)(x + 2)}$

101. $xy + 7x - 3y - 21 = x(y + 7) - 3(y + 7) = \boxed{(y + 7)(x - 3)}$

103. $2bx + 6x - 5b - 15 = 2x(b + 3) - 5(b + 3) = \boxed{(b + 3)(2x - 5)}$

105. $3ax - 4a - 6x + 8 = a(3x - 4) - 2(3x - 4) = \boxed{(3x - 4)(a - 2)}$

107. $y^2 - 3y + 4by - 12b = y\,(y - 3) + 4b(y - 3) = \boxed{(y - 3)(y + 4b)}$

109. $2xy + y^2 - 2x - y = y(2x + y) - (2x + y) = \boxed{(2x + y)(y - 1)}$

111. $y^2 - 9y + 3y - 27 = y(y - 9) + 3(y - 9) = \boxed{(y - 9)(y + 3)}$

113. $\boxed{D}$ is true.

115. area = length × width

$55x - x^2 = x(\text{width})$

$\text{width} = \dfrac{55x - x^2}{x} = \dfrac{55x}{x} - \dfrac{x^2}{x}$

$\boxed{55 - x}$

117. $\boxed{C}$ is true;

$\pi b^2 - \dfrac{1}{2}\,bh$

$= b(\pi b) - b\left(\dfrac{1}{2}h\right)$

$= b\left(\pi b - \dfrac{1}{2}h\right)$

119. area of shaded region = area of large circle − area of small circle

$A = \pi(3x)^2 - \pi x^2 = \pi x^2(9 - 1) = \boxed{8\pi x^2}$

121. area of shaded region = area of rectangle − area of semicircle

$A = (2x + x)(2x) - \dfrac{1}{2}\pi x^2 = 6x^2 - \dfrac{1}{2}\,\pi x^2 = x^2\left(6 - \dfrac{\pi}{2}\right)$

123. area of shaded region = area of square − 2 × area of circle

$A = (4r)^2 - 2\pi r^2 = \boxed{2r^2(8 - \pi)}$

125. Let x, y, z be the unknown numbers.

$x + y + z = 30;\ xyz = 840$

$840 = 2^3 \cdot 3 \cdot 5 \cdot 7 = 8 \cdot 15 \cdot 7;\ 8 + 7 + 15 = 30$

Therefore, $\boxed{x = 7, y = 8, z = 15}$

Review Problems

127.   Let $x =$ amount invested at 6%.
$x - 350 =$ amount invested at 8%.
$x(0.06) + (x - 350)(0.08) = 147$
$0.14x - 28 = 147$
$0.14x = 175$
$x = \$1250$
$x - 350 = 1250 - 350 = 900$
$\boxed{\$1250 \text{ at } 6\%; \$900 \text{ at } 8\%}$

128. $2x - y = -4$
$x - 3y = 3$
Lines intersect at $x = -3$, $y = -2$: $(-3, -2)$
$\boxed{\{(-3, -2)\}}$

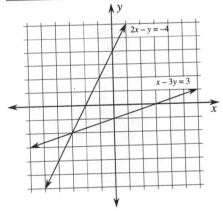

129. $Ax + By = C$
$A(-7) + B(2) = C$ (multiply by 5) $\rightarrow$ $-35A + 10B = 5C$
$A(-4) + B(5) = C$ (multiply by –2) $\rightarrow$ $\underline{8A - 10B = -2C}$
                                     $-27A = 3C$
                                     $A = -\dfrac{1}{9}C$

$\dfrac{4}{9}C + 5B = C; \; 5B = \dfrac{5}{9}C; \; B = \dfrac{1}{9}C$

Let $C = 9$. Then $A = -1$ and $B = 1$. (any nonzero value of C may be used.)
Equation is $-x + y = 9$, or $\boxed{x - y = -9}$.

Section 7.2   Factoring Trinomials Whose Leading Coefficient is 1

Problem Set 7.2, pp. 462-463

1. $(x^2 + 3x + 2) = (x + 2)(x +)$
Using FOIL, we must find a number y such that $2y = 2$ and $2 + y = 3$. The value of y is 1. Hence, the missing factor is $\boxed{x + 1}$.
Check: $(x + 2)(x + 1) = x^2 + 3x + 1$

3. $y^2 + y - 6 = (y + 3)(\quad) = (y + 3)\,\boxed{(y - 2)}$

Pairs of integers whose products are –6 are:
$(-6, 1)$
$(6)(-1)$
$(3)(-2)$ ← $3 - 2 = 1$
$(-3)(2)$
Only 3 and –2 have a sum of 1.

5. $x^2 + x - 12 = (x - 3)(\quad) = (x - 3)\,\boxed{(x + 4)}$

Pairs of integers whose products are –12:
$(1)(-12)$
$(-1)(12)$
$(2)(-6)$
$(-2)(6)$
$(3)(-4)$
$(-3)(4)$ ← only pair with a sum of 1. $-3 + 4 = 1$

7. $y^2 - 5y + 4 = (y - 1)(\quad) = (y - 1)\,\boxed{(y - 4)}$

Pairs of integers whose products are 4:
$(-1)(-4)$ ← only pair with a sum of –5. $-1 + (-4) = -5$
$(1)(4)$
$(-2)(-2)$
$(2)(2)$

9. $y^2 - 2y - 3 = (y + 1)(\quad) = (y + 1)\,\boxed{(y - 3)}$

Pairs of integers whose products are –3:
$(3)(-1)$
$(-3)(1)$ ← only pair with a sum of –2. $(-3) + 1 = -2$

11. $r^2 - 6r + 8 = (r - 2)(\quad) = (r - 2)\,\boxed{(r - 4)}$

Pairs of integers whose products are 8:
$(1)(8)$
$(-1)(-8)$
$(2)(4)$
$(-2)(-4)$ ← only pair with a sum of –6. $-2 + (-4) = -6$

13. $x^2 + 5x + 6 = (x + 2)\,\boxed{(x + 3)}$
$(2)(3) = 6,\ 2 + 3 = 5$

15. $y^2 - 2y - 15 = \boxed{(y - 5)(y + 3)}$
$(-5)(3) = -15,\ -5 + 3 = -2$

17. $r^2 + 7r + 12 = \boxed{(r + 4)(r + 3)}$
$(4)(3) = 12,\ 4 + 3 = 7$

19. $x^2 + 9x + 8 = \boxed{(x + 1)(x + 8)}$
$(1)(8) = 8,\ 1 + 8 = 9$

21. $x^2 - 5x - 6 = \boxed{(x - 6)(x + 1)}$
$(-6)(1) = -6,\ -6 + 1 = -5$

23. $y^2 - 14y + 45 = \boxed{(y - 5)(y - 9)}$
$(-5)(-9) = 45;\ -5 + (-9) = -14$

25. $r^2 + 12r + 27 = \boxed{(r + 3)(r + 9)}$
$3(9) = 27,\ 3 + 9 = 12$

27. $n^2 - 11n - 42 = \boxed{(n + 3)(n - 14)}$
$3 + (-14) = 11,\ 3 - 14 = -11$

29. $y^2 - 9y - 36 = \boxed{(y + 3)(y - 12)}$
$3(-12) = -36,\ 3 - 12 = -9$

31. $x^2 + 10x - 75 = \boxed{(x + 15)(x - 5)}$
$15(-5) = -75,\ 15 - 5 = 10$

33. $x^2 - 8x + 32$ is $\boxed{\text{prime}}$
cannot be factored

35. $y^2 + 30y + 200 = \boxed{(y + 10)(y + 20)}$
$(10)(20) = 200,\ 10 + 20 = 30$

37. $x^2 - 6x + 8 = \boxed{(x - 2)(x - 4)}$
$(-2)(-4) = 8,\ -2 + (-4) = -6$

39. $r^2 + 17r + 16 = \boxed{(r + 1)(r + 16)}$
$(16)(1) = 16 + 1 + 16 = 17$

41. $m^2 - 15m + 36 = \boxed{(m - 3)(m - 12)}$
$(-3)(-12) = 36,\ -3 + (-12) = -15$

43. $y^2 + y - 56 = \boxed{(y - 7)(y + 8)}$
$(-7)(8) = -56,\ -7 + 8 = 1$

45. $r^2 + 4r + 12$ is $\boxed{\text{prime}}$
cannot be factored

47. $y^2 - 4y - 21 = \boxed{(y - 7)(y + 3)}$
$(-7)(3) = -21$

49. $x^2 + 8x - 105 = \boxed{(x + 15)(x - 7)}$
$(15)(-7) = -105,\ 15 - 7 = 8$

51. $r^2 + 27r + 72 = \boxed{(r+3)(r+24)}$
$(3)(24) = 72,\ 3 + 24 = 27$

53. $x^2 + 5xy + 6y^2 = \boxed{(x+2y)(x+3y)}$
$(2)(3) = 6,\ 2 + 3 = 5$

55. $x^2 - 9xy + 20y^2 = \boxed{(x-4y)(x-5y)}$
$(-4)(-5) = 20,\ -4 + (-5) = -9$

57. $x^2 + 2xy - 8y^2 = \boxed{(x+4y)(x-2y)}$
$(4)(-2) = -8,\ 4 + (-2) = 2$

59. $x^2 - 6xy + 9y^2 = \boxed{(x-3y)^2}$
$(-3)-3) = 9,\ -3 + (-3) = -6$

61. $a^2 - 9ab + 20b^2 = \boxed{(a-5b)(a-4b)}$
$(-5)(-4) = 20,\ -5 + (-4) = -9$

63. $r^2 - 13rs + 42s^2 = \boxed{(r-6s)(r-7s)}$
$(-6)(-7) = 42,\ -6 + (-7) = -13$

65. $z^2 - 16zb + 15b^2 = \boxed{(z-b)(z-15b)}$
$(-1)(-15) = 15,\ -1 + (-15) = -16$

67. $a^2 - 8ad - 33d^2 = \boxed{(a-11d)(a+3d)}$
$(-11)(3) = -33,\ -11 + 3 = -8$

69. $x^2 - 9xy + 9y^2$ is $\boxed{\text{prime}}$

71. $3x^2 + 15x + 18 = 3(x^2 + 5x + 6) = \boxed{3(x+2)(x+3)}$

73. $4y^2 - 4y - 8 = 4(y^2 - y - 2) = \boxed{4(y+1)(y-2)}$

75. $xy^2 + 2xy - 15x = x(y^2 + 2y - 15) = \boxed{x(y+5)(y-3)}$

77. $ab^2 - 5ab + 6a = a(b^2 - 5b + 6) = \boxed{a(b-2)(b-3)}$

79. $2r^3 + 6r^2 + 4r = 2r(r^2 + 3r + 2) = \boxed{2r(r+1)(r+2)}$

81. $4x^3 + 12x^2 - 72x = 4x(x^2 + 3x - 18) = \boxed{4x(x+6)(x-3)}$

83. $2r^3 + 8r^2 - 64r = 2r(r^2 + 4r - 32) = \boxed{2r(r+8)(r-4)}$

85. $y^4 + 2y^3 - 80y^2 = y^2(y^2 + 2y - 80) = \boxed{y^2(y+10)(y-8)}$

87. $x^4 - 3x^3 - 10x^2 = x^2(x^2 - 3x - 10) = \boxed{x^2(x-5)(x+2)}$

89. $2w^4 - 26w^3 - 96w^2 = 2w^2(w^2 - 13w - 48) = \boxed{2w^2(w-16)(w+3)}$

91. $15xy^2 + 45xy - 60x = 15x(y^2 + 3y - 4) = \boxed{15x(y+4)(y-1)}$

93. $x^5 + 3x^4y - 4x^3y^2 = x^3(x^2 + 3xy - 4y^2) = \boxed{x^3(x+4y)(x-y)}$

95. $6a^3b + 30a^2b^2 + 36ab^3 = 6ab(a^2 + 5ab + 6b^2) = \boxed{6ab(a+2b)(a+3b)}$

97. Statement $\boxed{B}$ is true.
$x^2 - 10xy + 9y^2$
$= (x-y)(x-9y)$

$(-1)(-9) = 9,\ -1 + (-9) = -10$

99. area = width × length
$x^2 - 10x + 16 = (\text{width}) \times (x-2)$
width $= \dfrac{x^2 - 10x + 16}{x - 2}$
$w = \dfrac{(x-8)(x-2)}{x-2} = x - 8$ meters

101. $A = (x^2 + 7x + 12) = (x+3)(x+4)$
length of shaded area $= x + 4 - x = 4$
width of shaded area $= x + 3 - x = 3$
Area of shaded region $= (x + 3 - x)(x + 4 - x) = 3 \cdot 4 = \boxed{12}$.

103. $y^2 + by + 10$
Factors of 10 are (1, 10), (2, 5), (–1, –10), (–2, –5)
Possible values of b are $\boxed{11, 7, -11, -7}$.

Review Problems

107. Let t equal the tens' digit of Grandpa Drac. The units' digit is 3.
Grandpa Drac's age $= 10t + 3$ $(t\,3)$
Grandpa Drac's age reversed is $10(3) + t$ $(3\,t)$
$t^2 = 3(10) + t$
$t^2 - t - 30 = 0$
$(t - 6)(t + 5) = 0$
$t - 6 = 0$
$t + 5 = 0$
$x = 6$
$\boxed{\text{age} = 63}$

108. $\dfrac{9}{8} - \dfrac{6y}{5} = -\dfrac{y}{5} - \dfrac{1}{8}$

$\dfrac{10}{8} = \dfrac{5y}{5}$

$40y = 50$

$\boxed{y = \dfrac{5}{4}}$

109. $\dfrac{12(-3-1)}{-15(4-1)-(-5-2)(4+2)} = \dfrac{12(-4)}{-15(3)-(-7)(6)} = \dfrac{-48}{-45+42} = \dfrac{-48}{-3} = \boxed{16}$

Section 7.3 Factoring Trinomials Whose Leading Coefficient is Not 1

Problem Set 7.3, pp. 471-472

1. $5x^2 + 6x + 1 = (5x + 1)\boxed{(x + 1)}$ The factors of 5 are 5 and 1 and of 1 are 1 and 1.

3. $5y^2 + 29y - 6 = (y + 6)\boxed{(5y - 1)}$
Possible factors:
$(5y + 6)(y - 1), -5y + 6y = xy$
$(5y - 6)(y + 1), 5y - 6y = -y$
$(y + 6)(5y - 1), -y + 30y = 29y$ ← only correct middle
$(y - 6)(5y + 1), y - 30y = -29y$ term
$(5y - 3)(y + 2), 10y - 3y = 7y$
$(5y + 3)(y - 2), 10y + 3y = -7y$
$(5y - 2)(y + 3), 15y - 2y = 13y$
$(5y + 2)(y - 3), -15y + 2y = -13y$

5. $24r^2 - 22r - 35 = (6r + 5)\boxed{(4r - 7)}$
Possible factors of $24r^2$ are 24 and r, $2r$ and $12r$,
possible factors of –35 are 35 and –1, 3r and 8r,
–1 and 35, 4r and 6r, 5 and –7, –5 and 7,
$6r(-7) + 5(4r) = -42r + 20r = -22r$ ← only correct
                                      possibility

7. $6y^2 - 31y + 5 = (y - 5)\boxed{(6y - 1)}$
$-y - 30y = -31y$

9. $7y^2 - 40y - 63 = (y - 7)\boxed{(7y + 9)}$
$9y - 49y = -40y$ ← only correct middle term

11. $15m^2 + 7m - 22 = (m - 1)\boxed{(15m + 22)}$
$22m - 15m = 7m$

13. $2x^2 + 7x + 3 = \boxed{(2x + 1)(x + 3)}$
$6x + x = 7x$

15. $2x^2 + 17x + 35 = \boxed{(2x + 7)(x + 5)}$
$10x + 7x = 17x$

17. $2y^2 - 17y + 30 = \boxed{(2y - 5)(y - 6)}$
$-12y - 5y = -17y$

19. $3x^2 - x - 2 = \boxed{(3x + 2)(x - 1)}$
$-3x + 2x = -x$

21. $3y^2 + y - 10 = \boxed{(3y - 5)(y + 2)}$
$6y - 5y = y$

23. $3r^2 - 25r - 28 = \boxed{(3r - 28)(r + 1)}$
$3r - 28r = -25r$

25. $6y^2 - 11y + 4 = \boxed{(2y - 1)(3y - 4)}$
$-8y - 3y = -11y$

27. $8t^2 + 33t + 4 = \boxed{(8t + 1)(t + 4)}$
$32t + t - 33t$

29. $5x^2 + 33x - 14 = \boxed{(5x - 2)(x + 7)}$
$35x - 2x = 33x$

31. $14y^2 + 15y - 9 = \boxed{(7y - 3)(2y + 3)}$
$21y - 6y = 15y$

33. $25r^2 - 30r + 9 = (5r - 3)(5r - 3) = \boxed{(5r - 3)^2}$
$(-15r) + (-15r) = -30r$

35. $6x^2 - 7x + 3 = \boxed{\text{prime}}$

37. $10y^2 + 43y - 9 = \boxed{(5y - 1)(2y + 9)}$
$45y - 2y = 43y$

39. $8r^2 - 38r - 21 = \boxed{(4r - 21)(2r + 1)}$
$4r - 42r = -38r$

41. $15y^2 - y - 2 = \boxed{(5y - 2)(3y + 1)}$
$5y - 6y = -y$

43. $8m^2 - 2m - 1 = \boxed{(4m + 1)(2m - 1)}$
$-4m + 2m = -2m$

45. $35z^2 + 43z - 10 = \boxed{(7z + 10)(5z - 1)}$
$-7z + 50z = 43z$

47. $9y^2 - 9y + 2 = \boxed{(3y - 2)(3y - 1)}$
$-3y - 6y = -9y$

49. $20x^2 - 41x + 20 = \boxed{(4x - 5)(5x - 4)}$
$-16x - 25x = -41x$

51. $2x^2 + 3xy + y^2 = \boxed{(2x + y)(x + y)}$
$2xy + xy = 3xy$

53. $15x^2 + 11xy - 14y^2 = \boxed{(3x - 2y)(5x + 7y)}$
$21xy - 10xy = 11xy$

55. $2x^2 - 9xy + 9y^2 = \boxed{(2x - 3y)(x - 3y)}$
$-6xy - 3xy = -9xy$

57. $2x^2 + 7xy + 5y^2 = \boxed{(2x + 5y)(x + y)}$
$2xy + 5xy = 7xy$

59. $6a^2 - 5ab - 6b^2 = \boxed{(3a + 2b)(2a - 3b)}$
$-9ab + 4ab = -5ab$

61. $3a^2 - ab - 14b^2 = \boxed{(3a - 7b)(a + 2b)}$
$6ab - 7ab = -ab$

63. $12r^2 - 25rs + 12s^2 = \boxed{(4r - 3s)(3r - 4s)}$
$-16rs - 9rs = -25rs$

65. $-4x^2 - x + 3 = \boxed{(-4x + 3)(x + 1)}$
$-4x + 3x = -x$

67. $-4y^2 + 5y + 6 = \boxed{(-4y - 3)(y - 2)}$
$8y - 3y = 5y$

69. $2 + 7y + 6y^2 = \boxed{(2 + 3y)(1 + 2y)}$
$4y + 3x = -x$

71. $38 - 67x + 15x^2 = \boxed{(2 - 3x)(19 - 5x)}$
$-10x - 57y = -67x$

73. $18x^2 + 48x + 32 = 2(9x^2 + 24x + 16) = 2(3x + 4)(3x + 4) = \boxed{2(3x + 4)^2}$

75. $4y^2 + 2y - 30 = 2(2y^2 + y - 15) = \boxed{2(2y - 5)(y + 3)}$
$6y - 5y = y$

77. $9r^2 + 33r - 60 = 3(3r^2 + 11r - 20) = \boxed{3(3r - 4)(r + 5)}$
$15r - 4r = 11r$

79. $2y^3 - 3y^2 - 5y = y(2y^2 - 3y - 5) = \boxed{y(2y - 5)(y + 1)}$
$2y - 5y = -3y$

81. $9r^3 - 39r^2 + 12r = 3r(3r^2 - 13r + 4) = \boxed{3r(r - 4)(3r - 1)}$
$-r - 12r = -13r$

83. $14m^3 + 94m^2 - 28m = 2m(7m^2 + 47m - 14) = \boxed{2m(m + 7)(7m - 2)}$
$-2m + 49m = 47m$

85. $15x^4 - 39x^3 + 18x^2 = 3x^2(5x^2 - 13x + 6) = \boxed{3x^2(5x - 3)(x - 2)}$
$-10x - 3x = -13x$

87. $10x^5 - 17x^4 + 3x^3 = x^3(10x^2 - 17x + 3) = \boxed{x^3(5x - 1)(2x - 3)}$
$-15x - 2x = -17x$

89. $18x^4 - 37x^3y + 15x^2y^2 = x^2(18x^2 - 37xy + 15y^2) = \boxed{x^2(2x - 3y)(9x - 5y)}$
$-10x - 27x = -37x$

91. $40x^2 - 38xy - 56y^2 = 2(20x^2 - 19xy - 28y^2) = \boxed{2(4x - 7y)(5x + 4y)}$
$16xy - 35xy = -19xy$

93. $12a^2b - 46ab^2 + 14b^3 = 2b(6a^2 - 23ab + 7b^2) = \boxed{2b(3a - b)(2a - 7b)}$
$-21ab - 2ab = -23ab$

95. $-32x^2y^4 + 20xy^4 + 12y^4 = -4y^4(8x^2 - 5x - 3) = \boxed{-4y^4(8x + 3)(x - 1)}$
$-8x + 3x = -5x$

97. Statement $\boxed{A}$ is true.
$18y^2 - 6y + 6 = -8x + 3x = -5x$
$6(3y^2 - y + 1)$
$3y^2 - y + 1$ is prime

99. Let $w =$ width.
$(x - 5)w = 2x^2 - 11x + 5$
$w = \dfrac{(2x - 1)(x - 5)}{x - 5} = 2x - 1 \quad (x \neq 5)$
The width is $\boxed{2x - 1 \text{ meters}}$.

101. $V = lwh$
$28y^5 - 58y^4 - 30y^3 = (2y^3)(2y - 5)h$
$$h = \frac{28y^5 - 58y^4 - 30y^3}{(2y^3)(2y - 5)}$$
$$= \frac{2y^3(14y^2 - 29y - 15)}{2y^3(2y - 5)} = \frac{(7y + 3)(2y - 5)}{2y - 5} = 7y + 3 \qquad \left(y \neq \frac{5}{2}\right)$$
The height is $\boxed{7y + 3 \text{ meters}}$.

103. $3x^2 + bx + 2 = (3x + 2)(x + 1)$ if $\boxed{b = 5}$ or $(3x + 1)(x + 2)$ if $\boxed{b = 7}$
$= (3x - 2)(x - 1)$ if $\boxed{b = -5}$ or $(3x - 1)(x - 2)$ if $\boxed{b = -7}$

105. $3(x + 2)^2 - (x + 2) - 4 = 3(x^2 + 4x + 4) - x - 6 = 3x^2 + 12x + 12 - x - 6 = 3x^2 + 11x + 6 = \boxed{(3x + 2)(x + 3)}$

Review Problems

108. $-7 \leq 3x - 1 < 8$
$-6 \leq 3x < 9$
$-2 < x < 3$
$\boxed{\{x \mid -2 < x < 3\}}$

109.

$\begin{array}{llll}
\frac{2x}{3} - \frac{3y}{2} = -1 & (\times 6) \to & 4x - 9y = -6 & \text{(no change)} \to \quad 4x - 9y = -6 \\
\frac{x}{6} + \frac{5y}{2} = -6 & (\times 6) \to & x + 15y = -36 & (\times -4) \to \quad \underline{-4x - 60y = 144}
\end{array}$

$-69y = 138$
$y = -2$
$\frac{x}{6} + \frac{5(-2)}{2} = -6$
$\frac{x}{6} = -6 + 5$
$x = 6(-1) = -6$
$\boxed{\{(-6, -2)\}}$

110. Let $x =$ the cost to manufacture 144 can openers.
$$\frac{x}{144} = \frac{0.75}{15}$$
$$x = \frac{0.75}{15}(144) = \boxed{\$7.20}$$

Section 7.4 Factoring Special Forms

Problem Set 7.4, pp. 480-482

1. $x^2 - 25 = x^2 - 5^2 = \boxed{(x + 5)(x - 5)}$

3. $y^2 - 1 = y^2 - 1^2 = \boxed{(y + 1)(y - 1)}$

5. $4x^2 - 1 = (2x)^2 - 1^2 = \boxed{(2x + 1)(2x - 1)}$

7. $x^2 - 7$ is $\boxed{\text{prime}}$

9. $9y^2 - 4 = (3y)^2 - 2^2 = \boxed{(3y + 2)(3y - 2)}$

11. $9x^2 + 4$ is $\boxed{\text{prime}}$

13. $1 - 49x^2 = 1^2 - (7x)^2 = \boxed{(1 + 7x)(1 - 7x)}$

15. $25a^2 - 16b^2 = (5a)^2 - (4b)^2 = \boxed{(5a - 4b)(5a + 4b)}$

17. $x^2 + 9$ is $\boxed{\text{prime}}$

19. $16z^2 - y^2 = (4z)^2 - y^2 = \boxed{(4z + y)(4z - y)}$

21. $9 - 121a^2 = 3^2 - (11a)^2 = \boxed{(3 + 11a)(3 - 11a)}$

23. $(x + 1)^2 - 16 = (x + 1)^2 - 4^2 = [(x + 1) + 4][(x + 1) - 4] = \boxed{(x + 5)(x - 3)}$

25. $(2x + 3)^2 - 49 = (2x + 3)^2 - 7^2 = [(2x + 3) + 7][(2x + 3) - 7] = (2x + 10)(2x - 4) = \boxed{4(x + 5)(x - 2)}$

27. $(3x - 1)^2 - 64 = (3x - 1)^2 - 8^2 = [(3x - 1) + 8][(3x - 1) - 8] = (3x + 7)(3x - 9) = \boxed{3(3x + 7)(x - 3)}$

29. $25 - (x + 3)^2 = 5^2 - (x + 3)^2 = [5 - (x + 3)][5 + (x + 3)] = \boxed{(2 - x)(8 + x)}$

31. $2y^2 - 18 = 2(y^2 - 9) = 2(y^2 - 3^2) = \boxed{2(y + 3)(y - 3)}$

33. $2x^3 - 72x = 2x(x^2 - 36) = 2x(x^2 - 6^2) = \boxed{2x(x - 6)(x + 6)}$

35. $50 - 2y^2 = 2(25 - y^2) = 2(5^2 - y^2) = \boxed{2(5 - y)(5 + y)}$

37. $8y^3 - 2y = 2y(4y^2 - 1) = 2y[(2y)^2 - 1^2] = \boxed{2y(2y - 1)(2y + 1)}$

39. $8a^2 b - 18b = 2b(4a^2 - 9) = 2b[(2a)^2 - 3^2] = \boxed{2b(2a - 3)(2a + 3)}$

41. $2ab^3 - 8a^3 b = 2ab(b^2 - 4a^2) = 2ab[b^2 - (2a)^2] = \boxed{2ab(b - 2a)(b + 2a)}$

43. $x^4 - y^2 = (x^2)^2 - y^2 = \boxed{(x^2 - y)(x^2 + y)}$

45. $x^4 - 16 = x^2 - 4^2 = (x^2 - 4)(x^2 + 4) = (x^2 - 2^2)(x^2 + 4) = \boxed{(x - 2)(x + 2)(x^2 + 4)}$

47. $16y^4 - 81 = (4y^2)^2 - 9^2 = (4y^2 - 9)(4y^2 + 9) = [(2y)^2 - 3^2](4y^2 + 9) = \boxed{(2y - 3)(2y + 3)(4y^2 + 9)}$

49. $1 - y^4 = (1 - y^2)(1 + y^2) = \boxed{(1 - y)(1 + y)(1 + y^2)}$ **51.** $x^2 + 2x + 1 = \boxed{(x + 1)^2}$

53. $x^2 - 14x + 49 = x^2 - 2(7)x + 7^2 = \boxed{(x - 7)^2}$ **55.** $4y^2 + 4y + 1 = (2y)^2 + 2(2y) + 1 = \boxed{(2y + 1)^2}$

57. $9r^2 - 6r + 1 = (3r)^2 - 2(3r) + 1 = \boxed{(3r - 1)^2}$

59. $16t^2 + 1 + 8t = 16t^2 + 8t + 1 = (4t)^2 + 2(4t) + 1 = \boxed{(4t + 1)^2}$

61. $9b^2 - 42b + 49 = (3b)^2 - 2(3b)(7) + 7^2 = \boxed{(3b - 7)^2}$

63. $25x^2 + 30xy + 9y^2 = (5x)^2 + 2(5x)(3y) + (3y)^2 = \boxed{(5x + 3y)^2}$

65. $16x^2 - 24xy + 9y^2 = (4x)^2 - 2(4x)(3y) + (3y)^2 = \boxed{(4x - 3y)^2}$

67. $x^2 - 10x + 100$ is $\boxed{\text{prime}}$

69. $12k^2 - 12k + 3 = 3(4k^2 - 4k + 1) = 3[(2k)^2 - 2(2k) + 1] = \boxed{3(2k-1)^2}$

71. $x^2 y^2 - 6xy^2 + 9y^2 = y^2(x^2 - 6x + 9) = y^2[x^2 - 2(3)x + 3^2] = \boxed{y^2(x-3)^2}$

73. $32ab^2 - 48ab + 18a = 2a(16b^2 - 24b + 9) = 2a[(4b)^2 - 2(4b)(3) + 3^2] = \boxed{2a(4b-3)^2}$

75. $9a^4 b^2 + 24a^3 b^2 + 16a^2 b^2 = a^2 b^2(9a^2 + 24a + 16) = a^2 b^2[(3a)^2 + 2(3a)(4) + 4^2] = \boxed{a^2 b^2(3a+4)^2}$

77. $x^2 + 8x + 16 - y^2 = (x+4)^2 - y^2 = \boxed{(x+4-y)(x+4+y)}$

79. $9a^2 - 12a - 4 - y^2 = (3a-2)^2 - y^2 = \boxed{(3a-2-y)(3a-2+y)}$

81. $a^2 + 2ab + b^2 - 49 = (a+b)^2 - 49 = \boxed{(a+b-7)(a+b+7)}$

83. $x^3 + 27 = x^3 + 3^3 = \boxed{(x+3)(x^2 - 3x + 9)}$ **85.** $x^3 - 64 = x^3 - 4^3 = \boxed{(x-4)(x^2 + 4x + 16)}$

87. $8y^3 - 1 = (2y)^3 - 1 = (2y-1)[(2y)^2 + 2y + 1] = \boxed{(2y-1)(4y^2 + 2y + 1)}$

89. $64x^3 + 125 = (4x)^3 + 5^3 = (4x+5)[(4x)^2 - (4x)(5) + 5^2] = \boxed{(4x+5)(16x^2 - 20x + 25)}$

91. $125x^3 - 27y^3 = (5x)^3 - (3y)^3 = (5x-3y)[(5x)^2 + (5x)(3y) + (3y)^2] = \boxed{(5x-3y)(25x^2 + 15xy + 9y^2)}$

93. $2x^4 + 16x = 2x(x^3 + 8) = 2x(x^3 + 2^3) = \boxed{2x(x+2)(x^2 - 2x + 4)}$

95. Statement $\boxed{D}$ is true.   **97.** $A = x^2 - y^2 = \boxed{(x-y)(x+y)}$
$2x^2 - 18 = 2(x^2 - 9) = 2(x-3)(x+3)$
Three distinct factors.

99. $A = x^2 - 4y^2 = \boxed{(x-2y)(x+2y)}$

101. $1000^2 - 999^2 = (1000 - 999)(1000 + 999) = (1)(1999) = \boxed{1999}$

103. $1000^2 - 990^2 = (1000 - 990)(1000 + 990) = 10(1990) = \boxed{19{,}900}$

105. $x^2 + bx + 25 = x^2 + 2(5)x + 5^2 = (x+5)^2$ if $b = 2(5) = \boxed{10}$
or $x^2 + 2(-5)x + (-5)^2 = (x-5)^2$ if $b = 2(-5) = \boxed{-10}$

107. $9x^2 + bx + 4 = (3x)^2 + 2(3x) + 4 = (3x+2)^2$ if $b = 2(6) = \boxed{12}$

109. $4x^2 + kx + 1 = (2x+1)^2$ if $\boxed{k=4}$ **111.** $25x^2 + kxy + 9y^2 = (5x+3y)^2$ if $\boxed{k=30}$
$ = (2x-1)^2$ if $\boxed{k=-4}$ $ = (5x-3y)^2$ if $\boxed{k=-30}$

113. $64x^2 - 16x + k = (8x-1)^2$ if $\boxed{k=1}$ **115.** $x^5 - 1 = \boxed{(x-1)(x^4 + x^3 + x^2 + x + 1)}$

117.
$$0 = 0^2 - 0^2$$
$$1 = 1^2 - 0^2$$
$$3 = 2^2 - 1^2$$
$$4 = 2^2 - 0^2$$
$$5 = 3^2 - 2^2$$
$$7 = 4^2 - 3^2$$
$$8 = 3^2 - 1^2$$
$$9 = 5^2 - 4^2$$
$$11 = 6^2 - 5^2$$
$$12 = 4^2 - 2^2$$
$$13 = 7^2 - 6^2$$
$$15 = 8^2 - 7^2$$
$$16 = 5^2 - 3^2$$
$$17 = 9^2 - 8^2$$
$$19 = 10^2 - 9^2$$
$$20 = 6^2 - 4^2$$
$$21 = 11^2 - 10^2$$

The numbers that cannot be written as the difference of two squares are 2, 6, 10, 14, 18, ... are $2 + 4n$, where $n = 0, 1, 2, ...$

Review Problems

119. $\dfrac{4x^3 + 17x^2 + 6x - 27}{x + 3} = \boxed{4x^2 + 5x - 9}$

$$x + 3 \overline{\smash{\big)}\, 4x^3 + 17x^2 + 6x - 27}$$

quotient $4x^2 + 5x - 9$

$$\underline{4x^3 + 12x^2}$$
$$5x^2 + 6x$$
$$\underline{5x^2 + 15x}$$
$$-9x - 27$$
$$\underline{-9x - 27}$$
$$0$$

120. $\left(\dfrac{10x^{-4}}{2x^2}\right)^{-1} = \left(\dfrac{2x^2}{10x^{-4}}\right) = \dfrac{x^{2+4}}{5} = \boxed{\dfrac{x^6}{5}}$

121.
$$7(3 - x) + 8 \geq 4(3x - 1) - 8x$$
$$21 - 7x + 8 \geq 12x - 4 - 8x$$
$$33 \geq 11x$$
$$x \leq 3$$
$$\boxed{\{x \mid x \leq 3\}}$$

Section 7.5 A General Factoring Strategy

Problem Set 7.5, pp. 484-485

1. $6x^2 + 8xy = \boxed{2x(3x + 4y)}$

3. $2y^2 - 2y - 112 = 2(y^2 - y - 56) = \boxed{2(y - 8)(y + 7)}$

5. $7y^4 + 14y^3 + 7y^2 = 7y^2(y^2 + 2y + 1) = \boxed{7y^2(y+1)^2}$

7. $y^2 + 8y - 16$ is $\boxed{\text{prime}}$

9. $xy - 7x + 3y - 21 = x(y-7) + 3(y-7) = \boxed{(x+3)(y-7)}$

11. $16y^2 - 4y - 2 = 2(8y^2 - 2y - 1) = \boxed{2(2y-1)(4y+1)}$

13. $r^2 - 25r = \boxed{r(r-25)}$

15. $4w^2 + 8w - 5 = \boxed{(2w+5)(2w-1)}$

17. $x^3y - 4xy = xy(x^2 - 4) = \boxed{xy(x+2)(x-2)}$ **19.** $x^2 + 64$ is $\boxed{\text{prime}}$

21. $9y^2 + 13y + 4 = \boxed{(9y+4)(y+1)}$

23. $y^3 + 2y^2 - 4y - 8 = y^2(y+2) - 4(y+2) = (y+2)(y^2-4) = (y+2)(y-2)(y+2) = \boxed{(y-2)(y+2)^2}$

25. $9y^2 + 24y + 16 = \boxed{(3y+4)^2}$ **27.** $3xyz^2 - 12xy = 3xy(z^2 - 4) = \boxed{3xy(z-2)(z+2)}$

29. $x^{12} - 64 = (x^6 - 8)(x^6 + 8) = \boxed{(x^2-2)(x^4+2x^2+4)(x^2+2)\,(x^4-2x^2+4)}$

31. $5x^2 - 30x + 4$ is $\boxed{\text{prime}}$

33. $x^2 - 3xy - 4y^2 = \boxed{(x+y)(x-4y)}$

35. $5y^3 - 45y^2 + 70y = 5y(y^2 - 9y + 14) = \boxed{5y(y-2)(y-7)}$

37. $72a^3b^2 + 12a^2 - 24a^4b^2 = \boxed{12a^2(6ab^2 + 1 - 2a^2b^2)}$

39. $x^2 + 4x + ax + 4a = x(x+4) + a(x+4) = \boxed{(x+a)(x+4)}$

41. $3a^2 + 27ab + 54b^2 = 3(a^2 + 9ab + 18b^2) = \boxed{3(a+3b)(a+6b)}$

43. $y^5 - 81y = y(y^4 - 81) = y(y^2 - 9)(y^2 + 9) = \boxed{y(y-3)(y+3)(y^2+9)}$

45. $20a^4 - 45a^2 = 5a^2(4a^2 - 9) = \boxed{5a^2(2a-3)(2a+3)}$

47. $25x^7 - 16x^5 = \boxed{x^5(5x-4)(5x+4)}$

49. $y^4 + 7y^3 - 30y^2 = y^2(y^2 + 7y - 30) = \boxed{y^2(y+10)(y-3)}$

51. $10x^2 - 5xy - 15y^2 = 5(2x^2 - xy - 3y^2) = \boxed{5(2x-3y)(x+y)}$

53. $ax^2 - bx^2 - a + b = x^2(a-b) - (a-b) = (x^2-1)(a-b) = \boxed{(x-1)(x+1)(a-b)}$

55. $7x^5y - 7xy^5 = 7xy(x^4 - y^4) = 7xy(x^2 - y^2)(x^2 + y^2) = \boxed{7xy(x-y)(x+y)(x^2+y^2)}$

57. $24a^2b + 6a^3b - 45a^4b = 3a^2b(8 + 2a - 15a^2) = \boxed{3a^2b(4-5a)(2+3a)}$

59. $2bx^2 + 44bx + 242b = 2b(x^2 + 22x + 121) = \boxed{2b(x+11)^2}$

61. $12y^2 - 11y + 2 = \boxed{(3y-2)(4y-1)}$

63. $xz - bx + az - ab = x(z-b) + a(z-b) = \boxed{(x+a)(z-b)}$

65. $9y^2 - 64 = \boxed{(3y-8)(3y+8)}$ **67.** $9y^2 + 64$ is $\boxed{\text{prime}}$

69. $15a^2 + 11ab - 14b^2 = \boxed{(3a-2b)(5a+7b)}$

71. $y^2 + 14y + 49 - a^2 = (y+7)^2 - a^2 = \boxed{(y+7-a)(y+7+a)}$

73. $4y^3 - 14y^2 - 60y = \boxed{2y(y-6)(2y+5)}$

75. $2y^3 + 3y^2 - 50y - 75 = y^2(2y+3) - 25(2y+3) = (y^2-25)(2y+3) = \boxed{(y-5)(y+5)(2y+3)}$

77. $-6x^2 - x + 1 = \boxed{-(2x+1)(3x-1)}$

79. $36x^3y - 62x^2y^2 + 12xy^3 = 2xy(18x^2 - 31xy + 6y^2) = \boxed{2xy(2x-3y)(9x-2y)}$

81. $21y^2 - 11y^3 - 2y^4 = y^2(21 - 11y - 2y^2) = \boxed{y^2(3-2y)(7+y)}$

83. $r^2 + 15r - 34 = \boxed{(r+17)(r-2)}$

85. $42y^2 + 133y - 49 = 7(6y^2 + 19y - 7) = \boxed{7(3y-1)(2y+7)}$

87. $16x^2 - 40x + 25 - 81y^2 = (4x-5)^2 - 81y^2 = \boxed{(4x-5-9y)(4x-5+9y)}$

89. $8x^5 - 2x^3 = 2x^3(4x^2 - 1) = \boxed{2x^3(2x-1)(2x+1)}$ **91.** $3x^2 + 243 = \boxed{3(x^2+81)}$

93. $a^2y - b^2y - a^2x + b^2x = y(a^2 - b^2) - x(a^2 - b^2) = (a^2 - b^2)(y-x) = \boxed{(a-b)(a+b)(y-x)}$

95. $9ax^3 + 15ax^2 - 14ax = ax(9x^2 + 15x - 14) = \boxed{ax(3x+7)(3x-2)}$

97. $x^4 + 8x = x(x^3 + 8) = \boxed{x(x+2)(x^2-2x+4)}$

99. $2y^5 - 2y^2 = 2y^2(y^3 - 1) = \boxed{2y^2(y-1)(y^2+y+1)}$

Review Problems

103.
$$5x - 2y \geq 10$$
$$3x + y \leq 6$$
The lines intersect at $x = 2$, $y = 0$. The region satisfying both inequalities is shown.

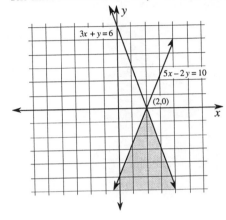

104. $f(x) = x^3 - 3x - 2$
$f(-3) = (-3)^3 - 3(-3) - 2 = -27 + 9 - 2 = \boxed{-20}$

105. Let t = unknown time.
$70t + 65t = 405$
$135t = 405$
$t = \boxed{3 \text{ hours}}$

Section 7.6 Solving Quadratic Equations by Factoring

Problem Set 7.6, pp. 491-493

1. $(x - 2)(x + 3) = 0$
$x - 2 = 0$ or $x - 3 = 0$ (Apply zero – product principle)
$x = 2$ or $x = -3$
Check: $x = 2$: $(2 - 2)(2 + 3) = 0$
$0(5) = 0$
$0 = 0$ True
$x = -3$: $(-3 - 2)(-3 + 3) = 0$
$(-5)(0) = 0$
$0 = 0$ True
The solution set is $\boxed{\{-3, 2\}}$.

3. $(3x + 4)(2x - 1) = 0$
$3x + 4 = 0$ or $2x - 1 = 0$
$x = -\dfrac{4}{3}$ or $x = \dfrac{1}{2}$
solution set: $\boxed{\left\{ -\dfrac{4}{3}, \dfrac{1}{2} \right\}}$

Note: **Checking will be omitted in the following problems unless steps other than simple factoring are used.**

5. $(2y - 1)(4y + 1) = 0$

$y = \dfrac{1}{2}$ or $y = -\dfrac{1}{4}$

$2y - 1 = 0$ or $4y + 1 = 0$

$\boxed{\left\{ -\dfrac{1}{4}, \dfrac{1}{2} \right\}}$

7. $(2y + 7)(3y - 1) = 0$

$y = -\dfrac{7}{2}$ or $y = \dfrac{1}{3}$

$2y + 7 = 0$ or $3y - 1 = 0$

$\boxed{\left\{ -\dfrac{7}{2}, \dfrac{1}{3} \right\}}$

9. $(z - 2)(3z + 7) = 0$

$z = 2$ or $z = -\dfrac{7}{3}$

$z - 2 = 0$ or $3z + 7 = 0$

$\boxed{\left\{ -\dfrac{7}{3}, -2 \right\}}$

11. $(4w - 9)(2w + 5) = 0$

$w = \dfrac{9}{4}$ or $w = -\dfrac{5}{2}$

$4w - 9 = 0$ or $2w + 5 = 0$

$\boxed{\left\{ -\dfrac{5}{2}, \dfrac{9}{4} \right\}}$

13. $x^2 + 8x + 15 = 0$

$(x + 3)(x + 5) = 0$ (Factor the left-hand side and apply the zero – product principle)

$x + 3 = 0$ or $x + 5 = 0$ (Apply the zero – product principle)

$x = -3$ or $x = -5$

Check: $x = -3$: $(-3)^2 + 8(-3) + 15 = 9 - 24 + 15 = 0$ True

$\qquad\qquad x = -5$: $(-5)^2 + 8(-5) + 15 = 25 - 40 + 15 = 0$ True

solution set: $\boxed{\{-5, -3\}}$

15. $y^2 - 2y - 15 = 0$

$(y - 5)(y + 3) = 0$

$y - 5 = 0$ or $y + 3 = 0$

$y = 5$ or $y = -3$

$\boxed{\{-3, 5\}}$

17. $m^2 - 4m - 21 = 0$

$(m - 7)(m + 3) = 0$

$m - 7 = 0$ or $m + 3 = 0$

$m = 7$ or $m = -3$

$\boxed{\{-3, 7\}}$

19. $z^2 = -9z - 8$
$z^2 + 9z + 8 = 0$
$(z + 1)(z + 8) = 0$
$z = -1 \text{ or } z = -8$
$z + 1 = 0 \text{ or } z + 8 = 0$
$z = -1 \text{ or } z = -8$
Check: $z = -1:\ (1)^2 = -9(-1) - 8$
$1 = 9 - 8$
$1 = 1\quad$ True
$z = -8:\ (-8)^2 = -9(-8) - 8$
$64 = 72 - 8$
$64 = 64\quad$ True
solution set: $\boxed{\{-8, -1\}}$

21. $2x^2 = 7x + 4$
$2x^2 - 7x - 4 = 0$
$(2x + 1)(x - 4) = 0$
$2x + 1 = 0 \text{ or } x - 4 = 0$
$x = -\dfrac{1}{2} \text{ or } x = 4$
$$\boxed{\left\{-\frac{1}{2}, 4\right\}}$$

23. $5x^2 + x = 18$
$5x^2 + x - 18 = 0$
$(5x - 9)(x + 2) = 0$
$5x - 9 = 0 \text{ or } x + 2 = 0$
$x = \dfrac{9}{5} \text{ or } x = -2$
$$\boxed{\left\{-2, \frac{9}{5}\right\}}$$

25. $x(6x + 23) + 7 = 0$
$6x^2 + 23x + 7 = 0$
$(2x + 7)(3x + 1) = 0$
$2x + 7 = 0 \text{ or } 3x + 1 = 0$
$x = -\dfrac{7}{2} \text{ or } x = -\dfrac{1}{3}$
$$\boxed{\left\{-\frac{7}{2}, -\frac{1}{3}\right\}}$$

27. $3s^2 + 4s = -1$
$3s^2 + 4s + 1 = 0$
$(3s + 1)(s + 1) = 0$
$3s + 1 = \text{ or } s + 1 = 0$
$s = -\dfrac{1}{3} \text{ or } s = -1$
$$\boxed{\left\{-1, \frac{1}{3}\right\}}$$

29. $4x(x + 1) = 15$
$4x^2 + 4x - 15 = 0$
$(2x + 5)(2x - 3) = 0$
$2x + 5 = 0$ or $2x - 3 = 0$
$x = -\dfrac{5}{2}$ or $x = \dfrac{3}{2}$
Check: $x = -\dfrac{5}{2}$: $4\left(-\dfrac{5}{2}\right)\left(-\dfrac{5}{2} + 1\right)$
$\qquad\qquad -10(-\dfrac{3}{2}) = 15$
$\qquad\qquad 15 = 15$ True
$\qquad x = \dfrac{3}{2}$: $4\left(\dfrac{3}{2}\right)\left(\dfrac{3}{2} + 1\right)$
$\qquad\qquad 6\left(\dfrac{5}{2}\right) = 15$
$\qquad\qquad 15 = 15$ True
solution set: $\boxed{\left\{-\dfrac{5}{2}, \dfrac{3}{2}\right\}}$

31. $12r^2 + 31r + 20 = 0$
$(3x + 4)(4x + 5) = 0$
$3x + 4 = 0$ or $4x + 5 = 0$
$r = -\dfrac{4}{3}$ or $r = -\dfrac{5}{4}$
$\boxed{\left\{-\dfrac{4}{3}, -\dfrac{5}{4}\right\}}$

33. $12s^2 + 28s - 24 = 0$
$4(3s^2 + 7s - 6) = 0$
$4(3s - 2)(s + 3) = 0$
$3s - 2 = 0$ or $s + 3 = 0$
$s = \dfrac{2}{3}$ or $s = -3$
$\boxed{\left\{-3, \dfrac{2}{3}\right\}}$

35. $w^2 - 5w = 18 + 2w$
$w^2 - 7w - 18 = 0$
$(w - 9)(w + 2) = 0$
$w - 9 = 0$ or $w + 2 = 0$
$w = 9$ or $w = -2$
$\boxed{\{-2, 9\}}$

37. $z(z + 8) = 16(z - 1)$
$z^2 + 8z - 16z + 16 = 0$
$z^2 - 8z + 16 = 0$
$(z - 4)^2 = 0$
$z - 4 = 0$ or $z - 4 = 0$
$z = 4$
$\boxed{\{4\}}$

39. $y^2 + 4y = 0$
   $y = 0$ or $y + 4 = 0$
   $y(y + 4) = 0$
   $y = 0$ or $y = -4$

   Check: $y = 0$: $0^2 + 4(0) = 0$
   $\qquad\qquad\quad 0 = 0$ True
   $\qquad\qquad y = -4$: $(-4)^2 + 4(-4) = 0$
   $\qquad\qquad\quad 16 - 16 = 0$
   $\qquad\qquad\qquad 0 = 0$ True
   $\boxed{\{-4, 0\}}$

41. $16x^2 - 49 = 0$
   $(4x - 7)(4x + 7) = 0$
   $4x - 7 = 0$ or $4x + 7 = 0$
   $x = \dfrac{7}{4}$ or $x = -\dfrac{7}{4}$
   $\boxed{\left\{ -\dfrac{7}{4}, \dfrac{7}{4} \right\}}$

43. $2x^2 = 6x$
   $2x^2 - 6x = 0$
   $2x(x - 3) = 0$
   $2x = 0$ or $x - 3 = 0$
   $x = 0$ or $x = 3$
   $\boxed{\{0, 3\}}$

45. $(y - 3)(y + 8) = -30$
   $y^2 + 5y - 24 = -30$
   $y^2 + 5y + 6 = 0$
   $(y + 2)(y + 3) = 0$
   $y + 2 = 0$ or $y + 3 = 0$
   $y = -2$ or $y = -3$
   $\boxed{\{-3, -2\}}$

47. $(z + 1)(2z + 5) = -1$
   $2z^2 + 7z + 5 = -1$
   $2z^2 + 7z + 6 = 0$
   $(2z + 3)(z + 2) = 0$
   $2z + 3 = 0$ or $z + 2 = 0$
   $z = -\dfrac{3}{2}$ or $z = -2$
   $\boxed{\left\{ -2, -\dfrac{3}{2} \right\}}$

49. $4y^2 + 20y + 25 = 0$
   $(2y + 5)^2 = 0$
   $2y + 5 = 0$
   $y = -\dfrac{5}{2}$
   $\boxed{\left\{ -\dfrac{5}{2} \right\}}$

51. $64w^2 - 48w + 9 = 0$
$(8w - 3)^2 = 0$
$8w - 3 = 0$
$w = \dfrac{3}{8}$

$$\boxed{\left\{\dfrac{3}{8}\right\}}$$

53. $2y^2 - 6y - 11 = 3y^2 + 2y + 5$
$y^2 + 8y + 16 = 0$
$(y + 4)^2 = 0$
$y + 4 = 0$
$y = -4$
Check: $2(-4)^2 - 6(-4) - 11 = 45$
$3(-4)^2 + 2(-4) + 5 = 45$
$45 = 45$ True

$$\boxed{\{-4\}}$$

55. $y(y - 6) + 2(y - 6) = 0$
$(y + 2)(y - 6) = 0$
$y + 2 = 0$ or $y - 6 = 0$
$y = -2$ or $y = 6$

$$\boxed{\{-2, 6\}}$$

57. $(x + 1)^2 = 2(x + 5)$
$x^2 + 2x + 1 = 2x + 10$
$x^2 - 9 = 0$
$(x - 3)(x + 3) = 0$
$x - 3 = 0$ or $x + 3 = 0$
$x = 3$ or $x = -3$

$$\boxed{\{-3, 3\}}$$

59. $(x + 1)(x - 3)(2x + 5) = 0$
$x + 1 = 0$ or $x - 3 = 0$ or $2x + 5 = 0$
$x = -1$ or $x = 3$ or $x = -\dfrac{5}{2}$

$$\boxed{\left\{-\dfrac{5}{2}, -1, 3\right\}}$$

61. $2x(x - 4)(3x + 8) = 0$
$2x = 0$ or $x - 4 = 0$ or $3x + 8 = 0$
$x = 0$ or $x = 4$ or $x = -\dfrac{8}{3}$

$$\boxed{\left\{-\dfrac{8}{3}, 0, 4\right\}}$$

63. $(x - 4)(x^2 + 5x + 6) = 0$
$(x - 4)(x + 2)(x + 3) = 0$
$x - 4 = 0$ or $x + 2 = 0$ or $x + 3 = 0$
$x = 4$ or $x = -2$ or $x = -3$

$$\boxed{\{-3, -2, 4\}}$$

65. $(y-5)(y^2-y-12)=0$
$(y-5)(y-4)(y+3)=0$
$y-5=0$ or $y-4=0$ or $y+3=0$
$y=5$ or $y=4$ or $y=-3$
$\boxed{\{-3,4,5\}}$

67. $x^3-36x=0$
$x(x^2-36)=0$
$x(x-6)(x+6)=0$
$x=0$ or $x-6=0$ or $x+6=0$
$x=0$ or $x=6$ or $x=-6$
$\boxed{\{-6,0,6\}}$

69. $16z^3-9z=0$
$z(16z^2-9)=0$
$z(4z-3)(4z+3)=0$
$z=0$ or $4z-3=0$ or $4z+3=0$
$z=0$ or $z=\dfrac{3}{4}$ or $z=-\dfrac{3}{4}$
$\boxed{\left\{-\dfrac{3}{4},0,\dfrac{3}{4}\right\}}$

71. $y^3+3y^2+2y=0$
$y(y^2+3y+2)=0$
$y(y+1)(y+2)=0$
$y=0$ or $y+1=0$ or $y+2=0$
$y=0$ or $y=-1$ or $y=-2$
$\boxed{\{-2,-1,0\}}$

73. $x^3=-6x^2-9x$
$x^3+6x^2+9x=0$
$x(x^2+6x+9)=0$
$x(x+3)^2=0$
$x=0$ or $x+3=0$
$x=0$ or $x=-3$
$\boxed{\{-3,0\}}$

75. $y^3-y^2-16y+16=10$
$y^2(y-1)-16(y-1)=0$
$(y^2-16)(y-1)=0$
$(y-4)(y+4)(y-1)=0$
$y-4=0$ or $y+4=0$ or $y-1=0$
$y=4$ or $y=-4$ or $y=1$
$\boxed{\{-4,4,1\}}$

77. $2(y-4)^2 + y^2 = y(50+y) - 46y$
$2y^2 - 16y + 32 + y^2 = 50y + y^2 - 46y$
$2y^2 - 20y + 32 = 0$
$y^2 - 10y + 16 = 0$
$(y-2)(y-8) = 0$
$y-2 = 0$ or $y-8 = 0$
$y = 2$ or $y = 8$
Check: $y = 2$: $2(2-4)^2 + 2^2 = 2(50+2) - 46(2)$
$\quad\quad\quad\quad 12 = 12$ True
$\quad\quad\quad y = 8$: $2(8-4)^2 + 8^2 = 8(50+8) - 46(8)$
$\quad\quad\quad\quad 96 = 96$ True
$\boxed{\{2,8\}}$

79. $d = t^2 + 2t$
$d = 8$ meters
$8 = t^2 + 2t$
$t^2 + 2t - 8 = 0$
$(t+4)(t-2) = 0$
$t+4 = 0$ or $t-2 = 0$
$t = 2$ sec   (Reject $t = -4$)

The time it takes a shock wave to travel 8 miles from an explosion site is $\boxed{2 \text{ seconds}}$.

81. $C = 600 + 1000x - 100x^2$
$C = \$2200$
$2200 = 600 + 1000x - 100x^2$
$100x^2 - 1000x + 1600 = 0$
$x^2 - 10x + 16 = 0$
$(x-2)(x-8) = 0$; $x = 2$ or $x = 8$
$x-2 = 0$ or $x-8 = 0$

Either $\boxed{200 \text{ or } 800 \text{ calculators}}$ can be manufactured.

Note: The total cost *drops* for $x > 5$.

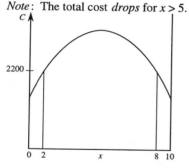

83. $P = -\dfrac{1}{50}A^2 + 2A + 22$

$72 = -\dfrac{1}{50}A^2 + 2A + 22$

$\dfrac{1}{50}A^2 - 2A + 50 = 0$

$A^2 - 100A + 2500 = 0$
$(A-50)^2 = 0$
$A - 50 = 0$
$A = 50$

The arousal level should be $\boxed{50}$.

85. $h = -16t^2 + 128t$

    a. $h = 0 = -16t^2 + 128t$

        $t^2 - 8t = 0$

        $t(t - 8) = 0$

        $t = 0$ or $t - 8 = 0$

        $t = \boxed{8 \text{ seconds}}$ (reject $t = 0$)

b.

Time t	Height $h = -16t^2 + 128t$
0	0
1	112
2	192
3	240
4	256
5	240
6	192
7	112
8	0

    c.

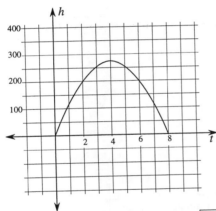

    d. The maximum height is reached at $t = \boxed{4 \text{ seconds}}$, and the height is $\boxed{256 \text{ ft}}$.

    e. The height is 0 at $t = 8$ seconds when the projectile is dropping.

        $\boxed{(8, 0) \text{ is a point on the graph}}$

87. Statement $\boxed{D}$ is true.

    $x(x + \pi) = 0$

    $x = 0$ or $x + \pi = 0$

    $x = 0$ or $x = -\pi$

    0 and $-\pi$ are solutions to the equation $x(x + \pi) = 0$

89. Statement $\boxed{\text{D}}$ is true.

    A. $2x(x^2 + 25) = 0$
        $2x = 0 \text{ or } x^2 + 25 = 0$
        $x = 0$ only

    B. If $y = -4$
        $7(16) + (2k - 5) - 20$
        $2k - 5 = 20 - 112$
        $2k = -92 + 5$
        $k = \dfrac{-87}{2} \neq 14$

    C. Check:
        $x = 1$: $(1 + 6)(2 - 4) = 6$
                   $7(-2) = 6$
                 $-14 = 6$ False
        $x = 5$: $10(6) = 6$
                 $60 = 6$ False

    D. None of the above is true.

91. $(x^2 - 5x + 5)^3 = 1$
    $x^2 - 5x + 5 = 1$
    since $1^3 = 1$
    $x^2 - 5x + 4 = 0$
    $(x - 1)(x - 4) = 0$
    $x - 4 = 0 \text{ or } x - 4 = 0$
    $x = 1 \text{ or } x = 4$
    $\boxed{\{1, 4\}}$

Review Problems

94.

Possible combinations

	5 nickels		3 dimes		2 quarters		$0.45 total
	number	amount	number	amount	number	amount	amount
1			2	0.20	1	0.25	0.45
2	2	0.10	1	0.10	1	0.25	0.45
3	4	0.20	0		1	0.25	0.45
	1	0.05	4	0.40	0		0.45
4	3	0.15	3	0.30			0.45
5	5	0.25	2	0.20			0.45
	7	0.35	1	0.10			0.45
	9	0.45	0				0.45

not possible; only 3 dimes

not possible; only 5 nickels
not possible; only 5 nickels

$\boxed{5}$ possible combinations

The possible combinations are 5 nickels and 2 dimes; 1 quarter, 1 dime, and 2 nickels; 1 quarter and 4 nickels; 3 nickels and 3 dimes; 1 quarter and 2 dimes.

95. Let $x =$ the measure of the third angle.
    $4x - 2 =$ the measure of the second angle
       $38 + 4x - 2 + x \;=\; 180$
               $5x \;=\; 144$
                $x \;=\; 28.8°$
        $4x - 2 \;=\; 4(28.8) - 2 = 113.2°$

The measure of the $\boxed{\text{second angle is } 113.2°}$ and the measure of the $\boxed{\text{third angle is } 28.80°}$.

96.

$$
\begin{array}{r}
y^2 - 2y + 5 \\
3y - 5 \overline{\smash{\big)}\ 3y^3 - 11y^2 + 25y - 25} \\
\underline{3y^3 - 5y^2} \\
-6y^2 + 25y \\
\underline{-6y^2 + 10y} \\
15y - 25 \\
\underline{15y - 25} \\
0
\end{array}
$$

$(3y^3 - 11y^2 + 25y - 25) \div (3y - 5) = \boxed{y^2 - 2y + 5}$

Section 7.7 Problem Solving

Problem Set 7.7, pp. 499-502

1. Let $x =$ the unknown number.
$x(x - 5) = 14$
$x^2 - 5x - 14 = 0$
$(x - 7)(x + 2) = 0$
$x - 7 = 0$ or $x + 2 = 0$
$x = 7$ or $x = -2$
The number is $\boxed{7 \text{ or } -2}$.

3. Let $x =$ the unknown number.
$(x + 1)(x - 1) = 8$
$x^2 - 1 = 8$
$x^2 - 9 = 0$
$(x - 3)(x + 3) = 0$
$x = 3$ or $x = -3$
$x - 3 = 0$ or $x + 3 = 0$
The number is $\boxed{3 \text{ or } -3}$.

5. Let $x =$ the unknown number.
$x^2 + 15 = 8x$
$x^2 - 8x + 15 = 0$
$(x - 3)(x - 5) = 0$
$x - 3 = 0$ or $x - 5 = 0$
$x = 3$ or $x = 5$
The number is $\boxed{3 \text{ or } 5}$.

7. Let $x =$ smaller integer.
$x + 1 =$ next consecutive integer
$x(x + 1) = 56$
$x^2 + x - 56 = 0$
$(x + 8)(x - 7) = 0$
$x + 8 = 0$ or $x - 7 = 0$
$x = 7$ or $x = -8$
$x + 1 = 8$ or $x + 1 = -7$
The integers are $\boxed{7 \text{ and } 8 \text{ or } -8 \text{ and } -7}$.

9. Let x = the smaller odd integer.

$x + 2$ = next consecutive odd integer

$$
\begin{aligned}
x(x + 2) &= 99 \\
x^2 + 2x - 99 &= 0 \\
(x + 11)(x - 9) &= 0 \\
x &= 9 \text{ or } x = -11 \\
x + 2 &= 11 \text{ or } x + 2 = -9
\end{aligned}
$$

The integers are $\boxed{9 \text{ and } 11}$ or $\boxed{-11 \text{ and } -9}$.

11. Let x = the even integer.

$x + 2$ = next consecutive even integer

$$
\begin{aligned}
x^2 + (x + 2)^2 &= 52 \\
x^2 + x^2 + 4x + 4 &= 52 \\
2x^2 + 4x - 48 &= 0 \\
x^2 + 2x - 24 &= 0 \\
(x + 6)(x - 4) &= 0 \\
x &= 4 \text{ or } x = -6 \\
x + 2 &= 6 \text{ or } x + 2 = -4
\end{aligned}
$$

The integers are $\boxed{4 \text{ and } 6}$ or $\boxed{-6 \text{ and } -4}$.

13. Let x = one number.

$13 - x$ = the other number

$$
\begin{aligned}
x(13 - x) &= 40 \\
-x^2 + 13x - 40 &= 0 \\
x^2 - 13x + 40 &= 0 \\
(x - 5)(x - 8) &= 0 \\
x &= 5 \text{ or } x = 8 \\
13 - x &= 8 \text{ or } 5
\end{aligned}
$$

The numbers are $\boxed{5 \text{ and } 8}$.

15. Let x = one number.

$9 - x$ = the other number

$$
\begin{aligned}
x^2 + (9 - x)^2 &= 41 \\
x^2 + 81 - 18x + x^2 &= 41 \\
2x^2 - 18x + 40 &= 0 \\
x^2 - 9x + 20 &= 0 \\
(x - 4)(x - 5) &= 0 \\
x &= 4 \text{ or } x = 5 \\
9 - x &= 5 \text{ or } 4
\end{aligned}
$$

The numbers are $\boxed{4 \text{ and } 5}$.

17. Let x = the smaller face value.

$x + 3$ = face value on other die

$$
\begin{aligned}
x(x + 3) &= 2(x + x + 3) \\
x^2 + 3x &= 4x + 6 \\
x^2 - x - 6 &= 0 \\
(x - 3)(x + 2) &= 0 \\
x - 3 &= 0 \text{ or } x + 2 = 0 \\
x &= 3 \quad \text{(reject } x = -2 \text{ since face value cannot be negative)} \\
x + 3 &= 6
\end{aligned}
$$

The numbers are $\boxed{3 \text{ and } 6}$.

19. Let x = the unknown number.

$$
\begin{aligned}
x^3 &= 4x \\
x^3 - 4x &= 0 \\
x(x^2 - 4) &= 0 \\
x(x - 2)(x + 2) &= 0 \\
x &= 0 \text{ or } x - 2 = 0 \text{ or } x + 2 = 0 \\
x &= 0 \text{ or } x = 2 \text{ or } x = -2
\end{aligned}
$$

The numbers are $\boxed{0, 2 \text{ or } -2}$

21. Let $x =$ the one number.
$x - 5 = 0$ the other number.

$$
\begin{aligned}
x(x-5) &= 84 \\
x^2 - 5x - 84 &= 0 \\
(x-12)(x+7) &= 0 \\
x &= 12 \text{ or } x = -7 \\
x - 5 &= 0 \text{ or } -2
\end{aligned}
$$

The numbers are $\boxed{12 \text{ and } 7}$ or $\boxed{-7 \text{ and } -12}$.

23. Let $x =$ the width of the rectangle.
$3x + 1 =$ the length of the rectangle
width $\times$ length $=$ area

$$
\begin{aligned}
x(3x+1) &= 52 \\
3x^2 + x - 52 &= 0 \\
(3x+13)(x-4) &= 0 \\
3x+13 &= 0 \text{ or } x - 4 = 0 \\
x &= 4 \quad \left(\text{reject } x = -\frac{13}{3} \text{ since width cannot be negative} \right)
\end{aligned}
$$

length $= 3x + 1 = 13$
The width is $\boxed{4 \text{ meters}}$ and the length is $\boxed{13 \text{ meters}}$.

25. Let $x =$ base of the triangle.
$x + 7 =$ height of the triangle

$$
A = \frac{1}{2}bh = \frac{1}{2}x(x+7)
$$

$$
\begin{aligned}
30 &= \frac{1}{2}(x^2 + 7x) \\
x^2 + 7x &= 60 \\
x^2 + 7x - 60 &= 0 \\
(x+12)(x-5) &= 0 \\
x &= 5 \quad (\text{reject } x = -12 \text{ since length cannot be negative}) \\
x+7 &= 5 + 7 = 12
\end{aligned}
$$

The height of the triangle is $\boxed{12 \text{ yards}}$ and the base is $\boxed{5 \text{ yards}}$.

27. Let $x =$ length of a side of the square.
perimeter $=$ area $+ 3$

$$
\begin{aligned}
4x &= x^2 + 3 \\
x^2 - 4x + 3 &= 0 \\
(x-3)(x-1) &= 0 \\
x-3 &= 0 \text{ or } x - 1 = 0 \\
x &= 1 \text{ or } x = 3
\end{aligned}
$$

The length of a side of the square can be $\boxed{1 \text{ or } 3}$.

29. Let x = the width of the rectangle.

$2x + 2$ = length of the rectangle

area = perimeter + 26

$$
\begin{aligned}
x(2x + 2) &= [2x + 2(2x + 2)] + 26 \\
2x^2 + 2x &= 2x + 4x + 4 + 26 \\
2x^2 + 2x &= 6x + 30 \\
2x^2 - 4x - 30 &= 0 \\
x^2 - 2x - 15 &= 0 \\
(x - 5)(x + 3) &= 0 \\
x - 5 &= 0 \text{ or } x + 3 = 0 \\
x &= 5 \quad (\text{reject } x = -3) \\
2x + 2 &= 12
\end{aligned}
$$

The width of the rectangle is $\boxed{5 \text{ meters}}$ and the length is $\boxed{12 \text{ meters}}$.

31. Let x = the length of the side of the original square.

$x + 3$ = length of side of new square

$$
\begin{aligned}
(x + 3)^2 &= 81 \\
x^2 + 6x + 9 &= 81 \\
x^2 + 6x - 72 &= 0 \\
(x + 12)(x - 6) &= 0 \\
x &= 6 \quad (\text{reject } x = -12)
\end{aligned}
$$

The length of the side of the original square is $\boxed{6 \text{ yards}}$.

33. Let x = the length of the longer leg of the right triangle.

$x + 1$ = length

$$
\begin{aligned}
3^2 + x^2 &= (x + 1)^2 \\
9 + x^2 &= x^2 + 2x + 1 \\
2x &= 8 \\
x &= 4
\end{aligned}
$$

The length of the longer leg is $\boxed{4 \text{ meters}}$.

35. Let x = the length of the shorter leg of the right triangle.

$x + 7$ = length of longer leg

$(x + 7) + 1 = x + 8$ = length of hypotenuse

$$
\begin{aligned}
x^2 + (x + 7)^2 &= (x + 8)^2 \\
x^2 + x^2 + 14x + 49 &= x^2 + 16x + 64 \\
x^2 - 2x - 15 &= 0 \\
(x - 5)(x + 3) &= 0 \\
x - 5 &= 0 \text{ or } x + 3 = 0 \\
x &= 5 \\
x + 7 &= 12 \\
x + 8 &= 13
\end{aligned}
$$

$\boxed{\text{shorter leg, 5 yards; longer leg, 12 yards; hypotenuse, 13 yards}}$

37. Let x = the width of the rectangle.

$x + 3$ = length of the rectangle

$x + 2$ = width of new rectangle

$x + 3 + 2 = x + 5$ = length of new rectangle

$$
\begin{aligned}
(x + 2)(x + 5) &= 54 \\
x^2 + 7x + 10 &= 54 \\
x^2 + 7x - 44 &= 0 \\
(x + 11)(x - 4) &= 0 \\
x &= 4 \quad (\text{reject } x = -11) \\
x + 3 &= 4 + 3 = 7
\end{aligned}
$$

dimensions of original rectangle: $\boxed{\text{width, 4 feet; length, 7 feet}}$

39. Let x = the length of the side of the smaller square room.

$x + 3$ = length of side of large square room

$$
\begin{aligned}
x^2 + (x+3)^2 &= 65 \\
x^2 + x^2 + 6x + 9 &= 65 \\
2x^2 + 6x - 56 &= 0 \\
x^2 + 3x - 28 &= 0 \\
(x+7)(x-4) &= 0 \\
x &= 4 \quad \text{(reject } x = -7) \\
x + 3 &= 0 \text{ or } x + 3 = 7
\end{aligned}
$$

dimensions of each room: length of side of smaller square, 4 yards; length of side of larger square, 7 yards

41. Let $2x - 1$ = unknown number of paces.

$$
\begin{aligned}
(x+1)^2 + x^2 &= (2x-1)^2 = y^2 \\
x^2 + 2x + 1 + x^2 &= 4x^2 - 4x + 1 \\
2x^2 - 6x &= 0 \\
x^2 - 3x &= 0 \\
x(x - 3) &= 0 \\
x &= 0 \text{ or } x = 3 \quad \text{(reject } x = 0) \\
2x - 1 &= 2(3) - 1 = 5
\end{aligned}
$$

Al Capone's treasure is buried   5 paces   from Video Rock.

43. Let $x =$ the measure of the smaller angle.

$90 - x$ = measure of complementary angle

$$
\begin{aligned}
x(90 - x) &= 2000 \\
90x - x^2 &= 2000 \\
x^2 - 90x + 2000 &= 0 \\
(x - 40)(x - 50) &= 0 \\
x &= 40° \text{ or } x = 50°
\end{aligned}
$$

The measure of the smaller angle is   40° .

45. Let x = width.

$x + 1$ = length

$$
\begin{aligned}
\text{Volume} &= lwh = 24 \\
(x+1)(x)(2) &= 24 \\
2x^2 + 2x &= 24 \\
x^2 + x &= 12 \\
x^2 + x - 12 &= 0 \\
(x+4)(x-3) &= 0 \\
x &= 3 \quad \text{(reject } x = -4)
\end{aligned}
$$

dimensions of base: width, 3 meters; length, 4 meters

47. Let x = radius of the smaller circle.

$x + 2$ = radius of larger circle

$$
\begin{aligned}
\pi(x+2)^2 - \pi x^2 &= 20\pi \\
x^2 + 4x + 4 - x^2 &= 20 \\
4x + 4 &= 20 \\
4x &= 16 \\
x &= 4 \\
x + 2 &= 4 + 2 = 6
\end{aligned}
$$

radius of smaller circle, 4 meters; radius of lager circle, 6 meters

49. Let h = height of the trapezoid.

$h + 3$ = shorter base of the trapezoid

$2h + 2$ = larger base of the trapezoid

$$\frac{1}{2}h(b_1 + b_2) \;=\; 50$$

$$\frac{1}{2}h(h + 3 + 2h + 2)$$

$$\begin{aligned}
\frac{h(3h + 5)}{2} &= 50 \\
3h^2 + 5h &= 100 \\
3h^2 + 5h - 100 &= 0 \\
(3h + 20)(h - 5) &= 0 \\
h &= 5 \quad \left(\text{reject } h = -\frac{20}{3}\text{ feet}\right) \\
h + 3 &= 5 + 3 = 8 \\
2h + 2 &= 2(5) + 2 = 12
\end{aligned}$$

$\boxed{\text{height, 5 feet; shorter base, 8 feet; larger base, 12 feet}}$

51.
$$\begin{aligned}
(12 + 2x)(15 + 2x) &= 378 = \text{total area} \\
180 + 54x + 4x^2 &= 378 \\
4x^2 + 54x - 198 &= 0 \\
2x^2 + 27x - 99 &= 0 \\
(2x + 33)(x - 3) &= 0 \\
x &= 3 \quad \left(\text{reject } x = -\frac{33}{2}\right)
\end{aligned}$$

$\boxed{\text{width of path, 3 meters}}$

53. Let x = the current age.

$x - 2$ = age 2 years ago

$$\begin{aligned}
4x^2 - 25(x - 2) &= 25 \\
4x^2 - 25x + 50 &= 0 \\
4x^2 - 25x + 25 &= 0 \\
(4x - 5)(x - 5) &= 0 \\
4x - 5 &= 0 \text{ or } x - 5 = 0 \\
x &= 5 \text{ or } x = \frac{5}{4}\text{ years}
\end{aligned}$$

Reject $\frac{5}{4}$, since Stella's age 2 years ago becomes meaningless.

Stella's age now: $\boxed{5 \text{ years old}}$

55. Let x = Stanley's present age.

$(x - 1)^2$ = Stanley's aunt's present age

$x + 2$ = Stanley's age 2 years from now

$$\begin{aligned}
(x - 1)^2 + 2 &= 6(x + 2) \\
x^2 - 2x + 1 + 2 &= 6x + 12 \\
x^2 - 8x - 9 &= 0 \\
(x - 9)(x + 1) &= 0 \\
x - 9 &= 0 \text{ or } x + 1 = 0 \\
x &= 9 \quad (\text{reject } x = -1)
\end{aligned}$$

Aunt's age $(x - 1)^2 = (9 - 1)^2 = 64$

$\boxed{\text{Stanley's age, 9 years old; Aunt's age, 64 years old}}$

57. Let x = the number of apples in each row.

$x - 3$ = number of rows

$$
\begin{aligned}
(x - 3)(x) &= 54 \\
x^2 - 3x - 54 &= 0 \\
(x - 9)(x + 6) &= 0 \\
x &= 9 \quad \text{(reject } x = -6) \\
x - 3 &= 9 - 3 = 6
\end{aligned}
$$

$\boxed{6 \text{ rows}}$

59. Let x = the shortest length.

Let $x + 2$ and $x + 4$ represent the remaining lengths of the sides of the right triangle.

$$
\begin{aligned}
x^2 + (x + 2)^2 &= (x + 4)^2 \\
x^2 + x^2 + 4x + 4 &= x^2 + 8x + 16 \\
x^2 - 4x - 12 &= 0 \\
(x - 6)(x + 2) &= 0 \\
x &= 6 \quad \text{(reject } x = -2)
\end{aligned}
$$

$\boxed{\text{Lengths of the sides are 6, 8, and 10}}$.

61.   demand = supply

$$
\begin{aligned}
100000 - 750x &= 5x^2 + 50000 \\
5x^2 + 750x - 50000 &= 0 \\
x^2 + 150x - 10000 &= 0 \\
(x + 200)(x - 50) &= 0 \\
x &= 50 \text{ metric tons} \quad \text{(reject } -200 \text{ tons)}
\end{aligned}
$$

Price = $150 + x =$ $\boxed{200\cancel{c}\text{ per bushel, or \$2.00 per bushel}}$

Review Problems

62. $y > \dfrac{2}{3}x + 1$

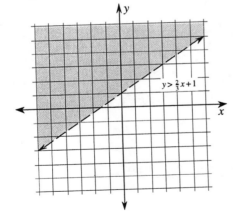

63. $\left(\dfrac{8x^4y^{-2}}{4x^7}\right)^{-3} = (2x^{4-7}y^{-2})^{-3} = (2x^{-3}y^{-2})^{-3} = 2^{-3}x^9y^6 = \boxed{\dfrac{x^9y^6}{8}}$

64. $\dfrac{8.4 \times 10^{-8}}{3.5 \times 10^{-6}} = \dfrac{8.4 \times 10^{-2}}{3.5} = \dfrac{0.084}{3.5} = \boxed{0.024}$

Chapter 7 Review Problems

Chapter 7 Review, pp. 504-506

1. $9y^2 - 18y = \boxed{9y(y-2)}$

2. $x^2 - 11x + 28 = \boxed{(x-4)(x-7)}$

3. $y^3 - 8y^2 + 7y = \boxed{y(y-7)(y-1)}$

4. $10r^2 + 9r + 2 = \boxed{(5r+2)(2r+1)}$

5. $15z^2 - z - 2 = \boxed{(5z-2)(3z+1)}$

6. $x^2 - 144 = \boxed{(x+12)(x-12)}$

7. $64 - y^2 = \boxed{(8-y)(8+y)}$

8. $9r^2 + 6r + 1 = \boxed{(3r+1)^2}$

9. $20a^7b^2 - 36a^3b^4 = \boxed{4a^3b^2(5a^4 - 9b^2)}$

10. $5(a-b) - c(a-b) = \boxed{(5-c)(a-b)}$

11. $4x^2 - 20x + 2xy - 10y = 4x(x-5) + 2y(x-5) = (4x+2y)(x-5) = \boxed{2(2x+y)(x-5)}$

12. $12y^2 + 11y - 5 = \boxed{(4y+5)(3y-1)}$

13. $16x^2 - 40x + 25 = \boxed{(4x-5)^2}$

14. $r^2 + 16$ is $\boxed{\text{prime}}$

15. $2x^3 + 19x^2 + 35x = x(2x^2 + 19x + 35) = \boxed{x(2x+5)(x+7)}$

16. $2x^4y - 2x^2y = 2x^2y(x^2-1) = \boxed{2x^2y(x-1)(x+1)}$

17. $10z^2 + 37z + 7 = \boxed{(5z+1)(2z+7)}$

18. $39a^2b - 52a + 13ab^4 = \boxed{13a(3ab - 4 + b^4)}$

19. $48 - 6a + 8b - ab = 6(8-a) + b(8-a) = \boxed{(6+b)(8-a)}$

20. $36y^2 - 59y - 7 = \boxed{(9y+1)(4y-7)}$

21. $12x^2 + 44xy + 35y^2 = \boxed{(2x+5y)(6x+7y)}$

22. $9r^2 + 8r - 3$ is $\boxed{\text{prime}}$

23. $10x^5y^4 - 44x^4y^5 + 16x^3y^6 = 2x^3y^4(5x^2 - 22xy + 8y^2) = \boxed{2x^3y^4(5x-2y)(x-4y)}$

24. $40x^2 + 17xy - 12y^2 = \boxed{(5x+4)(8x-3y)}$

25. $486z^2 - 24 = 6(81z^2 - 4) = \boxed{6(9z-2)(9z+2)}$

26. $48r^2 - 120r + 75 = 3(16r^2 - 40r + 25) = \boxed{3(4r-5)^2}$

27. $3y^4 - 9y^3 - 30y^2 = 3y^2(y^2 - 3y - 10) = \boxed{3y^2(y-5)(y+2)}$

28. $100y^2 - 49z^2 = \boxed{(10y - 7z)(10y + 7z)}$

29. $256x^4 - 1 = (16x^2 - 1)(16x^2 + 1) = \boxed{(4x - 1)(4x + 1)(16x^2 + 1)}$

30. $9x^5y^2 - 18x^4y^5 = \boxed{9x^4y^2(x - 2y^3)}$

31. $3w^2 + w - 5$ is $\boxed{\text{prime}}$

32. $64y^2 - 144y + 81 = \boxed{(8y - 9)^2}$

33. $x^2 + xy + y^2$ is $\boxed{\text{prime}}$

34. $a^2q + a^2z - p^2q - p^2z = a^2(q + z) - p^2(q + z) = (a^2 - p^2)(q + z) = \boxed{(a - p)(a + p)(q + z)}$

35. $y^3 - 8 = \boxed{(y - 2)(y^2 + 2y + 4)}$

36. $x^3 + 64 = \boxed{(x + 4)(x^2 - 4x + 16)}$

37. $4y^2 + 4y + 1 - 36x^2 = (2y + 1)^2 - 36x^2 = \boxed{(2y + 1 - 6x)(2y + 1 + 6x)}$

38. $x^2y^2 - 16x^2 - 4y^2 + 64 = x^2(y^2 - 16) - 4(y^2 - 16) = (x^2 - 4)(y^2 - 16) = \boxed{(x - 2)(x + 2)(y - 4)(y + 4)}$

39. $3x^4y^2 - 12x^2y^4 = 3x^2y^2(x^2 - 4y^2) = \boxed{3x^2y^2(x - 2y)(x + 2y)}$

40. $3r^4 + 12r^2 = \boxed{3r^2(r^2 + 4)}$

41. $56y^3 - 70y^2 + 21y = 7y(8y^2 - 10y + 3) = \boxed{7y(2y - 1)(4y - 3)}$

42. $a^2 + 4ab + 16b^2$ is $\boxed{\text{prime}}$

43. $s^2 - s - 90 = \boxed{(s - 10)(s + 9)}$

44. $125x^3 - 8y^3 = \boxed{(5x - 2y)(25x^2 + 10xy + 4y^2)}$

45. $8y^2 - 14y - 5$ is $\boxed{\text{prime}}$

46. $25x^2 + 25xy + 6y^2 = \boxed{(5x + 2y)(5x + 3y)}$

47. $p^4 + 125p = p(p^3 + 125) = \boxed{p(p + 5)(p^2 - 5p + 25)}$

48. $y^2 - 24y + 144 - 100a^2 = (y - 12)^2 - 100a^2 = \boxed{(y - 12 - 10a)(y - 12 + 10a)}$

49. $64x^2 + 48xy + 9y^2 = \boxed{(8x + 3y)^2}$

50. $6x^4 + 3x^3y^2 - 2xy^3 - y^5 = 3x^3(2x + y^2) - y^3(2x + y^2) = \boxed{(3x^3 - y^3)(2x + y^2)}$

51.
$$\begin{aligned} y^2 + 5y &= 14 \\ y^2 + 5y - 14 &= 0 \\ (y+7)(y-2) &= 0 \\ y &= -7 \text{ or } y = 2 \end{aligned}$$
$\boxed{\{-7, 2\}}$

52.
$$\begin{aligned} x(x-4) &= 32 \\ x^2 - 4x - 32 &= 0 \\ (x-8)(x+4) &= 0 \\ x &= 8 \text{ or } x = -4 \end{aligned}$$
$\boxed{\{-4, 8\}}$

53. $8w^2 - 37w + 20 = 0$
$(8w - 5))(w - 4) = 0$
$w = \dfrac{5}{8}$ or $w = 4$
$\boxed{\left\{\dfrac{5}{8}, 4\right\}}$

54.
$$\begin{aligned} (4x+7)(x-2) &= x(x-42) \\ 4x^2 - x - 14 &= x^2 - 42x \\ 3x^2 + 41x - 14 &= 0 \\ (3x-1)(x+14) &= 0 \\ x &= \frac{1}{3} \text{ or } x = -14 \end{aligned}$$
$\boxed{\left\{-14, \dfrac{1}{3}\right\}}$

55.
$$\begin{aligned} 6w^3 &= 17w^2 - 12w \\ w(6w^2 - 17w + 12) &= 0 \\ w(3w - 4)(2w - 3) &= 0 \\ w &= 0 \text{ or } w = \frac{4}{3} \text{ or } w = \frac{3}{2} \end{aligned}$$
$\boxed{\left\{0, \dfrac{4}{3}, \dfrac{3}{2}\right\}}$

56.
$$\begin{aligned} (x+6)(81x^2 - 90x + 25) &= 0 \\ (x+6)(9x-5)^2 &= 0 \\ x &= -6 \text{ or } x = \frac{5}{9} \end{aligned}$$
$\boxed{\left\{-6, \dfrac{5}{9}\right\}}$

57. Let x = the unknown number.
$$\begin{aligned} 3x(x+2) &= 9 \\ 3x^2 + 6x - 9 &= 0 \\ x^2 + 2x - 3 &= 0 \\ (x+3)(x-1) &= 0 \\ x &= -3 \text{ or } x = 1 \end{aligned}$$
The number is $\boxed{-3 \text{ or } 1}$.

58. Let x = the smaller number.
$x + 1$ = next consecutive integer
$$\begin{aligned} 3(x+1)^2 + 2(x^2) &= 35 \\ 3x^2 + 6x + 3 + 2x^2 &= 35 \\ 5x^2 + 6x - 32 &= 0 \\ (x-2)(5x+16) &= 0 \\ x &= 2 \quad \left(\text{reject } x = -\frac{16}{5} \text{ since } -\frac{16}{5} \text{ is not an integer}\right) \\ x + 1 &= 3 \end{aligned}$$
The integers are $\boxed{2 \text{ and } 3}$.

59. Let x = the smaller odd integer.
$x + 2$ = next consecutive integer
$$
\begin{aligned}
x(x + 2) &= 63 \\
x^2 + 2x - 63 &= 0 \\
(x + 9)(x - 7) &= 0 \\
x &= -9 \text{ or } x = 7 \\
x + 2 &= -7 \text{ or } x + 2 = 9
\end{aligned}
$$
The integers are $\boxed{-9 \text{ and } -7 \text{ or } 7 \text{ and } 9}$.

60. Let x and $21 - x$ be the numbers.
$$
\begin{aligned}
xy &= 108 \\
x(21 - x) &= 108 \\
-x^2 + 21x - 108 &= 0 \\
x^2 - 21x + 108 &= 0 \\
(x - 9)(x - 12) &= 0 \\
x &= 9 \text{ or } x = 12 \\
21 - x &= 12 \text{ or } 9
\end{aligned}
$$
The numbers are $\boxed{9 \text{ and } 12}$.

61. Let x = the number.
$$
\begin{aligned}
x^2 &= 48 - 2x \\
x^2 + 2x - 48 &= 0 \\
(x + 8)(x - 6) &= 0 \\
x &= -8 \text{ or } x = 6
\end{aligned}
$$
The number is $\boxed{-8 \text{ or } 6}$.

62. Let x = the unknown width of the rectangle.
$2x + 3$ = length of rectangle
$$
\begin{aligned}
x(2x + 3) &= 65 \\
2x^2 + 3x - 65 &= 0 \\
(2x + 13)(x - 5) &= 0 \\
x &= 5 \quad \left(\text{reject } x = -\frac{13}{2}\right)
\end{aligned}
$$
$2x + 3 = 2(5) + 3 = 13$
$\boxed{\text{width, 5 yards; length, 13 yards}}$

63. Let x = the length of the side of the smaller original square.
$x + 3$ = length of side of larger square
$$
\begin{aligned}
(x + 3)^2 &= 2x^2 - 7 \\
x^2 + 6x + 9 &= 2x^2 - 7 \\
x^2 - 6x - 16 &= 0 \\
(x - 8)(x + 2) &= 0 \\
x &= 8 \quad (\text{reject } x = -2)
\end{aligned}
$$
The length of the original square: $\boxed{8 \text{ inches}}$.

64. Let x = the width of the rectangle.
$x + 8$ = length of the rectangle
area = perimeter $- 11$
$$
\begin{aligned}
x(x + 8) &= 2(x + x + 8) - 11 \\
x^2 + 8x &= 4x + 16 - 11 \\
x^2 + 4x - 5 &= 0 \\
(x + 5)(x - 1) &= 0 \\
x &= 1 \quad (\text{reject } x = -5)
\end{aligned}
$$
$x + 8 = 1 + 8 = 9$ cm
$\boxed{\text{length, 9 cm; width, 1 cm}}$

65. Let x = the length of the shorter leg.
$x + 9$ = length of hypotenuse
$x + 9 - 2 = 2x + 7$ = length of the longer leg
$$
\begin{aligned}
x^2 + (x + 7)^2 &= (x + 9)^2 \\
x^2 + x^2 + 14x + 49 &= x^2 + 18x + 81 \\
x^2 - 4x - 32 &= 0 \\
(x - 8)(x + 4) &= 0 \\
x &= 8 \quad (\text{reject } x = -4)
\end{aligned}
$$
$\boxed{\text{length of shorter leg, 8 yards}}$

66. Let x equal the distance the car has traveled.

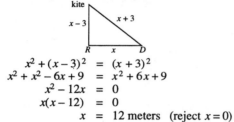

$$x^2 + (x+20)^2 = 100^2$$
$$x^2 + x^2 + 40x + 400 = 10000$$
$$2x^2 + 40x - 9600 = 0$$
$$x^2 + 20x - 4800 = 0$$
$$(x-60)(x+80) = 0$$
$$x = 60 \quad \text{(reject } x = -80)$$
$$x+20 = 60 + 20 = 80$$

distance car traveled, 60 miles; distance bus traveled, 80 miles

67. Let x = distance between Roseanne and Dan.
$x - 3$ = kite's height
$x + 3$ = length of kite's string

$$x^2 + (x-3)^2 = (x+3)^2$$
$$x^2 + x^2 - 6x + 9 = x^2 + 6x + 9$$
$$x^2 - 12x = 0$$
$$x(x-12) = 0$$
$$x = 12 \text{ meters} \quad \text{(reject } x = 0)$$

The length of the kite string, $x + 3 =$ 15 meters

68. $h = -16t^2 + 16t + 32$

   a. $h = 0$:
$$t^2 - t - 2 = 0$$
$$(t-2)(t+1) = 0$$
$$t = \boxed{2 \text{ seconds}} \quad \text{(reject } t = -1)$$

   b.

Time t	Height $h = -16t^2 + 16t + 32$
0	32
$\frac{1}{2}$	36
1	32
$1\frac{1}{2}$	20
2	0

c.

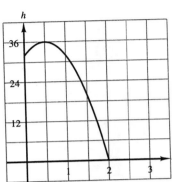

d. Maximum height of $\boxed{36 \text{ ft}}$ is reached at $t = \frac{1}{2}$ second.

e. $(2, 0)$ is a point on the graph

69.
$$
\begin{aligned}
C &= 20x^2 - 120x + 400 \\
1200 &= 20x^2 - 120x + 400 \\
20x^2 - 120x - 800 &= 0 \\
x^2 - 6x - 40 &= 0 \\
(x - 10)(x + 4) &= 0 \\
x &= 10 \text{ dozen pairs} \quad (\text{reject } x = -4)
\end{aligned}
$$

$\boxed{120 \text{ pairs were manufactured}}$

70. Let $x =$ length of shorter base.
$x + 2 =$ length of larger base
$x + 1 =$ length of height

$\text{Area} = \frac{1}{2} b(h_1 + h_2)$

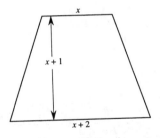

$$
\begin{aligned}
A &= \frac{1}{2}(x + 1)(x + x + 2) = 9 \\[4pt]
A &= \frac{1}{2}(x + 1)(2x + 2) = 9 \\[4pt]
x^2 + 2x + 1 &= 9 \\
x^2 + 2x - 8 &= 0 \\
(x + 4)(x - 2) &= 0 \\
x &= 2 \\
x + 2 &= 4 \\
x + 1 &= 3
\end{aligned}
$$

$\boxed{\text{length of shorter base, 2 feet; length of longer base, 4 feet; length of height, 3feet}}$

71. Let $x =$ base of triangle.
$2x + 1 =$ height of triangle

$\text{Area} = \frac{1}{2}bh = \frac{1}{2}x(2x + 1)$

$$
\begin{aligned}
x^2 + \frac{x}{2} &= 39 \\
2x^2 + x - 78 &= 0 \\
(2x + 13)(x - 6) &= 0 \\
x &= 6 \quad \left(\text{reject } x = -\frac{13}{2}\right) \\
2x + 1 &= 13
\end{aligned}
$$

$\boxed{\text{base, 6 meters; height, 13 meters}}$

72. Let x = Maria's age today.
$$\begin{aligned}
(x+2)^2 - 15(x-2) &= 10 \\
x^2 + 4x + 4 - 15x + 30 &= 10 \\
x^2 - 11x + 24 &= 0 \\
(x-8)(x-3) &= 0 \\
x &= 3 \text{ or } x = 8 \text{ years}
\end{aligned}$$
Maria is $\boxed{3 \text{ years old or 8 years old}}$.

Cumulative Review Problems (Chapters 1-7)

Chapter 7 Cumulative Review, pp. 506-507

1.
  a. Natural numbers: $\{1, 9\}$.
  b. Whole numbers: $\{0, 1, 9\}$.
  c. Integers: $\{-3, -2, 0, 1, 9\}$.
  d. Rational numbers: $\left\{ -3, -2, \frac{1}{7}, 0, 1, 9, 11.3 \right\}$.
  e. Irrational numbers: $\left\{ \sqrt{7}, 8\pi \right\}$.
  f. Real numbers: $\left\{ -3, -2, \frac{1}{7}, 0, 1, 9, 11.3, \sqrt{7}, 8\pi \right\}$.

2. $6[5 + 2(3-8) - 3] = 6[5 + 2(-5) - 3] = 6[5 - 10 - 3] = 6(-8) = \boxed{-48}$

3.
$$\begin{aligned}
(2y-3)(y-1) &= (2y+1)(y-4) - 1 \\
2y^2 - 5y + 3 &= 2y^2 - 7y - 5 \\
2y &= -8 \\
y &= -4
\end{aligned}$$
$\boxed{\{-4\}}$

4.
$$\begin{aligned}
\frac{5}{6}(y-4) &= \frac{7}{18} + \frac{2}{9}(y+4) \quad \text{LCM} = 18 \\
15(y-4) &= 7 + 4(y+4) \\
15y - 60 &= 7 + 4y + 16 \\
11y - 83 &= 0 \\
y &= \frac{83}{11}
\end{aligned}$$
$\boxed{\left\{ \frac{83}{11} \right\}}$

5. Let x = the measure of the base angles.
$3x - 10$ = the measure of third angle
$$\begin{aligned}
2x + 3x - 10 &= 180 \\
5x &= 190 \\
x &= 38°
\end{aligned}$$
$$3x - 10 = 3(38) - 10 = 104$$
The measures of the angles are $\boxed{30°, 38° \text{ and } 104°}$.

6.

$$5x + 6y \; > \; -30$$
$$6y \; > \; -5x - 30$$
$$y \; > \; -\frac{5}{6}x - 5$$

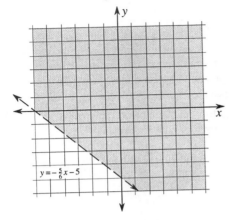

$y = -\frac{5}{6}x - 5$

7.

$$5x + 2y \; = \; 14$$
$$y \; = \; 2x - 11$$

(substitute for y)

$$5x + 2(2x - 11) \; = \; 14$$
$$5x + 4x - 22 \; = \; 14$$
$$9x \; = \; 36$$
$$x \; = \; 4$$
$$y \; = \; 2x - 11 = 2(4) - 11 = -3$$

$\boxed{\{(4, -3)\}}$

8. Let x = cost of pen.

y = cost of pad

$$
\begin{array}{llll}
4x + 7y \; = \; 6.40 & (\times -2) \to & -8x - 14y \; = \; -12.80 \\
19x + 2y \; = \; 5.40 & (\times 7) \to & \underline{133x + 14y \; = \; \quad 37.80} \\
& & 125x \; = \; 25.00
\end{array}
$$

$$x \; = \; 0.20$$
$$y = \frac{6.40 - 4x}{7} = \frac{6.40 - 0.80}{7} = 0.80$$

$\boxed{\text{cost of pen, \$0.20; cost of pad, \$0.80}}$

9. $\dfrac{6x^5 - 3x^4 + 9x^2 + 27}{-3x} = \boxed{-2x^4 + x^3 - 3x - \dfrac{9}{x}}$ $(x \neq 0)$

10. $\left(\dfrac{4y^{-1}}{2y^{-3}}\right)^3 = 8y^{-3-(-9)} = \boxed{8y^6}$

11. Let x = the length of the side of the painting.

$$x + 1 + 1 = x + 2 = \text{length of frame}$$
$$(x + 2)^2 - x^2 \; = \; 28$$
$$x^2 + 4x + 4 - x^2 \; = \; 28$$
$$4x \; = \; 24$$
$$x \; = \; 6$$

dimensions of painting: $\boxed{6 \text{ inches} \times 6 \text{ inches}}$

12.
$$y(5y + 17) = 12$$
$$5y^2 + 17y - 12 = 0$$
$$(5y - 3)(y + 4) = 0$$
$$y = \frac{3}{5} \text{ or } y = -4$$

$$\boxed{\left\{ -4, \frac{3}{5} \right\}}$$

13.
$$5(3y - 4) - 10y \leq 4(2y - 1) - 16$$
$$15y - 20 - 10y \leq 8y - 4 - 16$$
$$-3y \leq 0$$
$$y \geq 0$$

$$\boxed{\{y \mid y \geq 0\}}$$

14. Let x = hourly rate earned by Harpo.

$x + 0.60$ = hourly rate earned by Chico

$$4(x + 0.60) + 4x = 5(30)$$
$$8x + 2.40 = 150$$
$$8x = 147.60$$
$$x = \boxed{\$18.45 \text{ for Harpo}}$$
$$x + 0.60 = \boxed{\$19.05 \text{ for Chico}}$$

15.
$$\boxed{\text{gallons of white paint} = \frac{4}{5}(45) = 36}$$

$$\boxed{\text{gallons of black paint} = \frac{1}{5}(45) = 9}$$

16. $(2, 4), (3, 1)$

slope $= \dfrac{1 + 4}{3 - 2} = \dfrac{5}{1} = 5$

point-slope: $\boxed{y - 1 = 5(x - 3) \text{ or } y + 4 = 5(x - 2)}$

slope-intercept: $y - 1 = 5x - 15$

$$\boxed{y = 5x - 14}$$

standard: $\boxed{5x - y = 14}$

17.
$$f(x) = -(x - 2)^2$$
$$f(-1) = -((-1) - 2)^2 = -(-3)^2 = \boxed{-9}$$

18. $y < -\dfrac{2}{5}x + 2$

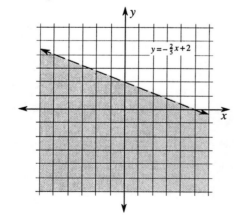

19.
$$\begin{aligned} 2x + 3y &= 5 & (\times 3) \rightarrow & & 6x + 9y &= 15 \\ 3x - 2y &= -4 & (\times -2) \rightarrow & & \underline{-6x - 4y} &= \underline{\ 8} \\ & & & & 13y &= 23 \end{aligned}$$

$$y = \frac{23}{13}$$

$$x = \frac{-4 + 2y}{3} = \frac{-4 + \frac{46}{13}}{3} = \frac{-52 + 46}{39} = -\frac{6}{39} = -\frac{2}{13}$$

$$\boxed{\left\{ \frac{-2}{13}, \frac{23}{13} \right\}}$$

20. $\dfrac{6x^3 + 5x^2 - 34x + 13}{3x - 5} = \boxed{2x^2 + 5x - 3 - \dfrac{2}{3x - 5}}$

$$\begin{array}{r} 2x^2 + 5x - 3 \\ 3x - 5\overline{\smash{\big)}\, 6x^3 + 5x^2 - 34x + 13} \\ \underline{6x^3 - 10x^2} \\ 15x^2 - 34x \\ \underline{15x^2 - 25x} \\ -9x + 13 \\ \underline{-9x + 15} \\ -2 \end{array}$$

quotient $= 2x^2 + 5x - 3$

remainder $= -\dfrac{2}{3x - 5}$

21. Let $x =$ the smaller number.
$x + 4 =$ larger number
$$\begin{aligned} (x + 4)^2 - x^2 &= 9x \\ x^2 + 8x + 16 - x^2 &= 9x \\ -x &= -16 \\ x &= 16 \\ x + 4 &= 20 \end{aligned}$$
The numbers are $\boxed{16 \text{ and } 20}$.

22. $\quad x\overline{\smash{\big)}\,\overset{\textstyle 1y}{133}}$

Since $7(19) = 133$, $\boxed{x = 7 \text{ and } y = 9}$.

23.
$$\begin{aligned} C &= 100 + 20x \\ R &= 24x \end{aligned}$$
At break–even point,
$$\begin{aligned} 100 + 20x &= 24x \\ 4x &= 100 \\ x &= \boxed{25 \text{ skateboards}} \end{aligned}$$

24.

$$b = \text{base of triangle}$$
$$h = 2b - 2 = \text{height of triangle}$$
$$A = \frac{1}{2}bh$$
$$30 = \frac{1}{2}b(2b - 2) = b(b - 1) = b^2 - b$$
$$b^2 - b - 30 = 0$$
$$(b - 6)(b + 5) = 0$$
$$b = 6 \quad (\text{reject } b = -5)$$
$$h = 12 - 2 = 10$$

base, 6 yards; height, 10 yards

25.
$$0.88y + 0.16(10 - y) = 0.40y$$
$$88y + 16(10 - y) = 40y$$
$$11y + 2(10 - y) = 5y$$
$$11y + 20 - 2y = 5y$$
$$4y = -20$$
$$y = -5$$

$\{-5\}$

26. Let x = the number of orchestra tickets.
y = number of mezzanine tickets.
$$x + y = 650$$
$$14x + 12y = 8390$$
$$y = 650 - x \quad (\text{substitute for } y)$$
$$14x + 12(650 - x) = 8390$$
$$2x = 8390 - 7800 = 590$$
$$x = 295$$
$$y = 650 - 295 = 355$$

295 orchestra tickets, 355 mezzanine tickets

27. Possible one–digit prime numbers are 2, 3, 5, 7.
Since $z > 3x$, x must be 2 and z must be 7. $y = 3$ or $y = 5$

28. Let x = the width of rectangle A.
$x + 2$ = width of rectangle B
$x + 4$ = length of rectangle B
$$x(3x) = (x + 2)(x + 4) \quad (\text{Areas are equal})$$
$$3x^2 = x^2 + 6x + 8$$
$$2x^2 - 6x - 8 = 0$$
$$x^2 - 3x - 4 = 0$$
$$(x - 4)(x + 1) = 0$$
$$x = 4 \quad (\text{reject } x = -1)$$
$$x + 2 = 6$$
$$x + 4 = 8$$

dimensions of rectangle A: $4 \text{ cm} \times 12 \text{ cm}$; dimensions of rectangle B: $6 \text{ cm} \times 8 \text{ cm}$

29. Let x = the liters of 60% acid solution.
y = liters of 30% (acid solution)
$$0.60x + 0.30y = 0.51(10) \quad \text{acid balance}$$
$$\underline{x + y = 10}$$
$$0.60x + 0.30(10 - x) = 5.1 \quad (\text{substitute for } y)$$
$$6x + 3(10 - x) = 51$$
$$6x + 30 - 3x = 51$$
$$3x = 21$$
$$x = 7$$
$$y = 3$$

7 liters of 60%; 3 liters of 30%

30. a. $3x^2 + 11x + 6 = \boxed{(x+3)(3x+2)}$

   b. $y^5 - 16y = y(y^4 - 16) = y(y^2 - 4)(y^2 + 4) = \boxed{y(y-2)(y+2)(y^2+4)}$

   c. $y^2 - 24y + 144 = \boxed{(y-12)^2}$

31. $7.9[18 - 2.6(-9.3)] = \boxed{333.22}$, rounded

32. $7.3(60 - 3.4^4) \div 5.2 = \boxed{-103.37}$, rounded

Chapter 8   Rational Expressions

Section 8.1 Rational Expressions and Their Simplification

Problem Set 8.1, pp. 523-526

1. $\dfrac{7}{2x}$

$2x = 0$

$x = 0$ Since 0 will make the denominator zero, the rational expression is undefined for $x = 0$.

Domain: $\boxed{\{x \mid x \neq 0\}}$

3. $\dfrac{x}{x-7}$

$x - 7 = 0$

$x = 7$

Domain: $\boxed{\{x \mid x \neq 7\}}$

5. $\dfrac{5y^2}{5y-15} = \dfrac{5y^2}{5(y-3)}$

$5(y-3) = 0$

$y = 3$

Domain: $\boxed{\{y \mid y \neq 3\}}$

7. $\dfrac{x+4}{(x+7)(x-3)}$

$(x+7)(x-3) = 0$

$x = -7$ or $x = 3$

Domain: $\boxed{\{x \mid x \neq -7 \text{ and } x \neq 3\}}$

9. $\dfrac{13z}{(3z-15)(z+2)} = \dfrac{13z}{3(z-5)(z+2)}$

$3(z-5)(z+2) = 0$

$z = 5$ or $z = -2$

Domain: $\boxed{\{z \mid z \neq -2 \text{ and } z = 5\}}$

11. $\dfrac{x+5}{x^2+x-12} = \dfrac{x+5}{(x+4)(x-3)}$

$(x+4)(x-3) = 0$

$x = -4$ or $x = 3$

Domain: $\boxed{\{x \mid x \neq -4 \text{ and } x \neq 3\}}$

13. $\dfrac{y+3}{4y^2+y-3} = \dfrac{y+3}{(4y-3)(y+1)}$

$(4y-3)(y+1) = 0$

$y = \dfrac{3}{4}$ or $y = -1$

Domain: $\boxed{\{y \mid y \neq -1 \text{ and } y \neq \dfrac{3}{4}\}}$

15. $\dfrac{7x}{x^2+4}$

$x^2+4 = 0$ No values of x for which the expression is undefined.

Domain: $\boxed{\{x \mid x \in R\}}$

17. $\dfrac{y^2-16}{8}$

$8 \neq 0$ No values of y for which the expression is undefined.

Domain: $\boxed{\{y \mid y \in R\}}$

19. $f(x) = \dfrac{130x}{100-x}$

   a. $f(40) = \dfrac{130(40)}{100-40} \approx \boxed{86.67}$ is $\boxed{\text{the cost in millions of dollars to inoculate 40\% of the population}}$

$f(80) = \dfrac{130(80)}{100-80} = \boxed{520}$ is $\boxed{\text{the cost to inoculate 80\% of the population}}$

$f(90) = \dfrac{130(90)}{100-90} = \boxed{1170}$ is $\boxed{\text{the cost to inoculate 90\% of the population}}$

b. Domain of $f = \boxed{\{x \mid 0 \le x < 100\}}$.

c. $\boxed{\text{The cost becomes infinitely large as } x \text{ approaches } 100\%.}$

21. a. $y = \dfrac{130x}{100-x}$

x	0	10	20	40	60	70	80	90	95	99	100
y	0	14	33	87	195	303	520	1170	2470	12870	undefined

b.

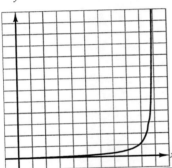

c. $\boxed{\begin{array}{l}\text{The values of the cost increase rapidly as the values of } x(\% \text{ inoculated}) \\ \text{approach } 100.\end{array}}$

23. $f(x) = \dfrac{0.2x}{x^2 + 4x + 4}$

$f(2) = \dfrac{0.2(2)}{4+8+4} = \boxed{0.025 \ \dfrac{\text{mg}}{\text{cm}^3} \text{ after 2 hours}}$

$f(4) = \dfrac{0.2(4)}{16+16+4} \approx \boxed{0.022 \ \dfrac{\text{mg}}{\text{cm}^3} \text{ after 4 hours}}$

$f(8) = \dfrac{0.2(8)}{64+32+4} = \boxed{0.016 \ \dfrac{\text{mg}}{\text{cm}^3} \text{ after 8 hours}}$

25. $\dfrac{3}{3x-9} = \dfrac{3}{3(x-3)} = \boxed{\dfrac{1}{x-3}}$

27. $\dfrac{16y^2}{7y} = \boxed{2y}$

29. $\dfrac{3y+9}{y+3} = \dfrac{3(y+3)}{y+3} = \boxed{3}$

31. $\dfrac{x+5}{x^2-25} = \dfrac{x+5}{(x+5)(x-5)} = \boxed{\dfrac{1}{x-5}}$

33. $\dfrac{2y-10}{3y-6} = \boxed{\dfrac{2(y-5)}{3(y-2)}}$

35. $\dfrac{s+1}{s^2-2s-3} = \dfrac{s+1}{(s-3)(s+1)} = \boxed{\dfrac{1}{s-3}}$

37. $\dfrac{4b-8}{b^2-4b+4} = \dfrac{4(b-2)}{(b-2)^2} = \boxed{\dfrac{4}{b-2}}$

39. $\dfrac{y^2-3y+2}{y^2+7y-18} = \dfrac{(y-1)(y-2)}{(y+9)(y-2)} = \boxed{\dfrac{y-1}{y+9}}$

41. $\dfrac{2y^2 - 7y + 3}{2y^2 - 5y + 2} = \dfrac{(2y-1)(y-3)}{(2y-1)(y-2)} = \boxed{\dfrac{y-3}{y-2}}$

43. $\dfrac{2x+3}{2x-5}$ is $\boxed{\text{irreducible}}$

45. $\dfrac{x^2 + 5x + 2xy + 10y}{x^2 - 25} = \dfrac{x(x+5) + 2y(x+5)}{x^2 - 25} = \dfrac{(x+5)(x+2y)}{(x+5)(x-5)} = \boxed{\dfrac{x+2y}{x-5}}$

47. $\dfrac{x^3 + 5x^2 - 6x}{x^3 - x} = \dfrac{x(x^2 + 5x - 6)}{x(x^2 - 1)} = \dfrac{(x+6)(x-1)}{(x+1)(x-1)} = \boxed{\dfrac{x+6}{x+1}}$

49. $\dfrac{2y^8 + y^7}{2y^6 + y^5} = \dfrac{y^7(2y+1)}{y^5(2y+1)} = \boxed{y^2}$

51. $\dfrac{x-5}{5-x} = \dfrac{-(5-x)}{5-x} = \boxed{-1}$

53. $\dfrac{2x-2y}{y-x} = \dfrac{2(x-y)}{y-x} = \boxed{-2}$

55. $\dfrac{-2x-8}{x^2 - 16} = \dfrac{-2(x+4)}{(x+4)(x-4)} = -\dfrac{2}{x-4} = \boxed{\dfrac{2}{4-x}}$

57. $\dfrac{4-6y}{3y^2 - 2y} = \dfrac{2(2-3y)}{y(3y-2)} = \dfrac{-2(3y-2)}{y(3y-2)} = \boxed{-\dfrac{2}{y}}$

59. $\dfrac{9-x^2}{x^2 - x - 6} = \dfrac{(3-x)(3+x)}{(x-3)(x+2)} = \dfrac{-(x-3)(3+x)}{(x-3)(x+2)} = \boxed{-\dfrac{3+x}{x+2}}$

61. $\dfrac{y^2 - 9y + 18}{y^3 - 27} = \dfrac{(y-3)(y-6)}{(y-3)(y^2 + 3y + 9)} = \boxed{\dfrac{y-6}{y^2 + 3y + 9}}$

63. $\dfrac{b^2 - b - 12}{4-b} = \dfrac{(b-4)(b+3)}{4-b} = \dfrac{(b-4)(b+3)}{-(b-4)} = -(b+3) \text{ or } \boxed{-b-3}$

65. $\dfrac{(2y+3) - (y+6)}{y^2 - 9} = \dfrac{y-3}{(y-3)(y+3)} = \boxed{\dfrac{1}{y+3}}$

67. $\dfrac{9-r^2}{r^2 - 3(2r-3)} = \dfrac{(3-r)(3+r)}{r^2 - 6r + 9} = \dfrac{(r-3)(3+r)}{(r-3)^2} = -\dfrac{3+r}{r-3} = \boxed{\dfrac{3+r}{3-r}}$

69. $\dfrac{ab + 2b + 3a + 6}{a^2 + 5a + 6} = \dfrac{b(a+2) + 3(a+2)}{(a+2)(a+3)} = \dfrac{(b+3)(a+2)}{(a+2)(a+3)} = \boxed{\dfrac{b+3}{a+3}}$

71. $\text{cost} = \dfrac{150}{250x - 50} = \dfrac{50(3)}{50(5x-1)} = \boxed{\dfrac{3}{5x-1}}$

$\text{cost (for } x = 70) = \dfrac{3}{5(70) - 1} = \dfrac{3}{349} \approx 0.0086 \text{ million dollars} = \boxed{\$8600}$

73. Statement $\boxed{\text{D}}$ is true, since $\dfrac{-3y-6}{y+2} = -\dfrac{3(y+2)}{y+2} = -3.$

75. $\boxed{y=}$ $\dfrac{x^2-25}{x-5} = \dfrac{(x+5)(x-5)}{x-5} =$ $\boxed{x+5 \quad (x\ne 5)}$

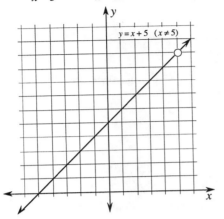

77. $\boxed{y=}$ $\dfrac{9x-18}{3x-6} = \dfrac{9(x-2)}{3(x-2)} =$ $\boxed{3 \quad (x\ne 2)}$

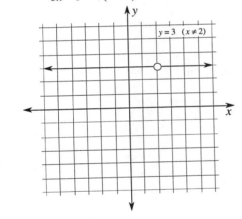

79. $\dfrac{1}{x+4}$ $\boxed{\text{approaches } 0}$ as x becomes large.

x	0	10	100
$\dfrac{1}{x+4}$	0.25	0.07	0.01

81. $\dfrac{2x}{x-1}$ $\boxed{\text{approaches } 2}$ as x becomes large.

x	0	10	100	1000
$\dfrac{2x}{x-1}$	0	2.22	2.02	2.00

83. $\dfrac{x+2}{x^2}$ $\boxed{\text{approaches } 0}$ as x becomes large.

x	0	10	100
$\dfrac{x+2}{x^2}$	undefined	0.12	0.01

85. $\dfrac{x}{x^2 - 1}$ ⬚ approaches 0 ⬚ as x becomes large.

x	0	10	100	1000
$\dfrac{x}{x^2 - 1}$	0	0.10	0.01	0.001

87. $\dfrac{x^2 + 7x + 12}{x + 3} = \dfrac{(x+3)(x+4)}{x+3} = \boxed{x+4}$

89. $\boxed{\dfrac{x^2 + 7x + 12}{x + 4}}$ is undefined when $x = -4$

91. Others are $\boxed{\dfrac{26}{65} = \dfrac{2}{5}}$ and $\boxed{\dfrac{49}{98} = \dfrac{4}{8}}$.

Review Problems

95. Let x equal the measure of the smaller angle. $2x + 6$ equals the measure of the larger angle.

$$x + 2x + 6 = 90$$
$$3x = 84$$
$$x = \boxed{28°}$$

The angles are complementary; measure of other angle $2x + 6 = \boxed{62°}$.

96.

$2x - 5y = -2$	$(\times 3)$	$6x - 15y = -6$
$3x + 4y = 20$	$(\times -2)$	$\underline{-6x - 8y = -40}$
		$23y = 46$
		$y = 2$

$$x = \dfrac{5y - 2}{2} = \dfrac{10 - 2}{2} = 4$$
$$y = 2$$
$$\boxed{\{(4, 2)\}}$$

97. $\dfrac{8.5 \times 10^{-3}}{1.7 \times 10^{-7}} = \dfrac{8.5}{1.7} \times 10^{-3 - (-7)} = \dfrac{8.5}{1.7} \times 10^4 = 5 \times 10^4 = \boxed{50000}$

Section 8.2 Multiplying and Dividing Rational Expressions

Problem Set 8.2, pp. 532-534

1. $\dfrac{5}{x+2} \cdot \dfrac{y-3}{7} = \boxed{\dfrac{5(y-3)}{7(x+2)}}$ $\boxed{x \neq -2}$

3. $\dfrac{x-2}{3x+9} \cdot \dfrac{2x+6}{2x-4} = \dfrac{x-2}{3(x+3)} \cdot \dfrac{2(x+3)}{2(x-2)} = \boxed{\dfrac{1}{3}}$

  $(x+3)(x-2) \neq 0$ $\boxed{x \neq -3, 2}$

5. $\dfrac{y-5}{y+2} \cdot \dfrac{6y-8}{2y+4} = \dfrac{y-5}{y+2} \cdot \dfrac{2(3y-4)}{2(y+2)} = \boxed{\dfrac{3y^2 - 19y + 20}{y^2 + 4y + 4}}$

  $(y+2)(y+2) \neq 0$ $\boxed{(y \neq -2)}$

7. $\dfrac{3}{x} \cdot \dfrac{2x}{9y} = \dfrac{6}{9y} = \boxed{\dfrac{2}{3y}}$

$x(9y) \neq 0$ $\boxed{(y \neq 0, x \neq 0)}$

9. $\dfrac{4y+30}{y^2-3y} \cdot \dfrac{y-3}{2y+15} = \dfrac{2(2y+15)}{y(y-3)} \cdot \dfrac{y-3}{2y+15} = \boxed{\dfrac{2}{y}}$

$y(y-3)(2y+15) \neq 0$ $\boxed{\left(y \neq 0, 3, -\dfrac{15}{2}\right)}$

11. $\dfrac{r^2-9}{r^2} \cdot \dfrac{r^2-3r}{r^2+r-12} = \dfrac{(r-3)(r+3)}{r^2} \cdot \dfrac{r(r-3)}{(r+4)(r-3)} = \boxed{\dfrac{(r-3)(r+3)}{r(r+4)}}$

$r^2(r+4)(r-3) \neq 0$ $\boxed{(r \neq 0, 3, -4)}$

13. $\dfrac{y^2-7y-30}{y^2-6y-40} \cdot \dfrac{2y^2+5y+2}{2y^2+7y+3} = \dfrac{(y-10)(y+3)}{(y-10)(y+4)} \cdot \dfrac{(2y+1)(y+2)}{(2y+1)(y+3)} = \boxed{\dfrac{y+2}{y+4}}$

$(y-10)(y+4)(2y+1)(y+3) \neq 0$

$\boxed{\left(y \neq -4, 10, -\dfrac{1}{2}, -3\right)}$

15. $(y^2-9)\left(\dfrac{4}{y-3}\right) = \dfrac{(y-3)(y+3)(4)}{y-3} = \boxed{4(y+3)}$

$\boxed{(y \neq 3)}$

17. $\dfrac{x^2-2x+4}{x^2-4} \cdot \dfrac{(x+2)^3}{2x+4} = \dfrac{x^2-2x+4}{(x-2)(x+2)} \cdot \dfrac{(x+2)^3}{2(x+2)} = \boxed{\dfrac{(x^2-2x+4)(x+2)}{2(x-2)}}$

$(x-2)(x+2)2(x+2) \neq 0$

$\boxed{(x \neq -2, 2)}$

19. $\dfrac{x^2-x-6}{3x-9} \cdot \dfrac{x^2-9}{x^2+6x+9} = \dfrac{(x-3)(x+2)}{3(x-3)} \cdot \dfrac{(x-3)(x+3)}{(x+3)^2} = \boxed{\dfrac{(x+2)(x-3)}{3(x+3)}}$

$3(x-3)(x+3)^2 \neq 0$

$\boxed{(x \neq -3, 3)}$

21. $\dfrac{y^2+10y+25}{y-4} \cdot \dfrac{y^2-y-12}{y+5} \cdot \dfrac{1}{y+3} = \dfrac{(y+5)^2}{y-4} \cdot \dfrac{(y-4)(y+3)}{y+5} \cdot \dfrac{1}{y+3} = \boxed{y+5}$

$(y-4)(y+5)(y+3) \neq 0$

$\boxed{(y \neq -3, 4, -5)}$

23. $\dfrac{(x-2)^3}{(x-1)^3} \cdot \dfrac{x-2x+1}{x^2-4x+4} = \dfrac{(x-2)^3}{(x-1)^3} \cdot \dfrac{(x-1)^2}{(x-2)^2} = \boxed{\dfrac{x-2}{x-1}}$

$(x-1)^3(x-2)^2 \neq 0$

$\boxed{(x \neq 1, 2)}$

25. $\dfrac{25 - y^2}{y^2 - 2y - 35} \cdot \dfrac{y^2 - 8y - 20}{y^2 - 3y - 10} = \dfrac{(5 - y)(5 + y)}{(y - 7)(y + 5)} \cdot \dfrac{(y - 10)(y + 2)}{(y - 5)(y + 2)} = \boxed{-\dfrac{y - 10}{y - 7}}$

$(y - 7)(y + 5)(y - 5)(y + 2) \neq 0$

$\boxed{(y \neq -2, -5, 5, 7)}$

27. $\dfrac{15}{x} \div \dfrac{3y}{2x} = \dfrac{15}{x} \cdot \dfrac{2x}{3y} = \boxed{\dfrac{10}{y}}$

$x(3y) \neq 0$

$\boxed{(x \neq 0, y \neq 0)}$

29. $\dfrac{y^2 - y}{15} \div \dfrac{y - 1}{5} = \dfrac{y(y - 1)}{15} \cdot \dfrac{5}{y - 1} = \boxed{\dfrac{y}{3}}$

$(y - 1) \neq 0$

$\boxed{(y \neq 1)}$

31. $\dfrac{x^2 + 2x + 1}{6x^2} \div \dfrac{x + 1}{12x^3} = \dfrac{(x + 1)^2}{6x^2} \cdot \dfrac{12x^3}{x + 1} = \boxed{2x(x + 1)}$

$6x^2(x + 1) \neq 0$

$\boxed{(x \neq 0, -1)}$

33. $\dfrac{y^3 + y}{y^2 - y} \div \dfrac{y^3 - y^2}{y^2 - 2y + 1} = \dfrac{y(y^2 + 1)}{y(y - 1)} \cdot \dfrac{(y - 1)^2}{y^2(y - 1)} = \boxed{\dfrac{y^2 + 1}{y^2}}$

$y(y - 1)y^2(y - 1) \neq 0$

$\boxed{(y \neq 0, 1)}$

35. $\dfrac{m^2 + 5m + 4}{m^2 + 12m + 32} \div \dfrac{m^2 - 12m + 35}{m^2 + 3m - 40} = \dfrac{(m + 4)(m + 1)}{(m + 4)(m + 8)} \cdot \dfrac{(m + 8)(m - 5)}{(m - 7)(m - 5)} = \boxed{\dfrac{m + 1}{m - 7}}$

$(m + 4)(m + 8)(m - 7)(m - 5) \neq 0$

$\boxed{(m \neq -4, -8, 7, 5)}$

37. $\dfrac{6y^2 + 31y + 18}{3y^2 - 20y + 12} \cdot \dfrac{2y^2 - 15y + 18}{6y^2 + 35y + 36} \div \dfrac{2y^2 - 13y + 15}{9y^2 + 15y + 4} = \dfrac{(3y + 2)(2y + 9)}{(3y - 2)(y - 6)} \cdot \dfrac{(2y - 3)(y - 6)}{(2y + 9)(3y + 4)} \cdot \dfrac{(3y + 1)(3y + 4)}{(2y - 3)(y - 5)}$

$= \boxed{\dfrac{(3y + 2)(3y + 1)}{(3y - 2)(y - 5)}}$

$(3y + 1)(3y + 4) \neq 0$

$(3y - 2)(y - 6) \neq 0$

$(2y + 9)(3y + 4) \neq 0$

$(2y - 3)(y - 5) \neq 0$

$\boxed{y \neq \left(\dfrac{2}{3}, 6, -\dfrac{9}{2}, -\dfrac{4}{3}, \dfrac{3}{2}, 5, \dfrac{1}{3}\right)}$

39. $\dfrac{32}{25x^3} \div \dfrac{16x^2}{50x^4y} \cdot \dfrac{x}{y^3} = \dfrac{32}{25x^3} \cdot \dfrac{50x^4y}{16x^2} \cdot \dfrac{x}{y^3} = \boxed{\dfrac{4}{y^2}}$

$\boxed{(x \neq 0, y \neq 0)}$

41. $\dfrac{2y^2 - 128}{y^2 + 16y + 64} \div \dfrac{y^2 - 6y - 16}{3y^2 + 30y + 48} = \dfrac{2(y - 8)(y + 8)}{(y + 8)^2} \cdot \dfrac{3(y + 8)(y + 2)}{(y - 8)(y + 2)} = \boxed{6}$

$(y + 8)^2(y - 8)(y + 2) \neq 0$

$\boxed{(y \neq -2, -8, 8)}$

43. $\dfrac{\dfrac{x^3}{6}}{\dfrac{x}{3}} = \dfrac{x^3}{6} \div \dfrac{x}{3} = \dfrac{x^3}{6} \cdot \dfrac{3}{x} = \boxed{\dfrac{x^2}{2}}$

$\boxed{(x \neq 0)}$

45. $\dfrac{\dfrac{3x+12}{4}}{\dfrac{x+4}{2}} = \dfrac{3x+12}{4} \div \dfrac{x+4}{2} = \dfrac{3(x+4)}{4} \cdot \dfrac{2}{x+4} = \boxed{\dfrac{3}{2}}$

$\boxed{(x \neq -4)}$

47. $\dfrac{\dfrac{x}{y-7}}{\dfrac{4}{7-y}} = \dfrac{x}{y-7} \div \dfrac{4}{7-y} = \dfrac{x}{y-7} \cdot \dfrac{7-y}{4} = \boxed{-\dfrac{x}{4}}$

$\boxed{(y \neq 7)}$

49. $\dfrac{\dfrac{x^2-9x+18}{x^2-9}}{\dfrac{2x^3-11x^2-6x}{2x^2+x}} = \dfrac{x^2-9x+18}{x^2-9} \div \dfrac{2x^3-11x^2-6x}{2x^2+x} = \dfrac{(x-6)(x-3)}{(x+3)(x-3)} \cdot \dfrac{x(2x+1)}{x(2x+1)(x-6)} = \boxed{\dfrac{1}{x+3}}$

$\boxed{\left(x \neq 3, -3, 0, -\dfrac{1}{2}, 6 \right)}$

51. $\dfrac{\left(\dfrac{7x}{3}\right)^2}{\left(\dfrac{7x}{2}\right)^3} = \left(\dfrac{7x}{3}\right)^2 \div \left(\dfrac{7x}{2}\right)^3 = \dfrac{(7x)^2}{9} \cdot \dfrac{8}{(7x)^3} = \boxed{\dfrac{8}{63x}}$

$\boxed{(x \neq 0)}$

53. $\dfrac{\dfrac{4}{x^2-3x-28}}{\dfrac{2}{x-7}} = \dfrac{4}{x^2-3x-28} \div \dfrac{2}{x-7} = \dfrac{4}{(x-7)(x+4)} \cdot \dfrac{x-7}{2} = \boxed{\dfrac{2}{x+4}}$

$\boxed{(x \neq -4, 7)}$

55. $\dfrac{y^2+6y+8}{y^2+y-2} \div \dfrac{y+4}{2y+4} \div \dfrac{y+3}{y-1} = \dfrac{(y+4)(y+2)}{(y-1)(y+2)} \cdot \dfrac{2(y+2)}{y+4} \cdot \dfrac{y+3}{y-1} = \boxed{\dfrac{2(y+2)(y+3)}{(y-1)^2}}$

$\boxed{(y \neq 1, -2, -3, -4)}$

57. Statement $\boxed{B}$ is true.

A. $5 \div x = \dfrac{5}{x} \text{ not } \dfrac{x}{5}$

B. $\dfrac{4}{x} \div \dfrac{x-2}{x} = \dfrac{4}{x-2}$

$\dfrac{4}{x} \cdot \dfrac{x}{x-2} = \dfrac{4}{x-2})$

$\dfrac{4}{x-2} = \dfrac{4}{x-2}$

true if $x \neq 0$ and $x \neq 2$

C. $\dfrac{x-5}{6} \cdot \dfrac{3}{5-x} = \dfrac{1}{2}$

$-\dfrac{1}{2} = \dfrac{1}{2}$ false

D.   Its value *remains the same*.

59. Area = length $\times$ width = $\dfrac{x+3}{2x-8} \cdot \dfrac{6x-24}{2x+1} = \dfrac{x+3}{2(x-4)} \cdot \dfrac{6(x-4)}{2x+1} = \boxed{\dfrac{3(x+3)}{2x+1}}$ square meters

$\boxed{\left(x \neq -\dfrac{1}{2}, 4\right)}$

61. $A = \dfrac{1}{2}bh = \dfrac{1}{2} \cdot \dfrac{2}{3x+3} \cdot \dfrac{x^2-1}{6} = \dfrac{1}{3(x+1)} \cdot \dfrac{(x+1)(x-1)}{6} = \boxed{\dfrac{x-1}{18}}$ square units

63. $w = \dfrac{A}{l} = \dfrac{y^2+y-2}{y^3} \div \dfrac{y-1}{y^2} = \dfrac{(y+2)(y-1)}{y^3} \cdot \dfrac{y^2}{y-1} = \boxed{\dfrac{y+2}{y}}$ meters

65. $\dfrac{\text{area of trapezoid}}{\text{area of rectangle}} = \dfrac{\dfrac{1}{2}(3x+x)\left(\dfrac{1}{8x+4}\right)}{\left(\dfrac{1}{14x+7}\right)(3)} = \dfrac{2x}{4(2x+1)} \cdot \dfrac{7(2x+1)}{3} = \dfrac{14x}{12} = \boxed{\dfrac{7x}{6}}$

67. $\left(1-\dfrac{1}{2}\right)\left(1-\dfrac{1}{3}\right)\left(1-\dfrac{1}{4}\right)\left(1-\dfrac{1}{5}\right)\cdots\left(1-\dfrac{1}{200}\right) = \dfrac{1}{2} \cdot \dfrac{2}{3} \cdot \dfrac{3}{4} \cdot \dfrac{4}{5} \cdots \dfrac{198}{199} \cdot \dfrac{199}{200} = \boxed{\dfrac{1}{200}}$

Review Problems

70. Let x equal the length of the longer leg. $\dfrac{1}{2}x - 1$ equals the length of the shorter leg. $x+1$ equals the length of the hypotenuse.

$x^2 + \left(\dfrac{1}{2}x-1\right)^2 = (x+1)^2$

$x^2 + \dfrac{1}{4}x^2 - x + 1 = x^2 + 2x + 1$

$\dfrac{1}{4x^2 - 3x} = 0$

$x\left(\dfrac{1}{4}x-3\right) = 0$

$x = 0$ or $\dfrac{1}{4}x - 3 = 0$

$x = 12$

$\dfrac{1}{2}x - 1 = 6 - 1 = 5$

$x + 1 = 13$

The lengths of the three sides are $\boxed{\text{5 meters, 12 meters, and 13 meters}}$.

71.
$$y(2y+9) = 5$$
$$2y^2 + 9y - 5 = 0$$
$$(2y-1)(y+5) = 0$$
$$2y - 1 = 0 \text{ or } y + 5 = 0$$
$$y = \frac{1}{2} \text{ or } y = -5$$

$$\boxed{\left\{\frac{1}{2}, -5\right\}}$$

72.
$$\frac{(2x^3)^{-1}(3x^{-3})^4}{2x^5} = \frac{3^4 x^{-12}}{(2x^3)(2x^5)}$$
$$= \frac{81x^{-12}}{4x^8} = \frac{81}{4} x^{-12-8} = \boxed{\frac{81}{4x^{20}}}$$

Section 8.3   Adding and Subtracting Rational Expressions with the Same Denominator

Problem Set 8.3, pp. 538-540

1. $\dfrac{4x}{9} + \dfrac{2x}{9} = \dfrac{6x}{9} = \boxed{\dfrac{2x}{3}}$

3. $\dfrac{5}{x} + \dfrac{3}{x} = \boxed{\dfrac{8}{x}}$ $(x \neq 0)$

5. $\dfrac{7}{9x} + \dfrac{5}{9x} = \dfrac{12}{9x} = \boxed{\dfrac{4}{3x}}$ $(x \neq 0)$

7. $\dfrac{m}{5} + \dfrac{2m}{5} = \boxed{\dfrac{3m}{5}}$

9. $\dfrac{7}{4-y} + \dfrac{3}{4-y} = \boxed{\dfrac{10}{4-y}}$ $(y \neq 4)$

11. $\dfrac{3x+2}{3x+4} + \dfrac{3x+6}{3x+4} = \dfrac{6x+8}{3x+4} = \dfrac{2(3x+4)}{3x+4} = \boxed{2}$ $\left(x \neq -\dfrac{4}{3}\right)$

13. $\dfrac{y^2-2y}{y^2+3y} + \dfrac{y^2+y}{y^2+3y} = \dfrac{2y^2-y}{y^2+3y} = \dfrac{y(2y-1)}{y(y+3)} = \boxed{\dfrac{2y-1}{y+3}}$ $(y \neq 0, -3)$

15. $\dfrac{y+2}{6y^3} + \dfrac{3y-2}{6y^3} = \dfrac{4y}{6y^3} = \boxed{\dfrac{2}{3y^2}}$ $(y \neq 0)$

17. $\dfrac{y^2+9y}{4y^2-11y-3} + \dfrac{3y-5y^2}{4y^2-11y-3} = \dfrac{-4y^2+12y}{4y^2-11y-3} = \dfrac{-4y(y-3)}{(4y+1)(y-3)} = \boxed{-\dfrac{4y}{4y+1}}$ $\left(y \neq 3, -\dfrac{1}{4}\right)$

19. $\dfrac{y}{2y+7} - \dfrac{2}{2y+7} = \boxed{\dfrac{y-2}{2y+7}}$ $\left(y \neq -\dfrac{7}{2}\right)$

21. $\dfrac{x}{x-1} - \dfrac{1}{x-1} = \dfrac{x-1}{x-1} = \boxed{1}$ $(x \neq 1)$

23. $\dfrac{2y+1}{3y-7} - \dfrac{y+8}{3y-7} = \dfrac{2y+1-y-8}{3y-7} = \boxed{\dfrac{y-7}{3y-7}}$ $\left(y \neq \dfrac{7}{3}\right)$

25. $\dfrac{y^3-3}{2y^4} - \dfrac{7y^3-3}{2y^4} = \dfrac{y^3-3-7y^3+3}{2y^4} = \dfrac{-6y^3}{2y^4} = \boxed{-\dfrac{3}{y}}$ $(y \neq 0)$

27. $\dfrac{2y+3}{3y-6} - \dfrac{3-y}{3y-6} = \dfrac{2y+3-3+y}{3y-6} = \dfrac{3y}{3y-6} = \dfrac{3y}{3(y-2)} = \boxed{\dfrac{y}{y-2}}$ $(y \neq 2)$

29. $\dfrac{y^2+3y}{y^2+y-12} - \dfrac{y^2-12}{y^2+y-12} = \dfrac{y^2+3y-y^2+12}{y^2+y-12} = \dfrac{3y+12}{y^2+y-12} = \dfrac{3(y+4)}{(y+4)(y-3)} = \boxed{\dfrac{3}{y-3}}$ $(y \neq -4, 3)$

31. $\dfrac{16r^2+3}{16r^2+16r+3} - \dfrac{3-4r}{16r^2+16r+3} = \dfrac{16r^2+3-3+4r}{16r^2+16r+3} = \dfrac{16r^2+4r}{16r^2+16r+3} = \dfrac{4r(4r+1)}{(4r+1)(4r+3)} = \boxed{\dfrac{4r}{4r+3}}$

$\left(r \neq -\dfrac{1}{4}, -\dfrac{3}{4} \right)$

33. $\dfrac{9x}{10} - \dfrac{7x}{10} + \dfrac{3x}{10} = \dfrac{5x}{10} = \boxed{\dfrac{x}{2}}$

35. $\dfrac{6y^2+y}{2y^2-9y+9} - \dfrac{2y+9}{2y^2-9y+9} - \dfrac{4y-3}{2y^2-9y+9} = \dfrac{6y^2+y-2y-9-4y+3}{2y^2-9y+9} = \dfrac{6y^2-5y-6}{2y^2-9y+9} = \dfrac{(2y-3)(3y+2)}{(2y-3)(y-3)}$

$= \boxed{\dfrac{3y+2}{y-3}}$ $\left(y \neq \dfrac{3}{2}, 3 \right)$

37. $\dfrac{6b^2-10b}{16b^2-48b+27} + \dfrac{7b^2-20b}{16b^2-48b+27} - \dfrac{6b-3b^2}{16b^2-48b+27} = \dfrac{6b^2-10b+7b^2-20b}{16b^2-48b+27} = \dfrac{16b^2-36b}{16b^2-48b+27}$

$= \dfrac{4b(4b-9)}{(4b-3)(4b-9)}$

$= \boxed{\dfrac{4b}{4b-3}}$ $\left(b \neq \dfrac{3}{4}, \dfrac{9}{4} \right)$

39. $\dfrac{2y}{y-5} - \left(\dfrac{2}{y-5} + \dfrac{y-2}{y-5} \right) = \dfrac{2y-2-y+2}{y-5} = \boxed{\dfrac{y}{y-5}}$ $(y \neq 5)$

41. $\dfrac{b}{a(c+d)-b(c+d)} - \dfrac{a}{a(c+d)-b(c+d)} = \dfrac{b-a}{(a-b)(c+d)} = \boxed{-\dfrac{1}{c+d}}$

43. $\dfrac{2y+7}{y-6} + \dfrac{3y}{6-y} = \dfrac{2y+7}{y-6} - \dfrac{3y}{y-6} = \boxed{\dfrac{7-y}{y-6}}$

45. $\dfrac{y^2}{y-2} + \dfrac{4}{2-y} = \dfrac{y^2}{y-2} - \dfrac{4}{y-2} = \dfrac{y^2-4}{y-2} = \dfrac{(y-2)(y+2)}{y-2} = \boxed{y+2}$ $(y \neq 2)$

47. $\dfrac{b-3}{b^2-25} + \dfrac{b-3}{25-b^2} = \dfrac{b-3}{b^2-25} - \dfrac{b-3}{b^2-25} = \boxed{0}$ $(b \neq -5, 5)$

49. $\dfrac{3(m-2)}{2m-3} + \dfrac{3(m-1)}{3-2m} + \dfrac{5(2m+1)}{2m-3} = \dfrac{3(m-2)}{2m-3} - \dfrac{3(m-1)}{2m-3} + \dfrac{5(2m+1)}{2m-3} = \dfrac{3m-6-3m+3+10m+5}{2m-3}$

$= \dfrac{10m+2}{2m-3}$

$= \boxed{\dfrac{2(5m+1)}{2m-3}}$ $\left(m \neq \dfrac{3}{2} \right)$

51. $\dfrac{y}{y-1} - \dfrac{1}{1-y} = \dfrac{y}{y-1} + \dfrac{1}{y-1} = \boxed{\dfrac{y+1}{y-1}}$ $(y \neq 1)$

53. $\dfrac{3-a}{a-7}-\dfrac{2a-5}{7-a}=\dfrac{3-a}{a-7}+\dfrac{2a-5}{a-7}=\boxed{\dfrac{a-2}{a-7}}$ $(a\neq 7)$

55. $\dfrac{z-2}{z^2-25}-\dfrac{z-2}{25-z^2}=\dfrac{z-2}{z^2-25}+\dfrac{z-2}{z^2-25}=\boxed{\dfrac{2(z-2)}{z^2-25}}$ $(z\neq -5, 5)$

57. $\dfrac{2(f-1)}{2f-3}-\dfrac{3(f+2)}{2f-3}-\dfrac{f-1}{3-2f}=\dfrac{2(f-1)}{2f-3}-\dfrac{3(f+2)}{2f-3}+\dfrac{f-1}{2f-3}=\dfrac{2f-2-3f-6+f-1}{2f-3}=\boxed{-\dfrac{9}{2f-3}}$

$\left(f\neq\dfrac{3}{2}\right)$

59. $\dfrac{x^2}{x-7}+\dfrac{6x+7}{7-x}=\dfrac{x^2}{x-7}-\dfrac{6x+7}{x-7}=\dfrac{x^2-6x-7}{x-7}=\dfrac{(x-7)(x+1)}{x-7}=\boxed{x+1}$ $(x\neq 7)$

61. $\dfrac{(y-3)(y+2)}{(y+1)(y-4)}-\dfrac{(y+2)(y+3)}{(y+1)(4-y)}-\dfrac{(y+5)(y-1)}{(y+1)(4-y)}$

$=-\dfrac{(y-3)(y+2)}{(y+1)(4-y)}-\dfrac{(y+2)(y+3)}{(y+1)(4-y)}-\dfrac{(y+5)(y-1)}{(y+1)(4-y)}$

$=\dfrac{-y^2+y+6-y^2-5y-6-y^2-4y+5}{(y+1)(4-y)}$

$=\boxed{\dfrac{-3y^2-8y+5}{(y+1)(4-y)}}$

63. Statement $\boxed{C}$ is true. $\dfrac{2y+1}{y-7}+\dfrac{3y+1}{y-7}-\dfrac{5y+2}{y-7}=\dfrac{2y+1+3y+1-5y+2}{y-7}=\dfrac{0}{y-7}=0$, true. $(y\neq 7)$

65. perimeter $=\dfrac{5}{y-1}+\dfrac{y^2-8}{y-1}+\dfrac{3y}{y-1}+\dfrac{y^2-2y}{y-1}=\dfrac{5+y^2-8+3y+y^2-2y}{y-1}=\dfrac{2y^2+y-3}{y-1}=\dfrac{(2y+3)(y-1)}{y-1}$

$=\boxed{2y+2}$ $(y\neq 1)$

Review Problems

69. $81y^4-1=(9y^2-1)(9y^2+1)=\boxed{(3y-1)(3y+1)(9y^2+1)}$

70. Let $t=$ the time (in hours)

$$\begin{aligned}550t+475t &= 2050\\ 1025t &= 2050\\ t &= 2\text{ hours}\\ \text{time} &= \text{noon}+2\text{ hours}=\boxed{2:00\text{ P.M.}}\end{aligned}$$

71. $\dfrac{3x^3+2x^2-26x-15}{x+3}=\boxed{3x^2-7x-5}$ $(x\neq -3)$

$$
\require{enclose}
\begin{array}{r}
3x^2-7x-5 \\
x+3\enclose{longdiv}{3x^3+2x^2-26x-15} \\
\underline{3x^3+9x^2} \\
-7x^2-26x \\
\underline{-7x^2-21x} \\
-5x-15 \\
\underline{-5x-15} \\
0
\end{array}
$$

Section 8.4 Adding and Subtracting Rational Expressions with Different Denominators

Problem Set 8.4, pp. 547-549

1. LCM (12, 10): $12 = 2^2 \cdot 3$
$10 = 2 \cdot 5$
LCM $= 2^2 \cdot 3 \cdot 5 = \boxed{60}$

3. LCM $(3x, x^3)$: $3x = 3 \cdot x$
$x^3 = x^3$
LCM $= \boxed{3x^3}$

5. LCM $(15x^2, 24x^5)$: $15x^2 = 3 \cdot 5 \cdot x^2$
$24x^5 = 2^3 \cdot 3 \cdot x^5$
LCM $= 2^3 \cdot 3 \cdot 5 \cdot x^5 = \boxed{120x^5}$

7. LCM $(100y, 120y)$: $100y = 2^2 \cdot 5^2 \cdot y$
$120y = 2^3 \cdot 3 \cdot 5 \cdot y$
LCM $= 2^3 \cdot 5^2 \cdot y = \boxed{600y}$

9. LCM $(15x^2y^3, 6x^5y)$: $15x^2y^3 = 3 \cdot 5 \cdot x^2 \cdot y^3$
$6x^5y = 2 \cdot 3 \cdot x^5 \cdot y$
LCM $= 2 \cdot 3 \cdot 5 \cdot x^5 y^3 = \boxed{30x^5y^3}$

11. LCM $(y - 3, y + 1)$: $y - 3 = y - 3$
$y + 1 = y + 1$
LCM $= \boxed{(y - 3)(y + 1) = y^2 - 2y - 3}$

13. LCM $(x, 7(x + 2))$: $x = x$
$7(x + 2) = 7 \cdot (x + 2)$
LCM $= \boxed{7x(x + 2)}$

15. LCM $(18x^2, 27x(x - 5))$: $18x^2 = 2 \cdot 3^2 \cdot x^2$
$27x(x - 5) = 3^3 \cdot x \cdot (x - 5)$
LCM $= 2 \cdot 3^3 \cdot x^2 \cdot (x - 5) = \boxed{54x^2(x - 5)}$

17. LCM $(x + 3, x^2 - 9)$: $x + 3 = x + 3$
$x^2 - 9 = (x + 3) \cdot (x - 3)$
LCM $= (x + 3)(x - 3) = \boxed{x^2 - 9}$

19. LCM $(y^2 - 4, y(y + 2))$: $y^2 - 4 = (y + 2) \cdot (y - 2)$
$y(y + 2) = y \cdot (y + 2)$
LCM $= y(y - 2)(y + 2) = \boxed{y(y^2 - 4)}$

21. LCM $(y^2 - 25, y^2 - 10y + 25)$: $y^2 - 25 = (y - 5) \cdot (y + 5)$
$y^2 - 10y + 25 = (y - 5)^2$
LCM $= \boxed{(y - 5)^2 \cdot (y + 5) \text{ or } y^3 - 5y^2 - 25y + 125}$

23. LCM $(2x^2 + 7x - 4, x^2 + 2x - 8)$: $\begin{aligned} 2x^2 + 7x - 4 &= (2x - 1) \cdot (x + 4) \\ x^2 + 2x - 8 &= (x + 4) \cdot (x - 2) \\ \text{LCM} &= \boxed{(2x - 1) \cdot (x - 2) \cdot (x + 4) \text{ or } 2x^3 + 3x^2 - 18x + 8} \end{aligned}$

25. $\dfrac{3}{x} + \dfrac{5}{x^2} = \dfrac{3x}{x^2} + \dfrac{5}{x^2} = \boxed{\dfrac{3x + 5}{x^2}}$ $(x \neq 0)$

27. $\dfrac{2}{9w} - \dfrac{11}{6w} = \dfrac{4}{18w} - \dfrac{33}{18w} = \boxed{\dfrac{-29}{18w}}$ $(w \neq 0)$

29. $\dfrac{x - 1}{6} - \dfrac{x + 2}{3} = \dfrac{x - 1}{6} - \dfrac{2x + 4}{6} = \dfrac{-x - 5}{6} = \boxed{-\dfrac{x + 5}{6}}$

31. $\dfrac{2}{x - 1} + \dfrac{3}{x + 2} = \dfrac{2(x + 2) + 3(x - 1)}{(x - 1)(x + 2)} = \boxed{\dfrac{5x + 1}{(x - 1)(x + 2)}}$ $(x \neq -2, 1)$

33. $\dfrac{2}{r + 5} + \dfrac{3}{4r} = \dfrac{2(4r) + 3(r + 5)}{4r(r + 5)} = \boxed{\dfrac{11r + 15}{4r(r + 5)}}$ $(r \neq 0, -5)$

35. $\dfrac{4y - 9}{3y} - \dfrac{3y - 8}{4y} = \dfrac{4(4y - 9) - 3(3y - 8)}{12y} = \boxed{\dfrac{7y - 12}{12y}}$ $(y \neq 0)$

37. $\dfrac{5a + 3b}{2a^2 b} - \dfrac{3a - 4b}{ab^2} = \dfrac{b(5a + 3b) - 2a(3a - 4b)}{2a^2 b^2} = \dfrac{5ab + 3b^2 - 6a^2 + 8ab}{2a^2 b^2} = \boxed{\dfrac{-6a^2 + 13ab + 3b^2}{2a^2 b^2}}$

$(a, b \neq 0)$

39. $\dfrac{7}{x + 5} - \dfrac{4}{x - 5} = \dfrac{7(x - 5) - 4(x + 5)}{(x + 5)(x - 5)} = \boxed{\dfrac{3x - 55}{(x + 5)(x - 5)}}$ $(x \neq -5, 5)$

41. $\dfrac{2z}{z^2 - 16} + \dfrac{z}{z - 4} = \dfrac{2z}{(z + 4)(z - 4)} + \dfrac{z}{z - 4} = \dfrac{2z + z(z + 4)}{(z + 4)(z - 4)} = \dfrac{z^2 + 6z}{(z + 4)(z - 4)} = \boxed{\dfrac{z(z + 6)}{(z + 4)(z - 4)}}$ $(z \neq -4, 4)$

43. $\dfrac{5y}{y^2 - 9} - \dfrac{4}{y + 3} = \dfrac{5y}{(y - 3)(y + 3)} - \dfrac{4}{y + 3} = \dfrac{5y - 4(y - 3)}{(y - 3)(y + 3)} = \boxed{\dfrac{y + 12}{(y - 3)(y + 3)}}$ $(y \neq -3, 3)$

45. $\dfrac{7}{y - 1} - \dfrac{3}{(y - 1)^2} = \dfrac{7(y - 1) - 3}{(y - 1)^2} = \boxed{\dfrac{7y - 10}{(y - 1)^2}}$ $(y \neq 1)$

47. $\dfrac{3r}{4r - 20} + \dfrac{9r}{6r - 30} = \dfrac{3r}{4(r - 5)} + \dfrac{9r}{6(r - 5)} = \dfrac{3r}{4(r - 5)} + \dfrac{3r}{2(r - 5)} = \dfrac{3r + 6r}{4(r - 5)} = \boxed{\dfrac{9r}{4(r - 5)}}$ $(r \neq 5)$

49. $\dfrac{y + 4}{y} - \dfrac{y}{y + 4} = \dfrac{(y + 4)(y + 4) - y(y)}{y(y + 4)} = \dfrac{y^2 + 8y + 16 - y^2}{y(y + 4)} = \boxed{\dfrac{8(y + 2)}{y(y + 4)}}$ $(y \neq 0, -4)$

51. $\dfrac{z}{z^2 + 2z + 1} + \dfrac{4}{z^2 + 5z + 4} = \dfrac{z}{(z + 1)^2} + \dfrac{4}{(z + 1)(z + 4)} = \dfrac{z(z + 4) + 4(z + 1)}{(z + 1)^2 (z + 4)} = \boxed{\dfrac{z^2 + 8z + 4}{(z + 1)^2 (z + 4)}}$

$(z \neq -1, -4)$

53. $\dfrac{y-5}{y+3} + \dfrac{y+3}{y-5} = \dfrac{(y-5)^2 + (y+3)^2}{(y+3)(y-5)} = \dfrac{2y^2 - 4y + 34}{(y+3)(y-5)} = \boxed{\dfrac{2(y^2 - 2y + 17)}{(y+3)(y-5)}}$ $(y \neq -3, 5)$

55. $\dfrac{5}{2y^2 - 2y} - \dfrac{3}{2y - 2} = \dfrac{5}{2y(y-1)} - \dfrac{3}{2(y-1)} = \boxed{\dfrac{5 - 3y}{2y(y-1)}}$ $(y \neq 0, 1)$

57. $\dfrac{4r+3}{r^2 - 9} - \dfrac{r+1}{r-3} = \dfrac{4r+3}{(r+3)(r-3)} - \dfrac{r+1}{r-3} = \dfrac{4r+3 - (r+1)(r+3)}{(r+3)(r-3)} = \dfrac{4r+3 - r^2 - 4r - 3}{(r+3)(r-3)}$

$= \boxed{\dfrac{-r^2}{(r+3)(r-3)}}$

$(r \neq -3, 3)$

59. $\dfrac{y^2 - 39}{y^2 + 3y - 10} - \dfrac{y-7}{y-2} = \dfrac{y^2 - 39}{(y-2)(y+5)} - \dfrac{y-7}{y-2} = \dfrac{y^2 - 39 - (y+5)(y-7)}{(y-2)(y+5)} = \dfrac{y^2 - 39 - y^2 + 2y + 35}{(y-2)(y+5)}$

$= \dfrac{2(y-2)}{(y-2)(y+5)}$

$= \boxed{\dfrac{2}{y+5}}$

$(y \neq -5, 2)$

61. $\dfrac{w^2 - 11}{3w^2 + 5w - 2} - \dfrac{w-5}{3w-1} = \dfrac{w^2 - 11}{(3w-1)(w+2)} - \dfrac{w-5}{3w-1} = \dfrac{w^2 - 11 - (w+2)(w-5)}{(3w-1)(w+2)} = \dfrac{w^2 - 11 - w^2 + 3w + 10}{(3w-1)(w+2)}$

$= \dfrac{3w-1}{(3w-1)(w+2)}$

$= \boxed{\dfrac{1}{w+2}}$

$(w \neq -2, 1)$

63. $4 + \dfrac{1}{x-3} = \dfrac{4(x-3) + 1}{x-3} = \boxed{\dfrac{4x - 11}{x-3}}$ $(x \neq 3)$

65. $3 - \dfrac{3y}{y+1} = \dfrac{3(y+1) - 3y}{y+1} = \boxed{\dfrac{3}{y+1}}$ $(y \neq -1)$

67. $\dfrac{r}{(r-2)^2} + \dfrac{r-4}{r^2 - 4} = \dfrac{r}{(r-2)^2} + \dfrac{r-4}{(r-2)(r+2)} = \dfrac{r(r+2) + (r-4)(r-2)}{(r+2)(r-2)^2} = \dfrac{r^2 + 2r + r^2 - 6r + 8}{(r+2)(r-2)^2}$

$= \dfrac{2r^2 - 4r + 8}{(r+2)(r-2)^2}$

$= \boxed{\dfrac{2(r^2 - 2r + 4)}{(r+2)(r-2)^2}}$

$(r \neq -2, 2)$

69. $\dfrac{9x+3}{x^2 - x - 6} + \dfrac{x}{3-x} = \dfrac{3(3x+1)}{(x-3)(x+2)} + \dfrac{x}{3-x} = \dfrac{3(3x+1)}{(x-3)(x+2)} - \dfrac{x}{x-3} = \dfrac{3(3x+1) - x(x+2)}{(x-3)(x+2)}$

$= \boxed{\dfrac{-x^2 + 7x + 3}{(x-3)(x+2)}}$

$(x \neq -2, 3)$

71. $\dfrac{y+3}{5y^2} - \dfrac{y-5}{15y} = \dfrac{3(y+3)}{15y^2} - \dfrac{y(y-5)}{15y^2} = \dfrac{3y + 9 - y^2 + 5y}{15y^2} = \dfrac{-y^2 + 8y + 9}{15y^2} = \boxed{\dfrac{(-y+9)(y+1)}{15y^2}}$ $(y \neq 0)$

73. $\dfrac{x+1}{4x^2+4x-15}-\dfrac{4x+5}{8x^2-10x-3}=\dfrac{x+1}{(2x+5)(2x-3)}-\dfrac{4x+5}{(4x+1)(2x-3)}=\dfrac{(x+1)(4x+1)-(4x+5)(2x+5)}{(2x+5)(2x-3)(4x+1)}$

$=\dfrac{4x^2+5x+1-8x^2-30x-25}{(2x+5)(2x-3)(4x+1)}$

$=\boxed{\dfrac{-4x^2-25x-24}{(2x+5)(2x-3)(4x+1)}}$

$\left(x\neq-\dfrac{5}{2},\dfrac{3}{2},-\dfrac{1}{4}\right)$

75. $\dfrac{4x}{x^2-1}-\dfrac{2}{x}-\dfrac{2}{x+1}=-\dfrac{4x}{(x-1)(x+1)}-\dfrac{2}{x}-\dfrac{2}{x+1}=\dfrac{4x(x)-2(x-1)(x+1)-2x(x-1)}{x(x-1)(x+1)}$

$=\dfrac{4x^2-2x^2+2-2x^2+2x}{x(x-1)(x+1)}$

$=\dfrac{2(x+1)}{x(x-1)(x+1)}$

$=\boxed{\dfrac{2}{x(x-1)}}$

$(x\neq0,-1,1)$

77. $\dfrac{3}{2y-1}-\dfrac{1}{y+2}-\dfrac{5}{2y^2+3y-2}=\dfrac{3}{2y-1}-\dfrac{1}{y+2}-\dfrac{5}{(2y-1)(y+2)}=\dfrac{3(y+2)-(2y-1)-5}{(2y-1)(y+2)}$

$=\dfrac{3y+6-2y+1-5}{(2y-1)(y+2)}$

$=\dfrac{y+2}{(2y-1)(y+2)}$

$=\boxed{\dfrac{1}{2y-1}}$

$\left(y\neq-2,\dfrac{1}{2}\right)$

79. $\dfrac{1}{y}+\dfrac{y}{2y+4}-\dfrac{2}{y^2+2y}=\dfrac{1}{y}+\dfrac{y}{2(y+2)}-\dfrac{2}{y(y+2)}=\dfrac{2(y+2)+y^2-4}{2y(y+2)}=\dfrac{y^2+2y}{2y(y+2)}=\dfrac{y(y+2)}{2y(y+2)}=\boxed{\dfrac{1}{2}}$

$(y\neq0,-2)$

81. $\dfrac{9w+3}{w^2-w-6}+\dfrac{w}{3-w}+\dfrac{w-1}{w+2}=\dfrac{3(3w+1)}{(w-3)(w+2)}-\dfrac{w}{w-3}+\dfrac{w-1}{w+2}=\dfrac{3(3w+1)-w(w+2)+(w-1)(w-3)}{(w-3)(w+2)}$

$=\dfrac{9w+3-w^2-2w+w^2-4w+3}{(w-3)(w+2)}$

$=\dfrac{3w+6}{(w-3)(w+2)}$

$=\dfrac{3(w+2)}{(w-3)(w+2)}$

$=\boxed{\dfrac{3}{w-3}}$

$(w\neq-2,3)$

83. $\dfrac{a-1}{a}+\dfrac{b+1}{b}=\dfrac{b(a-1)+a(b+1)}{ab}=\dfrac{ab-b+ab+a}{ab}=\boxed{\dfrac{2ab+a-b}{ab}}$ $(a,b\neq0)$

85. $\dfrac{7}{3x^2}+\dfrac{4}{x^2}-\dfrac{10}{7x}=\dfrac{7\cdot7+4\cdot21-10\cdot3\cdot x}{21x^2}=\boxed{\dfrac{-30x+133}{21x^2}}$

87. $\dfrac{1}{a-b} - \dfrac{a}{a^2-ab} + \dfrac{a^2}{a^3-a^2b} = \dfrac{1}{a-b} - \dfrac{a}{a(a-b)} + \dfrac{a^2}{a^2(a-b)} = \dfrac{1-1+1}{a-b} = \boxed{\dfrac{1}{a-b}}$ $(a \neq 0, b)$

89. $\dfrac{a}{b} + \dfrac{c}{d} = \boxed{\dfrac{ad+bc}{bd}}$ $(b, d \neq 0)$

91. $\dfrac{y^2+5y+4}{y^2+2y-3} \cdot \dfrac{y^2+y-6}{y^2+2y-3} - \dfrac{2}{y-1} = \dfrac{(y+1)(y+4)(y+3)(y-2)}{(y+3)(y-1)(y+3)(y-1)} - \dfrac{2}{y-1} = \dfrac{(y+1)(y+4)(y-2)}{(y-1)(y+3)(y-1)} - \dfrac{2}{y-1}$

$= \dfrac{(y+1)(y+4)(y-2) - 2(y-1)(y+3)}{(y-1)^2(y+3)}$

$= \boxed{\dfrac{y^3+y^2-10y-2}{(y-1)^2(y+3)}}$

93. Statement $\boxed{D}$ is true.

D. $\dfrac{3}{2xy^2} = \dfrac{3 \cdot 5x^2y^2}{2xy^2 \cdot 5x^2y^2} = \dfrac{15x^2y^2}{10x^3y^4}$ True

95. Statement $\boxed{C}$ is true.

C. $\dfrac{2}{y} + 1 = \dfrac{2+y}{y}$

$\dfrac{2+y}{y} = \dfrac{2+y}{y}$ $(y \neq 0)$ True

97. perimeter $= 2\left(\dfrac{x+5}{3x^2} + \dfrac{2x+1}{2x}\right) = 2\left(\dfrac{2(x+5) + 3x(2x+1)}{6x^2}\right) = 2\left(\dfrac{2x+10+6x^2+3x}{6x^2}\right)$

$= \boxed{\dfrac{6x^2+5x+10}{3x^2}}$ $(x \neq 0)$

99. Let $\dfrac{70}{x}$ = the measure of one angle.

measure of complement $= 90 - \dfrac{70}{x} = \boxed{\dfrac{90x-70}{x}}$ degrees

101. Fraction left to paint $= 1 - \dfrac{1}{x} - \dfrac{1}{x+3} = \dfrac{x(x+3) - (x+3) - x}{x(x+3)} = \dfrac{x^2+3x-x-3-x}{x(x+3)} = \boxed{\dfrac{x^2+x-3}{x(x+3)}}$

$(x \neq 0, -3)$

103. a. $\dfrac{a}{b} + \dfrac{a}{c} \neq \dfrac{a}{b+c}$

Example: $\dfrac{2}{5} + \dfrac{2}{10} = \dfrac{6}{10}$

$\dfrac{2}{5+10} = \dfrac{2}{15}$

$\dfrac{6}{10} \neq \dfrac{2}{15}$

b. $\dfrac{a}{b} + \dfrac{a}{c} = \boxed{\dfrac{ac+ab}{bc}}$

Review Problems

107. $7y^4(3y + 5)(2y - 7) = 7y^4(6y^2 - 11y - 35) = \boxed{42y^6 - 77y^5 - 245y^4}$

108. $\boxed{3x - y \;<\; 3}$
$\qquad\quad y \;>\; 3x - 3$

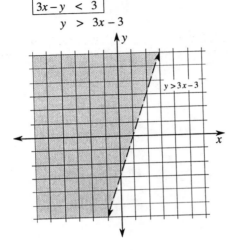

$y > 3x - 3$

109. $(-3, -4), (1, 0):\ m = \dfrac{-4 - 0}{-3 - 1} = 1$

using $(1, 0)$:
$$y - 0 \;=\; 1(x - 1)$$
$$y \;=\; x - 1$$
$$-x + y \;=\; -1$$
$$\boxed{x - y = 1}$$

Section 8.5 Complex Fractions

Problem Set 8.5, pp. 555-558

1. $\dfrac{\dfrac{1}{2} + \dfrac{1}{4}}{\dfrac{1}{2} + \dfrac{1}{3}} = \dfrac{\dfrac{4 + 2}{8}}{\dfrac{3 + 2}{6}} = \dfrac{3}{4} \cdot \dfrac{6}{5} = \dfrac{18}{20} = \boxed{\dfrac{9}{10}}$

3. $\dfrac{3 + \dfrac{1}{2}}{4 - \dfrac{1}{4}} = \dfrac{\dfrac{6 + 1}{2}}{\dfrac{16 - 1}{4}} = \dfrac{7}{2} \cdot \dfrac{4}{15} = \dfrac{28}{30} = \boxed{\dfrac{14}{15}}$

5. $\dfrac{\dfrac{2}{5} - \dfrac{1}{3}}{\dfrac{2}{3} - \dfrac{3}{4}} = \dfrac{\left(\dfrac{2}{5} - \dfrac{1}{3}\right)}{\left(\dfrac{2}{3} - \dfrac{3}{4}\right)} \cdot \dfrac{60}{60} = \dfrac{4(6 - 5)}{5(8 - 9)} = \dfrac{4}{-5} = \boxed{-\dfrac{4}{5}}$ (Using LCM of denominators)

7. $\dfrac{\frac{3}{4}-x}{\frac{3}{4}+x}=\dfrac{\frac{3-4x}{4}}{\frac{3+4x}{4}}=\boxed{\dfrac{3-4x}{3+4x}}$

9. $\dfrac{\frac{1}{y}-\frac{3}{2}}{\frac{1}{y}+\frac{3}{4}}=\dfrac{\frac{2-3y}{2y}}{\frac{4+3y}{4y}}=\dfrac{2-3y}{2y}\cdot\dfrac{4y}{4+3y}=\boxed{\dfrac{2(2-3y)}{4+3y}}\quad\left(y\neq0,-\dfrac{4}{3}\right)$

11. $\dfrac{5-\frac{2}{x}}{3+\frac{1}{x}}=\dfrac{\frac{5x-2}{x}}{\frac{3x+1}{x}}=\boxed{\dfrac{5x-2}{3x+1}}\quad\left(x\neq0,-\dfrac{1}{3}\right)$

13. $\dfrac{\frac{1}{3}-\frac{2}{x}}{4-\frac{1}{x^2}}=\dfrac{\frac{x-6}{3x}}{\frac{4x^2-1}{x^2}}=\dfrac{x-6}{3x}\cdot\dfrac{x^2}{4x^2-1}=\dfrac{x(x-6)}{3(4x^2-1)}=\boxed{\dfrac{x(x-6)}{3(2x-1)(2x+1)}}\quad\left(x\neq0,-\dfrac{1}{2},\dfrac{1}{2}\right)$

15. $\dfrac{\frac{2}{a}+\frac{7}{b}}{12}=\dfrac{2b+7a}{12ab}=\boxed{\dfrac{7a+2b}{12ab}}\quad(a,b\neq0)$

17. $\dfrac{3y-5}{\frac{5}{y}-3}=\dfrac{3y-5}{\frac{5-3y}{y}}=\dfrac{y(3y-5)}{5-3y}=\boxed{-y}\quad\left(y\neq0,\dfrac{5}{3}\right)$

19. $\dfrac{\frac{7}{r}+\frac{1}{3}}{\frac{2}{3}+\frac{5}{r}}=\dfrac{\frac{21+r}{3r}}{\frac{2r+15}{3r}}=\dfrac{21+r}{3r}\cdot\dfrac{3r}{2r+15}=\boxed{\dfrac{21+r}{2r+15}}$

21. $\dfrac{\frac{12}{y^2}-\frac{3}{y}}{\frac{15}{y}-\frac{9}{y^2}}=\dfrac{\frac{12-3y}{y^2}}{\frac{15y-9}{y^2}}=\dfrac{12-3y}{y^2}\cdot\dfrac{y^2}{15y-9}=\dfrac{12-3y}{15y-9}=\dfrac{(4-y)}{(5y-3)}=\boxed{\dfrac{4-y}{5y-3}}\quad\left(y\neq0,\dfrac{3}{5}\right)$

23. $\dfrac{\frac{1}{w}+\frac{2}{w^2}}{\frac{2}{w}+1}=\dfrac{\frac{w+2}{w^2}}{\frac{w(2+w)}{w^2}}=\dfrac{w+2}{w^2}\cdot\dfrac{w^2}{w(2+w)}=\dfrac{w+2}{w(2+w)}=\boxed{\dfrac{1}{w}}\quad(w\neq0,-2)$

25. $\dfrac{\frac{9}{5}+\frac{4}{5s}}{\frac{4}{s^2}+\frac{9}{5}}=\dfrac{9s+4}{5s}\cdot\dfrac{5s^2}{20+9s^2}=\boxed{\dfrac{s(9s+4)}{9s^2+20}}\quad(s\neq0)$

27. $\dfrac{\frac{7}{x^3}+\frac{11}{x^2}}{\frac{7}{x^4}+\frac{11}{x^3}}=\dfrac{7+11x}{x^3}\cdot\dfrac{x^4}{7+11x}=\boxed{x}\quad\left(x\neq0,-\dfrac{11}{7}\right)$

29. $\dfrac{\dfrac{5}{y+3}}{\dfrac{15}{y^2-9}} = \dfrac{5}{y+3} \cdot \dfrac{(y-3)(y+3)}{15} = \boxed{\dfrac{y-3}{3}}$ $(y \neq -3, 3)$

31. $\dfrac{\dfrac{1}{y+2}}{1+\dfrac{1}{y+2}} = \dfrac{\dfrac{1}{y+2}}{\dfrac{y+3}{y+2}} = \boxed{\dfrac{1}{y+3}}$ $(y \neq -2, -3)$

33. $\dfrac{\dfrac{x}{x+3}}{\dfrac{x}{x+3}+x} = \dfrac{\dfrac{x}{x+3}}{\dfrac{x+x^2+3x}{x+3}} = \dfrac{x}{x^2+4x} = \boxed{\dfrac{1}{x+4}}$ $(x \neq 0, -4)$

35. $\dfrac{2+\dfrac{2}{y-1}}{2-\dfrac{2}{y}} = \dfrac{\dfrac{2y-2+2}{y-1}}{\dfrac{2y-2}{y}} = \dfrac{2y}{y-1} \cdot \dfrac{y}{2(y-1)} = \boxed{\dfrac{y^2}{(y-1)^2}}$ $(y \neq 0, 1)$

37. $\dfrac{y+\dfrac{4}{y-1}}{y-\dfrac{y+1}{y-1}} = \dfrac{\dfrac{y^2-y+4}{y-1}}{\dfrac{y^2-y-y-1}{y-1}} = \boxed{\dfrac{y^2-y+4}{y^2-2y-1}}$ $(y \neq 1)$

39. $\dfrac{\dfrac{4}{y+1}+\dfrac{3}{y}}{\dfrac{2}{y+1}-\dfrac{3}{y}} = \dfrac{\dfrac{4y+3y+3}{y(y+1)}}{\dfrac{2y-3y-3}{y(y+1)}} = \dfrac{7y+3}{-y-3} = \boxed{-\dfrac{7y+3}{y+3}}$ $(y \neq 0, -1, -3)$

41. $\dfrac{\dfrac{1}{r+2}-\dfrac{3}{r+3}}{\dfrac{2}{r+3}+\dfrac{3}{r+2}} = \dfrac{\dfrac{r+3-3r-6}{(r+2)(r+3)}}{\dfrac{2r+4+3r+9}{(r+2)(r+3)}} = \dfrac{-2r-3}{5r+13} = \boxed{-\dfrac{2r+3}{5r+13}}$ $\left(r \neq -2, -3, -\dfrac{13}{5}\right)$

43. $\dfrac{\dfrac{3}{w-2}-\dfrac{4}{w+2}}{\dfrac{7}{w^2-4}} = \dfrac{\dfrac{3w+6-4w+8}{(w-2)(w+2)}}{\dfrac{7}{(w-2)(w+2)}} = \boxed{\dfrac{-w+14}{7}}$ $(w \neq -2, 2)$

45. $\dfrac{\dfrac{6}{x^2+2x-15}-\dfrac{1}{x-3}}{\dfrac{1}{x+5}+1} = \dfrac{\dfrac{6}{(x+5)(x-3)}-\dfrac{1}{x-3}}{\dfrac{1+x+5}{x+5}} = \dfrac{\dfrac{6-(x+5)}{(x+5)(x-3)}}{\dfrac{x+6}{x+5}} = \dfrac{1-x}{(x+5)(x-3)} \cdot \dfrac{x+5}{x+6} = \boxed{\dfrac{1-x}{(x-3)(x+6)}}$

$(x \neq -6, -5, 3)$

47. $\dfrac{\dfrac{y+2}{y}-\dfrac{3}{y+1}}{\dfrac{y-1}{y^2+y}+\dfrac{3}{y+1}} \cdot \dfrac{\dfrac{(y+2)(y+1)-3y}{y(y+1)}}{\dfrac{y^2-1+3y^2+3y}{y(y+1)(y+1)}} = \dfrac{y^2+3y+2-3y}{y(y+1)} \cdot \dfrac{y(y+1)^2}{4y^2+3y-1} = \dfrac{(y^2+2)(y+1)}{(4y-1)(y+1)} = \boxed{\dfrac{y^2+2}{4y-1}}$

49. $\dfrac{\dfrac{10}{y}+\dfrac{5}{y^2-y}}{\dfrac{5}{y^2-y}-\dfrac{15}{y-1}} = \dfrac{\dfrac{10}{y}+\dfrac{5}{y(y-1)}}{\dfrac{5}{y(y-1)}-\dfrac{15}{y-1}} = \dfrac{10(y-1)+5}{y(y-1)} \cdot \dfrac{y(y-1)}{5-15y} = \dfrac{10y-5}{5-15y} = \boxed{\dfrac{2y-1}{1-3y}}$ $\left(y \neq 0, 1, \dfrac{1}{3}\right)$

51. $\dfrac{2 + 3y^{-1}}{1 - 6y^{-1}} = \dfrac{2 + \dfrac{3}{y}}{1 - \dfrac{7}{y}} = \boxed{\dfrac{2y + 3}{y - 7}}$ $\left(\text{Multiply by } \dfrac{y}{y}\right)$ $(y \ne 0, 7)$

53. $\dfrac{1 + 4x^{-1}}{1 - 16x^{-2}} = \dfrac{1 + \dfrac{4}{x}}{1 - \dfrac{16}{x}} = \dfrac{x^2 + 4x}{x^2 - 16} = \dfrac{x(x + 4)}{(x + 4)(x - 4)} = \boxed{\dfrac{x}{x - 4}}$ $(x \ne 0, -4, 4)$

55. $\dfrac{y^{-1} - (y + 5)^{-1}}{5} = \dfrac{\dfrac{1}{y} - \dfrac{1}{y + 5}}{5} = \dfrac{(y + 5) - y}{5y(y + 5)} = \boxed{\dfrac{1}{y(y + 5)}}$ $(y \ne 0, -5)$

57. Statement $\boxed{\text{B}}$ is true.

B. $\dfrac{y - \dfrac{1}{2}}{y + \dfrac{3}{4}} = \dfrac{\dfrac{2y - 1}{2}}{\dfrac{4y + 3}{4}} = \dfrac{2y - 1}{2} \cdot \dfrac{4}{4y + 3} = \dfrac{2(y - 1)}{4y + 3} = \dfrac{4y - 2}{4y + 3}$

True for all values except $y = -\dfrac{3}{4}$.

59. $r_{\text{average}} = \dfrac{2d}{\dfrac{d}{r_1} + \dfrac{d}{r_2}} = \dfrac{2d}{\dfrac{dr_2 + dr_1}{r_1 r_2}} = \dfrac{2dr_1 r_2}{d(r_1 + r_2)} = \boxed{\dfrac{2r_1 r_2}{r_2 + r_2}}$

$r_1 = 30, \ r_2 = 20$

$r_{\text{average}} = \dfrac{2(30)(20)}{30 + 20} = \dfrac{1200}{50} = \boxed{24 \text{ miles per hour}}$

> The answer is not 25 miles per hour, which would be $\dfrac{r_1 + r_2}{2}$, but is the total distance divided by the total time.

61. $\dfrac{1 - \dfrac{9}{y^2}}{\dfrac{9}{y^2} + \dfrac{6}{y} + 1} - \dfrac{1 - \dfrac{4}{y^2}}{\dfrac{6}{y^2} + \dfrac{5}{y} + 1} = \dfrac{y^2 - 9}{9 + 6y + y^2} - \dfrac{y^2 - 4}{6 + 5y + y^2} = \dfrac{(y - 3)(y + 3)}{(y + 3)^2} - \dfrac{(y - 2)(y + 2)}{(y + 2)(y + 3)} = \dfrac{y - 3}{y + 3} - \dfrac{y - 2}{y + 3}$

$= \dfrac{y - 3 - y + 2}{y + 3}$

$= \boxed{-\dfrac{1}{y + 3}}$

$(y \ne 0, -2, -3)$

63. $\dfrac{R_1 R_2}{R_1 + R_2} = \dfrac{\dfrac{x + 3}{x^3} \cdot \dfrac{3}{x}}{\dfrac{x + 3}{x^3} + \dfrac{3}{x}} = \dfrac{\dfrac{3(x + 3)}{x^4}}{\dfrac{x + 3 + 3x^2}{x^3}} = \dfrac{3(x + 3)}{x^4} \cdot \dfrac{x^3}{3x^2 + x + 3} = \boxed{\dfrac{3(x + 3)}{x(3x^2 + x + 3)}}$

65. $1 + \dfrac{1}{1+1} = 1 + \dfrac{1}{2} = \dfrac{3}{2}$ or $1\dfrac{1}{2}$

$1 + \dfrac{1}{1+\dfrac{1}{1+1}} = 1 + \dfrac{1}{1+\dfrac{1}{2}} = 1 + \dfrac{1}{\dfrac{3}{2}} = 1 + \dfrac{2}{3} = \dfrac{5}{3}$ or $1\dfrac{2}{3}$

$1 + \dfrac{1}{1+\dfrac{1}{1+\dfrac{1}{1+1}}} = 1 + \dfrac{1}{\dfrac{5}{3}} = 1 + \dfrac{3}{5} = \dfrac{8}{5}$ or $1\dfrac{3}{5}$

> The pattern is $1 + \dfrac{1}{2}$, $1 + \dfrac{2}{3}$, $1 + \dfrac{3}{5}$, $1 + \dfrac{1}{\dfrac{8}{5}} = 1\dfrac{5}{8}$, $1 + \dfrac{1}{\dfrac{13}{8}} = 1\dfrac{8}{13}$, $1 + \dfrac{1}{\dfrac{21}{13}} = 1\dfrac{13}{21}$.

> If the nth term is $1 + \dfrac{a}{b}$, the $(n+1)$st term is $1 + \dfrac{b}{a+b}$.

Review Problems

68. Let $x =$ the number of nickels in the third pile.

$$2[15(0.05) + 21(0.05) + x(0.05)] = 4.50$$
$$0.75 + 1.05 + 0.05x = 2.25$$
$$1.80 + 0.05x = 2.25$$
$$0.05x = 0.45$$
$$x = \boxed{9 \text{ nickels}}$$

69.
$$h = -16t^2 + 32t \qquad h = 16 \text{ feet}$$
$$16 = -16t^2 + 32t$$
$$1 = -t^2 + 2t$$
$$t^2 - 2t + 1 = 0$$
$$(t-1)^2 = 0$$
$$t = \boxed{1 \text{ second}}$$

70.
$$f(x) = 4x - 3$$
$$f(-2) = 4(-2) - 3 = -8 - 3 = -11$$
$$3f((4) = 3[4(4) - 3] = 3(16 - 3) = 39$$
$$f(-2) + 3f(4) = -11 + 39 = \boxed{28}$$

Section 8.6 Equations Containing Rational Expressions

Problem Set 8.6, pp. 565-567

1.
$$2 - \dfrac{8}{x} = 6$$
$$x\left(2 - \dfrac{8}{x}\right) = 6x \qquad \text{Multiply each side of equation by LCM of denominators } (x \neq 0).$$
$$2x - 8 = 6x \qquad \text{Simplify.}$$
$$-8 = 4x$$
$$x = -2$$

$\boxed{\{-2\}}$

3.
$$\frac{2}{3} - \frac{5}{6} = \frac{1}{y}$$
$$6y\left(\frac{2}{3} - \frac{5}{6}\right) = 6y\left(\frac{1}{y}\right) \qquad (y \neq 0)$$
$$6y\left(\frac{-1}{6}\right) = 6$$
$$-y = 6$$
$$y = -6$$
$$\boxed{\{-6\}}$$

5.
$$\frac{4}{y} + \frac{1}{2} = \frac{5}{y}$$
$$\frac{8+y}{2y} = \frac{5}{y}$$
$$2y\left(\frac{8+y}{2y}\right) = 2y\left(\frac{5}{y}\right)$$
$$8+y = 10$$
$$y = 10 - 8 = 2$$
$$\boxed{\{2\}}$$

7.
$$\frac{2}{y} + 3 = \frac{5}{2y} + \frac{13}{4}$$
$$4y\left(\frac{2+3y}{y}\right) = 4y\left(\frac{10+13y}{4y}\right)$$
$$4(2+3y) = 10 + 13y$$
$$8 + 12y = 10 + 13y$$
$$y = -2$$
$$\boxed{\{-2\}}$$

9.
$$\frac{1}{z-1} + 5 = \frac{11}{z-1}$$
$$5 = \frac{11-1}{z-1} = \frac{10}{z-1}$$
$$5z - 5 = 10$$
$$z = 3$$
$$\boxed{\{3\}}$$

11.
$$\frac{4}{r^2-4} + \frac{2}{r-2} = \frac{1}{r+2} \qquad \text{LCM} = r^2 - 4$$
$$(r^2-4)\left[\frac{4}{r^2-4} + \frac{2}{r-2}\right] = (r^2-4)\left(\frac{1}{r+2}\right)$$
$$4 + 2(r+2) = r - 2$$
$$4 + 2r + 4 = r - 2$$
$$r = -8 - 2 = -10$$
$$\boxed{\{-10\}}$$

13.
$$\frac{4}{y-3} - \frac{2}{y-2} = \frac{7-y}{y^2-5y+6} \qquad \text{LCM} = y^2 - 5y + 6 = (y-2)(y-3)$$
$$(y-3)(y-2)\left[\frac{4}{y-3} - \frac{2}{y-2}\right] = (y^2-5y+6)\left[\frac{7-y}{y^2-5y+6}\right] \qquad (y \neq 2, 3)$$
$$4(y-2) - 2(y-3) = 7 - y$$
$$4y - 8 - 2y + 6 = 7 - y$$
$$3y = 9$$
$$y = 3 \qquad \text{but } y \neq 3$$
Since y is not permitted to be 3, there is no solution.
$$\boxed{\varnothing}$$

15.
$$\frac{5}{2x+6} - \frac{1}{x+3} = \frac{1}{x+1} \qquad \text{LCM} = 2(x+3)(x+1)$$
$$2(x+3)(x+1)\left(\frac{5}{2x+6} - \frac{1}{x+3}\right) = 2(x+3)(x+1)\left(\frac{1}{x+1}\right)$$
$$5(x+1) - 2(x+1) = 2(x+3)$$
$$3(x+1) = 2(x+3)$$
$$3x + 3 = 2x + 6$$
$$x = 3$$
$$\boxed{\{3\}}$$

17.
$$\frac{3y}{y-4} - 5 = \frac{12}{y-4}$$

$$(y-4)\left[\frac{3y}{y-4} - 5\right] = (y-4)\left(\frac{12}{y-4}\right)$$

$$3y - 5(y-4) = 12$$
$$-2y + 20 = 12$$
$$-2y = -8$$
$$y = 4 \qquad \text{but } y \neq 4$$

Since $y = 4$ is not permitted, there is no solution.

$$\boxed{\varnothing}$$

19.
$$\frac{4}{w} - \frac{w}{2} = \frac{7}{2}$$

$$(2w)\left(\frac{4}{w} - \frac{w}{2}\right) = 2w\left(\frac{7}{2}\right)$$

$$8 - w^2 = 7w$$
$$w^2 + 7w - 8 = 0$$
$$(w+8)(w-1) = 0$$
$$w = 1 \text{ or } w = -8$$

Check: $w = 1$: $\quad \dfrac{4}{1} - \dfrac{1}{2} = \dfrac{7}{2} \quad$ true

$\qquad w = -8$: $\quad \dfrac{4}{8} - \dfrac{-8}{2} = -\dfrac{1}{2} + \dfrac{8}{2} = \dfrac{7}{2} \quad$ true

Solution set = $\boxed{\{-8, 1\}}$

21.
$$\frac{5}{3y-8} = \frac{y}{y+2} \qquad\qquad \left(y \neq -2, \frac{8}{3}\right)$$

$$5(y+2) = 3y^2 - 8y$$

$$5y + 10 = 3y^2 - 8y$$
$$3y^2 - 13y - 10 = 0$$
$$(3y+2)(y-5) = 0$$
$$y = -\frac{2}{3} \text{ or } y = 5$$

Check: $y = -\dfrac{2}{3}$: $\quad \dfrac{5}{-2-8} = \dfrac{-\dfrac{2}{3}}{\dfrac{4}{3}}$

$$\qquad\qquad -\frac{1}{2} = -\frac{1}{2} \quad \text{true}$$

$\qquad y = 5$: $\quad \dfrac{5}{15-8} = \dfrac{5}{5+2}$

$$\qquad\qquad \frac{5}{7} = \frac{5}{7} \quad \text{true}$$

Solution set is $\boxed{\left\{-\dfrac{2}{3}, 5\right\}}$

23.
$$\frac{3}{z-1}+\frac{8}{z} = 3$$
$$z(z-1)\left(\frac{3}{z-1}+\frac{8}{z}\right) = z(z-1)(3)$$
$$3z+8(z-1) = 3z(z-1)$$
$$3z+8z-8 = 3z^2-3z$$
$$3z^2-14z+8 = 0$$
$$(3z-2)(z-4) = 0$$
$$z = \frac{2}{3} \text{ or } z=4$$

Check: $z=\frac{2}{3}$:
$$\frac{3}{-\frac{1}{3}}+\frac{8}{\frac{2}{3}} = -9+12 = 3$$
$$3 = 3 \quad \text{true}$$

$z=4$:
$$\frac{3}{3}+\frac{8}{4} = 1+2 = 3$$
$$3 = 3 \quad \text{true}$$

Solution set is $\boxed{\left\{\frac{2}{3},4\right\}}$

25.
$$\frac{2}{y-2}+\frac{y}{y+2} = \frac{y+6}{y^2-4} \qquad \text{LCM}=y^2-4$$
$$(y^2-4)\left(\frac{2}{y-2}+\frac{y}{y+2}\right) = (y^2-4)\left(\frac{y+6}{y^2-4}\right) \qquad (y\neq-2,2)$$
$$2(y+2)+y(y-2) = y+6$$
$$2y+4+y^2-2y = y+6$$
$$y^2-y-2 = 0$$
$$(y-2)(y+1) = 0$$
$$y = -1 \text{ or } y=2 \quad (\text{reject } y=2 \text{ since } y\neq2)$$

Solution set is $\boxed{\{-1\}}$

27.
$$\frac{4}{y-3}-\frac{3}{y+3} = \frac{1}{y^2-9} \qquad \text{LCM}=y^2-9$$
$$(y^2-9)\left(\frac{4}{y-3}-\frac{3}{y+3}\right) = (y^2-9)\left(\frac{1}{y^2-9}\right) \qquad (y\neq-3,3)$$
$$4(y+3)-3(y-3) = 1$$
$$4y+12-3y+9 = 1$$
$$y = -20$$

$\boxed{\{-20\}}$

29.
$$\frac{4x}{x-4}-\frac{3x}{x-2} = \frac{-3}{x^2-6x+8}$$
$$\frac{4x(x-2)-3x(x-4)}{(x-4)(x-2)} = \frac{-3}{(x-2)(x-4)}$$
$$4x^2-8x-3x^2+12x = -3$$
$$x^2+4x+3 = 0$$
$$(x+3)(x+1) = 0$$
$$x = -3 \text{ or } x=-1$$

$\boxed{\{-3,-1\}}$

31.

$$\frac{y}{y-3}+\frac{4}{y+1} = \frac{y^2-2y+2}{y^2-2y-3}$$

$$\frac{y(y+1)+4(y-3)}{(y-3)(y+1)} = \frac{y^2-2y+2}{(y-3)(y+1)}$$

$$y^2+y+4y-12 = y^2-2y+2$$

$$7y-14 = 0$$

$$y = 2$$

$$\boxed{\{2\}}$$

33.

$$\frac{x+2}{x^2-x}-\frac{6}{x^2-1} = 0$$

$$\frac{x+2}{x(x-1)}-\frac{6}{(x-1)(x+1)} = 0 \qquad (x \neq 0,-1,1)$$

$$(x+2)(x+1)-6x = 0$$

$$x^2+3x+2-6x = 0$$

$$x^2-3x+2 = 0$$

$$(x-1)(x-2) = 0$$

$$x = 2 \qquad (\text{reject } x=1 \text{ since } x \neq 1)$$

$$\boxed{\{2\}}$$

35.

$$\frac{w}{2w-2}+\frac{w+1}{2w+1} = \frac{-3w}{2w^2-w-1}$$

$$\frac{w}{2(w-1)}+\frac{w+1}{2w+1} = \frac{-3w}{(2w+1)(w-1)} \qquad \text{LCM}=2(w-1)(2w+1)$$

$$w(2w+1)+2(w+1)(w-1) = -6w$$

$$2w^2+w+2w^2-2 = -6w$$

$$4w^2+7w-2 = 0$$

$$(4w-1)(w+2) = 0$$

$$w = \frac{1}{4} \text{ or } w=-2$$

$$\boxed{\left\{-2,\frac{1}{4}\right\}}$$

37.

$$\frac{2}{y+3}+\frac{4}{y+1} = \frac{4}{3}$$

$$6(y+1)+12(y+3) = 4(y+3)(y+1)$$

$$6y+6+12y+36 = 4y^2+16y+12$$

$$4y^2-2y-30 = 0$$

$$2y^2-y-15 = 0$$

$$(2y+5)(y-3) = 0$$

$$y = -\frac{5}{2} \text{ or } y=3$$

$$\boxed{\left\{-\frac{5}{2},3\right\}}$$

39.

$$\frac{w+1}{2w^2-11w+5} = \frac{w-7}{2w^2+9w-5} - \frac{2w-6}{w^2-25}$$

$$\frac{w+1}{(2w-1)(w-5)} = \frac{w-7}{(2w-1)(w+5)} - \frac{2w-6}{(w-5)(w+5)} \qquad \text{LCM} = (2w-1)(w-5)(w+5)$$

$$(w+1)(w+5) = (w-7)(w-5) - (2w-6)(2w-1)$$
$$w^2+6w+5 = w^2-12w+35-4w^2+14w-6$$
$$4w^2+4w-24 = 0$$
$$w^2+w-6 = 0$$
$$(w+3)(w-2) = 0$$
$$w = -3 \text{ or } w = 2$$

$$\boxed{\{-3, 2\}}$$

41.

$$5y^{-2}+1 = 6y^{-1}$$
$$\frac{5}{y^2}+1 = \frac{6}{y} \qquad \text{LCM} = y^2$$
$$5+y^2 = 6y \qquad (y \neq 0)$$
$$y^2-6y+5 = 0$$
$$(y-1)(y-5) = 0$$
$$y = 1 \text{ or } y = 5$$

$$\boxed{\{1, 5\}}$$

43.

$$\frac{1}{y+5}+\frac{10}{y^2-25} = \frac{3}{y-5} \qquad \text{LCM} = (y+5)(y-5)$$
$$y-5+10 = 3(y+5)$$
$$y+5 = 3y+15 \qquad (y \neq -5, 5)$$
$$2y = -10$$
$$y = -5 \qquad \text{but } y \neq -5$$

Since y is not permitted to be -5, there is no solution.

$$\boxed{\varnothing}$$

45.

$$\frac{3}{y+1}-\frac{1}{1-y} = \frac{10}{y^2-1} \qquad \text{LCM} = (y+1)(y-1)$$
$$3(y-1)+(y+1) = 10 \qquad (y \neq -1, 1)$$
$$4y-2 = 10$$
$$4y = 12$$
$$y = 3$$

$$\boxed{\{3\}}$$

47.

$$P = \frac{500(1+3t)}{5+t} \qquad P = 1000$$
$$1000 = \frac{500(1+3t)}{5+t}$$
$$2(5+t) = 1+3t \qquad (t \neq -5)$$
$$10+2t = 1+3t$$
$$t = 9 \text{ hours}$$

It will take $\boxed{9 \text{ hours}}$ for the population to increase to 1000 insects.

49.

$$y = \frac{0.8x}{1 + 0.03x}$$

$$20 = \frac{0.8x}{1 + 0.03x}$$

$$20 + 0.6x = 0.8x$$

$$20 = 0.2x$$

$$x = \boxed{100 \text{ prey per unit area}}$$

51.

$$Q = \frac{PV}{P + V}$$

$$85 = \frac{80000V}{80000 + V}$$

$$85(80000) + 85V = 80000V$$

$$V = \frac{85(80000)}{80000 - 85} \approx 85.09$$

$$\boxed{\text{last year's volume, approximately } 85.09 \text{ thousands} \approx 85090}$$

53. Statement $\boxed{\text{B}}$ is true.

B.
$$\frac{y + 7}{2y - 6} = \frac{5}{y - 3} - 1$$

$$\frac{y + 7}{2(y - 3)} = \frac{5 - y + 3}{y - 3}$$

$$\frac{y + 7}{2(y - 3)} = \frac{8 - y}{(y - 3)}$$

$$y + 7 = 16 - 2y$$

$$3y = 9$$

$$y = 3 \quad \text{but } y \neq 3$$

no solution; true

55.

$$\frac{x - 17245}{x + 23927} = \frac{0.04301}{0.00526}$$

$$0.00526x - 90.7087 = 0.04301x + 1029.1003$$

$$0.03775x = -1119.809$$

$$x \approx -29663.8$$

$$\boxed{\{-29663.8\}}$$

Review Problems

58. Let x = the unknown calories

$$\frac{28.4}{110} = \frac{42.6}{x}$$

$$x = \frac{110(42.6)}{28.4} = \boxed{165 \text{ calories}}$$

59. $\boxed{\text{In C, } y \text{ is a function of } x \text{; it satisfies the vertical line test.}}$

60. Let x = the price of an orchestra seat and y = the price of a mezzanine seat.

$$4x + 2y = 22 \quad (\times -1) \qquad -4x - 2y = -22$$
$$2x + 3y = 16 \quad (\times 2) \qquad \underline{4x + 6y = 32}$$
$$4y = 10$$
$$y = 2.5$$

$$4x = 22 - 2(2.50) \qquad \leftarrow \text{(substitute for } y)$$
$$x = \frac{17}{4} = 4.25$$

Hence, $x = \$4.25$, $y = \$2.50$.

price of an orchestra seat: $\boxed{\$2.50}$

Section 8.7 Problem Solving

Problem Set 8.7, pp. 578-582

1. Let x = the numerator of the fraction. $x + 5$ = the denominator of the fraction.

$$\frac{x+1}{x+5+2} = \frac{1}{3}$$
$$3x + 3 = x + 7$$
$$2x = 4$$
$$x = 2$$
$$x + 5 = 7$$
$$\text{fraction} = \boxed{\frac{2}{7}}$$

Check: $\dfrac{2+1}{7+2} = \dfrac{3}{9} = \dfrac{1}{3}$

3. Let x = the numerator of the fraction. $2x + 1$ = the denominator of the fraction.

$$\frac{x-6}{2x+1-6} = \frac{5}{3}$$
$$3x - 18 = 10x - 25$$
$$7x = 7$$
$$x = 1$$
$$2x + 1 = 3$$
$$\text{fraction} = \boxed{\frac{1}{3}}$$

5. Let x = the number to be added.

$$\frac{7+x}{3+x} = \frac{5}{3}$$
$$21 + 3x = 15 + 5x$$
$$2x = 6$$
$$x = 3$$

The number to be added is $\boxed{3}$.

7. Let x = the unknown number. $\dfrac{1}{x}$ = the reciprocal of the number.

$$x + \frac{1}{x} = \frac{25}{12}$$

$$\frac{x^2 + 1}{x} = \frac{25}{12}$$

$$12x^2 + 12 = 25x$$

$$12x^2 - 25x + 12 = 0$$

$$(4x - 3)(3x - 4) = 0$$

$$x = \frac{3}{4} \text{ or } x = \frac{4}{3}$$

The number is $\boxed{\dfrac{3}{4} \text{ or } \dfrac{4}{3}}$.

9. Let x = the unknown number. $\dfrac{1}{x}$ = the reciprocal of the number.

$$1 - \frac{1}{x} = 3\left(\frac{1}{x}\right)$$

$$\frac{x - 1}{x} = \frac{3}{x}$$

$$x - 1 = 3 \qquad (x \neq 0)$$

$$x = 4$$

The number is $\boxed{4}$.

11. Let x = the smaller integer. $x + 1$ = the next consecutive integer.

$$\frac{1}{x} + \frac{1}{x + 1} = \frac{5}{6}$$

$$\frac{x + 1 + x}{x(x + 1)} = \frac{5}{6}$$

$$\frac{2x + 1}{x(x + 1)} = \frac{5}{6}$$

$$12x + 6 = 5x^2 + 5x$$

$$5x^2 - 7x - 6 = 0$$

$$(5x + 3)(x - 2) = 0$$

$$x = 2$$

$$\left(\text{Reject } x = -\frac{3}{5} \text{ since } -\frac{3}{5} \text{ is not an integer.}\right)$$

$$x + 1 = 3$$

The integers are $\boxed{2 \text{ and } 3}$.

13. Let x = the smaller number. $x + 14$ = the larger number.

$$\frac{x + 14}{x} = 3 + \frac{4}{x}$$

$$\frac{x + 14}{x} = \frac{3x + 14}{x} \qquad (x \neq 0)$$

$$x + 14 = 3x + 4$$

$$2x = 10$$

$$x = 5$$

$$x + 14 = 19$$

The numbers are $\boxed{5 \text{ and } 19}$.

15. Let x = the smaller number. $3x + 1$ = the larger number.

$$\frac{3x + 1}{x} = 3 + \frac{1}{x}$$

$$\frac{3x + 1}{x} = \frac{3x + 1}{x}$$

$$3x + 1 = 3x + 1$$

We have an identity, and so x can be $\boxed{\text{any real number except 0}}$. Also, to have a remainder, $3x + 1$ must not be divisible by x. Thus,

$$3x + 1 \neq kx$$
$$(k - 3)x \neq 1$$
$$x \neq \frac{1}{k - 3}$$
$$k \neq 3$$

k is an integer.

17. Let x = the smaller number. $11 - x$ = the larger number.

$$\frac{11 - x}{x} = 2 + \frac{2}{x}$$

$$\frac{11 - x}{x} = \frac{2x + 2}{x} \qquad (x \neq 0)$$

$$11 - x = 2x + 2$$
$$3x = 9$$
$$x = 3$$
$$11 - x = 8$$

The numbers are $\boxed{3 \text{ and } 8}$.

19. Let x = the units' digit. $x + 5$ = the tens' digit.

The number is $10(x + 5) + x$

$$\frac{10(x + 5) + x}{x + 5 + x} = 7 + \frac{6}{2x + 5}$$

$$\frac{11x + 50}{2x + 5} = \frac{14x + 35 + 6}{2x + 5}$$

$$11x + 50 = 14x + 41$$
$$3x = 9$$
$$x = 3 = \text{the units' digit}$$
$$x + 5 = 8 = \text{the tens' digit}$$

The number is $\boxed{83}$.

21. Let x = the tens' digit. $x + 2$ = the units' digit. $10x + (x + 2)$ = the number.

$$\frac{10x + (x + 2)}{x + (x + 2)} = 4 + \frac{3}{2x + 2}$$

$$\frac{11x + 2}{2x + 2} = \frac{8x + 8 + 3}{2x + 2}$$

$$11x + 2 = 8x + 11$$
$$3x = 9$$
$$x = 3 = \text{tens' digit}$$
$$x + 2 = 5 = \text{units' digit}$$

The number is $\boxed{35}$.

23. Let x equal the units' digit. $x + 3 =$ the tens' digit. $10(x + 3) + x =$ the number.

$$\frac{10(x + 3) + x}{9x} = 2 + \frac{2}{9x}$$

$$\frac{11x + 30}{9x} = \frac{18x + 2}{9x}$$

$$11x + 30 = 18x + 2$$

$$7x = 28$$

$$x = 4 = \text{the units' digit}$$

$$x + 3 = 7 = \text{the tens' digit}$$

The number is $\boxed{74}$.

25. Let $3x =$ the numerator. $4x =$ the denominator. $\dfrac{3x}{4x} = \dfrac{3}{4} =$ the fraction (reduced).

$$\frac{3x - 2}{4x + 4} = \frac{1}{2}$$

$$6x - 4 = 4x + 4$$

$$2x = 8$$

$$x = 4$$

$$3x = 12$$

$$4x = 16$$

$$\text{fraction} = \frac{12}{16}$$

Check: $$\frac{12}{16} = \frac{3}{4}$$

$$\frac{12 - 2}{16 + 4} = \frac{1}{2}$$

Original fraction $= \boxed{\dfrac{12}{16}}$

27. Let $x =$ the numerator of the fraction. $x + 8 =$ the denominator of the fraction.

$$\frac{x + 1}{x + 8 - 11} = \frac{x + 8}{x}$$

$$\frac{x + 1}{x - 3} = \frac{x + 8}{x}$$

$$x^2 + x = (x - 3)(x + 8)$$

$$x^2 + x = x^2 + 5x - 24 \qquad (x \neq 0, 3)$$

$$4x = 24$$

$$x = 6$$

$$x + 8 = 14$$

first fraction $= \boxed{\dfrac{6}{14}}$

Check: $$\frac{7}{3} = \frac{14}{6}$$

29. Let $x =$ the number.

$$\frac{1}{x^2} = 3\left(\frac{1}{x}\right) - 2 = \frac{3}{x} - 2 = \frac{3 - 2x}{x} \qquad (x \neq 0)$$

$$1 = x(3 - 2x) = 3x - 2x^2$$

$$2x^2 - 3x + 1 = 0$$

$$(2x - 1)(x - 1) = 0$$

$$x = 1 \text{ or } \frac{1}{2}$$

The number is $\boxed{1 \text{ or } \dfrac{1}{2}}$

31.
$$C = \frac{4x}{100 - x}$$
$$100C - Cx = 4x \qquad (x \neq 100)$$
$$(4 + C)x = 100C$$
$$\boxed{x = \frac{100C}{4 + C}}$$

If $C = 16$, $x = \dfrac{1600}{4 + 16} = \dfrac{1600}{20} = \boxed{80\%}$

33.
$$P = 30 - \frac{9}{t+1} = \frac{30t + 30 - 9}{t+1} = \frac{30t + 21}{t+1}$$
$$Pt + P = 30t + 21$$
$$(P - 30)t = 21 - P$$
$$t = \frac{21 - P}{P - 30}$$
$$\boxed{t = \frac{P - 21}{30 - P}}$$

If $P = 29$, $t = \dfrac{29 - 21}{30 - 29} = 8$ years from 1990, or the year $\boxed{1998}$.

35.
$$B = \frac{F}{S - V}$$

a.
$$BS - BV = F$$
$$S = \frac{BV + F}{B}$$
$$S = \boxed{V + \frac{F}{B}}$$

b.   If $F = 20000$, $V = 60$, and $B = 100$, then
$$S = 60 + \frac{20000}{100} = 60 + 200 = \boxed{\$260}$$

37.
$$\frac{1}{R} = \frac{1}{R_1} + \frac{1}{R_2}$$

a.   If $R = 4$ ohms, $R_1 = 12$ ohms, then,
$$\frac{1}{4} = \frac{1}{12} + \frac{1}{R_2}$$
$$\frac{1}{R_2} = \frac{1}{4} - \frac{1}{12}$$
$$\frac{1}{R_2} = \frac{1}{6}$$
$$R_2 = \boxed{6 \text{ ohms}}$$

b.   If $R_1 = 3R_2$, $R = 15$ ohms, then ,
$$\frac{1}{15} = \frac{1}{3R_2} + \frac{1}{R_2} = \frac{4}{3R_2}$$
$$\frac{3R_2}{4} = 15$$
$$\boxed{R_2 = 20 \text{ ohms}}$$
$$\boxed{R_1 = 3R_2 = 60 \text{ ohms}}$$

c.
$$\frac{1}{R} = \frac{R_2 + R_1}{R_1 R_2}$$
$$R = \boxed{\frac{R_1 R_2}{R_1 + R_2}}$$

39.
$$P = \frac{DN}{N+2}$$
$$PN + 2P = DN$$
$$N(P - D) = -2P$$
$$N = \boxed{\frac{2P}{D-P}}$$

41.
$$A = P + Prt = P(1 + rt)$$
$$P = \boxed{\frac{A}{1+rt}}$$

43.
$$A = \frac{rs}{r+s} \text{ for } s$$
$$Ar + As = rs$$
$$s(A - r) = -Ar$$
$$s = -\frac{Ar}{A-r} = \boxed{\frac{Ar}{r-A}}$$

45.
$$\frac{b}{y} = 1 + C \text{ for } y$$
$$b = y(1 + C)$$
$$y = \boxed{\frac{b}{1+C}}$$

47. Let x = the speed of current.

	D	R	$T = \dfrac{D}{R}$
with the current	33 miles	$18 + x$	$\dfrac{33}{18+x}$
against the current	21 miles	$18 - x$	$\dfrac{21}{18-x}$

$$\text{time} = \frac{\text{distance}}{\text{speed}} = \frac{33}{18+x} = \frac{21}{18-x}$$
$$33(18 - x) = 21(18 + x)$$
$$18(33 - 21) = x(21 + 33)$$
$$x = \frac{216}{54} = 4$$

The speed of the current is $\boxed{4 \text{ miles per hour}}$.

49. Let x = the jogger's walking speed. $x + 5$ = the running speed.

	D	R	$T = \dfrac{D}{R}$
running	8 miles	$x + 5$	$\dfrac{8}{x+5}$
walking	6 miles	x	$\dfrac{6}{x}$

$$\text{time} = \frac{\text{distance}}{\text{speed}}$$

$$\begin{aligned}
\frac{8}{x+5} + \frac{6}{x} &= 2 \\
8x + 6(x+5) &= 2x(x+5) \\
14x + 30 &= 2x^2 + 10x \\
2x^2 - 4x - 30 &= 0 \\
x^2 - 2x - 15 &= 0 \\
(x-5)(x+3) &= 0 \\
x &= 5
\end{aligned}$$

The jogger's walking speed is $\boxed{5 \text{ miles per hour}}$.

51. Let x = the speed of the boat in still water.

	D	R	$T = \dfrac{D}{R}$
with current	11 miles	$x + 2$	$\dfrac{11}{x+2}$
against current	9 miles	$x - 2$	$\dfrac{9}{x-2}$

$$\text{time} = \frac{\text{distance}}{\text{speed}}$$

$$\begin{aligned}
\frac{11}{x+2} &= \frac{9}{x-2} \\
11x - 22 &= 9x + 18 \\
2x &= 40 \\
x &= 20
\end{aligned}$$

The speed of the boat in still water is $\boxed{20 \text{ miles per hour}}$.

53. Let x = Beth's running speed. $x + 4$ = Jo's running speed.

	D	R	$T = \dfrac{D}{R}$
Beth	6 miles	x	$\dfrac{6}{x}$
Joe	10 miles	$x + 4$	$\dfrac{10}{x+4}$

$$\begin{aligned}
\text{time} = \frac{\text{distance}}{\text{speed}} = \frac{10}{x+4} &= \frac{6}{x} \\
10x &= 6x + 24 \\
4x &= 24 \\
x &= 6 \\
x + 4 &= 10
\end{aligned}$$

$\boxed{\text{Speeds: Jo, 10 miles per hour; Beth, 6 miles per hour.}}$

55. Let x = Mendel's speed on outgoing hike. $x + 1$ = speed on return hike.

	D	R	$T = \dfrac{D}{R}$
outgoing	6 miles	x	$\dfrac{6}{x}$
returning	6 miles	$x + 1$	$\dfrac{6}{x + 1}$

$$\text{time} \;=\; \frac{\text{distance}}{\text{speed}}$$

$$\begin{aligned}
\frac{6}{x} + \frac{6}{x+1} &= 5 \\
6x + 6 + 6x &= 5x(x+1) \\
12x + 6 &= 5x^2 + 5x \\
5x^2 - 7x - 6 &= 0 \\
(5x + 3)(x - 2) &= 0 \\
x &= 2
\end{aligned}$$

Mendel's speed on outgoing hike, $\boxed{\text{2 miles per hour}}$.

57. Let x = the race car's full speed.

$$\text{time} \;=\; \frac{\text{distance}}{\text{speed}}$$

$$\begin{aligned}
\frac{50}{x} &= \frac{30}{x} + 8 \\
50 &= 30 + 8x \\
x &= \frac{20}{8} = 2.5 \text{ miles per minute} \\
&= 2.5 \text{ miles (60) minutes minute hour} \\
&= \boxed{150 \text{ miles per hour}}
\end{aligned}$$

59. Let x = time for Esther to do the job. $2x$ = the time for Sid to do the job. Total rate = sum of individual rates.

	fractional part of job completed in one minute	time working together	fractional part of job completed when working together
Ester (x minutes)	$\dfrac{1}{x}$	40	$\dfrac{40}{x}$
Sid ($2x$ minutes)	$\dfrac{1}{2x}$	40	$\dfrac{40}{2x}$

$$\begin{aligned}
\frac{40}{x} + \frac{40}{2x} &= 1 \\
2x\left(\frac{40}{x} + \frac{40}{2x}\right) &= 2x(1) \\
80 + 40 &= 2x \\
120 &= 2x \\
x &= 60 \text{ minutes} \\
2x &= 120 \text{ minutes}
\end{aligned}$$

Time working alone: $\boxed{\text{Esther, 60 minutes; Sid, 120 minutes}}$

61. Let x = the time for B to do the job. $\frac{4}{5}x$ = the time for A to do the job. Total rate = sum of individual rates.

	fractional part of job completed in one hour	time working together (days)	fractional part of job completed when working together
A	$\dfrac{1}{\frac{4}{5}x}$	2	$\dfrac{2}{\frac{4}{5}x}$
B	$\dfrac{1}{x}$	2	$\dfrac{2}{x}$

$$\frac{2}{\frac{4}{5}x} + \frac{2}{x} = 1$$

$$\frac{1}{\frac{4}{5}x} + \frac{1}{x} = \frac{1}{2}$$

$$\frac{5}{4x} + \frac{4}{4x} = \frac{1}{2}$$

$$\frac{9}{4x} = \frac{1}{2}$$

$$4x = 18$$

$$x = 4.5 \text{ hours}$$

$$\frac{4}{5}x = \frac{4}{5}(4.5) = 3.6 \text{ hours}$$

Time working alone: $\boxed{\text{A, 3.6 hours; B, 4.5 hours}}$

63. Let x = the time for Mrs. Lovett to do the job. $x + 5$ = the time for Mr. Todd to do the job. Total rate = the sum of individual rates.

	fractional part of job completed in one day	time working together (days)	fractional part of job completed when working together
Mr. Todd	$\dfrac{1}{x+5}$	6	$\dfrac{6}{x+5}$
Mrs. Lovett	$\dfrac{1}{x}$	6	$\dfrac{6}{x}$

$$\frac{6}{x+5} + \frac{6}{x} = 1$$

$$\frac{1}{x+5} + \frac{1}{x} = \frac{1}{6}$$

$$\frac{x + x + 5}{x(x+5)} = \frac{1}{6}$$

$$6(2x + 5) = x(x+5)$$

$$12x + 30 = x^2 + 5x$$

$$x^2 - 7x - 30 = 0$$

$$(x - 10)(x + 3) = 0$$

$$x = 10 \quad (\text{reject } x = -3)$$

Mrs. Lovett, **time** : $\boxed{\text{10 days}}$

65. Let $x =$ additional consecutive hits.

$$\text{batter average} = \frac{\text{hits}}{\text{number of times at bat}}$$

$$\frac{12 + x}{40 + x} = 0.440$$

$$12 + x = 17.6 + 0.440x$$

$$0.560x = 5.6$$

$$x = \boxed{10}$$

The player must hit the ball $\boxed{10 \text{ additional consecutive times}}$.

67. $\frac{3}{4}$ perimeter of first square $=$ reciprocal of second square's area $+ \frac{1}{2}$ perimeter of third square

$$\frac{3}{4} \cdot 4\left(\frac{1}{4x - 8}\right) = \frac{1}{36} + \frac{1}{2} \cdot 4\left(\frac{1}{3x - 6}\right)$$

$$\frac{3}{4(x - 2)} = \frac{1}{36} + \frac{2}{3(x - 2)} \qquad \text{LCM} = 36(x - 2)$$

$$9 \cdot 3 = x - 2 + 2 \cdot 12$$

$$3 = x - 2$$

$$x = \boxed{5}$$

69. Let $x =$ the lower rate in percent. $x + 1 =$ the higher rate in percent.

$$\frac{Px}{100} = 175$$

$$\frac{P(x + 1)}{100} = 200$$

$$Px = 175 \cdot 100 = 17500$$

$$P(x + 1) = Px + P = 100 \cdot 200 = 20000$$

$$17500 + P = 20000$$

$$P = 2500$$

$$Px = 17500$$

$$x = \frac{17500}{2500} = 7\%$$

$$x + 1 = 8\%$$

The two interest rates are $\boxed{7\% \text{ and } 8\%}$.

71. $\dfrac{\text{cost}}{\text{movie}} = \dfrac{10 - 4}{10} = \dfrac{6}{10}$, or \$0.60

cost for 24 movies $= 24(0.60) + 4.00 = \boxed{\$18.40}$

73.
$$\text{product} = xy$$

$$x + y = \frac{1}{x} + \frac{1}{y}$$

$$x + y = \frac{x + y}{xy}$$

$$1 = \frac{1}{xy}$$

$$xy = 1$$

$$y = \frac{1}{x}$$

$$\text{product} = x\left(\frac{1}{x}\right) = \boxed{1}$$

75. Statement $\boxed{C}$ is true.

$$nt - 6n = 32 - 2t \quad \text{for } n$$
$$n(t - 6) = 32 - 2t \quad \text{first factor the left side.}$$
true

Review Problems

79. $(5, -2), (3, 8)$: $m = \dfrac{8 - (-2)}{3 - 5} = \dfrac{10}{-2} = \boxed{-5}$

80.
$$2x + y = -4$$
$$x + y = -3$$
lines intersect at $x = -1$, $y = -2$.

solution: $\boxed{\{(-1, -2)\}}$

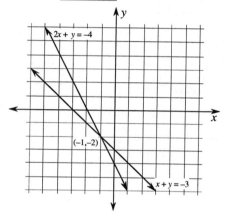

81. $16x^2y^2 - 40xy^2 + 25y^2 = y^2(16x^2 - 40x + 25) = \boxed{y^2(4x - 5)^2}$

Chapter 8 Review Problems

Chapter 8 Review, pp. 584-586

1.
$$\frac{5x}{6x - 24} = \frac{5x}{6(x - 4)}$$
$$6(x - 4) \neq 0$$
Domain: $\boxed{\{x \mid x \neq 4\}}$

2.
$$\frac{x + 3}{(x - 2)(x + 5)}$$
$$(x - 2)(x + 5) \neq 0$$
Domain: $\boxed{\{x \mid x \neq 2 \text{ and } x \neq -5\}}$

3.
$$\frac{x^2 + 3}{x^3 - 3x + 2} = \frac{x^2 + 3}{(x - 1)(x - 2)}$$
$$(x - 1)(x - 2) \neq 0$$
Domain: $\boxed{\{x \mid x \neq 1 \text{ and } x \neq 2\}}$

4. $\dfrac{5}{x^2 + 1}$

$x^2 + 1 \neq 0$

$x^2 + 1 \rightarrow$ never zero

Domain: $\boxed{\{x \mid x \in R\}}$

5. $f(x) = \dfrac{80000x}{100 - x}$

a. $\boxed{f(20) =}\ \dfrac{80000(20)}{100 - 20} = \boxed{\$20000}$ = cost to remove 20% of pollutants

$\boxed{f(50) =}\ \dfrac{80000(50)}{100 - 50} = \boxed{\$80000}$ = cost to remove 50% of pollutants

$\boxed{f(90) =}\ \dfrac{80000(90)}{100 - 90} = \boxed{\$720000}$ = cost to remove 90% of pollutants

$\boxed{f(98) =}\ \dfrac{80000(98)}{100 - 98} = \boxed{\$3920000}$ = cost to remove 98% of pollutants

b. domain of $f = \boxed{\{x \mid 0 \leq x < 100\}}$

c. $\boxed{\text{The cost increases rapidly as } x \text{ approaches 100\%. The difficulty of removing the pollutants}}$
$\boxed{\text{increases greatly as higher levels of purity are demanded.}}$

6. $f(x) = \dfrac{80000x}{100 - x}$

x	0	10	20	50	60	80	90	98	99	100
y	0	8889	20000	80000	120000	320000	720000	3920000	79200000	undefined

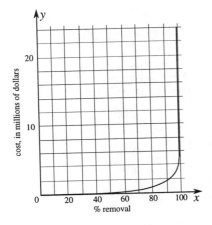

7. $\dfrac{x^2 - 4x - 5}{x^2 + 8x + 7} = \dfrac{(x - 5)(x + 1)}{(x + 7)(x + 1)} = \boxed{\dfrac{x - 5}{x + 7}} \quad (x \neq -1)$

8. $\dfrac{y^2 + 2y}{y^2 + 4y + 4} = \dfrac{y(y + 2)}{(y + 2)^2} = \boxed{\dfrac{y}{y + 2}} \quad (y \neq -2)$

9. $\dfrac{2y^2 - 8y + 8}{6y^2 - 12y} = \dfrac{2(y - 2)^2}{6y(y - 2)} = \boxed{\dfrac{y - 2}{3y}} \quad (y \neq 2)$

10. $\dfrac{3a^2 - 5a - 2}{4 - a^2} = \dfrac{(3a + 1)(a - 2)}{(2 - a)(2 + a)} = \boxed{-\dfrac{3a + 1}{a + 2}} \quad (a \neq 2)$

11. $\dfrac{2x^2 - 2xy + 3x - 3y}{2x + 3} = \dfrac{2x(x - y) + 3(x - y)}{2x + 3} = \dfrac{(2x + 3)(x - y)}{2x + 3} = \boxed{x - y}$ $\left(x \neq -\dfrac{3}{2}\right)$

12. $y = \dfrac{x^2 - 9}{x + 3} = \dfrac{(x + 3)(x - 3)}{x + 3} = \boxed{x - 3}$ $(x \neq -3)$

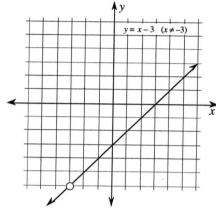

$y = x - 3 \quad (x \neq -3)$

13. $\dfrac{5y + 5}{6} \cdot \dfrac{3y}{y^2 + y} = \dfrac{5(y + 1)}{6} \cdot \dfrac{3y}{y(y + 1)} = \dfrac{15}{6} = \boxed{\dfrac{5}{2}}$ $(y \neq 0, -1)$

14. $\dfrac{x^2 + 6x + 9}{x^2 - 4} \cdot \dfrac{x + 3}{x - 2} = \dfrac{(x + 3)^2}{(x - 2)(x + 2)} \cdot \dfrac{x + 3}{x - 2} = \boxed{\dfrac{(x + 3)^3}{(x + 2)(x - 2)^2}}$ $(x \neq -2, 2)$

15. $\dfrac{2y^2 + y - 3}{4y^2 - 9} \cdot \dfrac{3y + 3}{5y - 5y^2} = \dfrac{(2y + 3)(y - 1)}{(2y + 3)(2y - 3)} \cdot \dfrac{3(y + 1)}{5y(1 - y)} = \boxed{-\dfrac{3(y + 1)}{5y(2y - 3)}}$ $\left(y \neq -\dfrac{3}{2}, \dfrac{3}{2}, 0, 1\right)$

16. $\dfrac{x^2 + x - 6}{x^2 + 6x + 9} \cdot \dfrac{x + 2}{x - 3} \cdot \dfrac{x^2 - 7x + 12}{x^2 - x - 2} = \dfrac{(x + 3)(x - 2)}{(x + 3)^2} \cdot \dfrac{x + 2}{x - 3} \cdot \dfrac{(x - 3)(x - 4)}{(x - 2)(x + 1)} = \boxed{\dfrac{(x + 2)(x - 4)}{(x + 3)(x + 1)}}$

$(x \neq -3, 3, 2, -1)$

17. $\dfrac{y^2 + y - 2}{10} \div \dfrac{2y + 4}{5} = \dfrac{(y + 2)(y - 1)}{10} \cdot \dfrac{5}{2(y + 2)} = \boxed{\dfrac{y - 1}{4}}$ $(y \neq -2)$

18. $\dfrac{6y + 2}{y^2 - 1} \div \dfrac{3y^2 + y}{y - 1} = \dfrac{2(3y + 1)}{(y - 1)(y + 1)} \cdot \dfrac{y - 1}{y(3y + 1)} = \boxed{\dfrac{2}{y(y + 1)}}$ $\left(y \neq 0, -\dfrac{1}{3}, -1, 1\right)$

19. $\dfrac{y^2 - 5y - 24}{2y^2 - 2y - 24} \div \dfrac{y^2 - 10y + 16}{4y^2 + 4y - 24} = \dfrac{(y - 8)(y + 3)}{2(y - 4)(y + 3)} \cdot \dfrac{4(y + 3)(y - 2)}{(y - 2)(y - 8)} = \boxed{\dfrac{2(y + 3)}{y - 4}}$ $(y \neq 4, -3, 2, 8)$

20. $\dfrac{z^2 - 10z + 21}{7 - z} \div (z + 3) = \dfrac{(z - 3)(z - 7)}{7 - z} \cdot \dfrac{1}{z + 3} = \boxed{\dfrac{3 - z}{3 + z}}$ $(z \neq 7, -3)$

21. $\dfrac{12x - 5}{3x - 1} \div \dfrac{1}{3x - 1} = \dfrac{12x - 5}{3x - 1} \cdot \dfrac{3x - 1}{1} = \boxed{12x - 5}$ $\left(x \neq \dfrac{1}{3}\right)$

22. $\dfrac{3y^2 + 2y}{y - 1} - \dfrac{10y - 5}{y - 1} = \dfrac{3y^2 - 8y + 5}{y - 1} = \dfrac{(3y - 5)(y - 1)}{y - 1} = \boxed{3y - 5}$ $\left(y \neq \dfrac{5}{3}\right)$

23. $\dfrac{2y-1}{y^2+5y-6} - \dfrac{2y-7}{y^2+5y-6} = \boxed{\dfrac{6}{(y+6)(y-1)}}$ $(y \neq 1, -6)$

24. $\dfrac{2x+7}{x^2-9} - \dfrac{x-4}{x^2-9} = \boxed{\dfrac{x+11}{(x-3)(x+3)}}$ $(x \neq 3, -3)$

25. $9x^3y = 3^2x^3y$
             $12xy = 2^2 \cdot 3xy$
      LCM $= \boxed{2^2 \cdot 3^2 x^3 y \text{ or } 36x^3 y}$

26. $8y^2(y-1)^2 = 2^3y^2(y-1)^2$
       $10y^3(y-1) = 2 \cdot 5y^3(y-1)$
       LCM $= \boxed{2^3 \cdot 5y^3(y-1)^2 \text{ or } 40y^3(y-1)^2}$

27. $x^2+4x+3 = (x+3)(x+1)$
        $x^2+10x+21 = (x+3)(x+7)$
          LCM $= \boxed{(x+1)(x+3)(x+7)}$

28. $\dfrac{3}{10y^2} + \dfrac{7}{25y} = \dfrac{15}{50y^2} + \dfrac{14y}{50y^2} = \boxed{\dfrac{15+14y}{50y^2}}$

29. $\dfrac{6y}{y^2-4} - \dfrac{3}{y+2} = \dfrac{6y}{y^2-4} - \dfrac{3(y-2)}{y^2-4} = \dfrac{3y+6}{y^2-4} = \dfrac{3(y+2)}{y^2-4} = \boxed{\dfrac{3}{y-2}}$ $(y \neq -2, 2)$

30. $\dfrac{2}{3x} + \dfrac{5}{x+1} = \dfrac{2x+2+15x}{3x(x+1)} = \boxed{\dfrac{17x+2}{3x(x+1)}}$

31. $\dfrac{2y}{y^2+2y+1} + \dfrac{y}{y^2-1} = \dfrac{2y}{(y+1)^2} + \dfrac{y}{(y+1)(y-1)} = \dfrac{2y(y-1)}{(y+1)^2(y-1)} + \dfrac{y(y+1)}{(y+1)^2(y-1)} = \dfrac{2y^2-2y+y^2+y}{(y+1)^2(y-1)}$
$= \dfrac{3y^2-y}{(y+1)^2(y-1)}$
$= \boxed{\dfrac{y(3y-1)}{(y+1)^2(y-1)}}$
$(y \neq -1, 1)$

32. $\dfrac{4z}{z^2+6z+5} - \dfrac{3}{z^2+5z+4} = \dfrac{4z}{(z+1)(z+5)} - \dfrac{3}{(z+1)(z+4)} = \dfrac{4z(z+4)-3(z+5)}{(z+1)(z+4)(z+5)} = \boxed{\dfrac{4z^2+13z-15}{(z+1)(z+4)(z+5)}}$
$(z \neq -1, -4, -5)$

33. $\dfrac{y}{y-2} - \dfrac{y-4}{2-y} = \dfrac{y}{y-2} + \dfrac{y-4}{y-2} = \dfrac{2y-4}{y-2} = \dfrac{2(y-2)}{y-2} = \boxed{2}$ $(y \neq 2)$

34. $\dfrac{4y-1}{2y^2+5y-3} - \dfrac{y+3}{6y^2+y-2} = \dfrac{4y-1}{(2y-1)(y+3)} - \dfrac{y+3}{(3y+2)(2y-1)}$
$= \dfrac{(4y-1)(3y+2)-(y+3)(y+3)}{(2y-1)(y+3)(3y+2)}$
$= \dfrac{12y^2+5y-2-y^2-6y-9}{(2y-1)(y+3)(3y+2)}$
$= \boxed{\dfrac{11y^2-y-11}{(2y-1)(y+3)(3y+2)}}$

35. $\dfrac{4}{y^2 - 3y + 2} - \dfrac{y}{y - 2} + \dfrac{2y}{y - 1} = \dfrac{4}{(y-2)(y-1)} - \dfrac{y(y-1) - 2y(y-2)}{(y-2)(y-1)} = \dfrac{4 - y^2 + y + 2y^2 - 4y}{(y-1)(y-2)}$

$= \boxed{\dfrac{y^2 - 3y + 4}{(y-1)(y-2)}}$

$(y \neq 1, 2)$

36. $\dfrac{\dfrac{1}{x} - \dfrac{1}{2}}{\dfrac{1}{3} - \dfrac{x}{6}} = \dfrac{2 - x}{2x} \cdot \dfrac{6}{2 - x} = \boxed{\dfrac{3}{x}} \quad (x \neq 2)$

37. $\dfrac{3 + \dfrac{12}{y}}{1 - \dfrac{16}{y^2}} = \dfrac{3y + 12}{y} \cdot \dfrac{y^2}{y^2 - 16} = \dfrac{3(y+4)y^2}{y(y+4)(y-4)} = \boxed{\dfrac{3y}{y - 4}} \quad (y \neq 0, -4, 4)$

38. $\dfrac{\dfrac{25}{y+5} + 5}{\dfrac{3}{y+5} - 5} = \dfrac{25 + 5y + 25}{3 - 5y - 25} = \boxed{\dfrac{50 + 5y}{-22 - 5y}} \quad (y \neq -5)$

39. $\dfrac{\dfrac{1}{y-2} - \dfrac{1}{y+2}}{\dfrac{1}{y+2} + \dfrac{1}{y^2 - 4}} = \dfrac{y + 2 - y + 2}{y - 2 + 1} = \boxed{\dfrac{4}{y - 1}} \quad (y \neq 1, -2, 2)$

40. $\dfrac{x^{-2}}{1 - x^{-2}} = \boxed{\dfrac{1}{x^2 - 1}} \quad (x \neq -1, 1, 0)$

41.
$$\begin{aligned}
\frac{13}{y-1} - 3 &= \frac{1}{y-1} \\
\frac{13 - 1}{y-1} &= 3 \\
12 &= 3y - 3 \\
3y &= 15 \\
y &= 5
\end{aligned}$$
$\boxed{\{5\}}$

42.
$$\begin{aligned}
\frac{3}{4x} - \frac{1}{x} &= \frac{1}{4} \\
\frac{3 - 4}{4x} &= \frac{1}{4} \\
-\frac{1}{4x} &= \frac{1}{4} \\
-4 &= 4x \\
x &= -1
\end{aligned}$$
$\boxed{\{-1\}}$

43.
$$\frac{5}{y+2} + \frac{y}{y+6} = \frac{24}{y^2+8y+12} = \frac{24}{(y+6)(y+2)}$$
$$5(y+6) + y(y+2) = 24$$
$$5y + 30 + y^2 + 2y = 24$$
$$y^2 + 7y + 6 = 0$$
$$(y+6)(y+1) = 0$$
$$y = -1 \qquad\qquad (y \neq -2, -6)$$

$$\boxed{\{-1\}}$$

44.
$$3 - \frac{6}{y} = y + 8$$
$$3y - 6 = y^2 + 8y$$
$$y^2 + 5y + 6 = 0$$
$$(y+2)(y+3) = 0$$
$$y = -2 \text{ or } y = -3 \qquad \text{(Both answers check.)}$$

$$\boxed{\{-2, -3\}}$$

45.
$$\frac{7y}{y-2} + 1 = y + \frac{14}{y-2}$$
$$1 - y = \frac{14 - 7y}{y-2} = \frac{7(2-y)}{y-2} = -7$$
$$y = 1 + 7 = 8$$

$$\boxed{\{8\}}$$

46.
$$\frac{2y}{y-3} - \frac{y}{y+3} = \frac{2y-6}{y^2-9} \qquad\qquad \text{LCM} = (y-3)(y+3)$$
$$2y(y+3) - y(y-3) = 2y - 6$$
$$2y^2 + 6y - y^2 + 3y = 2y - 6$$
$$y^2 + 7y + 6 = 0$$
$$(y+1)(y+6) = 0$$
$$y = -1 \text{ or } y = -6$$

$$\boxed{\{-1, -6\}}$$

47.
$$P = 30 - \frac{9}{t+1}$$
$$29 = 30 - \frac{9}{t+1}$$
$$-1 = -\frac{9}{t+1}$$
$$1 = \frac{9}{t+1}$$
$$t + 1 = 9$$
$$t = 8$$
$$1990 + 8 = 1998$$
corresponding to $\boxed{1998}$

48. Let x = numerator of fraction. $x + 6$ = denominator of fraction.

$$\frac{x+3}{x+6+3} = \frac{2}{5}$$
$$\frac{x+3}{x+9} = \frac{2}{5}$$
$$5x + 15 = 2x + 18$$
$$3x = 3$$
$$x = 1$$
$$x + 6 = 7$$
$$\text{fraction} = \boxed{\frac{1}{7}}$$

49. Let x = the number.

$$2x - \frac{3}{x} = 1$$
$$2x^2 - 3 = x$$
$$2x^2 - x - 3 = 0$$
$$(2x - 3)(x + 1) = 0$$
$$x = \frac{3}{2} \text{ or } x = -1$$

The number is $\boxed{\frac{3}{2} \text{ or } -1}$.

50. Let x = the smaller number. $x + 9$ = the larger number.

$$\frac{x+9}{x} = 2 + \frac{4}{x}$$
$$x + 9 = 2x + 4$$
$$x = 5$$
$$x + 9 = 14$$
The numbers are $\boxed{5 \text{ and } 14}$.

51. Let x = units' digit of number. $x + 3$ = the tens' digit of number. $10(x+3) + x$ = number.

$$\frac{10(x+3)+x}{x+(x+3)} = 6 + \frac{8}{2x+3}$$
$$\frac{10x+30+x}{2x+3} - \frac{8}{2x+3} = 6$$
$$11x + 22 = 6(2x+3)$$
$$11x + 22 = 12x + 18$$
$$x = 4 = \text{units' digit}$$
$$x + 3 = 7 = \text{tens' digit}$$
The number is $\boxed{74}$.

52.
$$\frac{x-1}{x+1} = \frac{y}{z} \quad \text{for } x$$
$$xz - z = yx + y$$
$$x(z-y) = y + z$$
$$x = \boxed{\frac{y+z}{z-y}} \quad (y \neq z)$$

53.
$$\frac{1}{a} + \frac{1}{b} = \frac{1}{c} \quad \text{for } a$$
$$\frac{b+a}{ab} = \frac{1}{c}$$
$$bc + ac = ab$$
$$a(c-b) = -bc$$
$$a = \boxed{\frac{bc}{b-c}}$$

54.
$$T = \frac{A-p}{pr} \quad \text{for } p$$
$$prT = A - p$$
$$p(rT + 1) = A$$
$$p = \boxed{\frac{A}{1 + rT}}$$

55.
$$\frac{1}{R} = \frac{1}{R_1} + \frac{1}{R_2}$$
If $R = 60$ ohms and $R_1 = 2R_2$, then
$$\frac{1}{60} = \frac{1}{2R_2} + \frac{1}{R_2}$$
$$\frac{1}{60} = \frac{3}{2R_2}$$
$$60 = \frac{2R_2}{3}$$
$$2R_2 = 180$$
$$\boxed{R_2 = 90 \text{ ohms}}$$
$$\boxed{R_1 = 180 \text{ ohms}}$$

56.
$$C = \frac{80000x}{100 - x}$$
$$100C - Cx = 80000x$$
$$x(80000 + C) = 100C$$
$$x = \boxed{\frac{100C}{80000 + C}}$$

If $C = 3{,}920{,}000$,
$$x = \frac{100(3920000)}{80000 + 3920000} = \frac{100(3920)}{80 + 3920} = \boxed{98\%} \text{ of pollutants eliminated}$$

57. Let $x =$ the speed of the boat in still water.
$$\text{time} = \frac{\text{distance}}{\text{speed}}$$
$$\frac{11}{x+3} = \frac{9}{x-3}$$
$$11x - 33 = 9x + 27$$
$$2x = 60$$
$$x = 30$$
The speed of the boat in still water is $\boxed{30 \text{ miles/hour}}$.

58. Let x = the speed on outgoing trip. $x + 1$ = speed on return trip.

$$\text{time} = \frac{\text{distance}}{\text{speed}}$$

$$\frac{6}{x} + \frac{6}{x+1} = 5$$
$$6(x+1) + 6x = 5(x^2 + x)$$
$$6x + 6 + 6x = 5x^2 + 5x$$
$$5x^2 - 7x + 6 = 0$$
$$(x-2)(5x+3) = 0$$
$$x = 2 \qquad \left(\text{reject } x = -\frac{3}{5}\right)$$

Rowing speed on outgoing trip: $\boxed{2 \text{ miles per hour}}$.

59. Let x = hours for Nora to do the job. $x + 10$ = hours for Nick to do the job.

$$\frac{1}{x} + \frac{1}{x+10} = \frac{1}{12}$$
$$\frac{x+10+x}{x(x+10)} = \frac{1}{12}$$
$$x^2 + 10x = 24x + 120$$
$$x^2 - 14x - 120 = 0$$
$$(x-20)(x+6) = 0$$
$$x = 20$$
$$x + 10 = 30$$

$\boxed{\text{Time for Nora, 20 hours; time for Nick, 30 hours}}$.

Cumulative Review Problems (Chapters 1-8)

Chapter 8 Cumulative Review, pp. 586-587

1. a. $8 + (7 + 3) = (8 + 7) + 3$
$\boxed{\text{Associative property of addition}}$

b. $-3(5 + 9) = (-3) \cdot 5 + (-3) \cdot 9$
$\boxed{\text{Distributive property}}$

c. $5(3 + 4) = 5(4 + 3)$
$\boxed{\text{Commutative property of additon}}$

2.
$$-10 \le 3y + 14 < 17$$
$$-24 \le 3y < 3$$
$$-8 \le y < 1$$

$\boxed{\{y \mid -8 \le y < 1\}}$

3. Let x = the number.
$$7x + 4 = 3x - 10$$
$$4x = -14$$
$$x = -\frac{7}{2}$$

The number is $\boxed{-\frac{7}{2}}$.

4.
$$3x - 4y \ < \ 12$$
$$y \ > \ -4$$

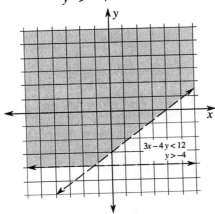

$$3x - 4y < 12$$
$$y > -4$$

5. $(2y^2 - y + 2)(3y^3 - 2y^2 + 3y + 1)$
$= 2y^2(3y^3 - 2y^2 + 3y + 1) - y(3y^3 - 2y^2 + 3y + 1) + 2(3y^3 - 2y^2 + 3y + 1)$
$= 6y^5 - 4y^4 + 6y^3 + 2y^2 - 3y^4 + 2y^3 - 3y^2 - y + 6y^3 - 4y^2 + 6y + 2$
$= \boxed{6y^5 - 7y^4 + 14y^3 - 5y^2 + 5y + 2}$

6. $2y^3 + 3y^2 - 2y - 3 = y^2(2y + 3) - (2y + 3) = (y^2 - 1)(2y + 3) = \boxed{(y + 1)(y - 1)(2y + 3)}$

7. $(y - 11)(2y^2 - y - 15) = \ 0$
$(y - 11)(2y + 5)(y - 3) \ = \ 0$
$y - 11 \ = \ 0 \ $ or $\ 2y + 5 = 0 \ $ or $\ y - 3 = 0$
$y \ = \ 11$
$y \ = \ -\dfrac{5}{2}$
$y \ = \ 3$

$$\boxed{\left\{ -\frac{5}{2}, 3, 11 \right\}}$$

8. Let x = the width of the rectangle. $2x + 1$ = the length of the rectangle.
$$\begin{aligned} \text{area} \ &= \ \text{perimeter} + 1 \\ x(2x + 1) \ &= \ 2(x + 2x + 1) + 1 \\ 2x^2 + x \ &= \ 2(3x + 1) + 1 \\ 2x^2 + x \ &= \ 6x + 3 \\ 2x^2 - 5x - 3 \ &= \ 0 \\ (x - 3)(2x + 1) \ &= \ 0 \\ x \ &= \ 3 \quad \left(\text{reject } x = -\frac{1}{2} \right) \\ 2x + 1 \ &= \ 7 \end{aligned}$$

$\boxed{\text{dimensions of rectangle: width, 3 meters; length, 7 meters}}$

9. $10y^3 - 17y^2 + 3y = y(10y^2 - 17y + 3) = \boxed{y(5y - 1)(2y - 3)}$

10. $\dfrac{27y^5 - 3y^3 + 6y^2 - 2y}{3y - 1} = \boxed{9y^4 + 3y^3 + 2y}$

$$
\begin{array}{r}
9y^4 \quad + \quad 3y^3 \quad + \quad 2y \\
3y-1\overline{\smash{)}27y^5 + 0y^4 - 3y^3 + 6y^2 - 2y} \\
\underline{27y^5 - 9y^4} \\
9y^4 - 3y^3 \\
\underline{9y^4 - 3y^3} \\
0 + 6y^2 - 2y \\
\underline{6y^2 - 2y} \\
0
\end{array}
$$

11. $(4xy^3)^2(-2xy^{-8}) = 4^2x^2y^6(-2)xy^{-8} = -32x^3y^{-2} = \boxed{\dfrac{-32x^3}{y^2}}$

12.
$$
\begin{array}{lll}
5x + 2y = -1 & (\times 5)\to & 25x + 10y = -5 \\
2x - 5y = 1 & (\times 2)\to & \underline{4x - 10y = 2} \\
& & 29x = -3 \\
& & x = -\dfrac{3}{29}
\end{array}
$$

$$
5\left(\dfrac{-3}{29}\right) + 2y = -1
$$
$$
2y = -1 + \dfrac{15}{29}
$$
$$
2y = -\dfrac{14}{29}
$$
$$
y = -\dfrac{7}{29}
$$

$$\boxed{\left\{\left(-\dfrac{3}{29}, -\dfrac{7}{29}\right)\right\}}$$

13. Let x = measure of angle. $90 - x$ = measure of its complement. $180 - x$ = measure of its supplement.
$$
\begin{aligned}
(180 - x) - 2(90 - x) &= 43 \\
180 - x - 180 + 2x &= 43 \\
x &= 43
\end{aligned}
$$
The angle's measure is $\boxed{43°}$.

14. $(1, -5), (-2, 10)$
$$
\begin{aligned}
\text{slope} &= \dfrac{10 + 5}{-2 - 1} = \dfrac{15}{-3} = -5 \\
y + 5 &= -5(x - 1) \\
y + 5 &= -5x + 5 \\
y &= -5x \\
\boxed{5x + y} &\boxed{= 0}
\end{aligned}
$$

15. x, y, z are prime, one is even
$x = 2$
$y, z = 1, 3, 5, 7$
difference between largest numbers: $7 - 5 = \boxed{2}$

16. Let x = the number of $5 bills. $90 - x$ = the number of $10 bills.

$$
\begin{aligned}
5x + 10(90 - x) &= 645 \\
5x + 900 - 10x &= 645 \\
-5x &= -255 \\
x &= 51 \\
90 - x &= 39
\end{aligned}
$$

51 five dollar bills; 39 ten dollar bills

17. Let x = my present age. $x + 30$ = the present age of my mother.

$$
\begin{aligned}
x + 30 &= 3x \\
30 &= 2x \\
15 &= x
\end{aligned}
$$

I am 15 years old .

18.

$$
\begin{aligned}
5(2y + 3) - 2(y - 4) &\le 3(2y + 2) + y \\
10y + 15 - 2y + 8 &\le 6y + 6 + y \\
8y + 23 &\le 7y + 6 \\
y &\le -17
\end{aligned}
$$

$\{y \mid y \le -17\}$

−17

19.

$$
\begin{aligned}
f(x) &= -x - x^2 \\
f(-3) &= -(-3) - (-3)^2 = 3 - 9 = \boxed{-6}
\end{aligned}
$$

20.

$$
\begin{aligned}
2x + y &= 4 \\
x + y &= 2
\end{aligned}
$$

Solution: $\{(2, 0)\}$

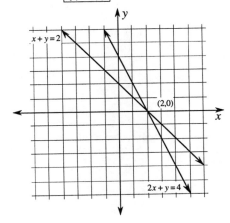

21. Let x = the number of records. y = the number of compact disks.

$$
\begin{aligned}
5x + 2y &= 65 \\
3x + 4y &= 81
\end{aligned}
$$

$(\times -2) \rightarrow$

$$
\begin{aligned}
-10x - 4y &= -130 \\
\underline{3x + 4y} &= \underline{81} \\
-7x &= -49 \\
x &= 7 \\
5(7) + 2y &= 65 \\
2y &= 30 \\
y &= 15
\end{aligned}
$$

record, $7 each; compact disk, $15 each

22. Let x = the length of the shorter leg. $x + 7$ = the length of the longer leg. $2x + 1$ = the hypotenuse.

$$
\begin{aligned}
x^2 + (x+7)^2 &= (2x+1)^2 \\
x^2 + x^2 + 14x + 49 &= 4x^2 + 4x + 1 \\
0 &= 2x^2 - 10x - 48 \\
x^2 - 5x - 24 &= 0 \\
(x - 8)(x + 3) &= 0 \\
x &= 8 \quad \text{(reject } x = -3) \\
x + 7 &= 15 \\
2x + 1 &= 17
\end{aligned}
$$

The lengths of the sides of the triangle are $\boxed{8 \text{ meters, 15 meters, and 17 meters}}$.

23.
$$
\begin{aligned}
\frac{7}{y+2} + \frac{5}{y-8} &= \frac{2}{y^2 - 6y - 16} \\
\frac{7(y-8) + 5(y+2)}{(y+2)(y-8)} &= \frac{2}{(y-8)(y+2)} \\
7y - 56 + 5y + 10 &= 2 \\
12y &= 48 \\
y &= 4
\end{aligned}
$$

$\boxed{\{4\}}$

24.
$$
\frac{y-1}{y^2 + 2y - 8} - \frac{2 - 3y}{y^2 + 3y - 4} = \frac{y-1}{(y+4)(y-2)} + \frac{3y-2}{(y+4)(y-1)} = \frac{(y-1)^2 + (y-2)(3y-2)}{(y+4)(y-2)(y-1)}
$$
$$
= \frac{y^2 - 2y + 1 + 3y^2 - 8y + 4}{(y+4)(y-2)(y-1)}
$$
$$
= \boxed{\frac{4y^2 - 10y + 5}{(y+4)(y-2)(y-1)}}
$$

25. Let x = the width of the rectangle. $2x - 7$ = the length of the rectangle.

$$
\begin{aligned}
x(2x - 7) &= 60 \\
2x^2 - 7x - 60 &= 0 \\
(2x - 15)(x + 4) &= 0 \\
x &= \frac{15}{2} \quad \text{(reject } x = -4) \\
x &= 7.5 \\
2x - 7 &= 8
\end{aligned}
$$

Dimensions: $\boxed{\text{width, 7.5 meters; length, 8 meters}}$.

26. Let x = the speed of the boat in still water. $x + 5$ = the speed of the boat with current. $x - 5$ = the speed of the boat against current.

$$
\begin{aligned}
\frac{240}{x+5} &= \frac{160}{x-5} \\
\frac{3}{x+5} &= \frac{2}{x-5} \\
3x - 15 &= 2x + 10 \\
x &= 25
\end{aligned}
$$

Speed of boat in still water, $\boxed{25 \text{ miles per hour}}$.

27.

	first floor	x	$x+2$	$x-1$
	second floor	y	$y-2$	$y+1$
	equation		$x+2=y-2$	$y+1=4(x-1)$

$$
\begin{aligned}
x+2 &= y-2 \\
x-y &= -4
\end{aligned}
\qquad\qquad
\begin{aligned}
y+1 &= 4x-4 \\
-4x+y &= -5
\end{aligned}
$$

$$
\begin{aligned}
x-y &= -4 \\
\underline{-4x+y} &= \underline{-5} \\
-3x &= -9 \\
x &= 3 \\
3-y &= -4 \\
y &= 7
\end{aligned}
$$

$\boxed{\text{3 downstairs; 7 upstairs}}$

28. $\dfrac{(9\times10^{-3})(1.2\times10^{6})}{(1.5\times10^{8})(9.6\times10^{-2})}=\left(\dfrac{9}{1.5}\right)\left(\dfrac{1.2}{9.6}\right)\dfrac{\times10^{-3+6}}{\times10^{8-2}}=\dfrac{6}{8}\times10^{3-6}=0.75\times10^{-3}$

$= \boxed{7.5\times10^{-4}}$

$= \boxed{0.00075}$

29.
$$
\begin{aligned}
y(y-4)-(y-2)(y+2) &= 1-4y \\
y^2-4y-y^2+4 &= 1-4y \\
4 &= 1 \quad \text{false}
\end{aligned}
$$

no solution; $\boxed{\varnothing}$

30.
$$
\begin{aligned}
\tfrac{1}{2}(y+5)-\tfrac{5}{6}(y+1) &= -\tfrac{1}{3}(y-5) \\
3(y+5)-5(y+1) &= -2(y-5) \quad (\times\,6) \\
3y+15-5y-5 &= -2y+10 \\
-2y+10 &= -2y+10 \\
10 &= 10
\end{aligned}
$$

True for all real values of y.

$\boxed{\{y\mid y\in R\}}$

Chapter 9   Roots and Radicals

Section 9.1 Finding Roots

Problem Set 9.1, pp. 596-598

1. The square roots of 36 are $\boxed{6 \text{ and } -6}$, since $6^2 = 36$ and $(-6)^2 = 36$.

3. Square roots of 144 are $\boxed{12 \text{ and } -12}$, since $12^2 = 144$ and $(-12)^2 = 144$.

5. Square roots of $\dfrac{9}{16}$ are $\boxed{\dfrac{3}{4} \text{ and } -\dfrac{3}{4}}$, since $\left(\dfrac{3}{4}\right)^2 = \dfrac{9}{16}$ and $\left(\dfrac{-3}{4}\right)^2 = \dfrac{9}{16}$.

7. Square roots of $\dfrac{49}{100}$ are $\boxed{\dfrac{7}{10} \text{ and } -\dfrac{7}{10}}$, since $\left(\dfrac{7}{10}\right)^2 = \dfrac{49}{100}$ and $\left(\dfrac{-7}{10}\right)^2 = \dfrac{49}{100}$.

9. $\sqrt{36} = \boxed{6}$

11. $-\sqrt{36} = \boxed{-6}$

13. $\sqrt{-36}$ does not exist, $\boxed{\text{not a real number}}$

15. $\sqrt{\dfrac{49}{25}} = \boxed{\dfrac{7}{5}}$

17. $-\sqrt{\dfrac{1}{25}} = \boxed{-\dfrac{1}{5}}$

19. $\sqrt{0.04} = \boxed{0.2}$

21. $\sqrt{33 - 8} = \sqrt{25} = \boxed{5}$

23. $\sqrt{2 \cdot 32} = \sqrt{64} = \boxed{8}$

25. $\sqrt{144 + 25} = \sqrt{169} = \boxed{13}$

27. $\sqrt{144} + \sqrt{25} = 12 + 5 = \boxed{17}$

29. $t = \sqrt{\dfrac{d}{16}}$, $d = 1411$ feet

$t = \sqrt{\dfrac{144}{16}} = \sqrt{9} = \boxed{3 \text{ seconds}}$

31. $H = (10.45 + \sqrt{100w} - w)(33 - t)$, $w = 4$ meters per second, $t = 0°C$

$H = \left(10.45 + \sqrt{100 \cdot 4} - w\right)(33 - 0) = (10.45 + 20 - 4)(33) = (26.45)(33) = 872.85$.

Since $H > 200$, the exposed flesh $\boxed{\text{will freeze in 1 minute or less}}$, $\boxed{\text{yes}}$.

33. $\sqrt{\dfrac{1}{4}} = \boxed{\dfrac{1}{2}}$, $\boxed{\text{rational}}$

35. $\sqrt{15} \approx \boxed{3.873}$, $\boxed{\text{irrational}}$

37. $\sqrt{400} = \boxed{20}$, $\boxed{\text{rational}}$

39. $-\sqrt{225} = \boxed{-15}$, $\boxed{\text{rational}}$

41. $\sqrt{-1}$ is $\boxed{\text{not a real number}}$

43. $-\sqrt{83} \approx \boxed{-9.110}$, $\boxed{\text{irrational}}$

45. $\sqrt{573} \approx \boxed{23.937}$, $\boxed{\text{irrational}}$

47. $-\sqrt{1369} = \boxed{-37}$, $\boxed{\text{rational}}$

49. $\dfrac{9 + \sqrt{144}}{3} = \dfrac{9 + 12}{3} = \dfrac{21}{3} = \boxed{7}$, $\boxed{\text{rational}}$

51. $\dfrac{12 + \sqrt{45}}{2} = \dfrac{12 + 3\sqrt{5}}{2} \approx \boxed{9.354}$, $\boxed{\text{irrational}}$

53. $\dfrac{12 + \sqrt{-45}}{2}$ is $\boxed{\text{not a real number}}$

55. $d = 1.22\sqrt{x} = 1.22\sqrt{25000} \approx \boxed{192.90 \text{ miles}}$

57. $r = \dfrac{\sqrt{A} - \sqrt{P}}{\sqrt{P}} = \dfrac{\sqrt{900} - \sqrt{800}}{\sqrt{800}} = \dfrac{\sqrt{900}}{\sqrt{800}} - 1 = \boxed{\dfrac{30}{20\sqrt{2}} - 1} \approx 0.061$

59.

x	0	1	2	4	6	8	9	10	12	14	16
y	0	1	1.41	2	2.45	2.83	3	3.16	3.46	3.74	4

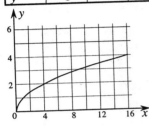

61. $\sqrt[4]{1} = 1$

63. $\sqrt[3]{64} = \boxed{4}$

65. $\sqrt[3]{-27} = \boxed{-3}$

67. $\sqrt[3]{125} = \boxed{5}$

69. $\sqrt[4]{16} = \boxed{2}$

71. $-\sqrt[4]{81} = \boxed{-3}$

73. $\sqrt[4]{-81}$ is $\boxed{\text{not a real number}}$

75. $\sqrt[4]{256} = \boxed{4}$

77. $\sqrt[5]{-32} = \boxed{-2}$

79. $\sqrt{\sqrt[3]{64}} = \sqrt{4} = \boxed{2}$

81. $f(x) = \sqrt[3]{2x + 1}$

$f(0) - f\left(-\dfrac{9}{2}\right) = \sqrt[3]{1} - \sqrt[3]{-9 + 1} = 1 - \sqrt[3]{-8} = 1 - (-2) = \boxed{3}$

83. a. $\left(\sqrt{3.1}\right)^2 = \boxed{3.1}$

  b. $\left(\sqrt{107}\right)^2 = \boxed{107}$

  c. $\left(\sqrt{29}\right)^2 = \boxed{29}$

  d. $\left(\sqrt{9216}\right)^2 = \boxed{9216}$

85. $\sqrt{16} = \boxed{4}$, $\sqrt{36} = \boxed{6}$, $\sqrt{64} = \boxed{8}$

87. $[(4 \uparrow 4) \downarrow 2] \uparrow 3 = [256 \downarrow 2] \uparrow 3 = 16 \uparrow 3 = \boxed{4096}$

  since $4 \uparrow 4 = 4^4 = 256$, $256 \downarrow 2 = \sqrt{256} = 16$, $16 \uparrow 3 = 16^3 = 4096$

Review Problems

90. $4x - 5y < 20$

$$\boxed{\text{Intercepts are } x = 5, \, y = -4}$$

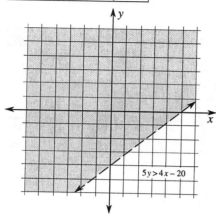

91. $\dfrac{x^2 - 2x - 3}{2x^2 - 5x - 3} \div \dfrac{2x^2 - 4x}{2x^2 - 3x - 2} = \dfrac{(x-3)(x+1)}{(x-3)(2x+1)} \div \dfrac{2x(x-2)}{(2x+1)(x-2)}$

$\quad = \dfrac{(x-3)(x+1)}{(x-3)(2x+1)} \cdot \dfrac{(2x+1)(x-2)}{2x(x-2)} = \boxed{\dfrac{x+1}{2x}} \quad x \neq -\dfrac{1}{2}, \, 2, \, 3$

92.
$$2(x-4) + 6 \leq 5x - 2(6-x)$$
$$2x - 8 + 6 \leq 5x - 12 + 2x$$
$$-2 \quad \leq 5x - 12$$
$$5x \quad \geq 10$$
$$x \quad \geq 2$$

$$\boxed{\{x \mid x \geq 2\}}$$

Section 9.2 Multiplying and Dividing Radicals

Problem Set 9.2, pp. 605-606

1. $\sqrt{7} \cdot \sqrt{6} = \boxed{\sqrt{42}}$

3. $\sqrt{6} \cdot \sqrt{6} = \sqrt{36} = \boxed{6}$

5. $\sqrt{3} \cdot \sqrt{5y} = \sqrt{15y}$

7. $\sqrt{50} = \sqrt{25 \cdot 2} = \boxed{5\sqrt{2}}$

9. $\sqrt{45} = \sqrt{9 \cdot 5} = \boxed{3\sqrt{5}}$

11. $\sqrt{80} = \sqrt{16 \cdot 5} = \boxed{4\sqrt{5}}$

13. $\sqrt{600} = \sqrt{100 \cdot 6} = \boxed{10\sqrt{6}}$

15. $2\sqrt{27} = 2\sqrt{9 \cdot 3} = \boxed{6\sqrt{3}}$

17. $7\sqrt{8} = 7\sqrt{4 \cdot 2} = \boxed{14\sqrt{2}}$

19. $\sqrt{27} \cdot \sqrt{18} = \sqrt{9 \cdot 3} \cdot \sqrt{9 \cdot 2} = \boxed{9\sqrt{6}}$

21. $\sqrt{15} \cdot \sqrt{21} = \sqrt{15 \cdot 21} = \sqrt{9 \cdot 35} = \boxed{3\sqrt{35}}$

23. $\sqrt{72} \cdot \sqrt{50} = \left(6\sqrt{2}\right)\left(5\sqrt{2}\right) = 30 \cdot 2 = \boxed{60}$

25. $\sqrt{3} \cdot \sqrt{6} \cdot \sqrt{18} = \sqrt{18} \cdot \sqrt{18} = \boxed{18}$

27. $\sqrt{y^3} = \sqrt{y^2 \cdot y} = \boxed{y\sqrt{y}}$.

29. $\sqrt{50x^2} = \sqrt{50} \cdot \sqrt{x^2} = \boxed{5\sqrt{2}x}$

31. $\sqrt{80x^4} = \sqrt{80} \cdot \sqrt{x^4} = \boxed{4\sqrt{5}x^2}$

33. $\sqrt{36x^4 \cdot 2x} = \sqrt{36x^4} \cdot \sqrt{2x} = \boxed{6x^2\sqrt{2x}}$

35. $\sqrt{2x^2} \cdot \sqrt{6x} = \sqrt{12x^2}\sqrt{x} = \sqrt{3} \cdot \sqrt{x}$, or $\boxed{2x\sqrt{3x}}$

37. $\sqrt{2y^3} \cdot \sqrt{10y} = \sqrt{20y^4} = \boxed{2y^2\sqrt{5}}$

39. $\sqrt{\dfrac{49}{16}} = \boxed{\dfrac{7}{4}}$

41. $\sqrt{\dfrac{35}{4}} = \boxed{\dfrac{\sqrt{35}}{2}}$

43. $\sqrt{\dfrac{7}{x^4}} = \boxed{\dfrac{\sqrt{7}}{x^2}}$

45. $\sqrt{\dfrac{72}{x^6}} = \boxed{\dfrac{6\sqrt{2}}{x^3}}$

47. $\dfrac{\sqrt{54}}{\sqrt{6}} = \sqrt{\dfrac{54}{6}} = \sqrt{9} = \boxed{3}$

49. $\dfrac{\sqrt{72}}{\sqrt{8}} = \sqrt{9} = \boxed{3}$

51. $\dfrac{15\sqrt{10}}{3\sqrt{2}} = \boxed{5\sqrt{5}}$

53. $\dfrac{30\sqrt{50}}{10\sqrt{5}} = \boxed{3\sqrt{10}}$

55. $\sqrt{\dfrac{28y}{81}} = \boxed{\dfrac{2\sqrt{7y}}{9}}$

57. $\dfrac{\sqrt{96y^5}}{\sqrt{8y}} = \sqrt{12y^4} = \boxed{2\sqrt{3}y^2}$

59. $\dfrac{\sqrt{24y^7}}{\sqrt{6}} = \sqrt{4y^7} = \boxed{2y^3\sqrt{y}}$

61. $A = \left(13\sqrt{2}\right)\left(5\sqrt{6}\right) = 65\sqrt{12} = \boxed{130\sqrt{3}\text{ square feet}}$

63. $h = \dfrac{s}{2}\sqrt{3}$

$h = \dfrac{\sqrt{18}}{2}\left(\sqrt{3}\right) = \dfrac{\left(3\sqrt{2}\right)\sqrt{3}}{2} = \boxed{\dfrac{3}{2}\sqrt{6}\text{ feet}}$

65. $\boxed{\text{Statement C}}$ is true, since $\sqrt{2x}\sqrt{6y} = \sqrt{12xy} = 2\sqrt{3xy}$.

67. $\sqrt[3]{32} = \sqrt[3]{8 \cdot 4}) = \boxed{2\sqrt[3]{4}}$

69. $\sqrt[3]{128} = \sqrt[3]{64 \cdot 2} = \boxed{4\sqrt[3]{2}}$

71. $\sqrt[4]{80} = \sqrt[4]{16 \cdot 5} = \boxed{2\sqrt[4]{5}}$

73. $\sqrt[3]{4} \cdot \sqrt[3]{2} = \sqrt[3]{8} = \boxed{2}$

75. $\sqrt[3]{9} \cdot \sqrt[3]{6} = \sqrt[3]{54} = \sqrt[3]{27 \cdot 2} = 3\sqrt[3]{2}$

77. $\sqrt[5]{16} \cdot \sqrt[5]{4} = \sqrt[5]{64} = \sqrt[5]{32 \cdot 2} = \boxed{2\sqrt[5]{2}}$

79. $\sqrt[3]{\dfrac{27}{8}} = \boxed{\dfrac{3}{2}}$

81. $\sqrt[4]{\dfrac{225}{81}} = \dfrac{\sqrt{15}}{3} = \dfrac{\sqrt{5} \cdot \sqrt{3}}{3} = \boxed{\sqrt{\dfrac{5}{3}}}$

83. $\sqrt[3]{\dfrac{3}{8}} = \boxed{\dfrac{\sqrt[3]{3}}{2}}$

85. $\sqrt[3]{y^3} = \boxed{y}$

87. $\sqrt[3]{2y^2} \cdot \sqrt[3]{4y} = \sqrt[3]{8y^3} = \boxed{2y}$

89. $\sqrt[3]{y^3} \cdot \sqrt[3]{y} = \boxed{y\sqrt[3]{y}}$

Review Problems

93.

$$\begin{array}{llll}
4x + 3y & = & 18 & (\times 5) \\
5x - 9y & = & 48 & (\times -4)
\end{array}$$

$$\begin{array}{rcl}
20x + 15y & = & 90 \\
-20x + 36y & = & -192 \\
\hline
51y & = & -102 \\
y & = & -2 \\
4x - 3(2) & = & 18 \\
4x & = & 24 \\
x & = & 6
\end{array}$$

$$\boxed{\{(6, -2)\}}$$

94. $\dfrac{2x}{3x + 9} + \dfrac{5}{2x - 4} - \dfrac{10x + 55}{6x^2 + 6x - 36} = \dfrac{2x(2x - 4) + 5(3x + 9)}{(3x + 9)(2x - 4)} - \dfrac{10x + 55}{(3x + 9)(2x - 4)}$

$= \dfrac{4x^2 - 8x + 15x + 45 - 10x - 55}{(3x + 9)(2x - 4)} = \dfrac{4x^2 - 3x - 10}{(3x + 9)(2x - 4)} = \dfrac{(4x + 5)(x - 2)}{6(x + 3)(x - 2)} = \boxed{\dfrac{4x + 5}{6(x + 3)}}$

95. Let b = base of triangle.
$b - 3$ = height of triangle

$$\begin{array}{rcl}
A & = & \dfrac{1}{2}bh \\
35 & = & \dfrac{1}{2}b(b - 3) = \dfrac{1}{2}(b^2 - 3b) \\
b^2 - 3b & = & 70 \\
b^2 - 3b - 70 & = & 0 \\
(b - 10)(b + 7) & = & 0 \\
b & = & 10 \quad (\text{reject } b = -7) \\
b - 3 & = & 7
\end{array}$$

$$\boxed{\text{base, 10 cm; height, 7 cm}}$$

Section 9.3 Adding and Subtracting Radicals

Problem Set 9.3, pp. 610-611

1. $7\sqrt{3} + 6\sqrt{3} = (7 + 6)\sqrt{3} = \boxed{13\sqrt{3}}$

3. $4\sqrt{13} - 6\sqrt{13} = \boxed{-2\sqrt{13}}$

5. $\sqrt{5} + \sqrt{5} = 2\sqrt{5}$

7. $\sqrt{13} + 2\sqrt{13} = 3\sqrt{13}$

9. $-4\sqrt{11} - 8\sqrt{11} = \boxed{-12\sqrt{11}}$

11. $5\sqrt{6} - \sqrt{6} = \boxed{4\sqrt{6}}$

13. $4\sqrt{2} - 5\sqrt{2} + 8\sqrt{2} = \boxed{7\sqrt{2}}$

15. $\sqrt{3} - 6\sqrt{7} - 12\sqrt{3} = \boxed{-6\sqrt{7} - 11\sqrt{3}}$

17. $6\sqrt[3]{4} - 5\sqrt[3]{4} = \boxed{\sqrt[3]{4}}$

19. $\sqrt{2} + \sqrt[3]{2}\ \boxed{\text{cannot be simplified}}$

21. $\sqrt[4]{5} + \sqrt[3]{6} + 8\sqrt[4]{5} - 2\sqrt[3]{6} = \boxed{9\sqrt[4]{5} - \sqrt[3]{6}}$

23. $\sqrt{8} + 3\sqrt{2} = 2\sqrt{2} + 3\sqrt{2} = \boxed{5\sqrt{2}}$

25. $6\sqrt{3} - \sqrt{27} = 6\sqrt{3} - 3\sqrt{3} = \boxed{3\sqrt{3}}$

27. $\sqrt{50} + \sqrt{18} = 5\sqrt{2} + 3\sqrt{2} = \boxed{8\sqrt{2}}$

29. $3\sqrt{18} - 5\sqrt{50} = 9\sqrt{2} - 25\sqrt{2} = \boxed{-16\sqrt{2}}$

31. $\frac{1}{4}\sqrt{12} - \frac{1}{2}\sqrt{48} = \frac{1}{4}\sqrt{12} - \frac{2}{2}\sqrt{12} = -\frac{3}{4}\sqrt{12} = \boxed{-\frac{3}{2}\sqrt{3}}$

33. $3\sqrt{75} + 2\sqrt{12} - 2\sqrt{48} = 15\sqrt{3} + 4\sqrt{3} - 8\sqrt{3} = \boxed{11\sqrt{3}}$

35. $6\sqrt{7} + 2\sqrt{28} - 3\sqrt{63} = 6\sqrt{7} + 4\sqrt{7} - 9\sqrt{7} = \boxed{\sqrt{7}}$

37. $\frac{1}{6}\sqrt{72} - \frac{3}{8}\sqrt{8} + \frac{1}{5}\sqrt{50} = \sqrt{2} - \frac{3}{4}\sqrt{2} + \sqrt{2} = \boxed{\frac{5}{4}\sqrt{2}}$

39. $3\sqrt{54} - 2\sqrt{20} + 4\sqrt{45} - \sqrt{24} = 9\sqrt{6} - 4\sqrt{5} + 12\sqrt{5} - 2\sqrt{6} = \boxed{7\sqrt{6} + 8\sqrt{5}}$

41. $\frac{1}{4}\sqrt{2} + \frac{2}{3}\sqrt{8} = \frac{1}{4}\sqrt{2} + \frac{4}{3}\sqrt{2} = \boxed{\frac{19}{12}\sqrt{2}}$

43. $\frac{\sqrt{45}}{4} - \sqrt{80} + \frac{\sqrt{20}}{3} = \frac{3}{4}\sqrt{5} - 4\sqrt{5} + \frac{2}{3}\sqrt{5} = \frac{9 - 48 + 8}{12}\sqrt{5} = \boxed{-\frac{31}{12}\sqrt{5}}$

45. $\sqrt[3]{81} + \sqrt[3]{24} = 3\sqrt[3]{3} + 2\sqrt[3]{3} = \boxed{5\sqrt[3]{3}}$

47. $5\sqrt[3]{54} + 2\sqrt[3]{16} = 15\sqrt[3]{2} + 4\sqrt[3]{2} = \boxed{19\sqrt[3]{2}}$

49. $5\sqrt[3]{16} - 2\sqrt[3]{54} = 10\sqrt[3]{2} - 6\sqrt[3]{2} = \boxed{4\sqrt[3]{2}}$

51. $7\sqrt{x} + 2\sqrt{x} = \boxed{9\sqrt{x}}$

53. $6\sqrt{x+1} - 2\sqrt{x+1} = \boxed{4\sqrt{x+1}}$

55. $\sqrt{36x} + \sqrt{9x} = 6\sqrt{x} + 3\sqrt{x} = \boxed{9\sqrt{x}}$

57. $\sqrt{18y} - \sqrt{8y} = 3\sqrt{2y} - 2\sqrt{2y} = \boxed{\sqrt{2y}}$

59. $7\sqrt{5x} - \sqrt{3x} + 6\sqrt{5x} + 2\sqrt{3x} = \boxed{13\sqrt{5x} - \sqrt{3x}}$

61. $5\sqrt{48x} + 2\sqrt{27x} = 20\sqrt{3x} + 6\sqrt{3x} = \boxed{26\sqrt{3x}}$

63. $5\sqrt{8y^2} - 4y\sqrt{2} = 10y\sqrt{2} - 4y\sqrt{2} = \boxed{6y\sqrt{2}}$

65. $6\sqrt{75x^2} - 4\sqrt{27x^2} = 30x\sqrt{3} - 12x\sqrt{3} = \boxed{18x\sqrt{3}}$

67. $\sqrt{500x^2} - 8x\sqrt{45} - \sqrt{180x^2} = 10x\sqrt{5} - 24x\sqrt{5} - 6x\sqrt{5} = \boxed{-20x\sqrt{5}}$

69. $\sqrt{y^3} + 2y\sqrt{y} = y\sqrt{y} + 2y\sqrt{y} = \boxed{3y\sqrt{y}}$

71. $5\sqrt{3y^2} - y\sqrt{3} = 5y\sqrt{3} - y\sqrt{3} = \boxed{4y\sqrt{3}}$

73. $5\sqrt{8x^3} + x\sqrt{50x} = 10x\sqrt{2x} + 5x\sqrt{2x} = \boxed{15x\sqrt{2x}}$

75. $6\sqrt{18y^3} - 2y\sqrt{48y} = 18y\sqrt{2y} - 8y\sqrt{3y} = \boxed{\left(18\sqrt{2} - 8\sqrt{3}\right)y\sqrt{y}}$

77. $\sqrt{3y^3} + 3y\sqrt{y} = \sqrt{3}y\sqrt{y} + 3y\sqrt{y} = \boxed{\left(\sqrt{3} + 3\right)y\sqrt{y}}$

79. $3\sqrt{3y} + \sqrt{27y} - 8\sqrt{75y} = 3\sqrt{3y} + 3\sqrt{3y} - 40\sqrt{3y} = \boxed{-34\sqrt{3y}}$

81. $7\sqrt{2} - 4\sqrt[3]{y} + 7\sqrt{2} = \boxed{14\sqrt{2} - 4\sqrt[3]{y}}$

83. $\sqrt[3]{8x^3} + \sqrt[3]{125x^3} = 2x + 5x = \boxed{7x}$

85. $\sqrt{x^5} + 2x^2\sqrt{x} = x^2\sqrt{x} + 2x^2\sqrt{x} = \boxed{3x^2\sqrt{x}}$

87. $\sqrt{3} \cdot \sqrt{2} + 5\sqrt{6} = \sqrt{6} + 5\sqrt{6} = \boxed{6\sqrt{6}}$

89. $\sqrt{6} \cdot \sqrt{2} + 4\sqrt{3} = 2\sqrt{3} + 4\sqrt{3} = \boxed{6\sqrt{3}}$

91. Statement $\boxed{D}$ is true.
A. $\sqrt{16} + \sqrt{9} = 4 + 3 = 7$ *not* 5
B. $7\sqrt[3]{3} - 4\sqrt{3}; \sqrt[3]{3} \neq \sqrt{3}$
C. $\sqrt{5} + 6\sqrt{6} = 7\sqrt{5}$ *not* $7\sqrt{10}$
D. None of the above is true.

93. perimeter $= 2l + 2w = 2\left(3\sqrt{75}\right) + 2\left(4\sqrt{18}\right) = 6\sqrt{75} + 8\sqrt{18} = \boxed{30\sqrt{3} + 24\sqrt{2} \text{ meters}}$

Review Problems

98. $64y^3 - y = y(64y^2 - 1) = \boxed{y(8y+1)(8y-1)}$

99. $(3y-2)(4y-3) - (2y-5)^2 = 12y^2 - 17y + 6 - 4y^2 + 20y - 25 = \boxed{8y^2 + 3y - 19}$

100. $y = x + 1$ From the graph, $x = 2$ and $y = 3$. Solution: $\boxed{\{(2, 3)\}}$
$y = 2x - 1$

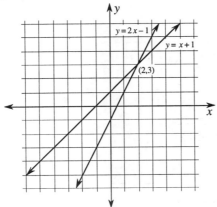

Section 9.4 Multiplying Radicals by Using the Distributive Property and the FOIL Method

Problem Set 9.4, pp. 616-617

1. $\sqrt{2}\left(\sqrt{3} + 4\right) = \boxed{\sqrt{6} + 4\sqrt{2}}$
Distributive property

3. $\sqrt{7}\left(\sqrt{6} - 5\right) = \boxed{\sqrt{42} - 5\sqrt{7}}$

5. $\sqrt{5}\left(\sqrt{3} + \sqrt{7}\right) = \boxed{\sqrt{15} + \sqrt{35}}$

7. $\sqrt{2}\left(\sqrt{5} - \sqrt{2}\right) = \sqrt{10} - \sqrt{4} = \boxed{\sqrt{10} - 2}$

9. $\sqrt{3}\left(5\sqrt{2} + \sqrt{3}\right) = \boxed{5\sqrt{6} + 3}$

11. $\sqrt{3}\left(4\sqrt{3} + \sqrt{5}\right) = 4 \cdot 3 + \sqrt{15} = \boxed{12 + \sqrt{15}}$

13. $5\sqrt{3}\left(4\sqrt{2} + 6\sqrt{5}\right) = \boxed{20\sqrt{6} + 30\sqrt{15}}$

15. $3\sqrt{10}\left(6\sqrt{2} - 4\sqrt{5}\right) = 18\sqrt{20} - 12\sqrt{50} = \boxed{36\sqrt{5} - 60\sqrt{2}}$

17. $4\sqrt{3}\left(\sqrt{32} - \sqrt{98}\right) = 4\sqrt{96} - 4\sqrt{294} = 16\sqrt{6} - 28\sqrt{6} = \boxed{-12\sqrt{6}}$

19. $\sqrt{x}\left(\sqrt{x} - \sqrt{y}\right) = \boxed{x - \sqrt{xy}}$

21. $\sqrt{3x}\left(\sqrt{3x} - \sqrt{3y}\right) = 3x - \sqrt{9xy} = 3x - 3\sqrt{xy} = \boxed{3\left(x - \sqrt{xy}\right)}$

23. $\sqrt[3]{2}\left(\sqrt[3]{4} + \sqrt[3]{5}\right) = \sqrt[3]{8} + \sqrt[3]{10} = \boxed{2 + \sqrt[3]{10}}$

25. $\sqrt{2}\left(\sqrt{10} + \sqrt{8} + \sqrt{6}\right) = \sqrt{20} + \sqrt{16} + \sqrt{12} = 2\sqrt{5} + 4 + 2\sqrt{3} = \boxed{4 + 2\left(\sqrt{3} + \sqrt{5}\right)}$

27. $(\sqrt{5}+2)(\sqrt{5}+3) = 5 + 5\sqrt{5} + 6 = \boxed{11 + 5\sqrt{5}}$

29. $(\sqrt{2}+6)(\sqrt{2}-5) = 2 + \sqrt{2} - 30 = \boxed{-28 + \sqrt{2}}$

31. $(\sqrt{2}+1)(\sqrt{3}-6) = \sqrt{6} - 6\sqrt{2} + \sqrt{3} - 6 = \boxed{\sqrt{6} + \sqrt{3} - 6(1+\sqrt{2})}$

33. $(\sqrt{3}+\sqrt{7})(\sqrt{3}+2\sqrt{7}) = 3 + 2\sqrt{21} + \sqrt{21} + 14 = \boxed{17 + 3\sqrt{21}}$

35. $(2\sqrt{7}+3)(4\sqrt{7}-5) = 56 - 10\sqrt{7} + 12\sqrt{7} - 15 = \boxed{41 + 2\sqrt{7}}$

37. $(4\sqrt{3}+7\sqrt{2})(5\sqrt{3}-6\sqrt{2}) = 60 - 24\sqrt{6} + 35\sqrt{6} - 84 = \boxed{-24 + 11\sqrt{6}}$

39. $(\sqrt{y}+5)(\sqrt{y}+2) = \boxed{y + 7\sqrt{y} + 10}$

41. $(2\sqrt{y}-4)(3\sqrt{y}-5) = 6y - 10\sqrt{y} - 12\sqrt{y} + 20 = \boxed{20 - 22\sqrt{y} + 6y}$

43. $(2\sqrt{7}-\sqrt{3})(\sqrt{14}-\sqrt{3}) = 2\sqrt{98} - 2\sqrt{21} - \sqrt{42} + 3 = \boxed{14\sqrt{2} - 2\sqrt{21} - \sqrt{42} + 3}$

45. $(\sqrt{5}+\sqrt{3})^2 = 5 + 2\sqrt{15} + 3 = \boxed{8 + 2\sqrt{15}}$

47. $(\sqrt{3}-1)^2 = 3 - 2\sqrt{3} + 1 = \boxed{4 - 2\sqrt{3}}$

49. $(2\sqrt{3}-4\sqrt{7})^2 = 12 - 16\sqrt{21} + 112 = \boxed{124 - 16\sqrt{21}}$

51. $(\sqrt{x}-4)^2 = \boxed{x - 8\sqrt{x} + 16}$ **53.** $(2\sqrt{x}+5)^2 = \boxed{4x + 20\sqrt{x} + 25}$

55. $(4+\sqrt{7})(4-\sqrt{7}) = 16 - 7 = \boxed{9}$ **57.** $(\sqrt{13}+\sqrt{5})(\sqrt{13}-\sqrt{5}) = 13 - 5 = \boxed{8}$

59. $(2\sqrt{3}-7)(2\sqrt{3}+7) = 12 - 49 = \boxed{-37}$ **61.** $(\sqrt{2x}-3)(\sqrt{2x}+3) = \boxed{2x - 9}$

63. $(2\sqrt{3}+\sqrt{5})(2\sqrt{3}-\sqrt{5}) = 12 - 5 = \boxed{7}$ **65.** $(3\sqrt{7}-2\sqrt{3})(3\sqrt{7}+2\sqrt{3}) = 63 - 12 = \boxed{51}$

67. $(\sqrt{x}+\sqrt{y})(\sqrt{x}-\sqrt{y}) = x - y$ **69.** $(\sqrt[3]{5}+3)(\sqrt[3]{2}-4) = \boxed{\sqrt[3]{10} - 4\sqrt[3]{5} + 3\sqrt[3]{2} - 12}$

71. Statement $\boxed{C}$ is correct. $(\sqrt{5}-\sqrt{3})^2 = (\sqrt{5})^2 - 2\sqrt{5}\sqrt{3} + (\sqrt{3})^2 = 5 - 2\sqrt{15} + 3 = 8 - 2\sqrt{15}$ True

73. area $= (5\sqrt{2}+3)(2\sqrt{3}-2) = 10\sqrt{6} - 10\sqrt{2} + 6\sqrt{3} - 6 = \boxed{10(\sqrt{6}-\sqrt{2}) + 6(\sqrt{3}-1)} \text{ cm}^2$

perimeter $= 2l + 2w = 2(5\sqrt{2}+3) + 2(2\sqrt{3}-2) = 10\sqrt{2} + 6 + 4\sqrt{3} - 4 = \boxed{2 + 10\sqrt{2} + 4\sqrt{3}} \text{ cm}$

Review Problems

76. Let x = the smallest number.

$x + 2$ and $x + 4$ are next two consecutive odd integers

$$
\begin{aligned}
4x - (x + 4) &= x + 10 \\
3x - 4 &= x + 10 \\
2x &= 14 \\
x &= 7 \\
x + 2 &= 9 \\
x + 4 &= 11
\end{aligned}
$$

The numbers are $\boxed{7, 9, \text{and } 11}$.

77.
$$
\begin{aligned}
(3x + 12)^\circ + (9x)^\circ &= 180^\circ \\
3x + 12 + 9x &= 180 \\
12x &= 168 \\
x &= 14 \\
(9x)^\circ &= (9 \cdot 14)^\circ = 126^\circ \\
(3x + 12)^\circ &= (3 \cdot 14 + 12)^\circ = 54^\circ
\end{aligned}
$$

The measures of the angles are $\boxed{54^\circ \text{ and } 126^\circ}$.

78. $a_n = a_1 + (n - 1)d$

$$d = \boxed{\dfrac{a_n - a_1}{n - 1}}$$

Section 9.5 Rationalizing Denominators; Simplified Radical Form

Problem Set 9.5, pp. 624-626

1. $\dfrac{2}{\sqrt{3}} = \dfrac{2}{\sqrt{3}} \cdot \dfrac{\sqrt{3}}{\sqrt{3}} = \boxed{\dfrac{2\sqrt{3}}{3}}$

3. $\dfrac{21}{\sqrt{7}} = \dfrac{21}{\sqrt{7}} \cdot \dfrac{\sqrt{7}}{\sqrt{7}} = \dfrac{21\sqrt{7}}{7} = \boxed{3\sqrt{7}}$

5. $\sqrt{\dfrac{2}{5}} = \dfrac{\sqrt{2}}{\sqrt{5}} \cdot \dfrac{\sqrt{5}}{\sqrt{5}} = \dfrac{\sqrt{2} \cdot \sqrt{5}}{5} = \boxed{\dfrac{\sqrt{10}}{5}}$

7. $\sqrt{\dfrac{7}{3}} = \dfrac{\sqrt{7}}{\sqrt{3}} \cdot \dfrac{\sqrt{3}}{\sqrt{3}} = \boxed{\dfrac{\sqrt{21}}{3}}$

9. $\sqrt{\dfrac{5}{18}} = \dfrac{\sqrt{5}}{\sqrt{18}} \cdot \dfrac{\sqrt{18}}{\sqrt{18}} = \dfrac{\sqrt{90}}{18} = \dfrac{3\sqrt{10}}{18} = \boxed{\dfrac{\sqrt{10}}{6}}$

11. $\sqrt{\dfrac{20}{3}} = \dfrac{\sqrt{20}}{\sqrt{3}} \cdot \dfrac{\sqrt{3}}{\sqrt{3}} = \dfrac{\sqrt{60}}{3} = \boxed{\dfrac{2\sqrt{15}}{3}}$

13. $\dfrac{12}{\sqrt{32}} = \dfrac{12}{4\sqrt{2}} \cdot \dfrac{\sqrt{2}}{\sqrt{2}} = \boxed{\dfrac{3\sqrt{2}}{2}}$

15. $\dfrac{15}{\sqrt{12}} = \dfrac{15}{2\sqrt{3}} \cdot \dfrac{\sqrt{3}}{\sqrt{3}} = \dfrac{15\sqrt{3}}{2(3)} = \boxed{\dfrac{5\sqrt{3}}{2}}$

17. $\dfrac{5}{\sqrt{x}} = \dfrac{5}{\sqrt{x}} \cdot \dfrac{\sqrt{x}}{\sqrt{x}} = \boxed{\dfrac{5\sqrt{x}}{x}}$

19. $\dfrac{2y^2}{\sqrt{y}} = \dfrac{2y^2}{\sqrt{y}} \cdot \dfrac{\sqrt{y}}{\sqrt{y}} = \dfrac{2y^2\sqrt{y}}{y} = \boxed{2y\sqrt{y}}$

21. $\sqrt{\dfrac{7x}{y}} = \dfrac{\sqrt{7x}}{\sqrt{y}} \cdot \dfrac{\sqrt{y}}{\sqrt{y}} = \boxed{\dfrac{\sqrt{7xy}}{y}}$

23. $\sqrt{\dfrac{72}{x}} = \dfrac{\sqrt{72}}{\sqrt{x}} \cdot \dfrac{\sqrt{x}}{\sqrt{x}} = \dfrac{\sqrt{72x}}{x} = \boxed{\dfrac{6\sqrt{2x}}{x}}$

25. $\dfrac{5}{\sqrt[3]{2}} = \dfrac{5}{\sqrt[3]{2}} \cdot \dfrac{\sqrt[3]{4}}{\sqrt[3]{4}} = \dfrac{5\sqrt[3]{4}}{\sqrt[3]{8}} = \boxed{\dfrac{5\sqrt[3]{4}}{2}}$

27. $\sqrt[3]{\dfrac{1}{4}} = \dfrac{1}{\sqrt[3]{4}} \cdot \dfrac{\sqrt[3]{2}}{\sqrt[3]{2}} = \dfrac{\sqrt[3]{2}}{\sqrt[3]{8}} = \boxed{\dfrac{\sqrt[3]{2}}{2}}$

29. $\sqrt[3]{\dfrac{2}{9}} = \dfrac{\sqrt[3]{2}}{\sqrt[3]{9}} \cdot \dfrac{\sqrt[3]{3}}{\sqrt[3]{3}} = \dfrac{\sqrt[3]{6}}{\sqrt[3]{27}} = \boxed{\dfrac{\sqrt[3]{6}}{3}}$

31. $\sqrt[3]{\dfrac{7}{3}} = \dfrac{\sqrt[3]{7}}{\sqrt[3]{3}} \cdot \dfrac{\sqrt[3]{9}}{\sqrt[3]{9}} = \dfrac{\sqrt[3]{63}}{\sqrt[3]{27}} = \boxed{\dfrac{\sqrt[3]{63}}{3}}$

33. $\dfrac{\sqrt{9}}{\sqrt{2x}} = \dfrac{\sqrt{9}}{\sqrt{2x}} \cdot \dfrac{\sqrt{2x}}{\sqrt{2x}} = \boxed{\dfrac{3\sqrt{2x}}{2x}}$

35. $\dfrac{\sqrt{2y}}{\sqrt{9x}} = \dfrac{\sqrt{2y}}{3\sqrt{x}} \cdot \dfrac{\sqrt{x}}{\sqrt{x}} = \boxed{\dfrac{\sqrt{2xy}}{3x}}$

37. $\dfrac{7}{\sqrt[3]{2x}} = \dfrac{7}{\sqrt[3]{2x}} \cdot \dfrac{\sqrt[3]{4x^2}}{\sqrt[3]{4x^2}} = \boxed{\dfrac{7\sqrt[3]{4x^2}}{2x}}$

39. $\dfrac{5}{\sqrt{3}-1} = \dfrac{5}{\sqrt{3}-1} \cdot \dfrac{\sqrt{3}+1}{\sqrt{3}+1}$ use conjugate

$= \dfrac{5\left(\sqrt{3}+1\right)}{3-1}$

$= \boxed{\dfrac{5\left(\sqrt{3}+1\right)}{2}}$

41. $\dfrac{15}{\sqrt{7}+2} = \dfrac{15}{\sqrt{7}+2} \cdot \dfrac{\sqrt{7}-2}{\sqrt{7}-2} = \dfrac{15\left(\sqrt{7}-2\right)}{7-4} = \boxed{5\left(\sqrt{7}-2\right)}$

43. $\dfrac{18}{3-\sqrt{3}} = \dfrac{18}{3-\sqrt{3}} \cdot \dfrac{3+\sqrt{3}}{3+\sqrt{3}} = \dfrac{18\left(3+\sqrt{3}\right)}{9-3} = \boxed{3\left(3+\sqrt{3}\right)}$

45. $\dfrac{\sqrt{2}}{\sqrt{2}+1} = \dfrac{\sqrt{2}}{\sqrt{2}+1} \cdot \dfrac{\sqrt{2}-1}{\sqrt{2}-1} = \dfrac{\sqrt{2}\left(\sqrt{2}-1\right)}{2-1} = \boxed{2-\sqrt{2}}$

47. $\dfrac{\sqrt{12}}{\sqrt{3}-1} = \dfrac{\sqrt{12}}{\sqrt{3}-1} \cdot \dfrac{\sqrt{3}+1}{\sqrt{3}+1} = \dfrac{2\sqrt{3}\left(\sqrt{3}+1\right)}{3-1} = \sqrt{3}\left(\sqrt{3}+1\right) = \boxed{3+\sqrt{3}}$

49. $\dfrac{3\sqrt{2}}{\sqrt{10}+2} = \dfrac{3\sqrt{2}}{\sqrt{10}+2} \cdot \dfrac{\sqrt{10}-2}{\sqrt{10}-2} = \dfrac{3\sqrt{2}\left(\sqrt{10}-2\right)}{10-4} = \dfrac{\sqrt{2}\left(\sqrt{10}-2\right)}{2} = \dfrac{\sqrt{20}-2\sqrt{2}}{2} = \boxed{\sqrt{5}-\sqrt{2}}$

51. $\dfrac{\sqrt{3}+1}{\sqrt{2}-1} = \dfrac{\sqrt{3}+1}{\sqrt{2}-1} \cdot \dfrac{\sqrt{2}+1}{\sqrt{2}+1} = \dfrac{\sqrt{6}+\sqrt{3}+\sqrt{2}+1}{2-1} = \boxed{\sqrt{6}+\sqrt{3}+\sqrt{2}+1}$

53. $\dfrac{\sqrt{2}-2}{2-\sqrt{3}} = \dfrac{\sqrt{2}-2}{2-\sqrt{3}} \cdot \dfrac{2+\sqrt{3}}{2+\sqrt{3}} = \dfrac{2\sqrt{2}+\sqrt{6}-4-2\sqrt{3}}{4-3} = \boxed{-4+2\sqrt{2}-2\sqrt{3}+\sqrt{6}}$

55. $\dfrac{2\sqrt{3}+1}{\sqrt{6}-\sqrt{3}} = \dfrac{2\sqrt{3}+1}{\sqrt{6}-\sqrt{3}} \cdot \dfrac{\sqrt{6}+\sqrt{3}}{\sqrt{6}+\sqrt{3}} = \dfrac{2\sqrt{18}+6+\sqrt{6}+\sqrt{3}}{6-3} = \dfrac{6\sqrt{2}+6+\sqrt{6}+\sqrt{3}}{3} = \boxed{\dfrac{6+6\sqrt{2}+\sqrt{3}+\sqrt{6}}{3}}$

57. $\dfrac{\sqrt{5}+\sqrt{6}}{\sqrt{5}+\sqrt{3}} = \dfrac{\sqrt{5}+\sqrt{6}}{\sqrt{5}+\sqrt{3}} \cdot \dfrac{\sqrt{5}-\sqrt{3}}{\sqrt{5}-\sqrt{3}} = \dfrac{5-\sqrt{15}+\sqrt{30}-\sqrt{18}}{5-3} = \boxed{\dfrac{5-\sqrt{15}+\sqrt{30}-3\sqrt{2}}{2}}$

59. $\dfrac{\sqrt{5}+\sqrt{2}}{\sqrt{5}-\sqrt{2}} = \dfrac{\sqrt{5}+\sqrt{2}}{\sqrt{5}-\sqrt{2}} \cdot \dfrac{\sqrt{5}+\sqrt{2}}{\sqrt{5}+\sqrt{2}} = \dfrac{5+2\sqrt{10}+2}{5-2} = \boxed{\dfrac{7+2\sqrt{10}}{3}}$

61. $\dfrac{8}{\sqrt{x}-1} = \dfrac{8}{\sqrt{x}-1} \cdot \dfrac{\sqrt{x}+1}{\sqrt{x}+1} = \boxed{\dfrac{8\left(\sqrt{x}+1\right)}{x-1}}$

63. $\sqrt{56} = \sqrt{4}\sqrt{14} = \boxed{2\sqrt{14}}$

65. $\sqrt[4]{32} = \sqrt[4]{16}\,\sqrt[4]{2} = \boxed{2\sqrt[4]{2}}$

67. $8\sqrt{27} - 3\sqrt{12} = 8 \cdot 3\sqrt{3} - 3 \cdot 2\sqrt{3} = \boxed{18\sqrt{3}}$

69. $7\sqrt{15} - 2\sqrt{5} \cdot \sqrt{3} = 7\sqrt{15} - 2\sqrt{15} = \boxed{5\sqrt{15}}$

71. $\dfrac{9}{\sqrt{18}} = \dfrac{9}{3\sqrt{2}} = \dfrac{3}{\sqrt{2}} \cdot \dfrac{\sqrt{2}}{\sqrt{2}} = \boxed{\dfrac{3\sqrt{2}}{2}}$

73. $\sqrt{\dfrac{3}{2}} \cdot \sqrt{\dfrac{5}{6}} = \sqrt{\dfrac{15}{12}} = \sqrt{\dfrac{5}{4}} = \boxed{\dfrac{\sqrt{5}}{2}}$

75. $\sqrt{\dfrac{1}{5}} \cdot \sqrt{\dfrac{3}{4}} = \sqrt{\dfrac{3}{20}} = \dfrac{\sqrt{3}}{2\sqrt{5}} \cdot \dfrac{\sqrt{5}}{\sqrt{5}} = \boxed{\dfrac{\sqrt{15}}{10}}$

77. $\dfrac{5xy}{\sqrt{x}} = \dfrac{5xy}{\sqrt{x}} \cdot \dfrac{\sqrt{x}}{\sqrt{x}} = \dfrac{5xy\sqrt{x}}{x} = \boxed{5\sqrt{x}\,y}$

79. $\left(2\sqrt{5}+\sqrt{3}\right)\left(\sqrt{2}+\sqrt{7}\right) = \boxed{2\sqrt{10}+2\sqrt{35}+\sqrt{6}+\sqrt{21}}$

81. $\dfrac{\sqrt{6}+1}{\sqrt{2}-4} = \dfrac{\sqrt{6}+1}{\sqrt{2}-4} \cdot \dfrac{\sqrt{2}+4}{\sqrt{2}+4} = \dfrac{\sqrt{12}+4\sqrt{6}+\sqrt{2}+4}{2-16} = \boxed{-\dfrac{4+\sqrt{2}+2\sqrt{3}+4\sqrt{6}}{14}}$

83. $\dfrac{1}{\sqrt[4]{2}} \cdot \dfrac{\sqrt[4]{8}}{\sqrt[4]{8}} = \dfrac{\sqrt[4]{8}}{\sqrt[4]{16}} = \boxed{\dfrac{\sqrt[4]{8}}{2}}$

85. $\dfrac{\sqrt{2}}{\sqrt{3}} + \dfrac{\sqrt{3}}{\sqrt{2}} = \dfrac{2+3}{\sqrt{6}} = \dfrac{5}{\sqrt{6}} \cdot \dfrac{\sqrt{6}}{\sqrt{6}} = \boxed{\dfrac{5\sqrt{6}}{6}}$

87. $\dfrac{2}{\sqrt{5}-1} \cdot \dfrac{\sqrt{5}+1}{\sqrt{5}+1} = \dfrac{2+2\sqrt{5}}{5-1} = \boxed{\dfrac{1+\sqrt{5}}{2} \approx 1.62}$

89. $\boxed{B}$ is true; $\dfrac{3xy}{x\sqrt{6y}} = \dfrac{3\sqrt{x}}{x\sqrt{6}} \dfrac{\sqrt{6}}{\sqrt{6}} = \dfrac{3\sqrt{6x}}{6x} = \dfrac{\sqrt{6x}}{2x}$ for $x>0$ and $y>0$, True.

91. $\sqrt{13 + \sqrt{2} + \dfrac{7}{3 + \sqrt{2}}} = \sqrt{13 + \sqrt{2} + \dfrac{7(3 - \sqrt{2})}{(3 + \sqrt{2})(3 - \sqrt{2})}} = \sqrt{13 + \sqrt{2} + \dfrac{21 - 7\sqrt{2}}{9 - 2}}$

$= \sqrt{13 + \sqrt{2} + \dfrac{21 - 7\sqrt{2}}{7}}$

$= \sqrt{13 + \sqrt{2} + 3 - \sqrt{2}}$

$= \sqrt{16}$

$= \boxed{4}$

Review Problems

94. $\qquad\quad x(x + 9)\ =\ 4(2x + 5)$

$\qquad\qquad\quad x^2 + 9x\ =\ 8x + 20$

$\qquad\quad x^2 + x - 20\ =\ 0$

$\qquad (x + 5)(x - 4)\ =\ 0$

$\qquad x = -5 \text{ or } x = 4$

$\qquad \boxed{\{-5, 4\}}$

95. $\quad \dfrac{1}{y - 1} + \dfrac{1}{y + 1}\ =\ \dfrac{3y - 2}{y^2 - 1}$

$\qquad \dfrac{y + 1 + y - 1}{y^2 - 1}\ =\ \dfrac{3y - 2}{y^2 - 1}$

$\qquad\qquad\qquad\quad 2y\ =\ 3y - 2 \quad (y \neq \pm 1)$

$\qquad\qquad\qquad\quad\ y\ =\ 2$

$\qquad\qquad \boxed{\{2\}}$

96. $81y^4 - 1 = (9y^2 - 1)(9y^2 + 1) = \boxed{(3y - 1)(3y + 1)(9y^2 + 1)}$

Section 9.6 Equations Containing Radicals

Problem Set 9.6, pp. 632-634

1. $\qquad\quad \sqrt{x}\ =\ 4$

$\qquad\qquad\ x\ =\ 4^2 = 16$

Squaring Property of Equality

$\boxed{\{16\}}$

3. $\qquad\quad \sqrt{x}\ =\ 5$

$\qquad\qquad\ x\ =\ 5^2 = 25$

$\boxed{\{25\}}$

5. $\qquad\quad \sqrt{x + 4}\ =\ 2$

$\qquad\qquad x + 4\ =\ 4$

$\qquad\qquad\quad\ x\ =\ 0$

$\boxed{\{0\}}$

7. $\qquad\quad \sqrt{3y - 2}\ =\ 4$

$\qquad\qquad 3y - 2\ =\ 16$

$\qquad\qquad\quad 3y\ =\ 18$

$\qquad\qquad\quad\ y\ =\ 6$

$\boxed{\{6\}}$

9.
$$\sqrt{3x+5} = 2$$
$$3x+5 = 4$$
$$3x = -1$$
$$x = -\frac{1}{3}$$

$$\boxed{\left\{-\frac{1}{3}\right\}}$$

11.
$$3\sqrt{z} = \sqrt{8z+16}$$
$$9z = 8z+16$$
$$z = \boxed{16}$$

Check: $3\sqrt{16} = 3\cdot4 = 12$

$$\sqrt{8\cdot16+16} = \sqrt{144} = 12 \checkmark$$

$$12 = 12 \text{ True}$$

$$\boxed{\{16\}}$$

13. $\sqrt{2y-3} = -5$

Since $\sqrt{2y-3}$ represents a nonnegative square root, there is no solution of the equation.

solution set: $\boxed{\varnothing}$

15.
$$\sqrt{3y+4}-2 = 3$$
$$\sqrt{3y+4} = 5$$
$$3y+4 = 25$$
$$3y = 21$$
$$y = 7$$

$$\boxed{\{7\}}$$

17.
$$\sqrt{6x-8}-3 = 1$$
$$\sqrt{6x-8} = 4$$
$$6x-8 = 16$$
$$6x = 24$$
$$x = 4$$

$$\boxed{\{4\}}$$

19.
$$3\sqrt{y-1} = \sqrt{3y+3}$$
$$9(y-1) = 3y+3$$
$$9y-3y = 3+9$$
$$6y = 12$$
$$y = 2$$

$$\boxed{\{2\}}$$

21.
$$\sqrt{y+3} = y-3$$
$$y+3 = y^2-6y+9$$
$$y^2-7y+6 = 0$$
$$(y-6)(y-1) = 0$$
$$y = 6 \text{ or } y = 1$$

Check: $y=6$: $\sqrt{6+3} = \sqrt{9} = 3;\ 6-3 = 3$

$y=1$: $\sqrt{1+3} = \sqrt{4} = 2;\ 1-3 = -2 \neq 2$

The solution is $\boxed{\{6\}}$.

23.
$$\sqrt{2x+13} = x+7$$
$$2x+13 = x^2+14x+49$$
$$x^2+12x+36 = 0$$
$$(x+6)^2 = 0$$
$$x = -6$$

$$\boxed{\{-6\}}$$

25.
$$\sqrt{y^2+5} = y+1$$
$$y^2+5 = y^2+2y+1$$
$$2y = 4$$
$$y = 2$$

$$\boxed{\{2\}}$$

27.
$$\sqrt{3y+3}+5 = y$$
$$\sqrt{3y+3} = y-5$$
$$3y+3 = y^2-10y+25$$
$$y^2-13y+22 = 0$$
$$(y-2)(y-11) = 0$$
$$y = 2 \text{ or } y = 11$$

Check: $y=2$: $\sqrt{6+3}+5 = 3+5 = 8 \neq 2$

$y=11$: $\sqrt{33+3}+5 = 6+5 = 11$

The solution set is $\boxed{\{11\}}$.

29.
$$\sqrt{3z + 7} - z = 3$$
$$\sqrt{3z + 7} = z + 3$$
$$3z + 7 = z^2 + 6z + 9$$
$$z^2 + 3z + 2 = 0$$
$$(z + 1)(z + 2) = 0$$
$$z = -1 \text{ or } z = -2$$

Check: $z = -1$: $\sqrt{-3 + 7} - (-1) = 3$
$\quad\quad\quad z = -2$: $\sqrt{-6 + 7} - (-2) = 3$

The solution set is $\boxed{\{-1, -2\}}$.

31.
$$\sqrt{3y} + 10 = y + 4$$
$$\sqrt{3y} = y - 6$$
$$3y = y^2 - 12y + 36$$
$$y^2 - 15y + 36 = 0$$
$$(y - 3)(y - 12) = 0$$
$$y = 3 \text{ or } y = 12$$

Check: $y = 3$: $\sqrt{9} + 10 = 13 \neq 3 + 4$
$\quad\quad\quad y = 12$: $\sqrt{36} + 10 = 16 = 12 + 4$

The solution set is $\boxed{\{12\}}$.

33.
$$\sqrt{4z^2 + 3z - 2} - 2z = 0$$
$$\sqrt{4z^2 + 3z - 2} = 2z$$
$$4z^2 + 3z - 2 = 4z^2$$
$$3z = 2$$
$$z = \frac{2}{3}$$

$$\boxed{\left\{\frac{2}{3}\right\}}$$

35.
$$\sqrt{3y^2 + 6y + 4} - 2 = 0$$
$$\sqrt{3y^2 + 6y + 4} = 2$$
$$3y^2 + 6y + 4 = 4$$
$$3y(y + 2) = 0$$
$$y = 0 \text{ or } y = -2$$

Check: $y = 0$: $\sqrt{4} = 2; 2 - 2 = 0$
$\quad\quad\quad y = -2$: $\sqrt{12 - 12 + 4} = \sqrt{4} = 2; 2 - 2 = 0$

The solution set is $\boxed{\{-2, 0\}}$.

37.
$$3\sqrt{y} + 5 = 2$$
$$3\sqrt{y} = -3$$
$$\sqrt{y} = -1$$

Since $\sqrt{y}$ represents a nonnegative value, there is no solution of the equation.

solution set: $\boxed{\varnothing}$

39.
$$\sqrt{y} = \sqrt{y+3} - 1$$
$$\sqrt{y} + 1 = \sqrt{y+3}$$
$$y + 2\sqrt{y} + 1 = y + 3$$
$$2\sqrt{y} = 2$$
$$\sqrt{y} = 1$$
$$y = 1$$

$\boxed{\{1\}}$

41.
$$\sqrt{2z} - \sqrt{z+7} = -1$$
$$\sqrt{2z} + 1 = \sqrt{z+7}$$
$$2z + 2\sqrt{2z} + 1 = z + 7$$
$$2\sqrt{2z} = 6 - z$$
$$4(2z) = 36 - 12z + z^2$$
$$z^2 - 20z + 36 = 0$$
$$(z - 18)(z - 2) = 0$$
$$z = 2 \text{ or } z = 18$$

Check: $z = 2$: $\sqrt{4} - \sqrt{9} = 2 - 3 = -1$
$\qquad\quad$ $z = 18$: $\sqrt{36} - \sqrt{25} = 6 - 5 = 1 \neq -1$

Therefore, the solution set is $\boxed{\{2\}}$.

43.
$$N = 5000\sqrt{100 - x}$$
$$40000 = 5000\sqrt{100 - x}$$
$$8 = \sqrt{100 - x}$$
$$64 = 100 - x$$
$$x = 100 - 64 = \boxed{36 \text{ years}}$$

45.
$$s = 30\sqrt{\frac{a}{p}}$$
$$90 = 30\sqrt{\frac{a}{100}}$$
$$3 = \sqrt{\frac{a}{100}}$$
$$9 = \frac{a}{100}$$
$$a = \boxed{900 \text{ feet}}$$

47.
$$T = \frac{11}{7}\sqrt{\frac{L}{2}}$$
$$2 = \frac{11}{7}\sqrt{\frac{L}{2}}$$
$$\frac{14}{11} = \sqrt{\frac{L}{2}}$$
$$\frac{196}{121} = \frac{L}{2}$$
$$L = \frac{392}{121} \text{ feet} \approx \boxed{3.24 \text{ feet}}$$

49. Let x be the number.
$$\sqrt{2x + 7} = 3$$
$$2x + 7 = 9$$
$$2x = 2$$
$$x = 1$$

The number is $\boxed{1}$.

51. Let $x =$ be the smaller integer.
$x + 1 =$ next consecutive integer

$$
\begin{aligned}
\sqrt{x + (x + 1)} &= x - 1 \\
2x + 1 &= x^2 - 2x + 1 \\
x^2 - 4x &= 0 \\
x &= 0 \text{ or } x = 4
\end{aligned}
$$

Check: $x = 0$: $1 \neq -1$
  $x = 4$: $\sqrt{9} = 3 = 4 - 1$

The integers are $\boxed{4 \text{ and } 5}$.

53. Let x be the number.

$$
\begin{aligned}
\sqrt{x + 8} &= \sqrt{x} + 2 \\
x + 8 &= x + 4\sqrt{x} + 4 \\
4 &= 4\sqrt{x} \\
\sqrt{x} &= 1 \\
x &= 1
\end{aligned}
$$

The number is $\boxed{1}$.

55.

$$
\begin{aligned}
\sqrt{1 + 24n} &= 5 \\
1 + 24n &= 25 \\
24n &= 24 \\
n &= \boxed{1}
\end{aligned}
$$

57. Statement $\boxed{B}$ is true.

The square root cannot be negative.
Solution set is $\varnothing$. True

59.

$$
\begin{aligned}
y &= \sqrt{x - 2} + 2 \\
z &= \sqrt{y - 2} + 2 \\
w &= \sqrt{z - 2} + 2
\end{aligned}
$$

Since $w = 2$, $2 = \sqrt{z - 2} + 2$; $\boxed{z = 2}$
  $2 = \sqrt{y - 2} + 2$; $\boxed{y = 2}$
  $2 = \sqrt{x - 2} + 2$; $\boxed{x = 2}$

Review Problems

63. Let x be the part that was invested at 12%.
$9000 - x =$ part invested at 8%

$$
\begin{aligned}
x(0.12) + (9000 - x)(0.08) &= 1000 \\
0.12x + 720 - 0.08x &= 1000 \\
0.04x &= 280 \\
x &= 7000 \\
9000 - x &= 2000
\end{aligned}
$$

$\boxed{\$7000 \text{ @ } 12\%}$, $\boxed{\$2000 \text{ @ } 8\%}$

64. Let $x =$ number of dimes.
$x + 2 =$ number of quarters
$x - 1 =$ number of nickels

$$
\begin{aligned}
0.10x + 0.05(x - 1) + 0.25(x + 2) &= 3.65 \\
0.10x + 0.05x - 0.05 + 0.25x + 0.5 &= 3.65 \\
0.4x &= 3.20 \\
x &= 8 \\
x + 2 &= 10 \\
x - 1 &= 7
\end{aligned}
$$

$\boxed{7 \text{ nickels, } 8 \text{ dimes, } 10 \text{ quarters}}$

Chapter 9 Review Problems

Chapter 9 Review Problems, pp. 642-643

1. $\sqrt{64} = 8$

   $-\sqrt{64} = \boxed{-8}$

2. $\sqrt{\dfrac{9}{25}} = \dfrac{3}{5}$

   $-\sqrt{\dfrac{9}{25}} = \boxed{-\dfrac{3}{5}}$

3. $\sqrt{121} = \boxed{11}$

4. $-\sqrt{121} = \boxed{-11}$

5. $\sqrt{-121}$ is $\boxed{\text{not a real number}}$

6. $\sqrt[3]{\dfrac{8}{125}} = \boxed{\dfrac{2}{5}}$

7. $\sqrt[5]{-32} = \boxed{-2}$

8. $-\sqrt[4]{81} = \boxed{-3}$

9. $\sqrt{\dfrac{8}{50}} = \dfrac{2\sqrt{2}}{5\sqrt{2}} = \boxed{\dfrac{2}{5}}$, $\boxed{\text{rational}}$

10. $\sqrt{1.21} = \boxed{1.1}$, $\boxed{\text{rational}}$

11. $\sqrt{75} = 5\sqrt{3} \approx \boxed{8.660}$, $\boxed{\text{irrational}}$

12. $\sqrt{-4}$ is $\boxed{\text{not a real number}}$

13. $\sqrt{300} = \sqrt{3} \cdot \sqrt{100} = \boxed{10\sqrt{3}}$

14. $6\sqrt{20} = 6\sqrt{4} \cdot \sqrt{5} = \boxed{12\sqrt{5}}$

15. $\sqrt{3} \cdot \sqrt{12} = \sqrt{36} = \boxed{6}$

16. $\sqrt{24} \cdot \sqrt{6} = \sqrt{144} = \boxed{12}$

17. $\sqrt{48} \cdot \sqrt{32} = 4\sqrt{3}\left(4\sqrt{2}\right) = \boxed{16\sqrt{6}}$

18. $\sqrt[3]{81} = \sqrt[3]{3 \cdot 27} = \boxed{3\sqrt[3]{3}}$

19. $\sqrt[4]{8} \cdot \sqrt[4]{10} = \sqrt[4]{80} = \sqrt[4]{5 \cdot 16} = \boxed{2\sqrt[4]{5}}$

20. $\sqrt{\dfrac{121}{4}} = \dfrac{\sqrt{121}}{\sqrt{4}} = \boxed{\dfrac{11}{2}}$

21. $\sqrt{\dfrac{7}{25}} = \boxed{\dfrac{\sqrt{7}}{5}}$

22. $\dfrac{6\sqrt{200}}{3\sqrt{2}} = \dfrac{2\sqrt{100 \cdot 2}}{\sqrt{2}} = \dfrac{20\sqrt{2}}{\sqrt{2}} = \boxed{20}$

23. $\sqrt{\dfrac{5}{2}} \cdot \sqrt{\dfrac{3}{8}} = \sqrt{\dfrac{15}{16}} = \boxed{\dfrac{\sqrt{15}}{4}}$

24. $\sqrt[3]{\dfrac{7}{64}} = \boxed{\dfrac{\sqrt[3]{7}}{4}}$

25. $\sqrt{63x^2} = x\sqrt{63} = \boxed{3x\sqrt{7}}$

26. $\sqrt{48y^3} = \boxed{4y\sqrt{3y}}$

27. $\sqrt{10x^3}\sqrt{8x^2} = \sqrt{80x^5} = \boxed{4x^2\sqrt{5x}}$

28. $\sqrt{\dfrac{7}{y^4}} = \boxed{\dfrac{\sqrt{7}}{y^2}}$

29. $\dfrac{\sqrt{75y^5}}{\sqrt{3y}} = \sqrt{25y^4} = \boxed{5y^2}$

30. $\sqrt[3]{16x^3} = \boxed{2x\sqrt[3]{2}}$

31. $\sqrt[3]{4y} \cdot \sqrt[3]{16y^2} = \sqrt[3]{64y^3} = \boxed{4y}$

32. $\dfrac{\sqrt{12x^5}}{\sqrt{3}} = \sqrt{4x^5} = \boxed{2x^2\sqrt{x}}$

33. $7\sqrt{5} + 13\sqrt{5} = \boxed{20\sqrt{5}}$

34. $\dfrac{1}{5}\sqrt{50} + \dfrac{3}{2}\sqrt{12} = \boxed{\sqrt{2} + 3\sqrt{3}}$

35. $\dfrac{5}{6}\sqrt{72} - \dfrac{3}{4}\sqrt{48} = \boxed{5\sqrt{2} - 3\sqrt{3}}$

36. $2\sqrt{18} + 3\sqrt{27} - \sqrt{12} = 6\sqrt{2} + 9\sqrt{3} - 2\sqrt{3} = \boxed{6\sqrt{2} + 7\sqrt{3}}$

37. $\sqrt[4]{7} + 3\sqrt[3]{5} - 2\sqrt[4]{7} - \sqrt[3]{5} = \boxed{-\sqrt[4]{7} + 2\sqrt[3]{5}}$

38. $4\sqrt[3]{16} + 5\sqrt[3]{2} = 8\sqrt[3]{2} + 5\sqrt[3]{2} = \boxed{13\sqrt[3]{2}}$

39. $6\sqrt{20x} - 2\sqrt{500x} = 12\sqrt{5x} - 20\sqrt{5x} = \boxed{-8\sqrt{5x}}$

40. $5\sqrt{27x^3} + x\sqrt{3x} = 15x\sqrt{3x} + x\sqrt{3x} = \boxed{16x\sqrt{3x}}$

41. $2\sqrt{75x^2} - 3x\sqrt{12} + 7\sqrt{27x^2} = 10x\sqrt{3} - 6x\sqrt{3} + 21x\sqrt{3} = \boxed{25x\sqrt{3}}$

42. $\sqrt{10}\left(\sqrt{5} + \sqrt{6}\right) = \sqrt{50} + \sqrt{60} = \boxed{5\sqrt{2} + 2\sqrt{15}}$

43. $\sqrt{3}\left(7\sqrt{2} + 4\sqrt{3}\right) = \boxed{7\sqrt{6} + 12}$

44. $7\sqrt{10}\left(6\sqrt{2} - 3\sqrt{5}\right) = 42\sqrt{20} - 21\sqrt{50} = \boxed{84\sqrt{5} - 105\sqrt{2}}$

45. $\left(\sqrt{2} + \sqrt{7}\right)\left(\sqrt{2} + 4\sqrt{7}\right) = 2 + 4\sqrt{14} + \sqrt{14} + 28 = \boxed{30 + 5\sqrt{14}}$

46. $\left(3\sqrt{6} - 2\sqrt{5}\right)\left(4\sqrt{6} + \sqrt{10}\right) = 72 + 3\sqrt{60} - 8\sqrt{30} - 2\sqrt{50} = \boxed{72 + 6\sqrt{15} - 8\sqrt{30} - 10\sqrt{2}}$

47. $\left(5\sqrt{7} - 3\right)^2 = 175 - 30\sqrt{7} + 9 = \boxed{184 - 30\sqrt{7}}$

48. $\left(\sqrt{11} - \sqrt{7}\right)\left(\sqrt{11} + \sqrt{7}\right) = 11 - 7 = \boxed{4}$

49. $\left(2\sqrt{3} + 7\sqrt{2}\right)\left(2\sqrt{3} - 7\sqrt{2}\right) = 12 - 98 = \boxed{-86}$

50. $\left(7\sqrt{3} - 2\sqrt{10}\right)^2 = 147 - 28\sqrt{30} + 40 = \boxed{187 - 28\sqrt{30}}$

51. $\left(3\sqrt{y} + \sqrt{13}\right)\left(2\sqrt{y} + 3\sqrt{13}\right)$ $(y \geq 0)$
$= 6y + 9\sqrt{13y} + 2\sqrt{13y} + 39 = \boxed{39 + 11\sqrt{13y} + 6y}$

52. $\dfrac{30}{\sqrt{5}} \cdot \dfrac{\sqrt{5}}{\sqrt{5}} = \dfrac{30\sqrt{5}}{5} = \boxed{6\sqrt{5}}$

53. $\dfrac{13}{\sqrt{50}} \cdot \dfrac{\sqrt{50}}{\sqrt{50}} = \dfrac{13\sqrt{50}}{50} = \dfrac{13 \cdot 5\sqrt{2}}{50} = \boxed{\dfrac{13\sqrt{2}}{10}}$

54. $\dfrac{7\sqrt{2}}{\sqrt{6}} = \dfrac{7}{\sqrt{3}} \cdot \dfrac{\sqrt{3}}{\sqrt{3}} = \boxed{\dfrac{7\sqrt{3}}{3}}$

55. $\sqrt{\dfrac{2}{3}} = \dfrac{\sqrt{2}}{\sqrt{3}} \cdot \dfrac{\sqrt{3}}{\sqrt{3}} = \boxed{\dfrac{\sqrt{6}}{3}}$

56. $\sqrt{\dfrac{1}{5}} \cdot \sqrt{\dfrac{7}{3}} = \dfrac{\sqrt{7}}{\sqrt{15}} \cdot \dfrac{\sqrt{15}}{\sqrt{15}} = \boxed{\dfrac{\sqrt{105}}{15}}$

57. $\sqrt{\dfrac{20}{y}} = \dfrac{2\sqrt{5}}{\sqrt{y}} \cdot \dfrac{\sqrt{y}}{\sqrt{y}} = \boxed{\dfrac{2\sqrt{5y}}{y}}$

58. $\dfrac{5}{\sqrt[3]{9}} = \dfrac{5}{\sqrt[3]{9}} \cdot \dfrac{\sqrt[3]{3}}{\sqrt[3]{3}} = \dfrac{5\sqrt[3]{3}}{\sqrt[3]{27}} = \boxed{\dfrac{5\sqrt[3]{3}}{3}}$

59. $\sqrt[3]{\dfrac{7}{4}} = \dfrac{\sqrt[3]{7}}{\sqrt[3]{4}} \cdot \dfrac{\sqrt[3]{2}}{\sqrt[3]{2}} = \dfrac{\sqrt[3]{14}}{\sqrt[3]{8}} = \boxed{\dfrac{\sqrt[3]{14}}{2}}$

60. $\dfrac{11}{\sqrt{5}+2} = \dfrac{11}{\sqrt{5}+2} \cdot \dfrac{\sqrt{5}-2}{\sqrt{5}-2} = \dfrac{-22+11\sqrt{5}}{5-4} = -22+11\sqrt{5} = \boxed{11\left(-2+\sqrt{5}\right)}$

61. $\dfrac{21}{4-\sqrt{3}} = \dfrac{21}{4-\sqrt{3}} \cdot \dfrac{4+\sqrt{3}}{4+\sqrt{3}} = \dfrac{21(4+\sqrt{3})}{16-3} = \boxed{\dfrac{21\left(4+\sqrt{3}\right)}{13}}$

62. $\dfrac{12}{\sqrt{5}+\sqrt{3}} = \dfrac{12}{\sqrt{5}+\sqrt{3}} \cdot \dfrac{\sqrt{5}-\sqrt{3}}{\sqrt{5}-\sqrt{3}} = \dfrac{12\left(\sqrt{5}-\sqrt{3}\right)}{5-3} = \boxed{6\left(\sqrt{5}-\sqrt{3}\right)}$

63. $\dfrac{\sqrt{3}+2}{\sqrt{6}-\sqrt{3}} = \dfrac{\sqrt{3}+2}{\sqrt{6}-\sqrt{3}} \cdot \dfrac{\sqrt{6}+\sqrt{3}}{\sqrt{6}+\sqrt{3}} = \dfrac{\sqrt{18}+3+2\sqrt{6}+2\sqrt{3}}{6-3} = \boxed{\dfrac{3+2\sqrt{3}+2\sqrt{6}+3\sqrt{2}}{3}}$

64.
$$\begin{aligned}
\sqrt{2y+3} &= 5 \\
2y+3 &= 25 \\
2y &= 22 \\
y &= 11
\end{aligned}$$
$\boxed{\{11\}}$

65.
$$\begin{aligned}
3\sqrt{x} &= \sqrt{6x+15} \\
9x &= 6x+15 \\
3x &= 15 \\
x &= 5
\end{aligned}$$
$\boxed{\{5\}}$

66.
$$\begin{aligned}
3\sqrt{z+3} &= \sqrt{2z+13} \\
9(z+3) &= 2z+13 \\
9z+27 &= 2z+13 \\
7z &= -14 \\
z &= -2
\end{aligned}$$
$\boxed{\{-2\}}$

67.

$$\sqrt{5x+1} = x+1$$
$$5x+1 = x^2+2x+1$$
$$x^2-3x = 0$$
$$x = 0 \text{ or } x = 3$$

Check: $x = 0$: $1 = 1$

$x = 3$: $\sqrt{16} = 4$

The solution set is $\boxed{\{0,3\}}$.

68.

$$\sqrt{y+1}+5 = y$$
$$\sqrt{y+1} = y-5$$
$$y+1 = y^2-10y+25$$
$$y^2-11y+24 = 0$$
$$(y-3)(y-8) = 0$$
$$y = 3 \text{ or } y = 8$$

Check: $y = 3$: $\sqrt{4}+5 = 7 \neq 3$

$y = 8$: $\sqrt{9}+5 = 8 = 8$

The solution set is $\boxed{\{8\}}$.

69. $y = \sqrt{y^2+4y+4}$

$$y^2 = y^2+4y+4$$
$$-4y = 4$$
$$y = -1$$

Check: $-1 = \sqrt{1-4+4}$

$-1 = 1$ False

no solution, $\boxed{\varnothing}$

70.

$$\sqrt{m+4}+\sqrt{m-4} = 4$$
$$m+4+2\left(\sqrt{m^2-16}\right)+m-4 = 16$$
$$2\sqrt{m^2-16} = 16-2m$$
$$m^2-16 = 64-16m+m^2$$
$$16m = 80$$
$$m = 5$$

$\boxed{\{5\}}$

71. $\sqrt{x-2}+5 = 1$

$$\sqrt{x-2} = -4$$

solution set is $\boxed{\{\varnothing\}}$ since $\sqrt{x-2} \geq 0$

72. $16^{1/2} = \sqrt{16} = \boxed{4}$

73. $25^{-1/2} = \dfrac{1}{\sqrt{25}} = \boxed{\dfrac{1}{5}}$

74. $125^{1/3} = \sqrt[3]{125} = \boxed{5}$

75. $27^{-1/3} = \dfrac{1}{27^{1/3}} = \boxed{\dfrac{1}{3}}$

76. $64^{2/3} = \left(\sqrt[3]{64}\right)^2 = 4^2 = \boxed{16}$

77. $27^{-4/3} = \dfrac{1}{27^{4/3}} = \dfrac{1}{\left(\sqrt[3]{27}\right)^4} = \dfrac{1}{3^4} = \boxed{\dfrac{1}{81}}$

78. $9^{-1} \cdot 9^{1/2} = 9^{-1/2} = \dfrac{1}{\sqrt{9}} = \boxed{\dfrac{1}{3}}$

79. $(2^{3/5})^{10} = 2^{30/5} = 2^6 = \boxed{64}$

80. $(16x^{12})^{1/4} = \boxed{2x^3}$

81. $\dfrac{3^{-3/4}}{3^{5/4}} = 3^{-2} = \boxed{\dfrac{1}{9}}$

82. $\dfrac{z^{1/2}z^{-3/2}}{z^2} = z^{(1-3-4)/2} = z^{-3} = \boxed{\dfrac{1}{z^3}}$

83. $r = \sqrt{\dfrac{A}{P}} - 1$

$r = \sqrt{\dfrac{144}{100}} - 1 = \dfrac{12}{10} - 1 = \boxed{0.20}$ or 20%

84. perimeter $= 2l + 2w = 2\left(4\sqrt{20} + 2\sqrt{8}\right) = \boxed{4\left(4\sqrt{5} + 2\sqrt{2}\right) \text{ meters}}$

area $= lw = \left(4\sqrt{20}\right)\left(2\sqrt{8}\right) = 8\sqrt{160} = \boxed{32\sqrt{10} \text{ square meters}}$

85.
$$\begin{aligned} C &= 200\sqrt{x} + 10 \\ 1010 &= 200\sqrt{x} + 10 \\ \frac{1000}{200} &= \sqrt{x} \\ 5 &= \sqrt{x} \\ x &= \boxed{25 \text{ tons}} \end{aligned}$$

86. Let x equal the unknown number.
$$\begin{aligned} \sqrt{1 + 2x} &= x - 7 \\ 1 + 2x &= x^2 - 14x + 49 \\ x^2 - 16x + 48 &= 0 \\ (x - 4)(x - 12) &= 0 \\ x &= 4 \text{ or } x = 12 \end{aligned}$$
Check: $x = 4$: $\sqrt{1 + 8} = 3 \neq 4 - 7 = -3$

$x = 12$: $\sqrt{1 + 24} = 5 = 12 - 7$

The number is $\boxed{12}$.

Cumulative Review Problems (Chapters 1-9)

Chapter 9 Cumulative Review, p. 644

1. $2x^2 - 5y^3 \div 12z$ $x = 2, y = -3, z = -5$
$$\begin{aligned} &= 2(2)^2 - 5(-3)^3 \div 12(-5) \\ &= 2(4) - 5(-27) \div 12(-5) \\ &= 8 + 135 \div 12(-5) \\ &= 8 + \frac{45}{4}(-5) \\ &= 8 - \frac{225}{4} \\ &= \boxed{\frac{-193}{4}} \end{aligned}$$

2.
$$23x - 13 = 19x + 15$$
$$4x = 28$$
$$x = 7$$
$$(23x - 13)° = (23 \cdot 7 - 13)° = (161 - 13)° = \boxed{148°}$$
$$(19x + 15)° = (19 \cdot 7 + 15)° = (133 + 15)° = \boxed{148°}$$

3. Let w = the width of the rectangle. $2w + 2$ = the length of the rectangle.
$$2(w + 2w + 2) = 40$$
$$2(3w + 2) = 40$$
$$3w + 2 = 20$$
$$3w = 18$$
$$w = 6$$
$$2w + 2 = 2 \cdot 6 + 2 = 14$$
dimensions: $\boxed{\text{width, 6 meters; length, 14 meters}}$

4.
$$8(5z - 7) - 4z = 9(4z - 6) - 3$$
$$40x - 56 - 4z = 36z - 54 - 3$$
$$36z - 56 = 36z - 57$$
$$-56 = -57 \quad \text{false}$$
no solution

solution set: $\boxed{\emptyset}$

5.
$$\frac{2}{3}y - 3 > \frac{5}{6}y - 1$$
$$\frac{4}{6}y - \frac{5}{6}y > -1 + 3$$
$$-\frac{y}{6} > 2$$
$$y < -12$$
$\boxed{\{y \mid y < -12\}}$

6. Let x = the present age of vase. $x + 15$ = the present age of the pitcher. $x - 10$ = the age of vase 10 years ago. $x + 15 - 10 = x + 5$ = the age of pitcher 10 years ago.
$$x + 5 = 4(x - 10)$$
$$x + 5 = 4x - 40$$
$$45 = 3x$$
$$15 = x$$
$$x + 15 = 30$$
$\boxed{\text{present age of vase, 15 years old; present age of pitcher, 30 years old}}$

7. $\dfrac{3y^3 - 25 + 16y^2}{y + 5} = \dfrac{3y^3 + 16y^2 - 25}{y + 5} = \boxed{3y^2 + y - 5}$

$$
\begin{array}{r}
3y^2 + y - 5 \\
y + 5 \overline{\smash{\big)}\ 3y^3 + 16y^2 + 0y - 25} \\
\underline{3y^3 + 15y^2} \\
y^2 + 0y \\
\underline{y^2 + 5y} \\
-5y - 25 \\
\underline{-5y - 25} \\
0
\end{array}
$$

8. $(2x - 3)(4x^2 + 6x + 9) = 2x(4x^2 + 6x + 9) - 3(4x^2 + 6x + 9) = 8x^3 + 12x^2 + 18x - 12x^2 - 18x - 27$
$= \boxed{8x^3 - 27}$

9.
$$
\begin{aligned}
(y - 2)(y - 5) + 2 &= (y - 3)(y + 3) \\
y^2 - 7y + 10 + 2 &= y^2 - 9 \\
-7y &= -21 \\
y &= 3
\end{aligned}
$$
$\boxed{\{3\}}$

10. $8y^3 + 72y^2 + 112y = 8y(y^2 + 9y + 14) = \boxed{8y(y + 2)(y + 7)}$

11.
$$
\begin{aligned}
x(3x + 11) &= 4 \\
3x^2 + 11x - 4 &= 0 \\
(3x - 1)(x + 4) &= 0 \\
x &= \tfrac{1}{3} \text{ or } x = -4
\end{aligned}
$$
$\boxed{\left\{ -4, \dfrac{1}{3} \right\}}$

12. Graph $5x + 3y \le -15$

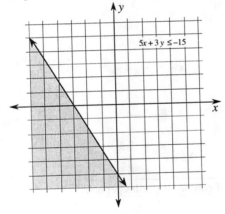

13.
$$
\begin{aligned}
8x - 5y &= -4 \\
2x + 15y &= -66
\end{aligned}
$$
$(\times 3) \rightarrow$
$$
\begin{aligned}
24x - 15y &= -12 \\
\underline{2x + 15y} &= \underline{-66} \\
26x &= -78 \\
x &= -3 \\
2(-3) + 15y &= -66 \\
15y &= -60 \\
y &= -4
\end{aligned}
$$

$\boxed{\{(-3, -4)\}}$

14. $(4, -2), (5, 1)$
$$
\begin{aligned}
\text{slope} &= \frac{1 + 2}{5 - 4} = \frac{3}{1} = 3 \\
y - 1 &= 3(x - 5) \\
y - 1 &= 3x - 15 \\
y &= 3x - 14 \\
\boxed{3x - y} &= \boxed{14}
\end{aligned}
$$

15.
$$y = x + 1$$
$$y = 2x - 1$$
solution: $\boxed{\{(2, 3)\}}$

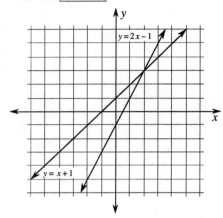

16. Let x = the speed of the boat in still water. c = the speed of current. $x + c$ = the rate of the boat with the current. $x - c$ = the rate of the boat against the current.

rate × time = distance

with the current: $(x + c)4\frac{1}{2} = 18$

against the current: $(x - c)6 = 18$

$$x + c = \frac{18}{4\frac{1}{2}} \quad \rightarrow \quad x + c = 4$$

$$x - c = \frac{18}{6} \quad \rightarrow \quad \underline{x - c = 3}$$

$$2x = 7$$
$$x = 3.5$$
$$3.5 + c = 4$$
$$c = 0.5$$

$\boxed{\text{rowing speed in still water, 3.5 miles per hour; speed of current, 0.5 miles per hour}}$.

17. Let u = the units' digit. t = the tens' digit. $10t + u$ = the number.
$$u = t - 2$$
$$10t + u = 8(u + t) - 1$$
$$10t + u = 8u + 8t - 1$$
$$2t - 7u = -1$$
$$2t - 7(t - 2) = -1 \quad \text{(substitute for } u)$$
$$2t - 7t + 14 = -1$$
$$-5t = -15$$
$$t = 3$$
$$u = t - 2 = 3 - 2 = 1$$
The number is $\boxed{31}$.

18. $\dfrac{8x^{-3}y^7}{2x^{-7}y^3} = 4x^{-3+7}y^{7-3} = \boxed{4x^4y^4}$

19. $(3x - 2)(4x - 3) - (2x - 5)^2 = 12x^2 - 8x - 9x + 6 - (4x^2 - 20x + 25) = 12x^2 - 17x + 6 - 4x^2 + 20x - 25$
$= \boxed{8x^2 + 3x - 19}$

20. $y^2 - 6y + 9 - x^2 = (y^2 - 6y + 9) - x^2 = (y-3)^2 - x^2 = [(y-3) - x][(y-3) + x] = \boxed{(y-3-x)(y-3+x)}$

21. Let $x =$ the smallest integer. $x + 1 =$ the next consecutive integer.

$$\begin{aligned}
x^2 + (x+1)^2 &= 9 + 8x \\
x^2 + x^2 + 2x + 1 &= 9 + 8x \\
2x^2 - 6x - 8 &= 0 \\
x^2 - 3x - 4 &= 0 \\
(x-4)(x+1) &= 0 \\
x &= 4 \text{ or } x = -1 \\
x + 1 &= 5 \\
x + 1 &= 0
\end{aligned}$$

The integers are $\boxed{4 \text{ and } 5 \text{ or } -1 \text{ and } 0}$.

22.

$$\begin{aligned}
D &= \frac{n(n-3)}{2} \quad D = 5 \\
5 &= \frac{n(n-3)}{2} \\
10 &= n^2 - 3n \\
n^2 - 3n - 10 &= 0 \\
(n-5)(n+2) &= 0 \\
n &= 5 \quad (\text{reject } n = -2)
\end{aligned}$$

The polygon with $\boxed{5}$ sides has 5 diagonals.

23. $\dfrac{3y}{y^2 + y - 2} - \dfrac{2}{y+2} = \dfrac{3y}{(y+2)(y-1)} - \dfrac{2}{y+2} = \dfrac{3y}{(y+2)(y-1)} - \dfrac{2(y-1)}{(y-2)(y-1)} = \dfrac{3y - 2y + 2}{(y+2)(y-1)}$

$= \dfrac{y+2}{(y+2)(y-1)}$

$= \boxed{\dfrac{1}{y-1}}$

$y \neq -2, 1$

24. $\dfrac{5x^2 - 6x + 1}{x^2 - 1} \div \dfrac{16x^2 - 9}{4x^2 + 7x + 3} = \dfrac{(5x-1)(x-1)}{(x-1)(x+1)} \div \dfrac{(4x-3)(4x+3)}{(4x+3)(x+1)} = \dfrac{5x-1}{x+1} \cdot \dfrac{x+1}{4x-3} = \boxed{\dfrac{5x-1}{4x-3}}$

$x \neq 1, -1, -\dfrac{3}{4}, \dfrac{3}{4}$

25.

$$\begin{aligned}
\frac{2y}{y+2} + \frac{3}{y^2 + 5y + 6} &= \frac{y}{y+3} \\
\frac{2y}{y+2} + \frac{3}{(y+3)(y+2)} &= \frac{y}{y+3} \\
\frac{2y(y+3)}{(y+2)(y+3)} + \frac{3}{(y+2)(y+2)} &= \frac{y(y+2)}{(y+3)(y+2)} \quad (y \neq -2, -3) \\
2y^2 + 6y + 3 &= y^2 + 2y \\
y^2 + 4y + 3 &= 0 \\
(y+3)(y+1) &= 0 \\
y &= -3 \text{ or } y = -1 \quad \text{but } y \neq -3
\end{aligned}$$

solution set: $\boxed{\{-1\}}$

26. $\dfrac{1 - \dfrac{1}{y} - \dfrac{6}{y^2}}{1 - \dfrac{4}{y^2}} = \dfrac{y^2\left(1 - \dfrac{1}{y} - \dfrac{6}{y^2}\right)}{y^2\left(1 - \dfrac{4}{y^2}\right)} = \dfrac{y^2 - y - 6}{y^2 - 4} = \dfrac{(y-3)(y+2)}{(y-2)(y+2)} = \boxed{\dfrac{y-3}{y-2}} \quad y \neq 2, -2, 0$

27. Let $x + 5 =$ the denominator. $x =$ the numerator.

$$\frac{x+4}{x+5+4} = \frac{2}{7}$$

$$\frac{x+4}{x+9} = \frac{2}{7}$$

$$7x + 28 = 2x + 18$$

$$5x = -10$$

$$x = -2$$

$$x + 5 = 3$$

original fraction: $\boxed{-\dfrac{2}{3}}$

28. $6\sqrt{75} - 4\sqrt{12} = 6\sqrt{25 \cdot 3} - 4\sqrt{4 \cdot 3} = 6 \cdot 5\sqrt{3} - 4 \cdot 2\sqrt{3} = 30\sqrt{3} - 8\sqrt{3} = \boxed{22\sqrt{3}}$

29. $\dfrac{5}{6 + \sqrt{11}} = \dfrac{5}{6 + \sqrt{11}} \cdot \dfrac{6 - \sqrt{11}}{6 - \sqrt{11}} = \dfrac{5\left(6 - \sqrt{11}\right)}{36 - 11} = \dfrac{5\left(6 - \sqrt{11}\right)}{25} = \boxed{\dfrac{6 - \sqrt{11}}{5}}$

30.
$$x = \sqrt{x-2} + 4$$

$$x - 4 = \sqrt{x-2}$$

$$(x-4)^2 = \left(\sqrt{x-2}\right)^2$$

$$x^2 - 8x + 16 = x - 2$$

$$x^2 - 9x + 18 = 0$$

$$(x - 6)(x - 3) = 0$$

$$x = 6 \text{ or } 3$$

Check: $x = 6$: $6 = \sqrt{6-2} + 4$

$$6 = \sqrt{4} + 4$$

$$6 = 2 + 4$$

$$6 = 6 \quad \text{True}$$

$x = 3$: $3 = \sqrt{3-2} + 4$

$$3 = 1 + 4$$

$$3 = 5 \quad \text{False}$$

The only solution is $x = 6$.

solution set: $\boxed{\{6\}}$

31. $13.5(-1.03)^8 \approx \boxed{17.10}$

Chapter 10   Quadratic Equations

Section 10.1   Solving Quadratic Equations by the Square Root Property

Problem Set 10.1, pp. 653-655

1.
$$x^2 = 36 \qquad \text{Original equation}$$
$$x = \pm\sqrt{36} \qquad \text{Apply the square root property.}$$
$$x = 6 \text{ or } x = -6 \qquad \text{Solutions}$$
$$\boxed{\{-6, 6\}}$$

3.
$$y^2 = 81 \qquad \text{Original equation}$$
$$y = \pm\sqrt{81} \qquad \text{Apply the square root property.}$$
$$y = 9 \text{ or } y = -9 \qquad \text{Solutions}$$
$$\boxed{\{-9, 9\}}$$

5.
$$x^2 = 7$$
$$x = \pm\sqrt{7}$$
$$x = \sqrt{7} \text{ or } x = -\sqrt{7}$$
$$\boxed{\left\{-\sqrt{7}, \sqrt{7}\right\}}$$

7.
$$x^2 = 50$$
$$x = \pm\sqrt{50} = \pm 5\sqrt{2}$$
$$\boxed{\left\{-5\sqrt{2}, 5\sqrt{2}\right\}}$$

9.
$$5y^2 = 20$$
$$y^2 = 4 \qquad \text{Divide both sides of the equation by 5.}$$
$$y = \pm 2$$
$$\boxed{\{-2, 2\}}$$

11.
$$4y^2 = 49$$
$$y^2 = \frac{49}{4}$$
$$y = \pm\frac{7}{2}$$
$$\boxed{\left\{-\frac{7}{2}, \frac{7}{2}\right\}}$$

13.
$$y^2 - 2y = 2(3 - y)$$
$$y^2 - 2y = 6 - 2y \qquad \text{Distributive property}$$
$$y^2 = 6 \qquad \text{Add } 2y \text{ to both sides of the equation.}$$
$$y = \pm\sqrt{6}$$
$$\boxed{\{-\sqrt{6}, \sqrt{6}\}}$$

15.
$$2z^2 + 2z - 5 = z(z + 2) - 3$$
$$2z^2 + 2z - 5 = z^2 + 2z - 3 \qquad \text{Distributive property}$$
$$z^2 = 2 \qquad \text{Simplify.}$$
$$z = \pm\sqrt{2}$$
$$\boxed{\left\{-\sqrt{2}, \sqrt{2}\right\}}$$

17.
$$11t^2 - 23 = 4t^2 + 33$$
$$7t^2 = 56$$
$$t^2 = 8$$
$$t = \pm 2\sqrt{2}$$

$$\boxed{\left\{-2\sqrt{2}, 2\sqrt{2}\right\}}$$

19.
$$3y^2 - 2 = 0$$
$$3y^2 = 2$$
$$y^2 = \frac{2}{3}$$
$$y = \pm\sqrt{\frac{2}{3}} = \pm\frac{\sqrt{2}}{\sqrt{3}} \cdot \frac{\sqrt{3}}{\sqrt{3}} = \pm\frac{\sqrt{6}}{3}$$

$$\boxed{\left\{-\frac{\sqrt{6}}{3}, \frac{\sqrt{6}}{3}\right\}}$$

21.
$$5m^2 - 7 = 0$$
$$5m^2 = 7$$
$$m^2 = \frac{7}{5}$$
$$m = \pm\sqrt{\frac{7}{5}} = \pm\frac{\sqrt{7}}{\sqrt{5}} \cdot \frac{\sqrt{5}}{\sqrt{5}}$$
$$= \pm\frac{\sqrt{35}}{5}$$

$$\boxed{\left\{-\frac{\sqrt{35}}{5}, \frac{\sqrt{35}}{5}\right\}}$$

23.
$$(y-3)^2 = 16$$
$$y - 3 = 4 \text{ or } y - 3 = -4 \qquad \text{Square root property}$$
$$y = 7 \text{ or } y = -1$$

$$\boxed{\{-1, 7\}}$$

25.
$$(2x+1)^2 = 64$$
$$2x + 1 = 8, \ 2x + 1 = -8$$
$$2x = 7 \text{ or } 2x = -9$$
$$x = \frac{7}{2} \text{ or } x = -\frac{9}{2}$$

$$\boxed{\left\{-\frac{9}{2}, \frac{7}{2}\right\}}$$

27.
$$(b+3)^2 = 5$$
$$b + 3 = \sqrt{5} \text{ or } b + 3 = -\sqrt{5}$$
$$b = \sqrt{5} - 3 \text{ or } b = -\sqrt{5} - 3$$

$$\boxed{\left\{\sqrt{5} - 3, -\sqrt{5} - 3\right\}}$$

29.
$$(y-2)^2 = 32$$
$$y - 2 = 4\sqrt{2} \text{ or } y - 2 = -4\sqrt{2}$$
$$y = 2 + 4\sqrt{2} \text{ or } y = 2 - 4\sqrt{2}$$

$$\boxed{\left\{2 - 4\sqrt{2}, 2 + 4\sqrt{2}\right\}}$$

31.
$$(3x-1)^2 = 12$$
$$3x - 1 = \pm 2\sqrt{3}$$
$$3x = 1 + 2\sqrt{3} \text{ or } 3x = 1 - 2\sqrt{3}$$
$$x = \frac{1}{3} + \frac{2\sqrt{3}}{3} \text{ or } x = \frac{1}{3} - \frac{2\sqrt{3}}{3}$$

$$\boxed{\left\{\frac{1}{3} - \frac{2\sqrt{3}}{3}, \frac{1}{3} + \frac{2\sqrt{3}}{3}\right\}}$$

33.
$$(6w+2)^2 = 27$$
$$6w + 2 = \pm 3\sqrt{3}$$
$$6w = -2 + 3\sqrt{3} \text{ or } 6w = -2 - 3\sqrt{3}$$
$$w = -\frac{1}{3} + \frac{\sqrt{3}}{2} \text{ or } w = -\frac{1}{3} - \frac{\sqrt{3}}{2}$$

$$\boxed{\left\{-\frac{1}{3} - \frac{\sqrt{3}}{2}, -\frac{1}{3} + \frac{\sqrt{3}}{2}\right\}}$$

35.
$$(3y + 6)^2 = 45$$
$$3y + 6 = \pm 3\sqrt{5}$$
$$y + 2 = \pm\sqrt{5} \quad \text{Simplify}$$
$$y = -2 + \sqrt{5} \text{ or } y = -2 - \sqrt{5}$$
$$\boxed{\left\{ -2 - \sqrt{5}, -2 + \sqrt{5} \right\}}$$

37.
$$(2a - 6)^2 = 98$$
$$2a - 6 = \pm 7\sqrt{2}$$
$$2a = 6 + 7\sqrt{2} \text{ or } 2a = 6 - 7\sqrt{2}$$
$$a = 3 + \frac{7\sqrt{2}}{2} \text{ or } a = 3 - \frac{7\sqrt{2}}{2}$$
$$\boxed{\left\{ 3 - \frac{7\sqrt{2}}{2}, 3 + \frac{7\sqrt{2}}{2} \right\}}$$

39.
$$\left(x - \frac{1}{3} \right)^2 = \frac{25}{9}$$
$$x - \frac{1}{3} = \pm\frac{5}{3}$$
$$x = 2 \text{ or } x = -\frac{4}{3}$$
$$\boxed{\left\{ -\frac{4}{3}, 2 \right\}}$$

41.
$$\left(y + \frac{1}{2} \right)^2 = \frac{5}{4}$$
$$y + \frac{1}{2} = \frac{\pm\sqrt{5}}{2}$$
$$y = -\frac{1}{2} + \frac{\sqrt{5}}{2} \text{ or } y = -\frac{1}{2} - \frac{\sqrt{5}}{2}$$
$$\boxed{\left\{ -\frac{1}{2} + \frac{\sqrt{5}}{2}, -\frac{1}{2} - \frac{\sqrt{5}}{2} \right\}}$$

43.
$$W = 3t^2, \ W = 108 \text{ grams}$$
$$108 = 3t^2$$
$$36 = t^2$$
$$t = 6 \quad (\text{reject } t = -6)$$
$$\boxed{6 \text{ weeks}}$$

45.
$$d = \frac{3}{50}v^2, \ d = 150 \text{ feet}$$
$$150 = \frac{3}{50}v^2$$
$$2500 = v^2 \quad \text{Simplify.}$$
$$v = \boxed{50 \text{ miles per hour}}$$

47.
$$A = P(1 + r)^2, \ A = 121, \ P = 100$$
$$121 = 100(1 + r)^2$$
$$11 = 10(1 + r) \qquad \text{Square root property}$$
$$1.1 = 1 + r \qquad \text{Divide by 10.}$$
$$r = \boxed{0.10} \text{ or } \boxed{10\%}$$

49.
$$A = \pi r^2$$
$$\pi r^2 = 49\pi$$
$$r^2 = 49$$
$$r = 7 \text{ meters}$$
$$C = 2\pi r = 2\pi(7)$$
$$= 14\pi$$

radius, 7 meters; circumference, 14π meters

51.
$$v = 1.2 - 2000\, r^2,\ v = 0.7 \text{ cm per second}$$
$$0.7 = 1.2 - 2000\, r^2$$
$$2000 r^2 = 0.5$$
$$r^2 = 0.00025$$
$$r = 0.01\sqrt{2.5} \approx 0.0158 \text{ cm}$$
$$= \boxed{0.02 \text{ cm, rounded}}$$

53. $d = \sqrt{60^2 + 60^2} = \sqrt{60^2(2)} = \boxed{60\sqrt{2} \text{ feet}}$

55.
length of diagonal $= \sqrt{30^2 + 40^2} = 50$ meters
length + width $= 40 + 30 = 70$ meters
distance saved $= 70 - 50 = \boxed{20 \text{ meters}}$

57. $AC = \sqrt{(5-1)^2 + (5-2)^2} = \sqrt{4^2 + 3^2} = \sqrt{16+9} = \sqrt{25} = \boxed{5}$

59. Let $x =$ number.
$$(7x+3)^2 = 9$$
$$7x + 3 = \pm 3$$
$$7x = 0 \text{ or } 7x = -6$$
$$x = 0 \text{ or } x = -\frac{6}{7}$$

The number is $\boxed{0 \text{ or } -\frac{6}{7}}$.

61. Let $x =$ length of side of larger square.
$x - 3 =$ length of side of smaller square
$$(x-3)^2 = 9$$
$$x - 3 = \pm 3$$
$$x = 6 \text{ or } x = 0$$

only solution is $x = 6$

length of side of large square, $\boxed{6 \text{ feet}}$

63. Let $x =$ length of side of smaller bed.
$x + 2 =$ length of side of larger flower bed
$$(x+2)^2 = 144$$
$$x + 2 = 12$$
$$x = 10 \text{ meters}$$

length of original square, $\boxed{10 \text{ meters}}$

65. Statement is $\boxed{C}$ is true.
$x^2 = -1$ has no solutions that are real numbers.

Statement A is equivalent to $x + 5 = \pm 2\sqrt{2}$
Statement B is incorrect because $x^2 = 0$ has
the solution $x = 0$

In Statement D, the solution set is $\left\{ -\sqrt{\dfrac{5}{3}}, \sqrt{\dfrac{5}{3}} \right\}$.

67.
$$ax^2 - b = 0 \ (a > 0 \text{ and } b > 0)$$
$$x^2 = \frac{b}{a}$$
$$x = \pm\sqrt{\frac{b}{a}} = \pm\frac{\sqrt{ab}}{a}$$
$$x = -\frac{\sqrt{ab}}{a} \text{ or } x = \frac{\sqrt{ab}}{a}$$

$$\boxed{\left\{ -\frac{\sqrt{ab}}{a}, \frac{\sqrt{ab}}{a} \right\}}$$

Review Problems

70. $6x^2 + 26x + 24$
$= 2(3x^2 + 13x + 12)$
$= \boxed{2(3x + 4)(x + 3)}$

71. $\dfrac{5}{y^2 + y - 6} + \dfrac{3y + 5}{y^2 + 4y + 3} - \dfrac{2y - 1}{y^2 - y - 2}$

$= \dfrac{5}{(y + 3)(y - 2)} + \dfrac{3y + 5}{(y + 3)(y + 1)} - \dfrac{2y - 1}{(y + 1)(y - 2)}$ Factor denominators.

$= \dfrac{5(y + 1) + (3y + 5)(y - 2) - (2y - 1)(y + 3)}{(y + 1)(y - 2)(y + 3)}$ Least common denominator

$= \dfrac{5y + 5 + 3y^2 - y - 10 - 2y^2 - 5y + 3}{(y + 1)(y - 2)(y + 3)}$ Carry out indicated multiplications.

$= \dfrac{y^2 - y - 2}{(y + 1)(y - 2)(y + 3)} = \dfrac{(y + 1)(y - 2)}{(y + 1)(y - 2)(y + 3)} = \boxed{\dfrac{1}{y + 3}}$ Simplify.

72. Let x = numerator.
$\dfrac{x + 6}{x + 29 + 22} = \dfrac{7}{12}$
$12(x + 6) = 7(x + 51)$
$5x = 285$
$x = 57$
original fraction $= \boxed{\dfrac{57}{86}}$

Section 10.2   Solving Quadratic Equations by Completing the Square

Problem Set 10.2, pp. 661-662

1. $x^2 + 12x$
$x^2 + 12x + \left(\dfrac{12}{2}\right)^2$ Add the square of one-half the coefficient of x
$x^2 + 12x + \boxed{36} = (x + 6)^2$ Simplify

3. $x^2 - 10x$
$x^2 - 10x + \left(\dfrac{10}{2}\right)^2$
$x^2 - 10x + \boxed{25} = (x - 5)^2$

5. $y^2 + 3y$
$y^2 + 3y + \left(\dfrac{3}{2}\right)^2$
$y^2 + 3y + \boxed{\dfrac{9}{4}} = \left(y + \dfrac{3}{2}\right)^2$

7. $y^2 - 7y$

$$y^2 - 7y + \left(-\frac{7}{2}\right)^2$$

$$y^2 - 7y + \boxed{\frac{49}{4}} = \left(y - \frac{7}{2}\right)^2$$

9. $x^2 - \frac{2}{3}x$

$$x^2 - \frac{2}{3}x + \left(-\frac{1}{3}\right)^2$$

$$x^2 - \frac{2}{3}x + \boxed{\frac{1}{9}} = \left(x - \frac{1}{3}\right)^2$$

11. $y^2 - \frac{1}{3}y$

$$y^2 - \frac{1}{3}y + \left(-\frac{1}{6}\right)^2$$

$$y^2 - \frac{1}{3}y + \boxed{\frac{1}{36}} = \left(y - \frac{1}{6}\right)^2$$

13.

$$\begin{aligned}
x^2 + 6x &= 7 & \text{Original equation} \\
x^2 + 6x + 9 &= 7 + 9 & \text{Complete the square.} \\
(x+3)^2 &= 16 \\
x + 3 &= \pm 4 & \text{Square root property} \\
x &= -7 \text{ or } x = 1
\end{aligned}$$

$$\boxed{\{-7, 1\}}$$

15.

$$\begin{aligned}
x^2 - 2x &= 2 \\
x^2 - 2x + 1 &= 2 + 1 \\
(x-1)^2 &= 3 \\
x - 1 &= \pm\sqrt{3} \\
x &= 1 + \sqrt{3} \text{ or } x = 1 - \sqrt{3}
\end{aligned}$$

$$\boxed{\left\{1 + \sqrt{3},\, 1 - \sqrt{3}\right\}}$$

17.

$$\begin{aligned}
y^2 - 6y - 11 &= 0 \\
y - 6y + 9 - 11 &= 9 \\
(y-3)^2 &= 20 \\
y - 3 &= \pm 2\sqrt{5} \\
x &= 3 + 2\sqrt{5} \text{ or } y = 3 - 2\sqrt{5}
\end{aligned}$$

$$\boxed{\left\{3 + 2\sqrt{5},\, 3 - 2\sqrt{5}\right\}}$$

19.

$$\begin{aligned}
r^2 + 4r + 1 &= 0 \\
r^2 + 4r + 4 &= 3 \\
(r+2)^2 &= 3 \\
r + 2 &= \pm\sqrt{3} \\
r &= -2 + \sqrt{3} \text{ or } r = -2 - \sqrt{3}
\end{aligned}$$

$$\boxed{\left\{-2 + \sqrt{3},\, -2 - \sqrt{3}\right\}}$$

21.

$$\begin{aligned}
x^2 + 3x - 1 &= 0 \\
x^2 + 3x + \frac{9}{4} - 1 &= \frac{9}{4} \\
\left(x + \frac{3}{2}\right)^2 &= \frac{13}{4} \\
x + \frac{3}{2} &= \pm\frac{\sqrt{13}}{2} \\
x &= -\frac{3}{2} - \frac{\sqrt{13}}{2} \text{ or } x = -\frac{3}{2} + \frac{\sqrt{13}}{2}
\end{aligned}$$

$$\boxed{\left\{-\frac{3}{2} - \frac{\sqrt{13}}{2},\, -\frac{3}{2} + \frac{\sqrt{13}}{2}\right\}}$$

23.

$$\begin{aligned}
y^2 &= 7y - 3 \\
y^2 - 7y + \left(\frac{7}{2}\right)^2 &= -3 + \left(\frac{7}{2}\right)^2 \\
\left(y - \frac{7}{2}\right)^2 &= -3 + \frac{49}{4} = \frac{37}{4} \\
y - \frac{7}{2} &= \pm\frac{\sqrt{37}}{2} \\
x &= \frac{7}{2} + \frac{\sqrt{37}}{2} \text{ or } y = \frac{7}{2} - \frac{\sqrt{37}}{2}
\end{aligned}$$

$$\boxed{\left\{\frac{7}{2} - \frac{\sqrt{37}}{2},\, \frac{7}{2} + \frac{\sqrt{37}}{2}\right\}}$$

25.
$$2z^2 - 7z + 3 = 0$$
$$2\left(z^2 - \frac{7}{2}z + \frac{3}{2}\right) = 0 \qquad \text{(Since } 2 \neq 0\text{, it may be canceled.)}$$
$$z^2 - \frac{7}{2}z + \left(-\frac{7}{4}\right)^2 + \frac{3}{2} = \left(-\frac{7}{4}\right)^2$$
$$\left(z - \frac{7}{4}\right)^2 = \frac{49}{16} - \frac{3}{2} = \frac{25}{16}$$
$$z - \frac{7}{4} = \pm\frac{5}{4}$$
$$z = \frac{7}{4} + \frac{5}{4} = 3 \text{ or } z = \frac{7}{4} - \frac{5}{4} = \frac{1}{2}$$

$$\boxed{\left\{\frac{1}{2}, 3\right\}}$$

27.
$$3y^2 = 3 + 8y$$
$$y^2 = 1 + \frac{8}{3}y \qquad \text{(Divide by 3.)}$$
$$y^2 - \frac{8}{3}y + \left(-\frac{4}{3}\right)^2 = 1 + \frac{16}{9} = \frac{25}{9}$$
$$\left(y - \frac{4}{3}\right)^2 = \frac{25}{9}$$
$$y - \frac{4}{3} = \pm\frac{5}{3}$$
$$y = \frac{4}{3} - \frac{5}{3} = -\frac{1}{3} \text{ or } y = \frac{4}{3} + \frac{5}{3} = 3$$

$$\boxed{\left\{-\frac{1}{3}, 3\right\}}$$

29.
$$4y^2 - 4y - 1 = 0$$
$$y^2 - y = \frac{1}{4}$$
$$y^2 - y + \left(-\frac{1}{2}\right)^2 = \frac{1}{4} + \frac{1}{4} = \frac{1}{2}$$
$$\left(y - \frac{1}{2}\right)^2 = \frac{1}{2}$$
$$y - \frac{1}{2} = \pm\frac{\sqrt{2}}{2}$$
$$y = \frac{1}{2} + \frac{\sqrt{2}}{2} \text{ or } y = \frac{1}{2} - \frac{\sqrt{2}}{2}$$

$$\boxed{\left\{\frac{1}{2} + \frac{\sqrt{2}}{2}, \frac{1}{2} - \frac{\sqrt{2}}{2}\right\}}$$

31.
$$3z^2 - 2z - 2 = 0$$
$$x^2 - \frac{2}{3}z = \frac{2}{3}$$
$$z^2 - \frac{2}{3}z + \left(-\frac{1}{3}\right)^2 = \frac{2}{3} + \frac{1}{9} = \frac{7}{9}$$
$$\left(z - \frac{1}{3}\right)^2 = \frac{7}{9}$$
$$z - \frac{1}{3} = \pm\frac{\sqrt{7}}{3}$$
$$z = \frac{1}{3} + \frac{\sqrt{7}}{3} \text{ or } z = \frac{1}{3} - \frac{\sqrt{7}}{3}$$

$$\boxed{\left\{\frac{1}{3} - \frac{\sqrt{7}}{3}, \frac{1}{3} + \frac{\sqrt{7}}{3}\right\}}$$

33.

$$
\begin{aligned}
2t^2 &= 3 - 10t \\
2t^2 + 10t &= 3 \\
t^2 + 5t &= \frac{3}{2} \\
t^2 + 5t + \left(\frac{5}{2}\right)^2 &= \frac{3}{2} + \left(\frac{5}{2}\right)^2 \\
\left(t + \frac{5}{2}\right)^2 &= \frac{31}{4} \\
t + \frac{5}{2} &= \frac{\pm\sqrt{31}}{2} \\
t &= \frac{-5}{2} - \frac{\sqrt{31}}{2} \text{ or } t = -\frac{5}{2} + \frac{\sqrt{31}}{2}
\end{aligned}
$$

$$\boxed{\left\{-\frac{5}{2} - \frac{\sqrt{31}}{2}, -\frac{5}{2} + \frac{\sqrt{31}}{2}\right\}}$$

35.

$$
\begin{aligned}
6y - y^2 &= 4 \\
y^2 - 6y &= -4 \\
y^2 - 6y + (-3)^2 &= -4 + 9 = 5 \\
(y - 3)^2 &= 5 \\
y - 3 &= \pm\sqrt{5} \\
y &= 3 + \sqrt{5} \text{ or } y = 3 - \sqrt{5}
\end{aligned}
$$

$$\boxed{\left\{3 - \sqrt{5}, 3 + \sqrt{5}\right\}}$$

37.

$$
\begin{aligned}
z(3z - 2) &= 6 \\
3z^2 - 2z &= 6 \\
z^2 - \frac{2}{3}z &= 2 \\
z^2 - \frac{2}{3}z + \left(-\frac{1}{3}\right)^2 &= 2 + \frac{1}{9} = \frac{19}{9} \\
\left(z - \frac{1}{3}\right)^2 &= \frac{19}{9} \\
z - \frac{1}{3} &= \pm\frac{\sqrt{19}}{3} \\
z &= \frac{1}{3} + \frac{\sqrt{19}}{3} \text{ or } z = \frac{1}{3} - \frac{\sqrt{19}}{3}
\end{aligned}
$$

$$\boxed{\left\{\frac{1}{3} - \frac{\sqrt{19}}{3}, \frac{1}{3} + \frac{\sqrt{19}}{3}\right\}}$$

39. Statement $\boxed{\text{D}}$ is correct.

$$x^2 - 7x + \left(-\frac{7}{2}\right)^2 = 5 + \left(-\frac{7}{2}\right)^2$$

Add: $\left(-\frac{7}{2}\right)^2 = \frac{49}{4}$ True

41.

$$
\begin{aligned}
x^2 + x + c &= 0 \\
x^2 + x + \left(\frac{1}{2}\right)^2 &= -c + \frac{1}{4} \\
\left(x + \frac{1}{2}\right)^2 &= \frac{1}{4} - c \\
x + \frac{1}{2} &= \pm\frac{\sqrt{1 - 4c}}{2} \\
x &= -\frac{1}{2} + \frac{\sqrt{1 - 4c}}{2} \text{ or } x = -\frac{\sqrt{1 - 4c}}{2}
\end{aligned}
$$

$$\boxed{\left\{-\frac{1}{2} - \frac{\sqrt{1 - 4c}}{2}, -\frac{1}{2} + \frac{\sqrt{1 - 4c}}{2}\right\}}$$

Review Problems

45. $3(x + 2) + 5y = -6$
            $2x - 3(y + 1) = 8$

(simplify) → $\quad 3x + 6 + 5y = -6$ (simplify) → $\quad 3x + 5y = -12$
$\qquad\qquad\quad 2x - 3y - 3 = 8$ (simplify) → $\quad 2x - 3y = 11$

Multiply the first equation by 2 and the second equation by –3. Then add the two equations. Go back and solve for x.

$$6x + 10y = -24$$
$$-6x + 9y = -33$$
$$19y = -57$$
$$y = -\frac{57}{19} = -3$$
$$x = \frac{1}{2}(11 + 3y) = \frac{1}{2}(11 - 9) = \frac{1}{2}(2) = 1$$

$$\boxed{\{(1, -3)\}}$$

46. $\dfrac{x^{-2}y^2}{x^4y^{-6}} = x^{-2-4}y^{2-(-6)} = x^{-6}y^8 = \boxed{\dfrac{y^8}{x^6}}$

47. $(x + 2)(3x - 1) - 3(x - 2)^2$
$= 3x^2 + 5x - 2 - 3(x^2 - 4x + 4)$
$= 3x^2 + 5x - 2 - 3x^2 + 12x - 12$
$= \boxed{17x - 14}$

Section 10.3   The Quadratic Formula

Problem Set 10.3, pp. 669-670

1. $x^2 - 3x - 18 = 0$

$x = \dfrac{-(-3) \pm \sqrt{(-3)^2 - 4(1)(-18)}}{2(1)}$

Original equation: $a = 1,\ b = -3,\ c = -18$

$x = \dfrac{-b \pm \sqrt{b^2 - 4ac}}{2a}$

$x = \dfrac{3 \pm \sqrt{9 + 72}}{2} = \dfrac{3 \pm \sqrt{81}}{2} = \dfrac{3 \pm 9}{2}$ Simplify

$x = 6$ or $x = -3$

$\boxed{\{-3, 6\}}$

3. $6x^2 - 5x - 6 = 0$

$x = \dfrac{5 \pm \sqrt{(-5)^2 - 4(6)(-6)}}{2(6)} = \dfrac{5 \pm \sqrt{25 + 144}}{12} = \dfrac{5 \pm \sqrt{169}}{12}$

$x = \dfrac{5 \pm 13}{12}$

$x = \dfrac{5 + 13}{12} = \dfrac{3}{2}$ or $x = \dfrac{5 - 13}{12} = -\dfrac{2}{3}$

$x = \dfrac{3}{2}$ or $x = -\dfrac{2}{3}$

$$\boxed{\left\{-\dfrac{2}{3}, \dfrac{3}{2}\right\}}$$

5. $x^2 - 2x - 10 = 0$

$$x = \frac{2 \pm \sqrt{4 - 4(-10)}}{2} = \frac{2 \pm \sqrt{44}}{2} = \frac{2 \pm 2\sqrt{11}}{2} = 1 \pm \sqrt{11}$$

$x = 1 + \sqrt{11}$ or $x = 1 - \sqrt{11}$

$$\boxed{\left\{ 1 - \sqrt{11}, 1 + \sqrt{11} \right\}}$$

7. $x^2 - x = 14$

$x^2 - x - 14 = 0$

$$x = \frac{1 \pm \sqrt{1 - 4(-14)}}{2} = \frac{1 \pm \sqrt{57}}{2}$$

$$x = \frac{1 + \sqrt{57}}{2} \text{ or } \frac{1 - \sqrt{57}}{2}$$

$$\boxed{\left\{ \frac{1 - \sqrt{57}}{2}, \frac{1 + \sqrt{57}}{2} \right\}}$$

9. $6y^2 + 6y + 1 = 0$

$$y = \frac{-6 \pm \sqrt{36 - 4(6)(1)}}{2(6)} = \frac{-6 \pm \sqrt{12}}{12} = \frac{-6 \pm 2\sqrt{3}}{12} = \frac{3 \pm \sqrt{3}}{6}$$

$$y = \frac{-3 + \sqrt{3}}{6} \text{ or } y = \frac{-3 - \sqrt{3}}{6}$$

$$\boxed{\left\{ \frac{-3 - \sqrt{3}}{6}, \frac{-3 + \sqrt{3}}{6} \right\}}$$

11. $4x^2 - 12x + 9 = 0$

$$x = \frac{12 \pm \sqrt{144 - 4(4)(9)}}{2(4)} = \frac{12 \pm 0}{8} = \frac{3}{2}$$

$$x = \frac{3}{2}$$

$$\boxed{\left\{ \frac{3}{2} \right\}}$$

13. $y^2 = 2(y + 1)$

$y^2 - 2y - 2 = 0$

$$y = \frac{2 \pm \sqrt{4 - 4(1)(-2)}}{2} = \frac{2 \pm \sqrt{12}}{2} = \frac{2 \pm 2\sqrt{3}}{2} = 1 \pm \sqrt{3}$$

$y = 1 - \sqrt{3}$ or $y = 1 + \sqrt{3}$

$$\boxed{\left\{ 1 - \sqrt{3}, 1 + \sqrt{3} \right\}}$$

15. $\dfrac{y^2}{4} + \dfrac{3y}{2} + 1 = 0$

$y^2 + 6y + 4 = 0$

$$y = \frac{-6 \pm \sqrt{36 - 4(1)(4)}}{2} = \frac{-6 \pm \sqrt{20}}{2} = \frac{-6 \pm 2\sqrt{5}}{2} = -3 \pm \sqrt{5}$$

$$y = -3 - \sqrt{5} \text{ or } y = -3 + \sqrt{5}$$

$$\boxed{\left\{ -3 - \sqrt{5}, -3 + \sqrt{5} \right\}}$$

17.
$$\frac{1}{x^2} + 3 = \frac{6}{x}$$
$$1 + 3x^2 = 6x$$
$$3x^2 - 6x + 1 = 0$$

$$x = \frac{6 \pm \sqrt{36 - 4(3)(1)}}{2(3)} = \frac{6 \pm \sqrt{24}}{6} = \frac{6 \pm 2\sqrt{6}}{6} = \frac{3 \pm \sqrt{6}}{3}$$

$$x = \frac{3 + \sqrt{6}}{3} \text{ or } x = \frac{3 - \sqrt{6}}{3}$$

$$\boxed{\left\{ \frac{3 - \sqrt{6}}{3}, \frac{3 + \sqrt{6}}{3} \right\}}$$

19. Statement $\boxed{C}$ is true.

21.
$$2x^2 - x = 1$$
$$2x^2 - x - 1 = 0$$
$$(2x + 1)(x - 1) = 0$$
$$x = -\frac{1}{2} \text{ or } x = 1$$

$$\boxed{\left\{ -\frac{1}{2}, 1 \right\}}$$

23.
$$5x^2 + 2 = 11x$$
$$5x^2 - 11x + 2 = 0$$
$$(5x - 1)(x - 2) = 0$$
$$x = \frac{1}{5} \text{ or } x = 2$$

$$\boxed{\left\{ \frac{1}{5}, 2 \right\}}$$

25.
$$y^2 = 20$$
$$y = \pm 2\sqrt{5}$$
$$y = -2\sqrt{5} \text{ or } y = 2\sqrt{5}$$

$$\boxed{\left\{ -2\sqrt{5}, 2\sqrt{5} \right\}}$$

27.
$$x^2 - 2x = 1$$
$$x^2 - 2x - 1 = 0$$

$$x = \frac{2 \pm \sqrt{4 + 4}}{2} = \frac{2 \pm \sqrt{8}}{2} = \frac{2 \pm 2\sqrt{2}}{2} = 1 \pm \sqrt{2}$$

$$x = 1 + \sqrt{2} \text{ or } x = 1 - \sqrt{2}$$

$$\boxed{\left\{ 1 - \sqrt{2}, 1 + \sqrt{2} \right\}}$$

29.
$$(2w + 3)(w + 4) = 1$$
$$2w^2 + 11w + 11 = 0$$

$$w = \frac{-11 \pm \sqrt{121 - 4(2)(11)}}{2(2)} = \frac{-11 \pm \sqrt{121 - 88}}{4} = \frac{-11 \pm \sqrt{33}}{4}$$

$$w = \frac{-11 + \sqrt{33}}{4} \text{ or } w = \frac{-11 - \sqrt{33}}{4}$$

$$\boxed{\left\{ \frac{-11 - \sqrt{33}}{4}, \frac{-11 + \sqrt{33}}{4} \right\}}$$

31.
$$(3r-4)^2 = 16$$
$$3r-4 = \pm 4$$
$$3r = 0 \text{ or } 3r = 8$$
$$r = 0 \text{ or } r = \frac{8}{3}$$
$$\left\{ 0, \frac{8}{3} \right\}$$

33.
$$3y^2 - 12y + 12 = 0$$
$$y^2 - 4y + 4 = 0$$
$$(y-2)^2 = 0$$
$$y = 2$$
$$\{2\}$$

35.
$$4w^2 - 16 = 0$$
$$w^2 - 4 = 0$$
$$w^2 = 4$$
$$w = \pm 2$$
$$w = -2 \text{ or } w = 2$$
$$\{-2, 2\}$$

37.
$$\frac{3}{4}y^2 - \frac{5}{2}y - 2 = 0$$
$$3y^2 - 10y - 8 = 0$$
$$(3y+2)(y-4) = 0$$
$$y = -\frac{2}{3} \text{ or } y = 4$$
$$\left\{ -\frac{2}{3}, 4 \right\}$$

39.
$$10x^2 - 11x + 2 = 0$$
$$x = \frac{11 \pm \sqrt{121 - 4(10)(2)}}{2(10)} = \frac{11 \pm \sqrt{41}}{20}$$
$$x = \frac{11 - \sqrt{41}}{20} \text{ or } x = \frac{11 + \sqrt{41}}{20}$$
$$\left\{ \frac{11 - \sqrt{41}}{20}, \frac{11 + \sqrt{41}}{20} \right\}$$

41.
$$\frac{y^2}{2} - 2y + \frac{3}{4} = 0$$
$$2y^2 - 8y + 3 = 0$$
$$y = \frac{8 \pm \sqrt{64 - 4(2)(3)}}{2(2)} = \frac{8 \pm \sqrt{40}}{4} = \frac{8 \pm 2\sqrt{10}}{4}$$
$$y = \frac{4 + \sqrt{10}}{2} \text{ or } y = \frac{4 - \sqrt{10}}{2}$$
$$\left\{ 2 - \frac{\sqrt{10}}{2}, 2 + \frac{\sqrt{10}}{2} \right\}$$

43.
$$(y-1)(y-2) = 2y(y+1)$$
$$y^2 - 3y + 2 = 2y^2 + 2y$$
$$y^2 + 5y - 2 = 0$$
$$y = \frac{-5 \pm \sqrt{25 - 4(1)(-2)}}{2} = \frac{-5 \pm \sqrt{33}}{2}$$
$$y = \frac{-5 + \sqrt{33}}{2} \text{ or } y = \frac{-5 - \sqrt{33}}{2}$$
$$\left\{ \frac{-5 - \sqrt{33}}{2}, \frac{-5 + \sqrt{33}}{2} \right\}$$

45.
$$(3x - 2)^2 = 10$$
$$3x - 2 = \pm\sqrt{10}$$
$$3x = 2 + \sqrt{10} \text{ or } 3x = 2 - \sqrt{10}$$
$$x = \frac{2 + \sqrt{10}}{3} \text{ or } x = \frac{2 - \sqrt{10}}{3}$$
$$\boxed{\left\{\frac{2 - \sqrt{10}}{3}, \frac{2 + \sqrt{10}}{3}\right\}}$$

47.
$$y^2 + 14y + 49 = 0$$
$$(y + 7)^2 = 0$$
$$y = -7$$
$$\boxed{\{-7\}}$$

49.
$$x^2 + 9x = 0$$
$$x(x + 9) = 0$$
$$x = 0 \text{ or } x = -9$$
$$\boxed{\{-9, 0\}}$$

51.
$$(x - 2)^2 - 49 = 0$$
$$(x - 2)^2 = 49$$
$$x = -5 \text{ or } x = 9$$
$$\boxed{\{-5, 9\}}$$

53.
$$\sqrt{2y + 3} = y - 1$$
$$2y + 3 = (y - 1)^2 = y^2 - 2y + 1$$
$$y^2 - 4y - 2 = 0$$
$$y = \frac{4 \pm \sqrt{16 - 4(1)(-2)}}{2} = \frac{4 \pm \sqrt{24}}{2} = \frac{4 \pm 2\sqrt{6}}{2} = 2 \pm \sqrt{6}$$
$$y = 2 + \sqrt{6} \text{ or } y = 2 - \sqrt{6}$$
$$\boxed{\left\{2 - \sqrt{6}, 2 + \sqrt{6}\right\}}$$

55.
$$\boxed{\frac{-b + \sqrt{b^2 - 4ac}}{2a} + \frac{-b - \sqrt{b^2 - 4ac}}{2a} = \frac{-2b}{2a} = -\frac{b}{a}}$$

57.
$$x^2 + 2\sqrt{3}x - 9 = 0$$
$$x = \frac{-2\sqrt{3} \pm \sqrt{(4)(3) - 4(1)(-9)}}{2} = \frac{-2\sqrt{3} \pm \sqrt{48}}{2} = \frac{-2\sqrt{3} \pm 4\sqrt{3}}{2} = -\sqrt{3} \pm 2\sqrt{3}$$
$$x = \sqrt{3} \text{ or } x = -3\sqrt{3}$$
$$\boxed{\left\{-3\sqrt{3}, \sqrt{3}\right\}}$$

59.
$$1.5x^2 - 17.6x + 8.03 = 0$$
$$x = \frac{17.6 \pm \sqrt{(-17.6)^2 - 4(1.5)(8.03)}}{2(1.5)} = \frac{17.6 \pm \sqrt{261.58}}{3} \approx \frac{17.6 \pm 16.17}{3}$$
$$x \approx 11.3 \text{ or } x \approx 0.48$$
$$\boxed{\{11.3, 0.48\}}$$

61.
$$d = 0.044v^2 + 1.1v, \, d = 550 \text{ feet}$$
$$550 = 0.044v^2 + 1.1v$$
$$0.044v^2 + 1.1v - 550 = 0$$
$$v = \frac{-1.1 + \sqrt{1.21 - 4(0.044)(-550)}}{2(0.044)} = \frac{-1.1 + \sqrt{98.01}}{0.088} = \frac{-1.1 + 9.9}{0.088}$$
$$v = \boxed{100 \text{ miles per hour}}$$

Review Problems

65. Let x = regular price.
$0.128x = 12.8\%$ reduction of regular price
$$\begin{aligned} x - 0.128x &= 287.76 \\ 0.872x &= 287.76 \\ x &= \frac{287.76}{0.872} = \boxed{\$330.00} \end{aligned}$$

66.
$$\begin{aligned} 4 - 7(x - 2) &> -3 \\ 4 - 7x + 14 &> -3 \\ -7x + 18 &> -3 \\ -7x &> -21 \\ x &< 3 \end{aligned}$$
$$\boxed{\{x \mid x < 3\}}$$

67. $\dfrac{\sqrt{5} - \sqrt{2}}{\sqrt{5} + 2\sqrt{2}} = \dfrac{\sqrt{5} - \sqrt{2}}{\sqrt{5} + 2\sqrt{2}} \cdot \dfrac{\sqrt{5} - 2\sqrt{2}}{\sqrt{5} - 2\sqrt{2}} = \dfrac{5 - 3\sqrt{10} + 4}{5 - 8} = \dfrac{9 - 3\sqrt{10}}{-3} = \boxed{-3 + \sqrt{10}}$

Section 10.4   Applications of Quadratic Equations

Problem Set 10.4, pp. 676-678

1. Let x = unknown number.
$$\begin{aligned} 4x + x^2 &= -1 \\ x^2 + 4x + 1 &= 0 \end{aligned}$$
$$\begin{aligned} x &= \frac{-4 \pm \sqrt{16 - 4}}{2} = \frac{-4 \pm \sqrt{12}}{2} = \frac{-4 \pm 2\sqrt{3}}{2} = -2 \pm \sqrt{3} \\ x &= -2 + \sqrt{3} \text{ or } x = -2 - \sqrt{3} \\ x &= -0.3 \text{ or } x = -3.7, \text{ rounded} \end{aligned}$$
The number is $\boxed{-2 + \sqrt{3} \text{ or} -2 - \sqrt{3}}$ or rounded, $\boxed{-0.3 \text{ or} -3.7}$.

3. Let x = one number.
$2 - x$ = other number
$$\begin{aligned} x(2 - x) &= -1 \\ 2x - x^2 &= -1 \\ x^2 - 2x - 1 &= 0 \end{aligned}$$
$$\begin{aligned} x &= \frac{2 \pm \sqrt{4 + 4}}{2} = \frac{2 \pm \sqrt{8}}{2} = \frac{2 \pm 2\sqrt{2}}{2} = 1 \pm \sqrt{2} \\ x &= 1 + \sqrt{2} \text{ or } x = 1 - \sqrt{2} \\ 2 - x &= 2 - 1 - \sqrt{2} \text{ or } 2 - 1 + \sqrt{2} \\ &= 1 - \sqrt{2} \text{ or } 1 + \sqrt{2} \end{aligned}$$
The numbers are $\boxed{1 + \sqrt{2} \text{ and } 1 - \sqrt{2}}$ or rounded, $\boxed{2.4 \text{ and} -0.4}$.

5. w = width of rectangle.
$l = w + 3$ = length of rectangle

$$w(w + 3) = 36$$
$$w^2 + 3w - 36 = 0$$
$$w = \frac{-3 + \sqrt{9 + 144}}{2} = \frac{-3 + \sqrt{153}}{2} = 4.7 \text{ meters, rounded; reject negative root as it}$$

would give negative length

$$w + 3 = \frac{3 + \sqrt{153}}{2} = 7.7 \text{ meters, rounded}$$

$$\boxed{\text{length}, \frac{3 + \sqrt{153}}{2} \approx 7.7 \text{ meters}}$$

$$\boxed{\text{width}, \frac{-3 + \sqrt{153}}{2} \approx 4.7 \text{ meters}}$$

7. Let b = base of triangle and h = height of triangle.

$$b = 2h - 1$$
$$A = \frac{bh}{2} = 9$$
$$\frac{(2h - 1)(h)}{2} = 9$$
$$2h^2 - h = 18$$
$$2h^2 - h - 18 = 0$$
$$h = \frac{1 + \sqrt{1 - 4(2)(-18)}}{2(2)} = \frac{1 + \sqrt{145}}{4}, b = \frac{1 + \sqrt{145}}{2} - 1 = \frac{-1 + \sqrt{145}}{2}; \text{ reject negative root}$$

$$\boxed{h = \frac{1 + \sqrt{145}}{4}} \quad \boxed{b = \frac{-1 + \sqrt{145}}{2}}$$

$$\boxed{h \approx 3.3 \text{ inches}, b \approx 5.5 \text{ inches, rounded}}$$

9.
$$(8 - x)(10 - x) = 47$$
$$80 - 18x + x^2 = 47$$
$$x^2 - 18x + 33 = 0$$
$$x = \frac{18 - \sqrt{324 - 4(33)}}{2} = \frac{18 - \sqrt{192}}{2} = \frac{18 - 8\sqrt{3}}{2} = 9 - 4\sqrt{3}$$

Note: The minus sign must be used, because $9 + 4\sqrt{3} > 10$.

$$x = \boxed{9 - 4\sqrt{3} \text{ meters} \approx 2.1 \text{ meters, rounded}}$$

11. Let x = length of side of original square.
$x - (2 + 2)$ = length of side of base of box

$$(x - 4)^2(2) = 108$$
$$(x - 4)^2 = 54$$
$$x - 4 = \sqrt{54}$$
$$x = \boxed{4 + 3\sqrt{6} \text{ cm} \approx 11.3 \text{ cm, rounded}}$$

13. Let x = length of shorter leg of triangle.

$x + 1$ = length of longer leg

$$x^2 + (x+1)^2 = 6^2 = 36$$
$$x^2 + x^2 + 2x + 1 = 36$$
$$2x^2 + 2x - 35 = 0$$

$$x = \frac{-2 + \sqrt{4 - 4(2)(-35)}}{2(2)} = \frac{-2 + \sqrt{4 + 280}}{4} = \frac{-2 + 2\sqrt{71}}{4}$$

$$x = \boxed{\frac{-1 + \sqrt{71}}{2}} \text{ mm} = 3.7 \text{ mm, rounded} \quad \boxed{\text{and}}$$

$$x + 1 = 1 + \frac{-1 + \sqrt{71}}{2} = \boxed{\frac{1 + \sqrt{71}}{2}} \approx 4.7 \text{ mm, rounded}$$

15.
$$x^2 + (400 - x)^2 = 300^2$$
$$x^2 + 160{,}000 - 800x + x^2 = 90{,}000$$
$$2x^2 - 800x + 70{,}000 = 0$$
$$x^2 - 400x + 35{,}000 = 0$$

$$x = \frac{400 - \sqrt{160{,}000 - 4(35{,}000)}}{2} = \frac{400 - \sqrt{20{,}000}}{2} = \frac{400 - 100\sqrt{2}}{2}$$

$$= \boxed{200 - 50\sqrt{2} \text{ yards}} \approx 129.3 \text{ yards, rounded}$$

$$\boxed{\text{and}} \quad 400 - x = \boxed{200 + 50\sqrt{2} \text{ yards}} \approx 270.7 \text{ yards, rounded}$$

17.
$$s = -16t^2 + 100t + 50$$
$$0 = -16t^2 + 100t + 50$$
$$8t^2 - 50t - 25 = 0$$

$$t = \frac{50 + \sqrt{2500 - 4(8)(-25)}}{2(8)} = \frac{50 + \sqrt{2500 + 800}}{16} = \frac{50 + 10\sqrt{33}}{16}$$

$$t = \boxed{\frac{25 + 5\sqrt{33}}{8} \text{ seconds}} \approx 6.7 \text{ seconds, rounded}$$

19.
$$i = t^2 - 4t + 16$$
$$50 = t^2 - 4t + 16$$
$$t^2 - 4t - 34 = 0$$

$$t = \frac{4 + \sqrt{16 - 4(-34)}}{2} = \frac{4 + \sqrt{16 + 136}}{2} = \frac{4 + 2\sqrt{38}}{2}$$

$$t = \boxed{2 + \sqrt{38} \text{ seconds}} \approx 6.7 \text{ second, rounded}$$

21. Let x = number of days for faster painter to paint house.

$x + 1$ = number of days for slower painter

Fraction painted in 1 day = $\dfrac{1}{x}$. For the painters working together,

$$\frac{2}{x} + \frac{2}{x+1} = 1$$

$$x(x+1)\left[\frac{2}{x} + \frac{2}{x+1}\right] = x(x+1)\cdot 1$$

$$2x + 2 + 2x = x^2 + x$$

$$4x + 2 = x^2 + x$$

$$x^2 - 3x - 2 = 0$$

$$x = \frac{3 + \sqrt{3^2 - 4(-2)}}{2}$$

$$= \boxed{\frac{3 + \sqrt{17}}{2}\text{ days} \approx 3.6\text{ days, rounded}}$$

$\boxed{\text{and}}$ Slower painter takes $x + 1 = 1 + \dfrac{3 + \sqrt{7}}{2} = \boxed{\dfrac{5 + \sqrt{17}}{2}\text{ days} \approx 4.6\text{ days, rounded}}$

23. Let x = hiker's speed going up the mountain.

$$\text{time going up} = \frac{4}{x}$$

$$\text{time going down} = \frac{4}{x+1}$$

$$\frac{4}{x} + \frac{4}{x+1} = 5$$

$$4x + 4 + 4x = 5x(x+1)$$

$$5x^2 - 3x - 4 = 0$$

$$x = \frac{3 + \sqrt{9 - 4(5)(-4)}}{2(5)} = \frac{3 + \sqrt{9 + 80}}{10}$$

$$x = \boxed{\frac{3 + \sqrt{89}}{10}\text{ miles per hr} \approx 1.2\text{ miles per hr, rounded}}$$

25. Let x = speed of boat in still water

$$\text{Speed upstream} = x - 2$$

$$\text{Speed downstream} = x + 2$$

$$\frac{6}{x-2} - \frac{20}{60} = \frac{6}{x+2}$$

$$\frac{6}{x-2} - \frac{6}{x+2} = \frac{1}{3}$$

$$3(x-2)(x+2)\left[\frac{6}{x-2} - \frac{6}{x+2}\right] = 3(x-2)(x+2)\cdot\frac{1}{3}$$

$$18x + 36 - 18x + 36 = x^2 - 4$$

$$72 = x^2 - 4$$

$$x^2 = 76$$

$$x = \sqrt{76} = \boxed{2\sqrt{19} \approx 8.7\text{ miles per hour, rounded}}$$

27. Let x = width of tile border.
$$9(2)(x) + 5(2)(x) + 4x^2 = 40$$
(length + width + corners)
$$4x^2 + 28x - 40 = 0$$
$$x^2 + 7x - 10 = 0$$
$$x = \frac{-7 + \sqrt{49 - 4(-10)}}{2} = \boxed{\frac{-7 + \sqrt{89}}{2} \text{ feet} \approx 1.2 \text{ feet, rounded}}$$

29. Let x = width of rectangle.
$x + 10$ = length of rectangle
$$\text{perimeter} = \text{area}$$
$$2x + 2(x + 10) = x(x + 10)$$
$$2x + 2x + 20 = x^2 + 10x$$
$$x^2 + 6x - 20 = 0$$
$$x = \frac{-6 + \sqrt{36 - 4(-20)}}{2} = \frac{-6 + \sqrt{116}}{2}$$

$\boxed{\text{width}, -3 + \sqrt{29} \text{ meters} \approx 2.4 \text{ meters, rounded}}$

$\boxed{\text{length}, \ 7 + \sqrt{29} \text{ meters} \approx 12.4 \text{ meters, rounded}}$

Review Problems

31.
$$7(y - 2) = 10 - 2(y + 3)$$
$$7y - 14 = 10 - 2y - 6$$
$$9y = 18$$
$$y = 2$$
$\boxed{\{2\}}$

32.
$$\frac{7}{y+2} + \frac{2}{y+3} = \frac{1}{y^2 + 5y + 6}$$
$$\frac{7(y+3) + 2(y+2)}{(y+2)(y+3)} = \frac{1}{(y+2)(y+3)}$$
$$7y + 21 + 2y + 4 = 1$$
$$9y = -24$$
$$y = -\frac{8}{3}$$
$\boxed{\left\{-\frac{8}{3}\right\}}$

33. $x - 2y > 2$

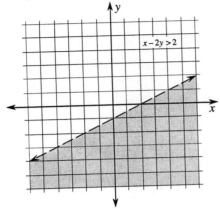

Section 10.5   Complex Numbers

Problem Set 10.5, pp. 687-688

1. $\sqrt{-16} = \sqrt{16}\sqrt{-1}$ Product rule for radicals

$= \boxed{4i}$ $i = \sqrt{-1}$

3. $\sqrt{-20} = \sqrt{20}\sqrt{-1} = \sqrt{20}i = \boxed{2\sqrt{5}i}$

5. $\sqrt{-45} = \sqrt{45}\sqrt{-1} = \boxed{3\sqrt{5}i}$ **7.** $\sqrt{-150} = \sqrt{150}i = \boxed{5\sqrt{6}i}$

9. $(4 + 9i) + (13 + 2i) = (4 + 13) + (9 + 2)i = \boxed{17 + 11i}$

11. $(-3 + i) + (7 - i) = (-3 + 7) + (1 - 1)i = \boxed{4}$

13. $(4 - 6i) - (2 + 5i) = (4 - 2) + (-6 - 5)i = \boxed{2 - 11i}$

15. $(-3 - 8i) - (-5 - i) = (-3 + 5) + (-8 + 1)i = \boxed{2 - 7i}$

17. $(7 - 3i) - (9 - 2i) = (7 - 9) + (-3 + 2)i = \boxed{-2 - i}$

19. $(7 - 8i) + (5 - i) - (2 - 7i) = (7 + 5 - 2) + (-8 - 1 + 7)i = \boxed{10 - 2i}$

21. $2(5 - 3i) + 4(2 + i) = (10 + 8) + (-6 + 4)i = \boxed{18 - 2i}$

23. $5i(6 - 3i) = 30i - 15i^2 = 30i + 15 = \boxed{15 + 30i}$

25. $-3i(2 + i) = -6i - 3i^2 = \boxed{3 - 6i}$

27. $(4 + 3i)(5 + 3i) = 20 + 12i + 15i + 9i^2 = 20 - 9 + 27i = \boxed{11 + 27i}$

29. $(4+3i)(5-i) = 20-4i+15i-3i^2 = \boxed{23+11i}$

31. $(5-2i)(4-3i) = 20-15i-8i+6i^2 = \boxed{14-23i}$

33. $(3+2i)(3-2i) = 9-4i^2 = 9+4 = \boxed{13}$

35. $(5-i)(5+i) = 25-i^2 = 25+1 = \boxed{26}$

37. $(2+i)^2 = 4+4i+i^2 = \boxed{3+4i}$

39. $(x+7i)(x-7i) = x^2-49i^2 = \boxed{x^2+49}$

41. $\dfrac{-3+i}{1-i} = \dfrac{-3+i}{1-i}\cdot\dfrac{1+i}{1+i} = \dfrac{-3-3i+i+i^2}{1-i^2} = \dfrac{-4-2i}{2} = \boxed{-2-i}$

43. $\dfrac{1-i}{2+i} = \dfrac{1-i}{2+i}\cdot\dfrac{2-i}{2-i} = \dfrac{2-3i+i^2}{2-i^2} = \boxed{\dfrac{1-3i}{3}}$ or $\boxed{\dfrac{1}{3}-i}$

45. $\dfrac{2}{1+4i} = \dfrac{2}{1+4i}\cdot\dfrac{1-4i}{1-4i} = \dfrac{2-8i}{1-16i^2} = \boxed{\dfrac{2-8i}{17}}$ or $\boxed{\dfrac{2}{17}-\dfrac{8}{17}i}$

47. $\dfrac{5i}{3-2i} = \dfrac{5i}{3-2i}\cdot\dfrac{3+2i}{3+2i} = \dfrac{15i+10i^2}{9-4i^2} = \boxed{\dfrac{-10+15i}{13}}$ or $\boxed{-\dfrac{10}{13}+\dfrac{15}{13}i}$

49. $\dfrac{4+3i}{4-3i} = \dfrac{4+3i}{4-3i}\cdot\dfrac{4+3i}{4+3i} = \dfrac{16+24i+9i^2}{16-9i^2} = \boxed{\dfrac{7+24i}{25}}$ or $\boxed{\dfrac{7}{25}+\dfrac{24}{25}i}$

51. $(x-3)^2 = -9$
$x-3 = \pm 3i$
$x = 3-3i$ or $x = 3+3i$
$\boxed{\{3-3i, 3+3i\}}$

53. $(x+7)^2 = -64$
$x+7 = \pm 8i$
$x = -7-8i$ or $x = -7+8i$
$\boxed{\{-7-8i, -7+8i\}}$

55. $(y-2)^2 = -7$
$y-2 = \pm\sqrt{7}i$
$y = 2-\sqrt{7}i$ or $y = 2+\sqrt{7}i$
$\boxed{\left\{2-i\sqrt{7}, 2+i\sqrt{7}\right\}}$

57. $(z+3)^2 = -18$
$z+3 = \pm 3\sqrt{2}i$
$z = -3-3\sqrt{2}i$ or $z = 3-3\sqrt{2}i$
$\boxed{\left\{-3-3i\sqrt{2}, -3+3i\sqrt{2}\right\}}$

59. $x^2+4x+5 = 0$
$$x = \dfrac{-4\pm\sqrt{16-20}}{2} = \dfrac{-4\pm 2i}{2} = -2\pm i$$
$x = -2-i$ or $x = -2+i$
$\boxed{\{-2-i, -2+i\}}$

61. $x^2-6x+13 = 0$
$$x = \dfrac{6\pm\sqrt{36-52}}{2} = \dfrac{6\pm 4i}{2} = 3\pm 2i$$
$x = 3-2i$ or $x = 3+2i$
$\boxed{\{3-2i, 3+2i\}}$

63. $x^2 - 12x + 40 = 0$

$$x = \frac{12 \pm \sqrt{144 - 160}}{2} = \frac{12 \pm 4i}{2} = 6 \pm 2i$$

$x = 6 - 2i$ or $x = 6 + 2i$

$\boxed{\{6 - 2i, 6 + 2i\}}$

65. $x^2 = 10x - 27$

$x^2 - 10x + 27 = 0$

$$x = \frac{10 \pm \sqrt{100 - 108}}{2} = \frac{10 \pm \sqrt{-8}}{2} = 5 \pm \sqrt{2}i$$

$x = 5 + \sqrt{2}i$ or $x = 5 - \sqrt{2}i$

$\boxed{\left\{5 - i\sqrt{2}, 5 + i\sqrt{2}\right\}}$

67. $5y^2 = 2y - 3$

$5y^2 - 2y + 3 = 0$

$$y = \frac{2 \pm \sqrt{4 - 4(5)(3)}}{2(5)} = \frac{2 \pm \sqrt{-56}}{10} = \frac{1 \pm \sqrt{14}i}{5}$$

$y = \frac{1}{5} + \frac{\sqrt{14}i}{15}$ or $y = \frac{1}{5} - \frac{\sqrt{14}i}{5}$

$\boxed{\left\{\frac{1}{5} - \frac{i\sqrt{14}}{5}, \frac{1}{5} + i\frac{\sqrt{14}}{4}\right\}}$

69. $5y^2 - y = y^2 + y - 5$

$4y^2 - 2y + 5 = 0$

$$y = \frac{2 \pm \sqrt{4 - 4(4)(5)}}{2(4)} = \frac{2 \pm \sqrt{-76}}{8} = \frac{1 \pm \sqrt{19}i}{4}$$

$y = \frac{1}{4} + \frac{\sqrt{19}i}{4}$ or $y = \frac{1}{4} - \frac{\sqrt{19}i}{4}$

$\boxed{\left\{\frac{1}{4} - \frac{i\sqrt{19}}{4}, \frac{1}{4} + i\frac{\sqrt{19}}{4}\right\}}$

71. $(5y^2 - 3y - 8) - (3y^2 + y - 12) = 0$

$2y^2 - 4y + 4 = 0$

$y^2 - 2y + 2 = 0$

$$y = \frac{2 \pm \sqrt{4 - 8}}{2} = \frac{2 \pm 2i}{2} = 1 + i \text{ or } y = 1 - i$$

$\boxed{\{1 - i, 1 + i\}}$

73. $3z(z + 3) = (z + 2)^2 - 10$

$3z^2 + 9z = z^2 + 4z + 4 - 10$

$2z^2 + 5z + 6 = 0$

$$z = \frac{-5 \pm \sqrt{25 - 4(2)(6)}}{4} = \frac{-5 \pm \sqrt{-23}}{4} = \frac{-5}{4} \pm \frac{\sqrt{23}}{4}i$$

$z = -\frac{5}{4} + \frac{\sqrt{23}}{4}i$ or $z = -\frac{5}{4} - \frac{\sqrt{23}}{4}i$

$\boxed{\left\{-\frac{5}{4} - \frac{i\sqrt{23}}{4}, -\frac{5}{4} + \frac{i\sqrt{23}}{2}\right\}}$

75.
$$\frac{1}{5}y^2 = -\frac{1}{20}y - \frac{1}{4}$$
$$4y^2 = -y - 5$$
$$4y^2 + y + 5 = 0$$

$$y = \frac{-1 \pm \sqrt{1 - 4(4)(5)}}{2(4)} = \frac{-1 \pm \sqrt{-79}}{8} = \frac{-1 \pm \sqrt{79}i}{8}$$

$$y = -\frac{1}{8} - \frac{\sqrt{79}}{8}i \text{ or } y = -\frac{1}{8} + \frac{\sqrt{79}}{8}i$$

$$\boxed{\left\{ -\frac{1}{8} - \frac{i\sqrt{79}}{8}, \ -\frac{1}{8} + \frac{i\sqrt{79}}{8} \right\}}$$

77. $\boxed{B}$ is true since $(2 - i)^2 - 4(2 - i) + 5 = 4 - 4i + i^2 - 8 + 4i + 5 = 0$.

79. $\boxed{D}$ is true.   **81.** $\sqrt{x - 5}$ will not be a real number if $x - 5 < 0$, or, if $\boxed{x < 5}$.

83. Let x = number we are looking for.
$$x^2 - 2x = -5$$
$$x^2 - 2x + 5 = 0$$
$$b^2 - 4ac = 4 - 4(1)(5) = -16 < 0$$
Since the discriminant < 0, there is $\boxed{\text{no real solution for } x}$.

85.
$$y = -50x^2 + 300x + 5400 = 7000$$
$$50x^2 - 300x + 1600 = 0$$
$$x^2 - 6x + 32 = 0$$
$$b^2 - 4ac = 36 - 4(32) = -92$$
Since the discriminant < 0, there is $\boxed{\text{no real value of } x}$ such that $y = 7000$.

87. $E = IR$
$$E = (2 - 3i)(3 + 5i) = 6 + 10i - 9i - 15i^2$$
$$= \boxed{21 + i}$$

Review Problems

92. Let x = number of liters of 8% alcohol solution.
$x + 32$ = number of liters of 12% alcohol solution
$$0.08x + 0.28(32) = 0.12(x + 32)$$
$$0.08x + 8.96 = 0.12x + 3.84$$
$$5.12 = 0.04x$$
$$x = 128$$
$\boxed{128 \text{ liters of 8\% alcohol solution}}$

93. Let x = first even integer in the set.
$x + 2$, $x + 4$ = next two even integers
$$3[(x + 2) + (x + 4)] = 7x + 4$$
$$3(2x + 6) = 7x + 4$$
$$6x + 18 = 7x + 4$$
$$x = 18 - 4 = 14$$
The integers are $\boxed{14, \ 16, \ \text{and } 18}$.

94. $(2\sqrt{3} + \sqrt{2})(2\sqrt{3} - 5\sqrt{2})$

$= (2\sqrt{3})^2 - 2\sqrt{3}\,(5\sqrt{2}) + \sqrt{2}\,(2\sqrt{3})$

$= 12 - 10\sqrt{6} + 2\sqrt{6} - 10$

$= \boxed{2 - 8\sqrt{6}}$

Section 10.6   Quadratic Functions and Their Graphs

Problem Set 10.6, pp. 699-703

1. $y = x^2 + 6x + 5$
x-intercepts: $x^2 + 6x + 5 = 0$
$\qquad\qquad\qquad (x + 1)(x + 5) = 0;\ x = -1,\ \text{and} -5$
y-intercepts: At $x = 0,\ y = 5$
Vertex:

x-coordinate, $x = -\dfrac{b}{2a} = -\dfrac{6}{2} = -3$

y-coordinate, $y = (-3)^2 + 6(-3) + 5 = 9 - 18 + 5 = -4$
At $x = -2,\ y = 4 - 12 + 5 = -3$
Additional point: $(-2, -3)$

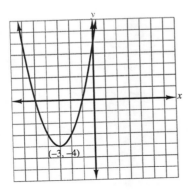
$(-3, -4)$

3. $y = x^2 + 4x + 3$

x-intercepts: $x^2 + 4x + 3 = (x + 1)(x + 3) = 0; x = -1, -3$

y-intercept: At $x = 0, y = 3$

Vertex:

x-coordinate: $x = -\dfrac{b}{2a} = -\dfrac{4}{2} = -2$

y-coordinate: $y = (-2)^2 + 4(-2) + 3 = 4 - 8 + 3 = -1$

At $x = -4, y = 16 - 16 + 3 = 3$

Additional point: $(-4, 3)$

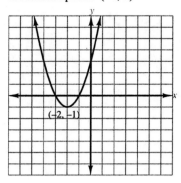

5. $y = -x^2 + 4x - 3$

x-intercepts: $x^2 - 4x + 3 = 0; (x - 1)(x - 3) = 0; x = 1, 3$

y-intercept: At $x = 0, y = -3$

Vertex:

x-coordinate: $x = -\dfrac{b}{2a} = \dfrac{4}{2} = 2$

y-coordinate: $y = -2^2 + 4(2) - 3 = 1$

At $x = -1, y = -6$

Additional point: $(-1, -6)$

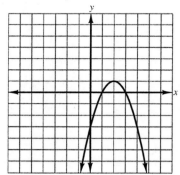

7. $y = -2x^2 + 16x - 30$

   x-intercepts: $x^2 - 8x + 15 = 0; (x-3)(x-5) = 0; x = 3, 5$

   y-intercept: At $x = 0, y = -30$

   Vertex:

   x-coordinate: $x = -\dfrac{b}{2a} = \dfrac{8}{2} = 4$

   y-coordinate: $y = -2(4^2) + 16(4) - 30 = 2$

   At $x = 2, y = -8 + 32 - 30 = -6$

   Additional point: $(2, -6)$

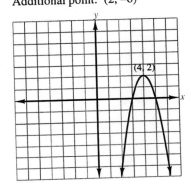

9. $y = x^2 + x$

   x-intercepts: $x(x+1) = 0; x = 0, -1$

   y-intercept: At $x = 0, y = 0$

   Vertex:

   x-coordinate: $x = -\dfrac{b}{2a} = -\dfrac{1}{2}$

   y-coordinate: $y = \left(-\dfrac{1}{2}\right)^2 - \dfrac{1}{2} = -\dfrac{1}{4}$

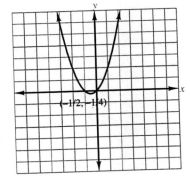

11. $y = x^2 + 4x + 4$

x-intercepts: $x^2 + 4x + 4 = 0; (x+2)^2 = 0, x = -2$

y-intercept: At $x = 0, y = 4$

Vertex:

x-coordinate: $x = -\dfrac{b}{2a} = -\dfrac{4}{2} = -2$

y-coordinate: $y = (-2)^2 + 4(-2) + 4 = 0$

At $x = 1, y = 1 + 4 + 4 = 9$

At $x = -1, y = 1$

Additional point: $(-1, 1)$

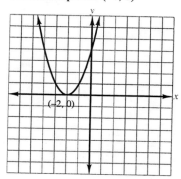

13. $y = x^2 - 4x + 6$

x-intercepts: $x^2 - 4x + 6 = 0; b^2 - 4ac = 16 - 24 = -8 < 0$; no x-intercepts

y-intercept: At $x = 0, y = 6$

Vertex:

x-coordinate: $x = -\dfrac{b}{2a} = \dfrac{4}{2} = 2$

y-coordinate: $y = 2^2 - 4(2) + 6 = 2$

At $x = 1, y = 1 - 4 + 6 = 3$

Additional point: $(1, 3)$

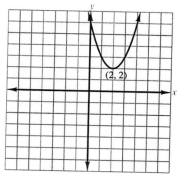

15. $y = -x^2 - 6x - 7$

x-intercepts: $x^2 + 6x + 7 = 0; x = \dfrac{-6 \pm \sqrt{36 - 28}}{2} = -3 \pm \sqrt{2}$

y-intercept: At $x = 0, y = -7$
Vertex:

x-coordinate: $x = -\dfrac{b}{2a} = -\dfrac{6}{2} = -3$

y-coordinate: $y = -(-3)^2 - 6(-3) - 7 = 2$
At $x = -1, y = -2$
Additional point: $(-1, -2)$

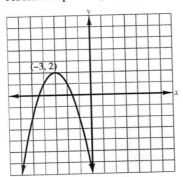

17. $y = x^2 - 4$
x-intercepts: $x^2 - 4 = 0; (x - 2)(x + 2) = 0; x = 2, -2$
y-intercept: At $x = 0, y = -4$
Vertex:

x-coordinate: $x = -\dfrac{b}{2a} = 0$

y-coordinate: $y = 0 - 4 = -4$
At $x = 1, y = -3$
Additional point: $(1, -3)$

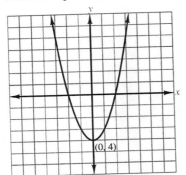

19. $y = -x^2 - 1$

   no x-intercepts: $x^2 + 1 = 0$

   y-intercept: At $x = 0$, $y = -1$

   Vertex:

   x-coordinate: $x = -\dfrac{b}{2a} = 0$

   y-coordinate: $y = 0 - 1 = -1$

   At $x = \pm 1$, $y = -2$; at $x = \pm 2$, $y = -5$

   Additional points: $(1, -2), (-1, -2), (2, -5), (-2, -5)$

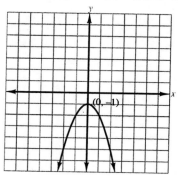

21. $y = -2x^2 + 4x$

   x-intercepts: $2x^2 - 4x = 0$; $x^2 - 2x = x(x - 2) = 0$; $x = 0, 2$

   y-intercept: At $x = 0$, $y = 0$

   Vertex:

   x-coordinate: $x = -\dfrac{b}{2a} = -\dfrac{4}{-4} = 1$

   y-coordinate: $y = -2(1)^2 + 4 = 2$

   At $x = 3$, $y = -2(9) + 12 = -6$

   Additional point: $(3, -6)$

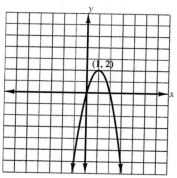

23. $y = -x^2 + 4x - 1$

  x-intercepts: $x^2 - 4x + 1 = 0$; $x = \dfrac{4 \pm \sqrt{16 - 4}}{2} = 2 \pm \sqrt{3}$

  y-intercept: At $x = 0$, $y = -1$
  Vertex:

  x-coordinate: $x = -\dfrac{b}{2a} = \dfrac{4}{2} = 2$

  y-coordinate: $y = -4 + 8 - 1 = 3$
  At $x = 1$, $y = -1 + 4 - 1 = 2$
  Additional point: $(1, 2)$

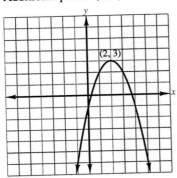

25. $y = 2x^2 + 8x + 1$

  x-intercepts: $2x^2 + 8x + 1 = 0$; $x = \dfrac{-8 \pm \sqrt{64 - 8}}{4} = \dfrac{-4 \pm \sqrt{14}}{2}$

  y-intercept: At $x = 0$, $y = 1$
  Vertex:

  x-coordinate: $x = -\dfrac{b}{2a} = -\dfrac{8}{4} = -2$

  y-coordinate: $y = 2(-2^2) + 8(-2) + 1 = -7$
  At $x = -1$, $y = 2 - 8 + 1 = -5$
  Additional point: $(-1, -5)$

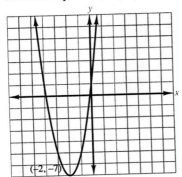

27. $y = -\frac{1}{5}x^2 + \frac{2}{5}x - 1$

no x-intercepts: $x^2 - 2x + 5 = 0$; $b^2 - 4ac = 4 - 20 = -16 < 0$

y-intercept: At $x = 0$, $y = -1$

Vertex:

x-coordinate: $-\frac{b}{2a} = \frac{2}{2} = 1$

y-coordinate: $-\frac{1}{5}(1) + \frac{2}{5}(1) - 1 = -\frac{4}{5}$

At $x = 3$, $y = -\frac{8}{5}$

Additional point: $\left(3, -\frac{8}{5}\right)$

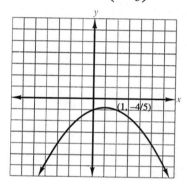

29. Statement $\boxed{\text{B}}$ is true.

31. Statement $\boxed{\text{C}}$ is true.

$-\frac{b}{2a} = \frac{96}{32} = 3$

At $t = 3$, $d = 144$ feet True

33. $h = -x^2 + 40x$

a. The x-coordinate of the vertex: $x = -\frac{b}{2a} = -\frac{40}{-2} = 20$ days

b. At $x = 20$, $y = -400 + 40(20) = \boxed{400 \text{ people}}$

c. $y = 0$ at $x^2 - 40x = x(x - 40) = 0$

$x = \boxed{40 \text{ days}}$

d.

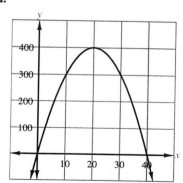

35. $y = -0.02x^2 + x + 1$

The x-coordinate of vertex: $x = -\dfrac{b}{2a} = -\dfrac{1}{2(-0.02)} = 25$ inches

$$y_{max} = -0.02(25^2) + 25 + 1 = \boxed{13.5 \text{ inches}}$$

37. $y = -0.00025x^2 + 0.105x - 1.025$

a. optimum rate of production = x-coordinate of vertex: $-\dfrac{b}{2a} = -\dfrac{0.105}{2(-0.00025)} = \boxed{210}$

b. At $x = 210$, $y = -0.00025(210^2) + 0.105(210) - 1.025 = \boxed{\$10}$ profit per lamp

c. At $y = 0$, $x^2 - 420x + 4100 = 0$
$(x - 10)(x - 410) = 0$
Break-even points are $x = 10$ and $x = 410$.

$\boxed{\text{The company loses money if it makes fewer than } \boxed{10} \text{ desk lamps or more than } \boxed{410} \text{ desk lamps.}}$

39. The matching is $\boxed{(a, 2), (b, 3), \text{ and } (c, 1)}$.

41. $y = -x$ and $y = -x^2 + 2x$
$-x = -x^2 + 2x$
$x^2 - 3x = 0$
$x = 0, x = 3$
Points of intersection: (0,0), (3, -3)

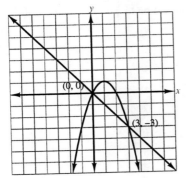

43. $R = 150x - 0.4x^2$
$C = 30x + 2000$

a. $P = R - C = \boxed{-0.4x^2 + 120x - 2000}$

b. optimum production rate = x-coordinate of vertex: $x = -\dfrac{b}{2a} = -\dfrac{120}{2(-0.4)} = \boxed{150 \text{ microwave ovens}}$

At $x = 150$, $P = -0.4(150^2) + 120(150) - 2000 = \boxed{\$7000}$

c.

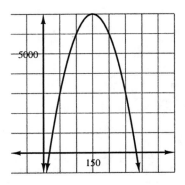

45.

> The shape of the lighted area is parabolic; the shape changes to an ellipse as the back end of the flashlight is raised.

47. a. $y = x^2$

  b. $y = \frac{1}{2}x^2$ The y-values are half those in part a.

  c. $y = -3x^2$ The parabola is inverted and lengthened in the y-direction by a factor of 3.

  d. $y = -\frac{1}{4}x^2$ The parabola in inverted and shortened in the y-direction to $\frac{1}{4}$ of the original value.

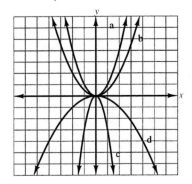

49. $y = 8 - \frac{1}{2}(x-3)^2$

  x-intercepts: $\boxed{-1.00, 7.00}$

  y-intercept: $\boxed{3.50}$

Review Problems

56. $0.00397 = \boxed{3.97 \times 10^{-3}}$

57. passing through $(5, -3)$
whose slope is -4
$y + 3 = -4(x - 5)$
$y + 3 = -4x + 20$
$y = -4x + 17$
$\boxed{4x + y = 17}$

58. $\dfrac{2y^2 + 5y - 3}{2y^2 + y - 36} \cdot \dfrac{2y^2 + 7y - 9}{3y^2 + 5y - 12} \div \dfrac{2y^2 + y - 3}{3y^2 - 16y + 16}$

$= \dfrac{(2y - 1)(y + 3)}{(y - 4)(2y - 9)} \cdot \dfrac{(2y + 9)(y - 1)}{(y + 3)(3y - 4)} \div \dfrac{(2y + 3)(y - 1)}{(3y - 4)(y - 4)}$

$= \dfrac{2y - 1}{y - 4} \cdot \dfrac{y - 1}{3y - 4} \cdot \dfrac{(3y - 4)(y - 4)}{(2y + 3)(y - 1)}$

$= \boxed{\dfrac{2y - 1}{2y + 3}}$

Chapter 10 Review Problems

Chapter 10 Review Problems pp. 705-706

1. $\quad x^2 = 64$
$\quad\quad x = \pm 8$
$\boxed{\{\pm 8\}}$

2. $\quad y^2 = 17$
$\quad\quad y = \pm\sqrt{17}$
$\boxed{\{\pm\sqrt{17}\}}$

3. $\quad r^2 = 75$
$\quad\quad r = \pm\sqrt{75} = \pm 5\sqrt{3}$
$\boxed{\{\pm 5\sqrt{3}\}}$

4. $\quad (y - 3)^2 = 9$
$\quad\quad (y - 3) = \pm 3$
$\quad\quad\quad y = 3 - 3 = 0 \text{ or } y = 3 + 3 = 6$
$\boxed{\{0, 6\}}$

5. $\quad (x + 4)^2 = 5$
$\quad\quad x + 4 = \pm\sqrt{5}$
$\quad\quad\quad x = -4 + \sqrt{5} \text{ or } x = -4 - \sqrt{5}$
$\boxed{\left\{-4 - \sqrt{5}, -4 + \sqrt{5}\right\}}$

6. $\quad (2x - 7)^2 = 25$
$\quad\quad 2x - 7 = \pm 5$
$\quad\quad\quad x = \dfrac{1}{2}(7 - 5) = 1 \text{ or } x = \dfrac{1}{2}(7 + 5) = 6$
$\boxed{\{1, 6\}}$

7.

$$(3x-4)^2 = 18$$

$$3x-4 = \pm 3\sqrt{2}$$

$$x = \frac{1}{3}\left(4-3\sqrt{2}\right) = \frac{4}{3}-\sqrt{2} \text{ or } x = \frac{1}{3}\left(4+3\sqrt{2}\right) = \frac{4}{3}+\sqrt{2}$$

$$\boxed{\left\{\frac{4}{3}-\sqrt{2}, \frac{4}{3}+\sqrt{2}\right\}}$$

8.

$$\left(x+\frac{1}{3}\right)^2 = \frac{5}{9}$$

$$x+\frac{1}{3} = \pm\frac{\sqrt{5}}{3}$$

$$x = -\frac{1}{3}-\frac{\sqrt{5}}{3} \text{ or } x = -\frac{1}{3}+\frac{\sqrt{5}}{3}$$

$$\boxed{\left\{-\frac{1}{3}-\frac{\sqrt{5}}{3}, -\frac{1}{3}+\frac{\sqrt{5}}{3}\right\}}$$

9.

$$A = P(1+r)^2$$

$$1.69 = 1(1+r)^2$$

$$1+r = \sqrt{1.69} = 1.30$$

$$r = 0.30 \text{ or } \boxed{30\%}$$

10. $d = \sqrt{10^2+10^2} = \boxed{10\sqrt{2} \text{ cm } \approx 14.14 \text{ cm, rounded}}$

11. $l = \sqrt{24^2+(49-42)^2} = \sqrt{576+49} = \sqrt{625} = \boxed{25 \text{ feet}}$

12.

$$x^2-12x+27 = 0$$

$$x^2-12x+\left(-\frac{12}{2}\right)^2 = -27+\left(-\frac{12}{2}\right)^2$$

$$x^2-12x+36 = (x-6)^2 = 9$$

$$x-6 = \pm 3$$

$$x = 6-3 = 3 \text{ or } x = 6+3 = 9$$

$$\boxed{\{3,9\}}$$

13.

$$x^2-6x+4 = 0$$

$$x^2-6x+9 = -4+9 = 5$$

$$(x-3)^2 = \pm\sqrt{5}$$

$$x = 3-\sqrt{5} \text{ or } x = 3+\sqrt{5}$$

$$\boxed{\left\{3-\sqrt{5}, 3+\sqrt{5}\right\}}$$

14.

$$3x^2 - 12x + 11 = 0$$

$$x^2 - 4x = -\frac{11}{3}$$

$$x^2 - 4x + 4 = 4 - \frac{11}{3} = \frac{1}{3}$$

$$(x-2)^2 = \frac{1}{3}$$

$$x - 2 = \frac{1}{\sqrt{3}}$$

$$x = 2 - \frac{1}{\sqrt{3}} \text{ or } x = 2 + \frac{1}{\sqrt{3}}$$

Since

$$\frac{1}{\sqrt{3}} = \frac{1}{\sqrt{3}} \cdot \frac{\sqrt{3}}{\sqrt{3}} = \frac{\sqrt{3}}{3}$$

$$x = 2 \pm \frac{\sqrt{3}}{3}$$

$$\boxed{\left\{2 - \frac{\sqrt{3}}{3},\ 2 + \frac{\sqrt{3}}{3}\right\}}$$

15.

$$2x^2 + 5x - 3 = 0$$

$$x = \frac{-5 \pm \sqrt{25 + 24}}{4} = \frac{-5 \pm 7}{4} = \frac{1}{2}, -3$$

$$x = \frac{1}{2} \text{ or } x = -3$$

$$\boxed{\left\{-3, \frac{1}{2}\right\}}$$

16.

$$3x^2 + 5 = 9x$$

$$3x^2 - 9x + 5 = 0$$

$$x = \frac{9 \pm \sqrt{81 - 60}}{6}$$

$$x = \frac{9 + \sqrt{21}}{6} \text{ or } x = \frac{9 - \sqrt{21}}{6}$$

$$\boxed{\left\{\frac{9 - \sqrt{21}}{6},\ \frac{9 + \sqrt{21}}{6}\right\}}$$

17.

$$4y^2 + 2y - 1 = 0$$

$$y = \frac{-2 \pm \sqrt{4 + 16}}{8} = \frac{-2 \pm 2\sqrt{5}}{8} = \frac{-1 \pm \sqrt{5}}{4}$$

$$y = \frac{-1 - \sqrt{5}}{4} \text{ or } y = \frac{-1 + \sqrt{5}}{4}$$

$$\boxed{\left\{\frac{-1 - \sqrt{5}}{4},\ \frac{-1 + \sqrt{5}}{4}\right\}}$$

18. $2x^2 - 11x + 5 = 0$
$(2x - 1)(x - 5) = 0$
$x = \dfrac{1}{2}$ or $x = 5$

$$\boxed{\left\{ \dfrac{1}{2}, 5 \right\}}$$

19. $(3x + 5)(x - 3) = 5$
$3x^2 - 9x + 5x - 15 = 5$
$3x^2 - 4x - 20 = 0$
$(3x - 10)(x + 2) = 0$
$3x - 10 = 0$ or $x + 2 = 0$
$x = \dfrac{10}{3}$ or $x = -2$

$$\boxed{\left\{ -2, \dfrac{10}{3} \right\}}$$

20. $3x^2 - 7x + 1 = 0$
$x = \dfrac{7 \pm \sqrt{49 - 12}}{6} = \dfrac{7 \pm \sqrt{37}}{6}$
$x = \dfrac{7 + \sqrt{37}}{6}$ or $x = \dfrac{7 - \sqrt{37}}{6}$

$$\boxed{\left\{ \dfrac{7 - \sqrt{37}}{6}, \dfrac{7 + \sqrt{37}}{6} \right\}}$$

21. $x^2 - 9 = 0$
$x^2 = 9$
$x = \pm 3$
$x = 3$ or $x = -3$

$\boxed{\{-3, 3\}}$

22. $(x - 3)^2 - 25 = 0$
$(x - 3)^2 = 25$
$x - 3 = \pm 5$
$x = 3 - 5 = -2$ or
$x = 3 + 5 = 8$

$\boxed{\{-2, 8\}}$

23. $5(x^2 - 1) = 2(x + 1) + 2(x - 1)$
$5x^2 - 5 = 2x + 2 + 2x - 2$
$5x^2 - 4x - 5 = 0$
$x = \dfrac{4 \pm \sqrt{16 + 100}}{10} = \dfrac{4 \pm 2\sqrt{29}}{10} = \dfrac{2 \pm \sqrt{29}}{5}$
$x = \dfrac{2 - \sqrt{29}}{5}$ or $x = \dfrac{2 + \sqrt{29}}{5}$

$$\boxed{\left\{ \dfrac{2 - \sqrt{29}}{5}, \dfrac{2 + \sqrt{29}}{5} \right\}}$$

24. Let x = unknown number.
$x^2 - 2x = 4$
$x^2 - 2x + 1 = 4 + 1 = 5$
$(x - 1)^2 = 5$
$x - 1 = \pm\sqrt{5}$
$x = 1 - \sqrt{5}$ or $x = 1 + \sqrt{5}$
$x = -1.2$ or $x = 3.2$, rounded

The number is $\boxed{1 - \sqrt{5} \text{ or } 1 + \sqrt{5}}$ or rounded, $\boxed{-1.2 \text{ or } 3.2}$.

25. Let x = unknown number.

$$x + \frac{1}{x} = 4$$
$$x^2 + 1 = 4x$$
$$x^2 - 4x + 1 = 0$$
$$x^2 - 4x + 4 = -1 + 4 = 3$$
$$(x-2)^2 = 3$$
$$x - 2 = \pm\sqrt{3}$$
$$x = 2 - \sqrt{3} \text{ or } x = 2 + \sqrt{3}$$
$$x = 0.3 \text{ or } x = 3.7, \text{ rounded}$$

The number is $\boxed{2 - \sqrt{3} \text{ or } 2 + \sqrt{3}}$ or rounded $\boxed{0.3 \text{ or } 3.7}$.

26. Let l = length.

$l - 2$ = width

$$(l-2)l = 16$$
$$l^2 - 2l = 16$$
$$l^2 - 2l + 1 = 16 + 1$$
$$(l-1)^2 = 17$$
$$l - 1 = \pm\sqrt{17}$$
$$l = 1 + \sqrt{17}, \ w = -1 + \sqrt{17}$$
$$l = 5.1, \ w = 3.1, \text{ rounded}$$

$$\boxed{l = 1 + \sqrt{17}, w = -1 + \sqrt{17}}$$

$$\boxed{l = 5.1, w = 3.1, \text{ rounded}}$$

27. $s = -16t^2 + 48t + 80$

If $s = 0$,

$$t^2 - 3t - 5 = 0$$
$$t = \frac{3 + \sqrt{9 + 20}}{2} = \frac{3 + \sqrt{29}}{2} \text{ seconds}$$
$$t = 4.2 \text{ seconds, rounded}$$

time to hit ground: $\boxed{\dfrac{3 + \sqrt{29}}{2} \text{ seconds} \approx 4.2 \text{ seconds}}$

28. Let x = width of tile.

$$(12 + 2x)(8 + 2x) - 12(8) = 124$$
$$40x + 4x^2 = 124$$
$$x^2 + 10x - 31 = 0$$
$$x = \frac{-10 + \sqrt{100 + 124}}{2} = \frac{-10 + \sqrt{224}}{2}$$

width of tile = $\boxed{-5 + 2\sqrt{14} \text{ feet} = 2.5 \text{ feet, rounded}}$

29. Let x = length of shorter leg of triangle.
$x + 2$ = length of longer leg

$$x^2 + (x+2)^2 = 6^2$$
$$x^2 + x^2 + 4x + 4 = 36$$
$$2x^2 + 4x - 32 = 0$$
$$x^2 + 2x - 16 = 0$$

$$x = \frac{-2 + \sqrt{4 + 64}}{2} = \frac{-2 + 2\sqrt{17}}{2} = \boxed{-1 + \sqrt{17};\ x + 2 = 1 + \sqrt{17}}$$

$$x = 3.1 \text{ feet}$$
$$\text{length of other leg} = 5.1 \text{ feet, rounded}$$

lengths of legs, $\boxed{-1 + \sqrt{17} \text{ feet and } 1 + \sqrt{17} \text{ feet}}$

or rounded, $\boxed{3.1 \text{ feet and } 5.1 \text{ feet}}$

30. Let x = time taken by faster worker, working alone.
$x + 2$ = time taken by slower worker

$$\frac{1}{x} + \frac{1}{x+2} = \frac{1}{5}$$

$$5x(x+2)\left[\frac{1}{x} + \frac{1}{x+2}\right] = 5x(x+2) \cdot \frac{1}{5}$$

$$5x + 10 + 5x = x^2 + 2x$$
$$x^2 - 8x - 10 = 0$$

$$x = \frac{8 + \sqrt{64 + 40}}{2} = \frac{8 + \sqrt{104}}{2}$$

time for faster worker = $\boxed{4 + \sqrt{26} \text{ hour} = 9.1 \text{ hours, rounded}}$

time for slower worker = $\boxed{6 + \sqrt{26} \text{ hours} = 11.1 \text{ hours, rounded}}$

31. Let x = speed of current.
$5 + x$ = rate of rowing with current
$5 - x$ = rate of rowing against current

$$\frac{10}{5-x} + \frac{10}{5+x} = 7$$
$$50 + 10x + 50 - 10x = 175 - 7x^2$$
$$7x^2 = 75$$
$$x^2 = \frac{75}{7}$$

$$x = \frac{5\sqrt{3}}{\sqrt{7}} \text{ miles per hour} = \frac{5\sqrt{3}}{\sqrt{7}} \cdot \frac{\sqrt{7}}{\sqrt{7}} = \frac{5\sqrt{21}}{7} = 3.3 \text{ miles per hour, rounded}$$

speed of current, $\boxed{\dfrac{5\sqrt{21}}{7} \text{ miles per hour; } 3.3 \text{ miles per hour, rounded}}$

32. $\sqrt{-81} = \boxed{9i}$ **33.** $\sqrt{-48} = \boxed{4\sqrt{3}i}$

34. $\sqrt{-17} = \boxed{\sqrt{17}i}$ **35.** $(-11 + 7i) + (-9 - 13i) = \boxed{-20 - 6i}$

36. $(-3 - 8i) - (5 - 7i) = (-3 - 5) + (-8 + 7)i = \boxed{-8 - i}$

37. $(-1 + i) - (-2 - i) = (-1 + 2) + (1 + 1)i = \boxed{1 + 2i}$

38. $(9 - 5i)(3 + 7i) = 27 + 63i - 15i - 35i^2 = \boxed{62 + 48i}$

39. $(7 + 2i)(7 - 2i) = 49 - 4i^2 = 49 + 4 = \boxed{53}$

40. $6i(3 - 7i) = 18i - 42i^2 = \boxed{42 + 18i}$

41. $\dfrac{4-i}{1+i} = \dfrac{4-i}{1+i} \cdot \dfrac{1-i}{1-i} = \dfrac{4 - 5i + i^2}{1 - i^2} = \dfrac{3 - 5i}{2} = \boxed{\dfrac{3}{2} - \dfrac{5}{2}i}$

42. $\dfrac{3-4i}{2+3i} = \dfrac{3-4i}{2+3i} \cdot \dfrac{2-3i}{2-3i} = \dfrac{6 - 17i + 12i^2}{4 - 9i^2} = \dfrac{-6 - 17i}{13} = \boxed{-\dfrac{6}{13} - \dfrac{17}{13}i}$

43. $\dfrac{2+i}{2-i} = \dfrac{2+i}{2-i} \cdot \dfrac{2+i}{2+i} = \dfrac{4 + 4i + i^2}{4 - i^2} = \boxed{\dfrac{3 + 4i}{5} = \dfrac{3}{5} + \dfrac{4}{5}i}$

44. $\dfrac{2-i}{4i} = \dfrac{2-i}{4i} \cdot \dfrac{-4i}{-4i} = \dfrac{-8i - 4}{16} = \dfrac{-1 - 2i}{4} = \boxed{-\dfrac{1}{4} - \dfrac{1}{2}i}$

45. $(3x - 4)^2 = -49$

$3x - 4 = \pm 7i$

$x = \dfrac{4}{3} + \dfrac{7}{3}i$ or $x = \dfrac{4}{3} - \dfrac{7}{3}i$

$\boxed{\left\{ \dfrac{4}{3} - \dfrac{7}{3}i, \dfrac{4}{3} + \dfrac{7}{3}i \right\}}$

46. $(7y + 1)^2 = -27$

$7y + 1 = \pm 3\sqrt{3}i$

$y = -\dfrac{1}{7} + \dfrac{3}{7}\sqrt{3}i$ or $y = -\dfrac{1}{7} - \dfrac{3}{7}\sqrt{3}i$

$\boxed{\left\{ -\dfrac{1}{7} - \dfrac{3}{7}\sqrt{3}i, -\dfrac{1}{7} + \dfrac{3}{7}\sqrt{3}i \right\}}$

47. $x^2 - 4x = -13$

$x^2 - 4x + 13 = 0$

$x = \dfrac{4 \pm \sqrt{16 - 52}}{2} = \dfrac{4 \pm \sqrt{-36}}{2} = 2 \pm 3i$

$x = 2 + 3i$ or $x = 2 - 3i$

$\boxed{\{2 - 3i, 2 + 3i\}}$

48. $x^2 + 4 = 3x$

$x^2 - 3x + 4 = 0$

$x = \dfrac{3 \pm \sqrt{9 - 16}}{2} = \dfrac{3 \pm \sqrt{7}}{2}$

$x = \dfrac{3}{2} + \dfrac{\sqrt{7}}{2}i$ or $x = \dfrac{3}{2} - \dfrac{\sqrt{7}}{2}i$

$\boxed{\left\{ \dfrac{3}{2} - \dfrac{\sqrt{7}}{2}i, \dfrac{3}{2} + \dfrac{\sqrt{7}}{2}i \right\}}$

49. $3y^2 - y + 2 = 0$

$y = \dfrac{1 \pm \sqrt{1 - 24}}{6} = \dfrac{1 \pm \sqrt{23}i}{6}$

$y = \dfrac{1}{6} + \dfrac{\sqrt{23}}{6}i$ or $y = \dfrac{1}{6} - \dfrac{\sqrt{23}}{6}i$

$\boxed{\left\{ \dfrac{1}{6} - \dfrac{\sqrt{23}}{6}i, \dfrac{1}{6} + \dfrac{\sqrt{23}}{6}i \right\}}$

50.
$$2y^2 = 3y - 5$$
$$2y^2 - 3y + 5 = 0$$
$$y = \frac{3 \pm \sqrt{9 - 40}}{4} = \frac{3 \pm \sqrt{31}i}{4}$$
$$y = \frac{3}{4} + \frac{\sqrt{31}}{4}i \text{ or } y = \frac{3}{4} - \frac{\sqrt{31}}{4}i$$

$$\left\{ \frac{3}{4} - \frac{\sqrt{31}}{4}i, \frac{3}{4} + \frac{\sqrt{31}}{4}i \right\}$$

51.
$$y = x^2 + 4x - 5$$

x-intercepts:
$$x^2 + 4x - 5 = 0$$
$$(x + 5)(x - 1) = 0$$
$$x = 1, -5$$

y-intercept: If $x = 0$, $y = -5$

Vertex:

x-coordinate:
$$-\frac{b}{2a} = -\frac{4}{2} = -2$$

y-coordinate: $(-2)^2 + 4(-2) - 5 = -9$

At $x = -1$, $y = 1 - 4 - 5 = -8$

Additional point: $(-1, -8)$

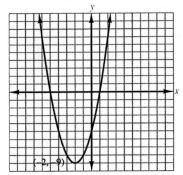

52.
$$y = -x^2 + 6x - 9$$
$$x^2 - 6x + 9 = 0$$
$$(x-3)^2 = 0$$

x-intercept: $\qquad x = 3$

y-intercept: $\qquad$ If $x = 0, y = -9$

Vertex:

x-coordinate: $\qquad -\dfrac{b}{2a} = \dfrac{6}{2} = 3$

y-coordinate: $\qquad -3^2 + 6(3) - 9 = 0$

At $x = 2, y = -4 + 12 - 9 = -1$

Additional point: $(2, -1)$

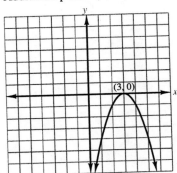

53.
$$y = x^2 - 6x + 7$$

x-intercepts: $\quad x^2 - 6x + 7 = 0$

y-intercept: $\qquad x = \dfrac{6 \pm \sqrt{36-28}}{2} = \dfrac{6 \pm 2\sqrt{2}}{2} = 3 \pm \sqrt{2}$

$\qquad\qquad\qquad x = 3 - \sqrt{2}, 3 + \sqrt{2}$

$\qquad\qquad$ If $x = 0, y = 7$

Vertex:

x-coordinate: $\qquad -\dfrac{b}{2a} = \dfrac{6}{2} = 3$

y-coordinate: $\qquad 3^2 + 6(3) + 7 = -2$

At $x = 1, y = 1 - 6 + 7 = 2$

Additional point: $(1, 2)$

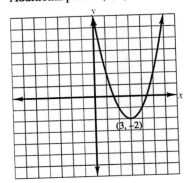

54.
$$y = -x^2 + 4x = x(-x + 4)$$

x-intercepts: $x = 0, 4$

y-intercept: If $x = 0$, $y = 0$

Vertex:

x-coordinate: $-\dfrac{b}{2a} = -\dfrac{4}{-2} = 2$

y-coordinate: $-2^2 + 4(2) = 4$

At $x = 1$, $y = 3$

Additional point: $(1, 3)$

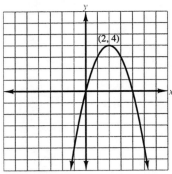

55.
$$y = x^2 - 4x + 10$$
$$x^2 - 4x + 10 = 0$$

no *x*-intercepts: $x = \dfrac{4 \pm \sqrt{16 - 40}}{2} = \dfrac{4 \pm 2\sqrt{6}i}{2} = 2 \pm \sqrt{6}i$

y-intercept: If $x = 0$, $y = 10$

Vertex:

x-coordinate: $-\dfrac{b}{2a} = \dfrac{4}{2} = 2$

y-coordinate: $2^2 - 4(2) + 10 = 6$

At $x = 1$, $y = 7$

Additional point: $(1, 7)$

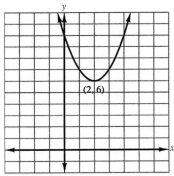

56.
$$y = -x^2 - 3$$
$$x^2 + 3 = 0$$

no x-intercepts: $x = \pm\sqrt{3}i$
y-intercept: If $x = 0$, $y = -3$
Vertex:

x-coordinate: $-\dfrac{b}{2a} = 0$

y-coordinate: -3
At $x = 1$, $y = -4$
Additional point: $(1, -4)$

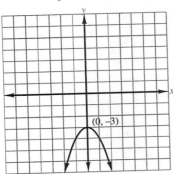

(0, –3)

57.
$$y = -2x^2 + 360x - 14{,}060$$

optimum production: $-\dfrac{b}{2a} = -\dfrac{360}{-4} = \boxed{90}$ tires

$$\boxed{\text{profit}} = -2(90^2) + 360(90) - 14{,}060 = \boxed{\$2140}$$

58. $d = -16t^2 + 160t$

a. $d = 0 \Rightarrow t^2 - 10t = 0$
$t = \boxed{10 \text{ seconds}}$

b. $-\dfrac{b}{2a} = -\dfrac{160}{-32} = \boxed{5 \text{ seconds}}$

maximum height $= -16(25) + 160(5) = \boxed{400 \text{ feet}}$

c.

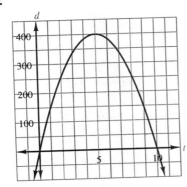

Appendix B

Review Problems Covering the Entire Book

Appendix B, pp. A4-A10

1.
$$-4(y+1)-3y = -7(2y-3)$$
$$-4y-4-3y = -14y+21$$
$$7y = 25$$
$$y = \frac{25}{7}$$
$$\boxed{\left\{\frac{25}{7}\right\}}$$

2.
$$\frac{2x}{3}+\frac{1}{5} = 1+\frac{3x}{5}-\frac{1}{3}$$
$$\frac{2x}{3}-\frac{3x}{5} = 1-\frac{1}{3}-\frac{1}{5}$$
$$\frac{10x-9x}{15} = 1-\frac{8}{15}$$
$$\frac{x}{15} = \frac{7}{15}$$
$$x = 7$$
$$\boxed{\{7\}}$$

3.
$$0.03(2x-3)+0.05 = 5$$
$$0.06x-0.09+0.05 = 5$$
$$0.06x-0.04 = 5$$
$$0.06x = 5.04$$
$$x = 84$$
$$\boxed{\{84\}}$$

4.
$$-2(y-5)+10 = -3(y+2)+y$$
$$-2y+10+10 = -3y-6+y$$
$$-2y+20 = -2y-6$$
$$20 = -6 \quad \text{False}$$
no solution
solution set: $\boxed{\varnothing}$

5.
$$5-2(3-x) \geq 2(2x+5)+1$$
$$5-6+2x \geq 4x+10+1$$
$$2x-1 \geq 4x+11$$
$$-2x \geq 12$$
$$x \leq -6$$
$$\boxed{\{x\mid x \leq -6\}}$$

6.
$$-10 \leq 2x-4 \leq 0$$
$$-6 \leq 2x \leq 4$$
$$-3 \leq x \leq 2$$
$$\boxed{\{x\mid -3 \leq x \leq 2\}}$$

7.
$$\frac{y+1}{2y+3} = \frac{2}{3}$$
$$3(y+1) = 2(2y+3)$$
$$3y+3 = 4y+6$$
$$-y = 3$$
$$y = -3$$
$$\boxed{\{-3\}}$$

8.
$$(x-2)(x-3) = 6(x-3)$$
$$(x-3)(x-2)-6(x-3) = 0$$
$$(x-3)(x-2-6) = 0$$
$$(x-3)(x-8) = 0$$
$$x-3 = 0 \quad \text{or} \quad x-8=0$$
$$x = 3 \quad \text{or} \quad x=8$$
$$\boxed{\{3,8\}}$$

9.

$$\frac{3}{y+5} - 1 = \frac{4-y}{2y+10}$$

$$\frac{3(2)}{2(y+5)} - \frac{2(y+5)}{2(y+5)} = \frac{4-y}{2(y+5)}$$

$$6 - 2y - 10 = 4 - y$$

$$-y = 8$$

$$y = -8$$

$$\boxed{\{-8\}}$$

10.

$$\frac{6}{y^2+y-2} = \frac{3}{y-1} - \frac{y}{y+2}$$

$$\frac{6}{(y+2)(y-1)} = \frac{3(y+2)}{(y-1)(y+2)} - \frac{y(y-1)}{(y+2)(y-1)}$$

$$6 = 3y + 6 - y^2 + y$$

$$y^2 - 4y = 0$$

$$y(y-4) = 0$$

$$y = 0 \quad \text{or} \quad y - 4 = 0$$

$$y = 0 \quad \text{or} \quad y = 4$$

$$\boxed{\{0, 4\}}$$

11.

$$\sqrt{4x} - \sqrt{2x+6} = 0$$

$$\sqrt{4x} = \sqrt{2x+6}$$

$$\left(\sqrt{4x}\right)^2 = \left(\sqrt{2x+6}\right)^2$$

$$4x = 2x + 6$$

$$2x = 6$$

$$x = 3$$

$$\boxed{\{3\}}$$

12.

$$\sqrt{2x-1} - x + 2 = 0$$

$$\sqrt{2x-1} = x - 2$$

$$\left(\sqrt{2x-1}\right)^2 = (x-2)^2$$

$$2x - 1 = x^2 - 4x + 4$$

$$0 = x^2 - 6x + 5$$

$$(x-5)(x-1) = 0$$

$$x - 5 = 0 \quad \text{or} \quad x - 1 = 0$$

$$x = 5 \quad \text{or} \quad x = 1$$

Check for extraneous roots:

$$x = 5: \quad \sqrt{2(5)-1} - 5 + 2 = 0$$

$$\sqrt{9} - 5 + 2 = 0$$

$$3 - 5 + 2 = 0$$

$$0 = 0$$

True, $x = 5$ is a solution.

$$x = 1: \quad \sqrt{2(1)-1} - 1 + 2 = 0$$

$$1 - 1 + 2 = 0$$

$$2 = 0$$

False, $x = 1$ is not a solution.

$$\boxed{\{5\}}$$

13.

$$(3x-2)^2 = 20$$

$$3x - 2 = \pm\sqrt{20}$$

$$3x = 2 \pm \sqrt{4 \cdot 5}$$

$$3x = 2 \pm 2\sqrt{5}$$

$$x = \frac{2 \pm 2\sqrt{5}}{3}$$

$$\boxed{\left\{\frac{2}{3} - \frac{2\sqrt{5}}{3}, \frac{2}{3} + \frac{2\sqrt{5}}{3}\right\}}$$

14.

$$3 + x(x+2) = 18$$

$$3 + x^2 + 2x = 18$$

$$x^2 + 2x - 15 = 0$$

$$(x+5)(x-3) = 0$$

$$x + 5 = 0 \quad \text{or} \quad x - 3 = 0$$

$$x = -5 \quad \text{or} \quad x = 3$$

$$\boxed{\{-5, 3\}}$$

15.
$$5x^2 - 1 = 0$$
$$5x^2 - x - 1 = 0$$
$$x = \frac{-(-1) \pm \sqrt{(-1)^2 - 4(5)(-1)}}{2(5)}$$
$$x = \frac{1 \pm \sqrt{1 + 20}}{10}$$
$$x = \frac{1 \pm \sqrt{21}}{10}$$
$$\boxed{\left\{ \frac{1}{10} - \frac{\sqrt{21}}{10}, \frac{1}{10} + \frac{\sqrt{21}}{10} \right\}}$$

16.
$$3x^2 - 6x + 2 = 0$$
$$x = \frac{-(-6) \pm \sqrt{(-6)^2 - 4(3)(2)}}{2(3)}$$
$$x = \frac{6 \pm \sqrt{36 - 24}}{6}$$
$$x = \frac{6 \pm \sqrt{12}}{6}$$
$$x = \frac{6 \pm 2\sqrt{3}}{6}$$
$$x = \frac{3 \pm \sqrt{3}}{3}$$
$$\boxed{\left\{ 1 - \frac{\sqrt{3}}{3}, 1 + \frac{\sqrt{3}}{3} \right\}}$$

17.
$$x^2 + 2x + 2 = 0$$
$$x = \frac{-2 \pm \sqrt{2^2 - 4(1)(2)}}{2(1)}$$
$$x = \frac{-2 \pm \sqrt{4 - 8}}{2}$$
$$x = \frac{-2 \pm \sqrt{-4}}{2}$$
$$x = \frac{-2 \pm 2i}{2}$$
$$x = -1 \pm i$$
$$\boxed{\{-1 - i, -1 + i\}}$$

18.
$$2z - \frac{14}{z - 2} = 1$$
$$2z(z - 2) - 14 = z - 2$$
$$2z^2 - 4z - 14 = z - 2$$
$$2z^2 - 5z - 12 = 0$$
$$(2z + 3)(z - 4) = 0$$
$$2z + 3 = 0 \quad \text{or } z = 4$$
$$z = \frac{-3}{2}$$
$$\boxed{\left\{ -\frac{3}{2}, 4 \right\}}$$

19.
$$y = 2x - 3$$
$$x + 2y = 9 \quad \text{(substitute for } y\text{)}$$
$$x + 2(2x - 3) = 9$$
$$x + 4x - 6 = 9$$
$$5x = 15$$
$$x = 3$$
$$y = 2x - 3 = 2(3) - 3 = 3$$
$$\boxed{\{(3, 3)\}}$$

20.

$$\begin{array}{rcl} 3x + 2y &=& -2 \\ \underline{-4x + 5y} &=& \underline{18} \end{array}$$

$(\times 4) \rightarrow$
$(\times 3) \rightarrow$

$$\begin{array}{rcl} 12x + 8y &=& -8 \\ \underline{-12x + 15y} &=& \underline{54} \\ 23y &=& 46 \\ y &=& 2 \\ 3x + 2(2) &=& -2 \\ 3x &=& -2 - 4 = -6 \\ x &=& -2 \end{array}$$

$$\boxed{\{(-2, 2)\}}$$

21.

$$\begin{array}{rcl} 3x - y &=& 4 \\ \underline{-9x + 3y} &=& \underline{-12} \end{array}$$

$(\text{no change}) \rightarrow$
$(\div 3) \rightarrow$

$$\begin{array}{rcl} 3x - y &=& 4 \\ \underline{-3x + y} &=& \underline{-4} \\ 0 &=& 0 \quad \text{true for all real values of } x \end{array}$$

$$\boxed{\{(x, y) \mid 3x - y = 4\}}$$

22.

$$\begin{array}{rcl} \frac{2}{3}x + \frac{5}{6}y &=& \frac{1}{3} \\[2mm] \underline{\frac{5}{4}x + \frac{3}{2}y} &=& \underline{\frac{3}{4}} \end{array}$$

$(\times 6) \rightarrow$
$(\times 4) \rightarrow$

$$\begin{array}{rcl} 4x + 5y &=& 2 \\ \underline{5x + 6y} &=& \underline{3} \end{array}$$

$(\times -6) \rightarrow$
$(\times 5) \rightarrow$

$$\begin{array}{rcl} -24x - 30y &=& -12 \\ \underline{25x + 30y} &=& \underline{15} \\ x &=& 3 \\ 4(3) + 5y &=& 2 \\ 5y &=& -10 \\ y &=& -2 \end{array}$$

$$\boxed{\{(3, -2)\}}$$

23. $4x^2 - 13x + 3 = \boxed{(4x - 1)(x - 3)}$

24. $4x^2 - 49y^2 = \boxed{(2x - 7y)(2x + 7y)}$

25. $4x^2 - 20x + 25 = \boxed{(2x - 5)^2}$

26. $3x - 6 + xy - 2y = 3(x - 2) + y(x - 2) = \boxed{(x - 2)(3 + y)}$

27. $20a^2b^2 - 45b^2 = 5b^2(4a^2 - 9) = \boxed{5b^2(2a - 3)(2a + 3)}$

28. $6x^2 - 3x + 2$ is $\boxed{\text{prime}}$, cannot be factored over integers

29. $12x^2 - 30x + 12 = 6(2x^2 - 5x + 2) = \boxed{6(2x - 1)(x - 2)}$

30. $x^3 - 1 = \boxed{(x - 1)(x^2 + x + 1)}$

31. $8y^3 + 125 = (2y)^3 + 5^3 = \boxed{(2y + 5)(4y^2 - 10y + 25)}$

32. $24 \div 8 \cdot 3 + 28 \div (-7) = 3 \cdot 3 - 4 = 9 - 4 = \boxed{5}$

33. $\dfrac{11 - (-9) + 6(10 - 4)}{2 + 3 \cdot 4} = \dfrac{11 + 9 + 6(6)}{2 + 12} = \dfrac{20 + 36}{14} = \dfrac{56}{14} = \boxed{4}$

34. $-21 - 16 - 3(2 - 8) = -21 - 16 - 3(-6) = -37 + 18 = \boxed{-19}$

35. $\dfrac{-8[9-(-10+3)]}{\left|3(-14)+(-11-6)(-5+3)\right|} = \dfrac{-8[9-(-7)]}{\left|-42+(-17)(-2)\right|} = \dfrac{-8(9+7)}{\left|-42+34\right|} = \dfrac{-8(16)}{\left|-8\right|} = \dfrac{-8(16)}{8} = \boxed{-16}$

36. $-(-3y+2)-4(6-5y)-3y-7 = 3y-2-24+20y-3y-7 = \boxed{20y-33}$

37. $(8x^5-7x^2+3x-6)-(-x^5+2x^2-x+5) = 8x^5-7x^2+3x-6+x^5-2x^2+x-5 = \boxed{9x^5-9x^2+4x-11}$

38. $(3x-5)^2-(2x-3)(4x+5) = 9x^2-30x+25-(8x^2-2x-15) = 9x^2-30x+25-8x^2+2x+15$
$= \boxed{x^2-28x+40}$

39. $(4y-3)(5y^2+6y-2) = 4y(5y^2+6y-2)-3(5y^2+6y-2) = 20y^3+24y^2-8y-15y^2-18y+6$
$= \boxed{20y^3+9y^2-26y+6}$

40. $\dfrac{-50w^6+8w^4-24w^2-72w}{-8w^3} = \dfrac{-50w^6}{-8w^3}+\dfrac{8w^4}{-8w^3}-\dfrac{24w^2}{-8w^3}-\dfrac{72w}{-8w^3} = \boxed{\dfrac{25w^3}{4}-w+\dfrac{3}{w}+\dfrac{9}{w^2}}$

41. $\dfrac{4y^4-4y^3+y^2+4y-3}{2y-1} = \boxed{2y^3-y^2+2-\dfrac{1}{2y-1}}$

$$\begin{array}{r}
2y^3-y^2\quad+\ \ 2 \\
2y-1\overline{\smash{)}4y^4-4y^3+y^2-4y-3} \\
\underline{4y^4-2y^3}\qquad\qquad \\
-2y^3+y^2\qquad \\
\underline{-2y^3+y^2}\qquad \\
0+4y-3 \\
\underline{4y-2} \\
-1
\end{array}$$

42. $\dfrac{-50x^2y}{150x^4y^5} = -\dfrac{1}{3}x^{2-4}y^{1-5} = -\dfrac{1}{3}x^{-2}y^{-4} = \boxed{\dfrac{-1}{3x^2y^4}}$

43. $\dfrac{(5x^4)^3}{x^{10}} = \dfrac{5^3x^{12}}{x^{10}} = 125x^{12-10} = \boxed{125x^2}$

44. $(-6xy^5)\left(-\dfrac{2}{3}xy^{-8}\right) = (-6)\left(-\dfrac{2}{3}\right)x^{1+1}y^{5-8} = 4x^2y^{-3} = \boxed{\dfrac{4x^2}{y^3}}$

45. $\dfrac{(x^4)^3x^{-2}}{(x^{-2})^5} = \dfrac{x^{12}x^{-2}}{x^{-10}} = x^{12-2+10} = \boxed{x^{20}}$

46. $\dfrac{y^2-y-12}{y^2-16}\cdot\dfrac{2y^2+7y-4}{y^2-4y-21} = \dfrac{(y-4)(y+3)}{(y+4)(y-4)}\cdot\dfrac{(2y-1)(y+4)}{(y-7)(y+3)} = \boxed{\dfrac{2y-1}{y-7}}$

47. $\dfrac{15-3y}{y+6}\div(y^2-9y+20) = \dfrac{-3(y-5)}{y+6}\div(y-5)(y-4) = \dfrac{-3(y-5)}{y+6}\cdot\dfrac{1}{(y-5)(y-4)} = \boxed{\dfrac{-3}{(y+6)(y-4)}}$

48. $\dfrac{x+6}{x-2}+\dfrac{2x+1}{x+3} = \dfrac{(x+6)(x+3)+(2x+1)(x-2)}{(x-2)(x+3)} = \dfrac{x^2+9x+18+2x^2-3x-2}{(x-2)(x+3)} = \boxed{\dfrac{3x^2+6x+16}{(x-2)(x+3)}}$

49. $\dfrac{x}{x^2+2x-3}-\dfrac{x}{x^2-5x+4}=\dfrac{x}{(x+3)(x-1)}-\dfrac{x}{(x-11)(x-1)}=\dfrac{x(x-4)-x(x+3)}{(x+3)(x-1)(x-4)}=\dfrac{x^2-4x-x^2-3x}{(x+3)(x-1)(x-4)}$

$=\boxed{\dfrac{-7x}{(x+3)(x-1)(x-4)}}$

50. $\dfrac{2}{y^2-9}+\dfrac{y}{y+3}+\dfrac{2y}{y-3}=\dfrac{2}{(y+3)(y-3)}+\dfrac{y(y-3)}{(y+3)(y-3)}+\dfrac{2y(y+3)}{(y-3)(y+3)}=\dfrac{2+y^2-3y+2y^2+6y}{(y+3)(y-3)}$

$=\boxed{\dfrac{3y^2+3y+2}{(y+3)(y-3)}}$

51. $\dfrac{\frac{1}{x}-2}{4-\frac{1}{x}}=\dfrac{\frac{1-2x}{x}}{\frac{4x-1}{x}}=\dfrac{1-2x}{x}\cdot\dfrac{x}{4x-1}=\boxed{\dfrac{1-2x}{4x-1}}$

52. $\dfrac{\frac{1}{x^2-9}+\frac{2}{x+3}}{\frac{3}{x-3}}=\dfrac{\frac{1}{(x+3)(x-3)}+\frac{2(x-3)}{(x+3)(x-3)}}{\frac{3}{x-3}}=\dfrac{1+2(x-3)}{(x+3)(x-3)}\cdot\dfrac{x-3}{3}=\dfrac{1+2x-6}{3(x+3)}=\boxed{\dfrac{2x-5}{3(x+3)}}$

53. $3\sqrt{20}+2\sqrt{45}-4\sqrt{80}=3\sqrt{4\cdot5}+2\sqrt{9\cdot5}-4\sqrt{16\cdot5}=3\left(2\sqrt{5}\right)+2\left(3\sqrt{5}\right)-4\left(4\sqrt{5}\right)$

$=6\sqrt{5}+6\sqrt{5}-16\sqrt{5}$

$=\boxed{-4\sqrt{5}}$

54. $2\sqrt[3]{16}+7\sqrt[3]{2}=2\sqrt[3]{8\cdot2}+7\sqrt[3]{2}=2\left(2\sqrt[3]{2}\right)+7\sqrt[3]{2}=4\sqrt[3]{2}+7\sqrt[3]{2}=\boxed{11\sqrt[3]{2}}$

55. $\sqrt{3y}\sqrt{6y}=\sqrt{18y^2}=\sqrt{9y^2\cdot2}=\boxed{3y\sqrt{2}}\qquad y\ge0$

56. $\sqrt{5}\left(\sqrt{2}+3\sqrt{7}\right)=\sqrt{5}\left(\sqrt{2}\right)+\sqrt{5}\left(3\sqrt{7}\right)=\boxed{\sqrt{10}+3\sqrt{35}}$

57. $\left(\sqrt{2}-3\sqrt{6}\right)\left(\sqrt{2}+\sqrt{6}\right)=\left(\sqrt{2}\right)^2+\sqrt{2}\left(\sqrt{6}\right)-3\sqrt{6}\left(\sqrt{2}\right)-3\sqrt{6}\left(\sqrt{6}\right)=2+\sqrt{12}-3\sqrt{12}-3(6)$

$=2-2\sqrt{12}-18$

$=-16-2\left(2\sqrt{3}\right)$

$=\boxed{-16-4\sqrt{3}}$

58. $\left(2+3\sqrt{5}\right)^2=4+12\sqrt{5}+\left(3\sqrt{5}\right)^2=4+12\sqrt{5}+9(5)=4+45+12\sqrt{5}=\boxed{49+12\sqrt{5}}$

59. $\sqrt{\dfrac{7}{8}}=\dfrac{\sqrt{7}}{\sqrt{8}}\cdot\dfrac{\sqrt{8}}{\sqrt{8}}=\dfrac{\sqrt{7}\cdot2\sqrt{2}}{8}=\boxed{\dfrac{\sqrt{14}}{4}}$

60. $\dfrac{6}{\sqrt[3]{4}}=\dfrac{6}{\sqrt[3]{4}}\cdot\dfrac{\sqrt[3]{2}}{\sqrt[3]{2}}=\dfrac{6}{2}\sqrt[3]{2}=\boxed{3\sqrt[3]{2}}$

61. $\dfrac{\sqrt{5}}{\sqrt{5}+\sqrt{6}} = \dfrac{\sqrt{5}}{\left(\sqrt{5}+\sqrt{6}\right)} \cdot \dfrac{\sqrt{5}-\sqrt{6}}{\sqrt{5}-\sqrt{6}} = \dfrac{\sqrt{5}\left(\sqrt{5}-\sqrt{6}\right)}{5-6} = \dfrac{5-\sqrt{30}}{-1} = \boxed{-5+\sqrt{30}}$

62. $\dfrac{11}{\sqrt{5}-3} = \dfrac{11}{\sqrt{5}-3} \cdot \dfrac{\sqrt{5}+3}{\sqrt{5}+3} = \dfrac{11\left(\sqrt{5}+3\right)}{5-3^2} = \dfrac{11\sqrt{5}+33}{-4} = \boxed{-\dfrac{11}{4}\sqrt{5}-\dfrac{33}{4}}$

63. $49^{-1/2} + 8^{-1/3} = \dfrac{1}{49^{1/2}} + \dfrac{1}{8^{1/3}} = \dfrac{1}{7} + \dfrac{1}{2} = \dfrac{2+7}{14} = \boxed{\dfrac{9}{14}}$

64. $\dfrac{2^{-3/4}}{2^{5/4}} = 2^{-3/4-5/4} = 2^{-8/4} = 2^{-2} = \dfrac{1}{2^2} = \boxed{\dfrac{1}{4}}$

65. $(3-2i)(5+3i) - 2i(4-3i) = 15 - i - 6i^2 - 8i + 6i^2 = \boxed{15-9i}$

66. $\dfrac{3-2i}{1+i} = \dfrac{3-2i}{1+i} \cdot \dfrac{1-i}{1-i} = \dfrac{3-3i-2i+2i^2}{1-i^2} = \dfrac{3-5i-2}{1+1} = \dfrac{1-5i}{2} = \boxed{\dfrac{1}{2}-\dfrac{5}{2}i}$

67. $y + 3 = 0$

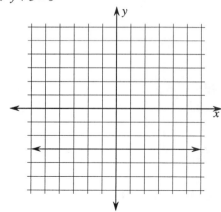

68. $3x - 2y = 6$

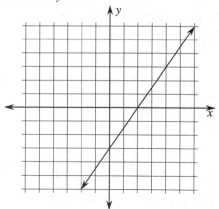

69. $y = -\dfrac{2}{3}x + 1$

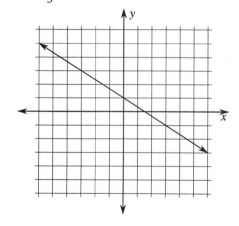

70. $5x + 2y < -10$

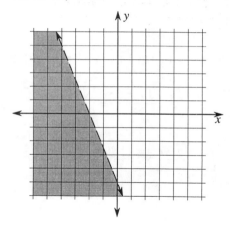

71. $y > -2x + 3$

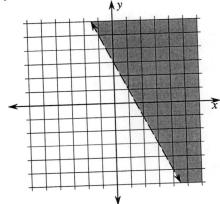

72. $y = x^3 - x$

x	-2	-1	0	1	2
y	-6	0	0	0	6

$x = -2$: $y = (-2)^3 - (-2) = -6$
$x = -1$: $y = (-1)^3 - (-1) = 0$
$x = 0$: $y = 0^3 - 0 = 0$
$x = 1$: $y = 1^3 - 1 = 0$
$x = 2$: $y = 2^3 - 2 = 6$

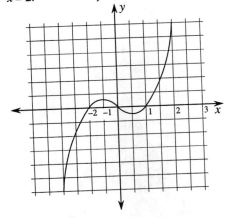

73. $y = x^2 - 2x - 8$

x–intercepts: $x^2 - 2x - 8 = 0$

$(x - 4)(x + 2) = 0$

$x = 4$ or $x = -2$

$(4, 0), (-2, 0)$

y–intercept: $y = 0 - 0 - 8 = -8$

$(0, -8)$

vertex: $-\dfrac{b}{2a} = -\dfrac{(-2)}{2(1)} = 1$

$y = 1^2 - 2 \cdot 1 - 8 = -9$

vertex $= (1, -9)$

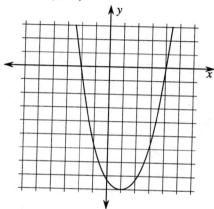

74. $2x + y = 6$

$\underline{-2x + y = 2}$

solution: $\boxed{\{(1, 4)\}}$

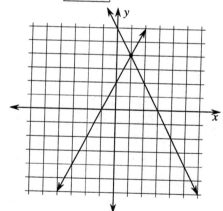

75. $2x + y < 4$

$y - 2x < 4$

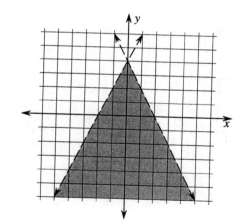

76. a. Natural numbers: $\left\{ 6, \sqrt{169} \right\}$.

  b. Whole numbers: $\left\{ 0, 6, \sqrt{169} \right\}$.

  c. Integers: $\left\{ -14, 0, 6, \sqrt{169} \right\}$.

  d. Rational numbers: $\left\{ -14, 0, 0.45, 6, 7\frac{1}{5}, \sqrt{169} \right\}$.

  e. Irrational numbers: $\left\{ -\pi, \sqrt{3} \right\}$.

  f. Real numbers: $\left\{ -14, -\pi, 0, 0.45, \sqrt{3}, 6, 7\frac{1}{5}, \sqrt{169} \right\}$.

77. $x^3 - 3x^3 y + 2y - 5$ when $x = -3$ and $y = -4$
$= (-3)^3 - 3(-3)^3(-4) + 2(-4) - 5$
$= -27 - 3(-27)(-4) - 8 - 5$
$= -27 + 81(-4) - 13$
$= -27 - 423 - 13$
$= \boxed{-364}$

78. $8 + (9 + 5) = 8 + (5 + 9)$
Commutative Property of Addition

79. $(13 \cdot 7) \cdot 3 = 13 \cdot (7 \cdot 3)$
Associative Property of Multiplication

80. $-5\left(-\frac{1}{5} \right) = 1$
Multiplicative Inverse Property

81. $3(2x - 6x) + 4(x + 5) = 3(-4x) + 4x + 20 = -12x + 4x + 20 = \boxed{-8x + 20}$

82. $7x - 4y \;=\; 15 \quad$ for y
$4y \;=\; 7x - 15$
$\boxed{y \;=\; \dfrac{7x}{4} - \dfrac{15}{4}}$

83. passing through $(-2, 6)$ and $(3, -4)$

slope $= \dfrac{-4 - 6}{3 + 2} = -\dfrac{10}{5} = -2$

point–slope form: $\boxed{y - 6 = -2(x + 2) \text{ or } y + 4 = -2(x - 3)}$

slope–intercept form: $y - 6 \;=\; -2x - 4$
$\boxed{y \;=\; -2x + 2}$

standard form: $\boxed{2x + y = 2}$

84. passing through $(-3, -7)$ and parallel to the line whose equation is $-6x + y = -2$

slope of line parallel: $y \;=\; 6x - 2$
$m \;=\; 6$

slope of line $= 6$

point–slope form: $\boxed{y + 7 = 6(x + 3)}$

slope–intercept form: $x + 7 \;=\; 6x + 18$
$\boxed{y \;=\; 6x + 11}$

standard form: $\boxed{6x - y = -11}$

85. (x, y):

Reading 1 $(1, 32)$

Reading 2 $(3, 96)$

$$\text{slope} = \frac{96 - 32}{3 - 1} = \frac{64}{2} = 32$$

$$\begin{aligned} y - 32 &= 32(x - 1) \\ y - 32 &= 32x - 32 \\ \boxed{y = 32x} \end{aligned}$$

If $x = 5$ seconds, $y = 32(5) = \boxed{160 \text{ feet per second}}$

86. $f(x) = 3x^2 - x + 2$

$f(-2) = 3(-2)^2 - (-2) + 2 = 3(4) + 2 + 2 = 12 + 4 = \boxed{16}$

87. C does not define y as a function of x since at $x = 0$ there are 2 values of y.

88. $(-1, 3), (2, 9)$

$y = mx + b$

$(-1, 3)$: $3 = -m + b$ $(\times -1) \rightarrow$ $-3 = m - b$

$(2, 9)$: $\underline{9 = 2m + b}$ $\underline{9 = 2m + b}$

                                                                            $6 = 3m$

                                                                            $2 = m$

                                                                            $3 = -2 + b$

                                                                            $5 = b$

$\boxed{m = 2, b = 5; \; y = 2x + 5}$

89. $(0.000\,73)(8{,}200{,}000) = (7.3 \times 10^{-4})(8.2 \times 10^6) = \boxed{5.986 \times 10^3}$

90. $\dfrac{1200}{0.006} = \dfrac{1.2 \times 10^3}{6 \times 10^{-3}} = 0.2 \times 10^{3+3} = 2 \times 10^{-1} \times 10^6 = \boxed{2 \times 10^5}$

91. $\dfrac{x + 3}{x^2 + 3x - 28}$

$x^2 + 3x - 28 \neq 0$

$(x + 7)(x - 4) \neq 0$

$x \neq -7 \text{ and } 4$

$\boxed{\text{Domain: } \{x \mid x \neq -7 \text{ and } x \neq 4\}}$

92. $\dfrac{3x^2 - 8x + 5}{4x^2 - 5x + 1} = \dfrac{(3x - 5)(x - 1)}{(4x - 1)(x - 1)} = \boxed{\dfrac{3x - 5}{4x - 1}}$

93.

$$\frac{1}{a} = \frac{1}{x} - \frac{1}{b} \quad \text{for } x$$

$$abx \left(\frac{1}{a}\right) = abx \left(\frac{1}{x} - \frac{1}{b}\right)$$

$$bx = ab - ax$$

$$ax + bx = ab$$

$$x(a + b) = ab$$

$$\boxed{x = \frac{ab}{a + b}}$$

94. $\sqrt{90x^3} = \sqrt{9x^2 \cdot 10x} = \boxed{3x\sqrt{10x}}$

95. $\sqrt[3]{16x^5} = \sqrt[3]{8x^3 \cdot 2x^2} = \boxed{2x\sqrt[3]{2x^2}}$

96. $\sqrt{-75} = \sqrt{(-25)(3)} = \boxed{5i\sqrt{3}}$

97. Let $x =$ the number.
$$\begin{aligned}
2x - 5 &= 3x - 6 \\
-x &= -1 \\
x &= 1
\end{aligned}$$
The number is $\boxed{1}$.

98. Let $x, x + 2, x + 4$ represent three consecutive even integers.
$$\begin{aligned}
(x + 4) - x &= (x + 2) + 6 \\
4 &= x + 8 \\
-4 &= x \\
-2 &= x + 2 \\
0 &= x + 4
\end{aligned}$$
The integers are $\boxed{-4, -2, \text{ and } 0}$.

99. Let $x =$ the number.
$$\begin{aligned}
0.30x &= 0.15 \\
x &= 0.5
\end{aligned}$$
The number is $\boxed{0.5}$.

100. Let $R =$ the number of Republicans. $3R =$ the number of Democrats. $R + 14 =$ the number of Independents.
$$\begin{aligned}
R + 3R + (R + 14) &= 79 \\
5R + 14 &= 79 \\
5R &= 65 \\
R &= 13 \\
3R &= 39 \\
R + 14 &= 13 + 14 = 27
\end{aligned}$$
$\boxed{39 \text{ Democrats}, 13 \text{ Republicans}, 27 \text{ Independents}}$

101. Let $x =$ the price of VCR before reduction.
$$\begin{aligned}
x - 0.20x &= 124 \\
0.80x &= 124 \\
x &= \boxed{\$155}
\end{aligned}$$

102. $M = \dfrac{2x}{1 - x}$
If $M = 3$, then
$$\begin{aligned}
3 &= \frac{2x}{1 - x} \\
3 - 3x &= 2x \\
-5x &= -3 \\
x &= \frac{3}{5} = 0.60 \text{ or } \boxed{60\%}
\end{aligned}$$

103. Let x = Rodney's present age. $x + 7$ = Allison's present age.

$$(x + 7 + 6) = 2(x + 6)$$
$$x + 13 = 2x + 12$$
$$1 = x$$
$$\underline{x + 7 = 8}$$

Allison, 8 years old; Rodney, 1 year old

104. Let x = the minimum amount of sales.

$$175 + 0.05x \geq 300$$
$$0.05x \geq 125$$
$$x \geq 2500$$

minimum sales at least \$2500

105. Let x = the amount of sales.

$$x + 0.08x = 648$$
$$1.08x = 648$$
$$x = 600$$

sales tax, $0.08x = 0.08(600) = $ \$48

106. Let x = the number of hours worked.

$$350 + 23x = 971$$
$$23x = 621$$
$$x = \boxed{27 \text{ hours}}$$

107. Let x = the number of nickels. $20 - x$ = the number of dimes.

$$0.05x + 0.10(20 - x) = 1.85$$
$$0.05x + 2 - 0.10x = 1.85$$
$$-0.05x = -0.15$$
$$x = 3$$
$$20 - x = 17$$

3 nickels, 17 dimes

108.

$$\frac{10 \text{ pounds of fertilizer}}{980 \text{ square feet}} = \frac{x}{1470 \text{ square feet}}$$
$$980x = 10(1470) \text{ pounds}$$
$$x = \boxed{15 \text{ pounds of fertilizer}}$$

109. Let x = the first number. y = the second number.

$$5x + 2y = 1 \qquad (\times 3) \rightarrow \qquad 15x + 6y = 3$$
$$\underline{-2x - 3y = 15} \qquad (\times 2) \rightarrow \qquad \underline{-4x - 6y = 30}$$
$$11x = 33$$
$$x = 3$$
$$5(3) + 2y = 1$$
$$2y = -14$$
$$y = -7$$

The numbers are 3 and –7.

110. Let u = units' digit. t = tens' digit. $10 + u$ = the number.

$$\begin{array}{rcl} 3t + 4u &=& 38 \\ 10t + u &=& 3u + 4 \end{array}$$

(simplify) $\rightarrow$
(no change) $\rightarrow$
($\times 2$) $\rightarrow$

$$\begin{array}{rcl} 3t + 4u &=& 38 \\ \underline{10t - 2u} &=& \underline{4} \\ 3t + 4u &=& 38 \\ \underline{20t - 4u} &=& \underline{8} \\ 23t &=& 46 \\ t &=& 2, \text{tens' digit} \\ 3(2) + 4u &=& 38 \\ 4u &=& 32 \\ u &=& 8, \text{units' digit} \end{array}$$

The number is $\boxed{28}$.

111. Let x = the cost of a pen. y = the cost of a pad.

$$\begin{array}{rcl} 10x + 12y &=& 42 \\ 5x + 10y &=& 29 \end{array}$$

($\div -2$) $\rightarrow$
(no change) $\rightarrow$

$$\begin{array}{rcl} -5x - 6y &=& -21 \\ \underline{5x - 10y} &=& \underline{29} \\ 4y &=& 8 \\ y &=& 2 \\ 5x + 10(2) &=& 29 \\ 5x &=& 9 \\ x &=& \dfrac{5}{9} = 1.80 \end{array}$$

$\boxed{\text{cost of pen, \$1.80; cost of pad, \$2}}$

112.

$$D = \frac{n(n-3)}{2}$$

If $D = 5$ diagonals, then

$$\begin{array}{rcl} 5 &=& \dfrac{n(n-3)}{2} \\ n(n-3) &=& 10 \\ n^2 - 3n - 10 &=& 0 \\ (n-5)(n+2) &=& 0 \\ n &=& 5 \end{array}$$

$\boxed{\text{number of sides, 5}}$

113.

$$\begin{array}{rcl} P &=& 30 - \dfrac{9}{t+1} \\ P &=& 27 \ (\text{thousands}) \\ 27 &=& 30 - \dfrac{9}{t+1} \\ \dfrac{9}{t+1} &=& 3 \\ 3(t+1) &=& 9 \\ t+1 &=& 3 \\ t &=& 2 \end{array}$$

The community will have a population of 27,000 in $1995 + 2 = \boxed{1997}$.

114. Let x = the numerator of a fraction. $x + 4$ = the denominator of a fraction.

$$\frac{x+1}{x+4+1} = \frac{3}{4}$$

$$\frac{x+1}{x+5} = \frac{3}{4}$$

$$4x + 4 = 3x + 15$$

$$x = 11$$

$$x + 4 = 15$$

original fraction: $\boxed{\dfrac{11}{15}}$

115. Let x = one number. $x + 5$ = another number.

$$\frac{x+5}{x} = 3 + \frac{1}{x}$$

$$x + 5 = 3x + 1$$

$$4 = 2x$$

$$2 = x$$

$$x + 5 = 7$$

The numbers are $\boxed{2 \text{ and } 7}$.

116. Let x = the number.

$$x + \frac{1}{x} = 4$$

$$x^2 + 1 = 4x$$

$$x^2 - 4x + 1 = 0$$

$$x = \frac{-(-4) \pm \sqrt{(-4)^2 - 4(1)(1)}}{2(1)}$$

$$= \frac{4 \pm \sqrt{16 - 4}}{2}$$

$$= \frac{4 \pm \sqrt{12}}{2}$$

$$= \frac{4 \pm 2\sqrt{3}}{2}$$

$$= 2 \pm \sqrt{3}$$

The number is $\boxed{2 + \sqrt{3} \text{ or } 2 - \sqrt{3}}$ or rounded, $\boxed{3.7 \text{ or } 0.3}$.

117.

$$h = -16t^2 + 96t + 80$$

If $h = 128$ feet, then

$$128 = -16t^2 + 96t + 80$$

$$16t^2 - 96t + 48 = 0$$

$$t^2 - 6t + 3 = 0$$

$$t = \frac{-(-6) \pm \sqrt{(-6)^2 - 4(1)(3)}}{2(1)}$$

$$= \frac{6 \pm \sqrt{36 - 12}}{2}$$

$$= \frac{6 \pm \sqrt{24}}{2}$$

$$= \frac{6 \pm 2\sqrt{6}}{2}$$

$$= 3 \pm \sqrt{6}$$

The time is $\boxed{3 + \sqrt{6} \text{ seconds or } 3 - \sqrt{16} \text{ second}}$ or rounded, $\boxed{5.4 \text{ seconds or } 0.6 \text{ seconds}}$.

118. $C = x^2 - 120x + 4200$

x–coordinate of the vertex x: $-\dfrac{b}{2a} = \dfrac{-(-120)}{2(1)} = 60$

cost (of $x = 60$) $= (60)^2 - 120(60) + 4200 = 3600 - 7200 + 4200 = 600$

$\boxed{60 \text{ desks}}$ produced for a daily minimum cost of $\boxed{\$600}$.

119. Let $x =$ the width of the rectangle. $2x - 7 =$ the length of the rectangle.

$$
\begin{aligned}
\text{perimeter} &= 58 \text{ meters} \\
2x + 2(2x - 7) &= 58 \\
2x + 4x - 14 &= 58 \\
6x &= 72 \\
x &= 12 \\
2x - 7 &= 24 - 7 = 17
\end{aligned}
$$

$\boxed{\text{width, 12 meters; length, 17 meters}}$

120. Let $x =$ the length of the side of an equilateral triangle. $x + 4 =$ the length of the side of a square.

perimeter of square = the perimeter of the triangle + 18

$$
\begin{aligned}
4(x + 4) &= 3x + 18 \\
4x + 16 &= 3x + 18 \\
x &= 2 \\
x + 4 &= 6
\end{aligned}
$$

$\boxed{\text{length of each side of equilateral triangle, 2 centimeters; length of each side of the square, 6 centimeters}}$

121. Let $x =$ the measure of angle. $90 - x =$ the measure of its complement. $180 - x =$ the measure of its supplement.

$$
\begin{aligned}
(180 - x) &= 25 + 2(90 - x) \\
180 - x &= 25 + 180 - 2x \\
x &= 25
\end{aligned}
$$

The angle measures $\boxed{25^\circ}$.

122.

$$
\begin{aligned}
(x + 10) + (4x + 20) + x &= 180 \\
6x + 30 &= 180 \\
6x &= 150 \\
x &= 25 \\
x + 10 &= 35 \\
4x + 20 &= 120
\end{aligned}
$$

The measures of the angles are $\boxed{25^\circ, 35^\circ, \text{ and } 120^\circ}$.

123.

$$
\begin{aligned}
10x - 48 &= 6x \\
4x &= 48 \\
x &= 12 \\
6x &= 6(12) = 72 \\
10x - 48 &= 10(12) - 48 = 120 - 48 = 72
\end{aligned}
$$

The measures of the angles are $\boxed{72^\circ \text{ and } 72^\circ}$.

124.

$$
\begin{aligned}
2x - 5 &= x + 22 \\
x &= 27 \\
x + 22 &= 27 + 22 = 49 \\
2x - 5 &= 2(27) - 5 = 54 - 5 = 49
\end{aligned}
$$

The measures of the angles are $\boxed{49^\circ \text{ and } 49^\circ}$.

125.

$$BA = BC$$
$$\underline{AC = BC + 4}$$

$\rightarrow$

$\rightarrow$

(simplify) $\rightarrow$

$$3x + 4y - 6 = 3y$$
$$\underline{-x + 4y + 2 = 3y + 4}$$
$$3x + y = 6$$
$$\underline{-x + y = 2} \quad (\times -1) \rightarrow$$

$$3x + y = 6$$
$$\underline{x - y = -2}$$
$$4x = 4$$
$$x = 1$$
$$-1 + y = 2$$
$$y = 3$$

$$BA = 3x + 4y - 6$$
$$= 3(1) + 4(B) - 6$$
$$= 3 + 12 - 6$$
$$= 9$$
$$BC = 3y = 3(3)$$
$$= 9$$
$$AC = -x + 4y + 2$$
$$= -1 + 4(3) + 2$$
$$= -1 + 12 + 2$$
$$= 13$$

The lengths of the sides are: | AB, 9 decimeters; BC, 9 decimeters; AC, 13 decimeters |

126. Let L = the length of the rectangular garden. W = the width of the rectangular garden.

$$2L + 2W = 32$$
$$3(2L) + 2(2W) = 84$$

$(\div -2) \rightarrow$

$$2L + 2W = 32$$
$$\underline{-3L - 2W = -42}$$
$$-L = -10$$
$$L = 10$$
$$2(10) + 2W = 32$$
$$2W = 12$$
$$W = 6$$

rectangular garden dimensions: | length, 10 yards; width, 6 yards |

127. Let x = the width of the rectangle. $x + 3$ = the length of the rectangle. $x + 2$ = the length of the side of the square.

$$\text{area of square} - \text{area of rectangle} = 11$$
$$(x + 2)^2 - x(x + 3) = 11$$
$$x^2 + 4x + 4 - x^2 - 3x = 11$$
$$x = 7$$
$$x + 3 = 10$$
$$x + 2 = 9$$

dimensions: | rectangle: width, 7 yards; length, 10 yards
square: side, 9 yards |

128. Let x = the width of the rectangle. $2x + 1$ = the length of the rectangle.

$$x(2x + 1) = 36$$
$$2x^2 + x - 36 = 0$$
$$(2x + 9)(x - 4) = 0$$
$$x = 4 \quad \left(\text{reject } x = -\frac{9}{2} \right)$$
$$2x + 1 = 2(4) + 1 = 9$$

dimensions: | width, 4 meters; length, 9 meters |

129.

$$\text{area} = \frac{1}{2}h(b_1 + b_2)$$

$$h = x - 1$$
$$b_1 = x$$
$$b_2 = x - 2$$
$$\text{area} = 32 \text{ square millimeters}$$

$$\frac{1}{2}(x-1)(x+x-2) = 32$$

$$\frac{(x-1)(2x-2)}{2} = 32$$

$$(x-1)(x-1) = 32$$
$$x^2 - 2x + 1 = 32$$
$$x^2 - 2x - 31 = 0$$

$$x = \frac{-(-2) \pm \sqrt{(-2)^2 - 4(1)(-31)}}{2(1)}$$

$$= \frac{2 \pm \sqrt{4 + 124}}{2}$$

$$= \frac{2 \pm \sqrt{128}}{2}$$

$$= \frac{2 \pm 8\sqrt{2}}{2}$$

$$= 1 \pm 4\sqrt{2}$$

$$= 1 + 4\sqrt{2} \quad \left(\text{reject } 1 - 4\sqrt{2} \text{ since length cannot be negative}\right)$$

$$b_1 = x = 1 + 4\sqrt{2}$$
$$b_2 = x - 2 = 1 + 4\sqrt{2} - 2 = -1 + 4\sqrt{2}$$
$$h = x - 1 = 1 + 4\sqrt{2} - 1 = 4\sqrt{2}$$

bases, $-1 + 4\sqrt{2}$ millimeters, and $1 + 4\sqrt{2}$ millimeters; height, $4\sqrt{2}$ millimeters

or rounded,

bases, 4.7 millimeters and 6.7 millimeters; height, 5.7 millimeters

130. Let $x =$ the length of the short leg. $x + 10 =$ the length of the hypotenuse. $x + 10 - 5 = x - 5 =$ the length of the long leg.

$$x^2 + (x+5)^2 = (x+10)^2$$
$$x^2 + x^2 + 10x + 25 = x^2 + 20x + 100$$
$$x^2 - 10x - 75 = 0$$
$$(x-15)(x+5) = 0$$
$$x = 15 \quad (\text{reject } x = -5)$$
$$x + 5 = 20$$
$$x + 10 = 25$$

The length of the sides of the right triangle are 15 decimeters, 20 decimeters, and **25 decimeters**.

131.

$$x^2 + 15^2 = 18^2$$
$$x^2 + 225 = 324$$
$$x^2 = 99$$
$$x = \sqrt{99}$$
$$x = 3\sqrt{11}$$

The base of the ladder is $3\sqrt{11}$ feet ≈ 9.9 feet from the wall.

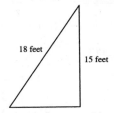

18 feet

15 feet

x

132. Let x = the width of the margin. $10 - 2x$ = the length of the side of print area.

$$
\begin{aligned}
\text{area of page with margin} - \text{area of page} &= \text{area of blue margin} \\
10^2 - (10 - x - x)^2 &= \text{area of blue margin} \\
100 - (10 - 2x)^2 &= (10 - 2x)^2 \\
100 &= 2(10 - 2x)^2 \\
(10 - 2x)^2 &= 50 \\
10 - 2x &= \pm\sqrt{50} \\
-2x &= -10 \pm 5\sqrt{2} \\
x &= 5 \pm \frac{5}{2}\sqrt{2} \\
x &= 5 - \frac{5}{2}\sqrt{2} \quad \left(\text{reject } 5 + \frac{5}{2}\sqrt{2} \text{ since } 2x > 10\right) \\
10 - 2x &= 10 - 2\left(5 - \frac{5}{2}\sqrt{2}\right) \\
&= 10 - 10 + 5\sqrt{2} \\
&= 5\sqrt{2}
\end{aligned}
$$

dimensions of portion of page on which the print appears are $\boxed{5\sqrt{2}\text{ inches} \times 5\sqrt{2}\text{ inches}}$ or rounded, $\boxed{7.1\text{ inches} \times 7.1\text{ inches}}$.

133. Let x = the amount invested at 5%. $4000 - x$ = the amount invested at 9%.

$$
\begin{aligned}
0.05x + 0.09(4000 - x) &= 311 \\
0.05x + 360 - 0.09x &= 311 \\
-0.04x &= -49 \\
x &= 1225 \\
4000 - x &= 2775
\end{aligned}
$$

$\boxed{\$1225 \text{ at } 5\%, \ \$2775 \text{ at } 9\%}$

134. Let t = the time (in hours) for boats to be 232 miles apart.

$$
\begin{aligned}
13t + 19t &= 232 \\
32t &= 232 \\
t &= \boxed{7.25 \text{ hours}}
\end{aligned}
$$

135. Let x = the amount of 80% acid solution. $10 - x$ = the amount of 65% acid solution.

$$
\begin{aligned}
0.80x + 0.65(10 - x) &= 0.75(10) \\
0.80x + 6.5 - 0.65x &= 7.5 \\
0.15x &= 1 \\
x &= 6\frac{2}{3} \\
10 - x &= 3\frac{1}{3}
\end{aligned}
$$

$\boxed{6\frac{2}{3} \text{ gallons of } 80\%, \ 3\frac{1}{3} \text{ gallons of } 65\%}$

136. Let x = the amount (in pounds) of macadamia nuts at \$5.00. $20 + x$ = the amount (in pounds) of mixture.

$$
\begin{aligned}
20(3.00) + x(5.00) &= (20 + x)(3.50) \\
60 + 5x &= 70 + 3.5x \\
1.5x &= 10 \\
x &= 6\frac{2}{3}
\end{aligned}
$$

$\boxed{6\frac{2}{3} \text{ pounds of macadamia nuts}}$

137. x = the time for painter and assistant to work together. $\frac{x}{4}$ = portion of job done by painter.

$\frac{x}{12}$ = portion of job done by assistant.

$$\frac{x}{4} + \frac{x}{12} = 1$$
$$12\left(\frac{x}{4} + \frac{x}{12}\right) = 12(1)$$
$$3x + x = 12$$
$$4x = 12$$
$$x = \boxed{3 \text{ days}}$$

138. Let v = the speed of the boat in still water. c = speed of the current. $r + c$ = the rate of boat with the current. $r - c$ = rate of the boat against the current.

$$
\begin{array}{lcll}
\text{rate} \times \text{time} & = & \text{distance} \\
(r + c)2 & = & 48 & (\div 2) \rightarrow \\
(r - c)3 & = & 48 & (\div 3) \rightarrow \\
\end{array}
$$

$$
\begin{array}{rcl}
r + c & = & 24 \\
r - c & = & 16 \\
\hline
2r & = & 40 \\
r & = & 20 \\
20 + c & = & 24 \\
c & = & 4 \\
\end{array}
$$

$\boxed{\text{speed of boat in still water, 20 miles per hour, rate of current, 4 miles per hour}}$

139. Let x = the time for Louis to do the job working alone.
$x - 9$ = the time for Prior to do the job working alone.

$\frac{20}{x}$ = fractional part of job done by Louis

$\frac{20}{x-9}$ = fractional part of job done by Prior

$$
\begin{array}{rcl}
\frac{20}{x} + \frac{20}{x-9} & = & 1 \\
20(x-9) + 20x & = & x(x-9) \\
20x - 180 + 20x & = & x^2 - 9x \\
0 & = & x^2 - 49x + 180 \\
(x-45)(x-4) & = & 0 \\
x & = & 45 \quad (\text{reject } x = 4 \text{ since } x - 9 = -5) \\
x - 9 & = & 36
\end{array}
$$

Time working alone: $\boxed{\text{Louis, 45 hours; Prior, 36 hours}}$

140. Let x = the rate of plane B. $x = 50$ = the rate of plane A.

	rate	$\times$	time	=	distance
plane A	$x + 50$		$\frac{500}{x+50}$		500 miles
plane B	x		$\frac{400}{x}$		400 miles

$$
\begin{array}{rcl}
\frac{500}{x+50} & = & \frac{400}{x} \\
500x & = & 400x + 20000 \\
100x & = & 20000 \\
x & = & 200 \\
x + 50 & = & 250
\end{array}
$$

$\boxed{\text{rate of plane } A, 250 \text{ miles per hour; rate of } B, 200 \text{ miles per hour}}$

141.

term (n)		pattern
3	$1 + 2 + 3 = 6 = 4 \cdot \dfrac{3}{2}$	$(3 + 1)\dfrac{3}{2} = 4 \cdot \dfrac{3}{2}$
5	$1 + 2 + 3 + 4 + 5 = 15 = 6 \cdot \dfrac{5}{2}$	$(5 + 1)\dfrac{5}{2} = 6 \cdot \dfrac{5}{2}$
8	$1 + 2 + 3 + 4 + 5 + 6 + 7 + 8 = 36 = 9 \cdot \dfrac{8}{2}$	$(8 + 1)\dfrac{8}{2} = 9 \cdot \dfrac{8}{2}$
10	$1 + 2 + \ldots + 9 + 10 = 55 = 11 \cdot \dfrac{10}{2}$	$(10 + 1)\dfrac{10}{2} = 11 \cdot \dfrac{10}{2}$
70	$1 + 2 + \ldots + 68 + 69 + 70 = ?$	$(70 + 1)\dfrac{70}{2} = 71 \cdot \dfrac{70}{2} = \boxed{2485}$
n	$1 + 2 + \ldots + (n - 2) + (n - 1) + n = ?$	$(n + 1)\dfrac{n}{2} = \boxed{\dfrac{n^2 + n}{2}}$

142. w, x, y, z represent natural numbers.

$$\begin{aligned} x &> w \\ y &= z - 1 \\ z &= x + 4 \\ x &> w \\ x + 4 &> w + 4 \\ \hline \boxed{z > w + 4} \end{aligned}$$

143.   **a.** $10 + 4 - 8 = 6$
            $x = 10,\ y = 4,\ z = 8$
        b. $4 \cdot 8 - 10 = 22$
            $a = 4,\ b = 8,\ c = 10$
        c. $10 \div 2 + 8 = 13$
            $r = 10,\ s = 2,\ w = 8$

144. $\boxed{C}$ is true. Let $X =$ any number.

$$\begin{aligned} X &> A \\ B &< X \\ C &= A \\ X &> C \quad \text{True} \end{aligned}$$

145.

	first term $n = 1$	second term $n = 2$	third term $n = 3$	fourth term $n = 4$	fifth term $n = 5$	rule for the nth term
a.	1	3	5	7	9	$2n - 1$
b.	5	8	11	14	17	$3(n + 1) - 1$
c.	1	4	9	16	25	n^2
d.	0	3	8	15	24	$n^2 - 1$
e.	1	8	27	64	125	n^3

146.   There are $\boxed{\text{five}}$ sizes of squares possible.

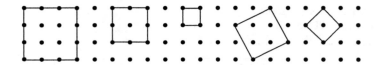